Molecular Aspects of Plant-Pathogen Interaction

Archana Singh • Indrakant K. Singh
Editors

Molecular Aspects of Plant-Pathogen Interaction

Springer

Editors
Archana Singh
Department of Botany
Hans Raj College, University of Delhi
New Delhi, India

Indrakant K. Singh
Molecular Biology Research Laboratory,
Department of Zoology
Deshbandhu College, University of Delhi
New Delhi, India

ISBN 978-981-10-7370-0 ISBN 978-981-10-7371-7 (eBook)
https://doi.org/10.1007/978-981-10-7371-7

Library of Congress Control Number: 2018930517

Printed on acid-free paper

This Springer imprint is published by the registered company Springer Nature Singapore Pte Ltd. part of Springer Nature.
The registered company address is: 152 Beach Road, #21-01/04 Gateway East, Singapore 189721, Singapore

Preface

Pathogen attack has been one of the chief constrains that reduce crop productivity worldwide. Plants have established sophisticated mechanisms to counter and acclimatize over these invading pathogens at physiological, biochemical as well as molecular levels. Due to severe crop losses by pathogen outbreak, it is mandatory to completely understand the resistance/defense mechanisms against pathogen and develop advanced tactics to improve biotic stress tolerance in crop plants.

We present this book with an objective to realize the plant defense against different pathogens better and to document fundamentals as well as recent findings. This book has an amalgamation of basic information about disease resistance along with current insights into plant-pathogen interaction. The book has 15 chapters to disseminate the most updated information and detailed overviews on the present knowledge on molecular aspects of plant responses and adaptation to biotic stresses. This book is an essential reading for researchers and professionals in plant pathology, cell biology, molecular biology and genetics. This is highly recommended for the ones who are involved in plant disease resistance and crop improvement and to all plant scientists and undergraduates.

Depending on their modes of nutrition, phytopathogens have been categorized as necrotrophs, biotrophs and hemibiotrophs. These pathogens can be bacterial and fungal and cause various diseases in plants. In addition, viruses are another important class of pathogens and are causal agents for many common plant diseases. Plants counter to pathogens by activating a cascade of genes, encoding different receptors, signaling and protective molecules. During biotic stress, first of all effector molecules i.e. pathogen-associated molecular patterns (PAMPs) are perceived by plant recognition receptors (PRRs), after which PRRs interact with additional trans-membrane proteins that act as signaling adapters or amplifiers to achieve full functionality and PAMP triggered immunity (PTI). Defense response by receptor-like protein is a complex strategy, characterized by specific interaction between disease resistance (*R*) genes of plants and corresponding avirulence (*avr*) genes of pathogen that induce effector-triggered immunity (ETI) through hypersensitive response.

The NBS-LRR genes are important class of resistance gene families and their products recognize factors secreted by pathogens, which activates downstream signaling pathways leading to defense. Mitogen-activated protein kinases (MAPKs), which are cell-signaling enzymes that also show vital functions in transmitting extracellular signals to the nucleus during biotic stress. To achieve defense against pathogen, transcription factors such as WRKY transcription factors bind to plant-specific *cis*-regulatory elements and activate gene expression thereby inducing transcriptional reprogramming and proteomic alterations to coordinate the perception and activation of pathways specific to the type of pathogen in question. Mainly phytohormones, small RNAs and other factors regulate this change at transcript level and protein level. Amongst all the targets, the induction and accumulation of pathogenesis-related (PR) proteins and biosynthesis of secondary metabolites are an integral component of innate immune responses in plants during pathogen attack.

Overall this volume will convey an overview of plant-pathogen interactions and it is a must read to understand this process for the genetic improvement of crops for disease resistance.

We are obliged to the authors of various chapters of this book for writing their chapters methodically and with great responsibility. We are extremely thankful to Dr. Rama, Principal, Hans Raj College, University of Delhi and Dr. Ajay Arora, Principal, Deshbandhu College, University of Delhi for providing overall support for our research and academic pursuits. We would also like to convey our gratitude to Dr. V. K. Kawatra, Mr. P. K. Singh and Dr. Vijay Rani Rajpal for always motivating us. We appreciate the beautiful ambiance created by our little angels Saumya and Kimaya, which allowed us to work tirelessly and gave us all emotional support. We are grateful to our parents for their constant support and blessings. Last but not the least, our sincere thanks to the handling editors and publisher.

We are optimistic that this book will be effective in broadcasting the latest knowledge about the plant-pathogen interaction.

New Delhi, India

Archana Singh
Indrakant K. Singh

Contents

1 *Arabidopsis thaliana* as a Model Organism to Study Plant-Pathogen Interactions 1
Shachi Agrawal

2 Fungal and Bacterial Biotrophy and Necrotrophy 21
Geeta and Reema Mishra

3 Plant–Virus Interactions 43
Sunita Yadav and Anju K. Chhibbar

4 Pathogen-Associated Molecular Patterns and Their Perception in Plants 79
Jitendra Singh Rathore and Chaitali Ghosh

5 Role of NBS-LRR Proteins in Plant Defense 115
Namo Dubey and Kunal Singh

6 Recent Advancement on Map Kinase Cascade in Biotic Stress 139
Monika Jaggi

7 Role of Phytohormones in Plant Defense: Signaling and Cross Talk 159
Vibha Gulyani Checker, Hemant Ritturaj Kushwaha, Pragati Kumari, and Saurabh Yadav

8 Insights into the Role of WRKY Superfamily of Protein Transcription Factor in Defense Response 185
Preeti Singh, Shashi Shekhar, Anjana Rustagi, Vinay Sharma, and Deepak Kumar

9 Small Noncoding RNA-Based Regulation of Plant Immunity 203
Om Prakash Gupta, Vanita Pandey, Nand Lal Meena, Suhas G. Karkute, Sagar Banerjee, and Anil Dahuja

10 Transcriptomic Studies Revealing Enigma of Plant-Pathogen Interaction 219
Zahoor Ahmed Wani and Nasheeman Ashraf

11 Proteomic Studies Revealing Enigma of Plant–Pathogen Interaction 239
Anjana Rustagi, Garvita Singh, Shachi Agrawal, and Prem Kumar Gupta

12 Role of Pathogenesis-Related (PR) Proteins in Plant Defense Mechanism 265
Deepti Jain and Jitendra Paul Khurana

13 Antimicrobial Compounds and Their Role in Plant Defense 283
Anupama Razdan Tiku

14 Explorations of Plant's Chemodiversity: Role of Nitrogen-Containing Secondary Metabolites in Plant Defense 309
Sanjay Kumar Singh

15 Plant Cell Wall: A Simple Physical Barrier or a Complex Defense Modulator – Exploring Its Dynamic Role at Plant-Fungus Interface 333
Sumanti Gupta and Amit Roy

Contributors

Shachi Agrawal Department of Botany, Gargi College, New Delhi, India

Nasheeman Ashraf Plant Biotechnology Division, CSIR—Indian Institute of Integrative Medicine, Srinagar, Jammu and Kashmir, India

Academy of Scientific and Innovative Research, CSIR—Indian Institute of Integrative Medicine, Jammu, Tawi, India

Sagar Banerjee Division of Biochemistry, ICAR-Indian Agricultural Research Institute, New Delhi, India

Vibha Gulyani Checker Department of Botany, Kirori Mal College, University of Delhi, New Delhi, India

Anju K. Chhibbar Department of Botany, Deshbandhu College, University of Delhi, New Delhi, India

Anil Dahuja Division of Biochemistry, ICAR-Indian Agricultural Research Institute, New Delhi, India

Namo Dubey Molecular Plant Pathology Laboratory, Division of Biotechnology, CSIR-Institute of Himalayan Bioresource Technology, Palampur, Himachal Pradesh, India

Geeta Department of Botany, Gargi College, University of Delhi, New Delhi, Delhi, India

Chaitali Ghosh Department of Zoology, Gargi College, University of Delhi, New Delhi, India

Om Prakash Gupta Division of Biochemistry, ICAR-Indian Agricultural Research Institute, New Delhi, India

Division of Quality and Basic Sciences, ICAR-Indian Institute of Wheat and Barley Research, Karnal, Haryana, India

Prem Kumar Gupta Ionics Clinical Laboratory, Gurgaon, India

Sumanti Gupta Department of Botany, Rabindra Mahavidyalaya, Tarakeswar, Hooghly, India

Monika Jaggi National Institute of Science Communication and Information Resources (NISCAIR), New Delhi, Delhi, India

Deepti Jain Interdisciplinary Centre for Plant Genomics, Department of Plant Molecular Biology, Delhi University-South Campus, New Delhi, Delhi, India

Suhas G. Karkute Division of Vegetable Improvement, ICAR-Indian Institute of Vegetable Research, Varanasi, Uttar Pradesh, India

Jitendra Paul Khurana Interdisciplinary Centre for Plant Genomics, Department of Plant Molecular Biology, Delhi University-South Campus, New Delhi, Delhi, India

Deepak Kumar Department of Botany, Central University of Jammu, Jammu, J&K, India

Pragati Kumari Department of Life Sciences, Singhania University, Rajasthan, India

Indian Agricultural Research Institute (IARI), New Delhi, India

Hemant Ritturaj Kushwaha School of Biotechnology, Jawaharlal Nehru University, New Delhi, India

Nand Lal Meena Plant Biochemistry, ICAR-Indian Institute of Millets Research, Hyderabad, Telangana, India

Reema Mishra Department of Botany, Gargi College, University of Delhi, New Delhi, Delhi, India

Vanita Pandey Division of Quality and Basic Sciences, ICAR-Indian Institute of Wheat and Barley Research, Karnal, Haryana, India

Jitendra Singh Rathore School of Biotechnology, Gautam Buddha University, Greater Noida, Uttar Pradesh, India

Amit Roy Department of Entomology, College of Agriculture, University of Kentucky, Lexington, KY, USA

Present Address: Faculty of Forestry and Wood Sciences, EXTEMIT-K, Czech University of Life Sciences, Suchdol, Czech Republic

Anjana Rustagi Department of Botany, Gargi College, New Delhi, India

Vinay Sharma Department of Bioscience and Biotechnology, Banasthali University, Rajasthan, India

Shashi Shekhar School of Life Sciences, Jawaharlal Nehru University, New Delhi, India

Garvita Singh Department of Botany, Gargi College, New Delhi, India

Kunal Singh Molecular Plant Pathology Laboratory, Division of Biotechnology, CSIR-Institute of Himalayan Bioresource Technology, Palampur, Himachal Pradesh, India

Preeti Singh School of Life Sciences, Jawaharlal Nehru University, New Delhi, India

Sanjay Kumar Singh Kentucky Tobacco Research and Development Center, University of Kentucky, Lexington, KY, USA

Anupama Razdan Tiku Department of Botany, Ramjas College, University of Delhi, Delhi, India

Zahoor Ahmed Wani Plant Biotechnology Division, CSIR—Indian Institute of Integrative Medicine, Srinagar, Jammu and Kashmir, India

Academy of Scientific and Innovative Research, CSIR—Indian Institute of Integrative Medicine, Jammu, Tawi, India

Saurabh Yadav Department of Biotechnology, Hemvati Nandan Bahuguna Garhwal (Central) University, Srinagar Garhwal, Uttarakhand, India

Sunita Yadav Department of Botany, HansRaj College, University of Delhi, New Delhi, India

About the Editors

Dr. Archana Singh is a Ph.D. in Plant Molecular Biology from the National Institute for Plant Genome Research (NIPGR). Since 2009, Dr. Archana has been working as an Assistant Professor in the Department of Botany, Hans Raj College, University of Delhi, where she teaches plant biotechnology, bioinformatics, genetics, cell biology and molecular biology. She has been awarded with Young Scientist Under Fast-Track Scheme (2013–2016) by Science and Engineering Research Board (SERB), Ministry of Science of Technology, Government of India. She has also been awarded with a prestigious Raman Post-Doctoral Fellowship by University Grant Commission, Ministry of Human Resources and Development, Government of India. Her quality research is evidenced by her publications in peer-reviewed journals of international repute like *Journal of Experimental Botany*, *Scientific Reports*, *Journal of Integrated Plant Biology*, *Journal of Asia-Pacific Entomology*, *Australasian Plant Pathology*, *Computational Biology and Chemistry* and *Plant Gene*. She has trained and supervised a few Junior Research Fellows and M.Sc. and B.Tech. students for their research projects. Over the years, she expanded her research interests from functional genomics of biotic stress to bioinformatics and proteomics.

Dr. Indrakant K. Singh is a Ph.D. in Zoology from the University of Delhi (DU). For last ten years, he has been working as an Assistant Professor in the Department of Zoology, Deshbandhu College, University of Delhi, where he teaches genetics, genomics, computation biology and immunology. He has been awarded with Young Scientist Award/Fellowship (2013–2016) by Science and Engineering Research Board (SERB), Ministry of Science of Technology, Government of India. He has also been awarded with a prestigious Raman Post-Doctoral Fellowship (2016–2017) by University Grant Commission, Ministry of Human Resources and Development, Government of India. He has published his quality research in peer-reviewed journals of international repute like *Journal of Experimental Botany*, *Scientific Reports*, *Journal of Biomedical Research*, *Journal of Asia-Pacific Entomology*, *Australasian Plant Pathology*, *Computational Biology and Chemistry* and *Plant Gene*. Presently, two Ph.D. scholars are working under his supervision. He has trained several B.Sc. and M.Sc. students for their research projects and now they are pursuing their doctoral research abroad. Over the years, he expanded his research interest from molecular basis of insect-plant interaction to functional genomics and proteomics.

Abbreviations

12-OPDA	12-oxophytodienoic acid
24-EpiBL	24-epibrassinolide
28-HomoBL	28-homobrassinolide
2D-DIGE	Two dimensional differential gel electrophoresis
2D-PAGE	Two dimensional polyacrylamide gel electrophoresis
5-HPT	5-hydroxytryptophan
AABL1	AB15 like1
ABA	Abscisic acid
ABC	ATP binding cassette
ACC	1-aminocyclopropane-1-carboxylic aid
ACMV	African cassava mosaic virus
ACS6	1-aminocyclopropane-1-carboxylic acid synthase
AGO	Argonaute
ALS	Amyotrophic lateral sclerosis
AMV	Alfaalfa mosaic virus
AP2	Apetala2
APAF	Apoptotic protease activating factor
APX	Ascorbate peroxidase
ARID	A/T rich interaction domain
as-1	Activation sequence 1
At	*Agrobacterium tumefaciens*
AtAGO1	*Arabidopsis* AGO1
AtEFR	*Arabidiopsis thaliana* EF-TU receptor
AtFSL2	*Arabidopsis thaliana* flagellin sensing 2
ATRM	*Arabidopsis* transcriptional regulatory map
ATX1	*Arabidopsis* homolog of trithorax
Aux	Auxin
Avr	Avirulence
BAK1	BRI1 associated receptor kinase 1
BAW	Beet armyworm
Bc	Botrytis cinerea
BECs	Blumeria effector candidates
BHNs	Broad host range necrotrophs
BIA	β-(Isoxazolin-5-on-2yl)-alanine

BIK1	BRI1 associated kinase
BIN2	Brassinosteroid insensitive 2
BL	Brassinolide
BMAA	β-methylamino-L-alanine
BPH	Brown plant hopper
BRI1	Brassinosteroid insensitive 1
BRs	Brassinosteroids
BWMK1	Blast and wounding activated map KINASE 1
bZIP	Basic domain leucine zipper
CaBD	Calmodulin binding domain
CaM	Calmodulin
CARD	Caspase recruitment domain
Cas	CRISPR associated
CAT	Catalase
CBB	Coomassie brilliant blue
CBD	Chitin binding domain
CC	Coiled coil
CD	C-terminal common docking
cDNA	Complementary DNA
CDPK	Calcium dependent protein kinase
CEBiP	Chitin elicitor binding protein
CED4	Cell death protein 4
CERK1	Chitin elicitor receptor kinase 1
CESA	Cellulose synthase cataylic subunit
CGs	Cyogenic glycosides
ChIP	Chromatin immunoprecipitation
CHS	Chalcone synthase
CKs	Cytokinins
CMV	Cucumber mosaic virus
CNR	Crinkler & necrosis
Col	Columbia
COR	Coronatine
COX5	Cytochrome oxidase subunit v
CP	Coat protein
CRISPR	Clustered regularly interspaced short palindromic repeats
CSEPs	Candidate secreted effector protein
CT	P-coumaroyl tyramine
CWDE	Cell wall degrading enzymes
D	Aspartate
DAMPs	Damage associated molecular patterns
DCL	DICER-like
DGE	Differential gene expression
DHAR	Dehydroascorbate reductase
DMT	DNA methyl transferase
DORN1	Does not respond to nucleotide 1

DRBs	Double stranded RNA binding proteins
DRE	Dehydration responsive element
dsDNA	Double stranded deoxyribonucleic acid
dsRNA	Double stranded Ribonucleic acid
DTI	Danger triggered immunity
eATP	Extracellular ATP
EBL	Epibrassinolide
ECC	*Erwinia carotovora*
ECM	Extra cellular matrix
ECS	Endocytosis cell signaling
EDS 1	Enhanced disease susceptibility 1
EF	Elongation factor
EFR	EF-Tu receptor
EF-Tu	Elongation factor Tu
EHC	Encasement of the haustorial complex
EHM	Extra haustorial membrane
EIX	Ethylene inducing xylanase
ELISA	Enzyme-linked immunosorbent assay
eLRR	Extra cytoplasmic leucine rich repeat
EPD	Eukaryotic promoter domain
EPSs	Extracellular polysaccharides
ER	Endoplasmic reticulum
ERE	Ethylene responsive elements
EREBP	Ethylene responsive element binding protein
ERF	Ethylene response factor
ESI	Electrospray ionization
ESTs	Expressed sequence tags
ET	Ethylene
ETI	Effector triggered immunity
ETR2	Ethylene resistant 2
ETS	Effector-triggered susceptibility
FISH	Fluorescence in-situ hybridization
FLAK	Phenylalanine, leucine, alanine, lysine
Flg	Flagellin
FLS 2	Flagellin sensing 2
FMs	Functional markers
FT	Feruloyl tyramine
GA	Gibberellic acids
GC-MS	Gas chromatography mass spectrometry
GE	Glucan elicitor
GEBP	GE-binding protein
GM	Genetically modified
GNA	*Galanthus nivalis* agglutinin
GR	Glutathine reductase
GST	Glutathione s-transferase

HCN	Hydrogen cynide
HCRSV	Hibiscus chlorotic ringspot carmovirus
hc-siRNAs	Hetrochromatic small interfering RNAs
HDA19	Histone deacetylase 19
HeMV	Hempane mosaic virus
HEN1	HUA ENHANCER
HGA	Homogalactouronam
HMGR2	3-hydroxy-3-methylglutaryl CoA reductase 2
HPLC	High performance liquid chromatography
HR	Hypersensitive response
hrc	HR and conserved
hrp	Hypersensitive reaction and pathogenicity
HSN	Host specific necrotroph
HSPs	Heat shock proteins
HSTs	Host specific toxins
HVMPK4	*Hordeum vulgare* signaling protein MAP kinase 4
HYL1	Hyponastic leaves 1
IAA26	Indole-3-acetic acid transcription factor
ICAT	Isotope coded affinity tag
IEF	Isoelectric focusing
IP-ELISA	Immune virus particle-ELISA
IPG	Immobilized pH gradient
IPP	Isopentenyl diphosphate
IPT	Isopentyl transferase
ISR	Induced systemic resistance
ITRAQ	Isobarric tagged for relative and absolute quantitation
JA	Jasmonic acid
JARE	Jasmonic acid responsive element
JS	Justamembrane
LC	Liquid chromatography
LC-MS/MS	Liquid chromatography-tandem mass spectrometry
L-DOPA	L-3,4-dihydoxyphenylalanine
LEA	Late embrogenesis abundant
LecRK	Lectin receptor kinase
Ler	Landsberg erecta
LGD1	Lagging growth development 1
LIR	Localised induced responses
LOS	Lipo-oligosaccharide
LPS	Lipopolysachharide
LRK	LRR kinase
LRR	Leucine rich repeat
lsiRNAs	Long small interfering RNAs
LTP	Lipid transfer protein
LYK3	LysM receptor like kinase 3
LZNBS-LRR	Leucine zipper nucleotide binding site leucine rich repeat

m/z	Mass to charge ratio
MALDI	Matrix assisted laser desorption/ionisation
MAMPs	Microbe associated molecular patterns
MAPK	Mitogen activated protein kinase
MCPs	Methyl accepting chemotaxis proteins
MDP	Muramyl dipeptide
MeJA	Methyl jasmonate
MIMPs	Microbe induced molecular patterns
miRNA	MicroRNA
MKK	MAP kinase kinase
MKS	MPK4 substrate
MLA 10	Mildew resistance locus A10
MoAGOs	M. oryzae genome encoded AGOs
MoDCL1	M. oryzae genome encoded DCL1
MoDCL2	M. oryzae genome encoded DCL2
MPs	Movement proteins
MPSS	Massively parallel signature sequencing
MS	Mass spectrometry
MSUD	Meiotic silencing of unpaired DNA
MTA	5′-methylthyoadenosine
MTI	MAMP triggered immunity
MTI	Microbial associated molecular pattern(MAMP) triggered immunity
MudPIT	Multidimensional protein identification technology
NAG	N-acetylglucosamine
NAM	N-acetyl muramic
NAMP	Nematode associated molecular pattern
NASBA	Nucleic acid sequence based amplification
nat-siRNAs	Natural antisense transcript-derived small interfering RNAs
NBS	Nucleotide binding site
NDPR	Non-pathogen derived resistance
NDR 1	Non race specific disease resistance 1
NEP1	Necrosis and ethylene inducing protein 1
NF	Nodulation factor
NFP	NOD factor perception
NGS	Next generation sequencing
NLPs	Necrosis and ethylene inducing peptide 1 like proteins
NLPs	Nep L like proteins
NLS	Nuclear localization signal
NO	Nitric oxide
NOD	Nucleotide binding oligomerization domain
NPAA	Nonprotein amino acids
NRG1	N-requirement gene
NtMKP1	Tobacco MAP kinase phosphatase 1
nTNL	Non TIR-NBS-LRR
O_2	Oxygen

OBF	Octopine synthase element binding factor
Obpv	Obuda pepper virus
OG	Oligogalacturonides
ORCAs	Octadecanoid responsive catharanthus APETALA2 domain proteins
Osa	*Oryza sativa*
osNramp6	osa-miRNA negative regulation of natural resistance associated macrophage protein 6
PAB	Plant associated bacteria
PAD3	Phytoalexin deficient 3
PAL	Phenylalanine ammonia lyase
PAMPs	Pathogen associated molecular pattern
PCD	Programmed cell death
PCR	Polymerase chain reaction
PD	Plasmodesmata
PDR	Pathogen derived resistance
Pel	Pectate lyase
PEN	Penetration genes
PepMV	Pepino mosaic virus
PEST	Pro-Glu-Ser-Thr
PG	Polygalactouronases
PGIP	Polygalactouronase inhibiting protein
PGN	Peptidoglycan
pI	Isoelectric point
PKs	Protein kinases
PLRV	Potato leafroll virus
PMEs	Pectin methyl esterases
PNPs	Plant natriuretic peptides
Pop	*Pseudomonas* outer protein
PPO	Poly phenol oxidase
PPP	Pentose phosphate pathway
PPV	Plum pox potyvirus
PR	Pathogenesis related
PR1	Pathogenesis related elements
pre-miRNA	Precursor miRNA
pri-miRNA	Primary-miRNA
PRLs	PR like proteins
PRRs	Pathogen recognition receptors
PRRs	Pattern recognition receptors
PRSV	Papaya ring spot virus
PS I	Photosystem I
PS II	Photosystem II
PSE1	Penetration specific effector 1
PstDC3000	Pathovar tomato strain DC3000
PTA-ELISA	Plate trapped antigen-ELISA
PTGS	Post transcriptional gene silencing

PTGS	Post translational gene silencing
PTI	PAMP triggered immunity
PTMs	Post translational modification
PVX	Potato virus X
PVY	Potato virus Y
PX	Peroxiredoxins
QAs	Quinolizidine alkaloids
qPCR	Real time quantitative PCR
QS	Quorum sensing
QTL	Quantitative trait loci
R	Resistance
RALF	Rapid alkalisation factor
ra-siRNAs	Repeat associated small interfering RNAs
RdDM	RNA-directedDNAmethylation
RDR	RNA dependent RNA polymerase
RDV	Rice dwarf virus
REL	Reticuloendotheliosis
REn	Replication enhancer
RenSeq	R gene enrichment and sequencing
RGC2	Resistance gene candidate 2
RIP	Repeat induced point
RISC	RNA induced silencing complex
RLCKs	Receptor like cytoplasmic kinases
RLK	Receptor like kinase
RLP	Receptor like protein
RMs	Random markers
RNAi	RNA interference
RNS	Reactive nitrogen species
ROMT	Resveratrol-o-methyltransferase
ROS	Reactive oxygen species
RP	Reverse phase
RPA	Reverse phase protein microarray
RPW8	Resistance to powdery mildew 8
Rsp1	Repetitive secreted protein 1
RSS	RNA silencing suppressors
RTD	Read through domain
RTP	Read through protein
RT-PCR	Reverse transcriptase-PCR
RUBISCO	Ribulose-1,5-biphosphate carboxylase oxygenase
RYMV	Rice yellow mottle virus
SA	Salicylic acid
SAGE	Serial analysis of gene expression
SAM	S-adenosyl methionine
SAR	Systemic acquired resistance
SARE	Salicylic acid responsive elements

SARE	SA-responsive element
SCF	SKP1-cullin-f-box protein
SCX	Strong cation exchange
SDS-PAGE	SDS- polyacrylamide gel electrophoresis
SE	Sieve element
See1	Seedling efficient effector 1
SEL	Size exclusion limit
SERK3	Somatic embryo receptor kinase 3
SIB1	Sigma factor binding protein 1
SIPK	Salicylic acid protein kinase
SIS	Sex induced silencing
SMV	Soybean mosaic virus
SNARE	Soluble N-Ethylmaleimide-sensitive factor attachment protein receptor
SNPs	Single-nucleotide polymorphisms
SOD	Superoxide dismutase
sRNAs	Small RNAs
ssDNA	Single stranded deoxyribonucleic acid
SSEM	Serologically specific electron microscopy
SSH	Suppression subtractive hybridization
SSITL	*Sclerotinia sclerotiorum* integrin like
SSPs	Small secretory proteins
SSR	Simple sequence tags
ssRNA	Single stranded ribonucleic acid
SSRs	Simple sequence repeats
STAND	Signal transduction ATPases with numerous domains
STB	Septoria tritici blotch
STS	Stilbone synthase gene
SYMRK	Symbiosis receptor kinase
ta-siRNAs	Trans-acting small interfering RNAs
TBP	TATA-box -binding protein
TCNL	TIR-CC-NBS-LRR
TCV	Turnip crinkle virus
TFs	Transcription factors
TGB	Triple gene block
TGS	Transcriptional gene silencing
THI2.1	Thionin 2.1
TIPK	Trichoderma induced MAPK
TIR	Toll interleukin 1 receptor
TLR 4	Toll like receptor 4
TMV	Tobacco mosaic virus
TNL	TIR-NBS-LRR
TOF-MS	Time of flight mass spectrometry
TrD	Transmembrane domain
tRNA	Transfer RNA

TSWV	Tomato spotted wilt virus
TTSS	Type three secretion system
TuMP	Turnip mosaic virus
TVMV	Tobacco vein molting virus
TYLCD	Tomato yellow leaf curl disease
VIP1	VirE2 interacting protein 1
VOCs	Volatile organic compounds
VRC	Viral replication complex
vRNP	Viral ribonucleoprotein complex
W	Tryptophan
WAK1	Wall associated kinase 1
WIPK	Wound induced protein kinase
Xop	Xanthomonas outer protein

Arabidopsis thaliana as a Model Organism to Study Plant-Pathogen Interactions

1

Shachi Agrawal

Abstract

Arabidopsis thaliana (*a crucifer*) provides a model system in every discipline of plant sciences including plant pathology with a varied array of molecular and genetic resources and biological information. Members of crucifer are widely distributed geographically and are well adapted to various plant pathogens such as fungi, bacteria, viruses, and nematodes. Besides small plant size, short life cycle, small genome size, availability of whole genome sequence, and easy genetic and mutational analysis, its response to the pathogen attack in a similar fashion as other higher plant species and an extensive collection of mutants available to determine defense pathway are the characteristics, which identify this plant as an indispensable research model in plant-pathogen interaction studies. This chapter mainly focuses on various existing model pathosystems of *Arabidopsis* with viral, bacterial, and fungal pathogens including an outlook on how this knowledge can be translated from *Arabidopsis*-pathogen model system to other crop plants. A general and brief overview of plant-pathogen interactions and how *A. thaliana* recognize and respond to pathogens is also portrayed.

Keywords

Effector molecules · Hypersensitive response (HR) · Plant defense · Plant defensin gene · PR proteins · Resistance genes · Signal molecules · Systemic acquired resistance (SAR)

S. Agrawal (✉)
Department of Botany, Gargi College, New Delhi, India
e-mail: gupta_shachi@yahoo.co.in

A. Singh, I. K. Singh (eds.), *Molecular Aspects of Plant-Pathogen Interaction*,
https://doi.org/10.1007/978-981-10-7371-7_1

1.1 Introduction

When a pathogen attacks a plant, either a pathogen can proliferate and can cause development of disease or the plant can resist the pathogen by means of active or passive form of resistance. During resistance, plants recognize a race-specific avirulence determinant produced by the pathogen (Keen 1990; Scofield et al. 1996; Tang et al. 1996); defense mechanisms are activated leading to hypersensitive response (HR) (Matthews 1991). At the same time, expressions of pathogenesis-related (PR) proteins as well as plant defensins are induced due to gene-for-gene interactions and rapid localized cell death (Narasimhan et al. 2001; Asano et al. 2012). Signaling molecules, salicylic acid (SA), jasmonic acid (JA), ethylene (ET), and reactive oxygen species (ROS) are directly involved in plant defense against pathogens (Clarke et al. 2000; Kunkel and Brooks 2002; Hossain et al. 2007; Asano et al. 2012). There is rich information on plant-pathogen interaction on many species. Advanced molecular tools are also accessible that can be used to study the function and evolution of genes that are important for plant defense such as those that control responses to wide range of pathogens. However, studies in molecular plant pathology require large initial investments in molecular technologies. It is cost-effective since these investments are shared among multiple laboratories by means of publications, bioinformatic tools, and databases such as TAIR. Moreover, researchers gain in-depth biological understanding when they compare and match previous studies from a research community that shares the tools and resources of model organisms. Although, it is imperative to study individual plant-pathogen interactions at species level to gain better knowledge. But, at the same time, *A. thaliana* serves as a model system to answer many basic questions related to plant-pathogen interaction due to availability of complete genome sequence and having a small genome size together with the extensive collection of new mutants and germplasm as well as the presence of specialized transformation techniques, its rapid growth, can be handled easily in the laboratory conditions, mutagenesis can also be done easily and the possibility of using microarrays for gene expression analysis. *Arabidopsis* is susceptible to only a limited number of pathogens including viruses, bacteria, fungi, nematodes, and insect pests, and it responds to the pathogen attack in a similar fashion to those of other higher plant species.

A. thaliana (L.) Heynh. is an annual flowering plant that belongs to mustard family (Cruciferae or Brassicaceae). It is a native of Eurasia, which has a broad natural distribution throughout Europe, Asia, and North America. Of late it has been introduced and naturalized worldwide. It is speculated that its spread was facilitated by the expansion of agriculture (Francois et al. 2008). *A. thaliana* is considered as a weed as it grows in open or recently disturbed habitats. *Arabidopsis* shows extensive natural variation for different developmental, abiotic, and biotic stress resistance traits (Koornneef et al. 2004; Alonso-Blanco et al. 2009; Atwell et al. 2010). Till date, over 750 different ecotypes (accessions) of *A. thaliana* have been collected from natural populations that are available for experimental analyses. The most commonly used ecotypes of *Arabidopsis* for genetic and molecular studies are Columbia (Col) and Landsberg erecta (L*er*). The entire life cycle of *A. thaliana* is completed in 6 weeks, which includes seed germination, rosette formation, bolting,

Fig. 1.1 An *Arabidopsis* plant grown under laboratory condition

flowering, and maturation of seeds. The plant is a small-sized herb with overall length of around 15–20 cm; leaves are 1.5–5 cm long and 2–10 mm broad. Flowers (2 mm length and 3 mm diameter) undergo self-pollination but can be crossed manually. The fruit is called a silique (5–20 mm long) that contains 20–30 seeds. On germination, the seed develops into a rosette plant (2–10 cm diameter), wherein the whorls of leaves are covered with trichomes (Fig. 1.1). Under laboratory conditions *Arabidopsis* can be grown easily in petri plates, pots, or hydroponics, either under fluorescent lights or in a greenhouse. Inflorescence is a corymb that appears as a result of bolting after 3 weeks of planting. Several hundred siliques are produced per plant, which account for more than 5000 total seeds. The plant has a single primary root that grows vertically downward and produces smaller lateral roots that are easy to study in culture.

1.2 Plant-Pathogen Interactions

An array of pathogens including fungi, bacteria, and viruses attack the plant kingdom. Different strategies have been devised by different pathogens to invade, feed on, and reproduce in the host plants. Pathogens can be broadly classified as biotrophs and necrotrophs based upon the strategy used by them to invade and infect a plant (Oliver and Ipcho 2004). Biotrophic pathogens are those that require a living host tissue for its growth and reproduction. In some cases wherein the tissue dies in the later stages of the infection, the pathogens are classified as hemibiotrophs. On the contrary, necrotrophic pathogens kill the host tissue as soon as they infect it and then grow and feed on the dead tissue. Viruses are classified as biotrophic pathogens, whereas bacteria and fungi follow both biotrophic as well as necrotrophic strategies of invasion. Plants respond to different kinds of pathogens differently. Pathogens can further be classified as those with different primary target tissues encountering different environmental conditions. Those pathogens that target the green, photosynthesizing, and assimilate-producing source tissues like leaves will encounter different kinds of defense responses in comparison to pathogens infecting the assimilate-importing tissue such as roots, flowers, and sink leaves (Berger et al. 2007).

Plant defense mechanism against pathogens can be either preformed (primary) or induced (secondary). The first and foremost step required for the activation of defense response is to recognize the presence of microorganisms. Elicitors are molecules that at very low concentrations induce plant defense response (Thakur and Sohal 2013). Recognition of microorganism-derived elicitors initiates the basal resistance in plants. This defense response involves activation of ion fluxes, phosphorylation/dephosphorylation of proteins by protein kinases and phosphatases, and production of signaling molecules such as adapter proteins, salicylic acid, jasmonic acid, ethylene, reactive oxygen species, and nitric oxide. These steps further initiate an array of signaling that leads to the regulation of expression of defense-related genes and the induction of defense responses. These responses include cell wall strengthening, accumulation of phytoalexins and pathogenesis-related (PR) proteins, and localized programmed cell death (PCD) (McDowell and Dangl 2000; Dangl and Jones 2001; Garcia-Brugger et al. 2006).

Plants also possess an innate immune system that perceives the presence of pathogens by recognition of molecules known as microbe- or pathogen-associated molecular patterns (MAMPs or PAMPs, respectively) or by sensing effector proteins secreted by the host during plant-pathogen interactions. Early interactions between PAMPs/MAMPs and cell surface receptors (pathogen recognition receptors or PRRs) lead to appropriate defenses by activating multicomponent and multilayered responses. The establishment of defense is triggered by several pathways that can involve Ca^{2+} influx, generation of reactive oxygen and nitrogen species (ROS and RNS, respectively), and synthesis of phytohormones such as jasmonic acid (JA), salicylic acid (SA), and ethylene (ET), which act as signal molecules (Pieterse et al. 2009). Plant immunity may be described at two levels (Jones and Dangl 2006). The first one involves cell surface pattern recognition receptors (PRRs) to detect pathogen-associated molecular patterns (PAMPs) and initiate PAMP-triggered immunity (PTI). The second involves nucleotide-binding leucine-rich repeat (NB-LRR) proteins, encoded by resistance (*R*) genes, which sense pathogen effectors and elicit a potent immune response called effector-triggered immunity (ETI). ETI is faster, longer, and stronger than PTI and usually leads to a local cell death, the hypersensitive response (HR), which stops the spread of the pathogen (Jones and Dangl 2006). In some cases, pathogens can evade such recognitions also and suppress host immunity with effectors, causing effector-triggered susceptibility

(ETS). R proteins recognize some effectors that enable the pathogen to overcome PTI, and the effectors are thus termed an avirulence (Avr) protein (Jones and Dangl 2006).

As per our current understanding, virulence of the virulent pathogenesis is contributed by the production of effector molecules which thereby suppress plant defense, and thus the compatible interactions allow the spread of the pathogen in the susceptible plant (Jones and Dangl 2006). Herein the pathogen proliferates at a rate in which the plant defense could not keep pace with that subsequently leads to the development of disease and necrosis. On the other hand, in resistant plants, the specific resistance is governed by the recognition of the activity of pathogen effector molecules (race-specific avirulence determinant) by plant receptor proteins (Keen 1990; Scofield et al. 1996; Tang et al. 1996; Berger et al. 2007). Hence, these incompatible interactions prevent the pathogen from spreading and impart resistance to the plant. The disease resistance "R" genes encode the microbe recognizing plant receptors. Those pathogens that cannot establish themselves in the host plant are called as avirulent strains of plant pathogens, and their early recognition combined with fast activation of plant defense mechanisms results in the inducible defense system (Jones and Dangl 2006). Moreover, the recognition of the avirulent strain determinant activates a hypersensitive response (HR) that is characterized by localized PCD resulting in small necrotic lesions that efficiently restrict the spread of biotrophic pathogens (Heath 2000; Narasimhan et al. 2001). In addition, plant defensins (PDF1.1, PDF1.2) mRNAs are expressed in response to gene for gene interaction (Narasimhan et al. 2001). As discussed earlier, various signaling molecules like jasmonic acid, salicylic acid, ethylene, and reactive oxygen species (ROS) are directly involved in such inducible defense systems (Clarke et al. 2000; Kunkel and Brooks 2002; Hossain et al. 2007). The jasmonate/ethylene signaling pathway seems to be the most important mechanism in defending against necrotrophic pathogens. On the other hand, in order to combat against the biotrophic pathogens, plants recruit the salicylic acid-dependent responses (Thomma et al. 2001).

1.3 How *Arabidopsis thaliana* Recognize and Respond to Pathogens?

In nature, plants are exposed to a large number of pathogens, but somehow they are susceptible to only a few of them. This may be due to the presence of different defense mechanisms exhibited by the plants (Nimchuk et al. 2003; Jones and Takemoto 2004). The disease resistance (*R*) genes that are involved in pathogen recognition show excessive polymorphism. This polymorphism has been speculated as a cause for plant resistance. In monoculture, loss of R gene polymorphism results in reduced resistance and increased susceptibility (Stahl and Bishop 2000). *Arabidopsis* is prone to infection by pathogens that includes viruses, bacteria, fungi, nematodes, and insects. As the mode of response to the pathogen attack is highly conserved in higher plant species, study of *Arabidopsis*-pathogen interactions have greatly helped the scientists to understand the molecular and cellular basis of host-pathogen interactions, disease resistance, and pathogen virulence (Andargie and Li 2016).

As stated earlier, R genes are important for parasite recognition and initiation of defense mechanism. A total of 150 different R genes have been identified in *Arabidopsis* genome that are located unevenly on chromosomes with 49, 2, 16, 28, and 55 R gene loci on chromosome number 1, 2, 3, 4, and 5, respectively (*Arabidopsis*-Genome-Initiative 2000). These R genes encode for proteins that contain nucleotide-binding (NB) domain(s) that binds to ATP or GTP along with a carboxy-terminal leucine-rich repeat (LRR) domain (S) that facilitate protein-protein interactions and ligand binding. They are further classified as those that contain toll interleukin 1 receptor domain (TIR) or coiled-coil (CC) domain at their amino terminal. Thus, broadly they can be classified as TIR-NB-LRR and CC-NB-LRR. *Arabidopsis* genome contains 85 TIR-NB-LRR resistance genes at 64 loci and 36 CC-NB-LRR resistance genes at 30 loci (*Arabidopsis*-Genome-Initiative 2000). Some of these R genes carry additional domains also, like WRKY transcription factor domain and protein kinase domain that have also been implicated in plant defense.

Studies were carried out to compare the defense mechanisms in plants and animals. Nitric oxide production seems to be a common response in both plants and mammals in conditions of biotic stress. But distinct homologue of nitric oxide synthase gene was not found in *Arabidopsis*. REL (reticuloendotheliosis) domain transcription factors or similar proteins that are involved in innate immunity in both *Drosophila* and mammals or their homologs were not detected in *Arabidopsis thaliana*. Moreover, no homologues were detected for genes like classical caspases, bcl2/ced9, and baculovirus p35 that are involved in apoptosis regulation in animal cells; however, eight homologues of metacaspase family protein and 36 cysteine proteases were found in *Arabidopsis* (*Arabidopsis*-Genome-Initiative 2000; Uren et al. 2000). The production of reactive oxygen intermediates is a primary response that is common to both plant and animal during pathogen recognition. This process involves transfer of electrons across the plasma membrane in mitochondria to make superoxide by a specialized respiratory burst oxidase. In mammals, gp91 is the subunit of NADH oxidase that catalyzes the final step of electron transfer to molecular oxygen (O_2), resulting in the generation of superoxide ion (O_2^-) (Yu et al. 1998). The *Arabidopsis* genome has eight functional homologues of *gp91*. These homologues are called *Atrboh* genes and have been implicated in plant defense response (Torres et al. 2002). In mammals, gp91 activity requires the action of Rac proteins, but no Rac or Ras proteins are found in *Arabidopsis*; however, a large family of rop genes that are related to G-proteins are present and may carry the same function. The various pathogens of *Arabidopsis thaliana*, gene associated with natural variation of response to those pathogens and their molecular functions, are summarized in Table 1.1.

Table 1.1 Various pathogens of *Arabidopsis thaliana* along with gene associated with natural variation of response to pathogen interactions and their molecular functions (Roux and Bergelson 2016)

Pathogens of *Arabidopsis*	Associated gene locus	Class of the associated gene
Viruses		
Turnip crinkle virus (TCV)	HRT	CC-NBS-LRR protein
Cucumber mosaic virus (CMV)	RCY1	CC-NBS-LRR protein
Tobacco ringspot virus (TRSV)	TTR1	TIR-NBS-LRR protein
Tobacco etch virus (TEV)	RTM1	Jacalin-like lectin protein
	RTM2	Small heat shock-like protein
	RTM3	MATH domain-containing protein
Plum pox virus (PPV)	RTM1	Jacalin-like lectin protein
	RTM2	Small heat shock-like protein
	RTM3, rwm1/rpv1	MATH domain-containing protein
		Nucleus-encoded chloroplast phosphoglycerate kinase
Lettuce mosaic virus (LMV)	RTM1	Jacalin-like lectin protein
	RTM2	Small heat shock-like protein
	RTM3	MATH domain-containing protein
Plantago asiatica mosaic virus (PAMV)	JAX1	Jacalin-like lectin protein
Watermelon mosaic virus (WMV)	rwm1/rpv1	Nucleus-encoded chloroplast phosphoglycerate kinase
Bacteria		
Pseudomonas syringae	RPM1/RPS3	CC-NBS-LRR protein
	RPS2	CC-NBS-LRR protein
	RPS5	CC-NBS-LRR protein
	RPS4	TIR-NBS-LRR protein
	RRS1	TIR-NBS-LRR WRKY protein
	ACD6	Ankyrin-repeat transmembrane protein
Xanthomonas campestris	RPS4	TIR-NBS-LRR protein
	RRS1	TIR-NBS-LRR WRKY protein
	RKS1	A typical kinase
	AT5G22540	Protein of unknown function
Ralstonia solanacearum	RPS4	TIR-NBS-LRR protein
	RRS1	TIR-NBS-LRR WRKY protein
	ERECTA	LRR receptor-like kinase
Fungi		
Elicitor from *Sclerotinia sclerotiorum*	RLP30	Receptor-like protein
Botrytis cinerea	RLP30	Receptor-like protein
	EGM1	Receptor-like kinase
	EGM2	Receptor-like kinase
	RLM3	TIR-NB protein

(continued)

Table 1.1 (continued)

Pathogens of *Arabidopsis*	Associated gene locus	Class of the associated gene
Fusarium oxysporum	RFO1	Wall-associated receptor-like kinase
	RFO2	Receptor-like protein
	RFO3	Receptor-like kinase
Alternaria brassicicola	RLM3	TIR-NB protein
Alternaria brassicae	RLM3	TIR-NB protein
Colletotrichum higginsianum	RPS4	TIR-NBS-LRR protein
	RRS1	TIR-NBS-LRR WRKY protein
Oomycetes		
Albugos candida	RAC1	TIR-NBS-LRR protein

1.4 *Arabidopsis thaliana*: An Important Model Host for Studying Plant-Pathogen Interactions

A. thaliana is an important model host for studying plant-pathogen interactions due to several reasons as described earlier. *Arabidopsis* is susceptible to only a limited number of pathogens including viruses, bacteria, fungi, nematodes, and insect pests. Diseases resulting from these pathogens have been reported in the wild (Holub et al. 1994, 1995; Tsuji and Somerville 1992) suggesting both the pathogen and the host share an ecological niche, and when the appropriate environmental conditions are present, disease can occur. Diseases have also been observed in a laboratory setting where the host is deliberately exposed to the pathogen. Regardless of the setting, nature, or the laboratory, *Arabidopsis* responds to the pathogen attack in a similar fashion as other higher plant species when exposed to viral, prokaryotic, or eukaryotic pathogens (Andargie and Li 2016). Since the 1990s till today, several plants have been recognized as model systems for plant-pathogen interactions such as tobacco, tomato, etc., but *A. thaliana* has been used extensively as a model plant to have an overview of the plant-pathogen interactions with a wide variety of pathogens. The *A. thaliana* genetic system is significantly more tractable than those of the other plant species, which were hampered by long generation times and large, polyploid, or repetitive genomes. Agriculturally important crucifers such as *Brassica napus*, *Brassica rapa* (oilseed rape, canola), *B. oleracea*, *Brassica* spp., European cabbage, cauliflower, Chinese cabbage, and radish (*Raphanus* spp.) are the closest relatives of *Arabidopsis*, so all the informations available can be useful for studying plant-pathogen interactions in these related spp. that are economically important crops. But molecular studies can be more complex in these spp. since they are mostly polyploids.

Besides, *A. thaliana* exhibits all of the major kinds of defense responses described in other plants. Furthermore, a large number of virulent and avirulent bacterial, fungal, and viral pathogens of *A. thaliana* have been deciphered (Glazebrook et al. 1997). Mutants defective in almost every aspect of plant growth and development have been identified and studied by the various research groups over the world. Novel insights into events subsequent to pathogen recognition in *A. thaliana* have been obtained from mutants altered in defense (Buell 1998). Several mutant groups in *A. thaliana* exist today: lesion mimic mutants, phytoalexin mutants, as well as enhanced susceptibility and resistance mutants. With the variety of mutants available, it is possible to determine which defense pathways are activated during pathogen attack and what leads to the subsequent resistance or susceptibility. As research progresses, the different mutants will be linked to specific genes finally leading to a better understanding of the various genes involved in plant response pathways (Glazebrook et al. 1997).

1.5 *A. thaliana*-Pathogen Interactions

Arabidopsis has been reported as a susceptible host to a range of pathogens and resistant to other pathogens. The findings related to defense mechanism in *Arabidopsis* have been successfully implemented in many model systems, which have been developed to better understand interactions between plants and pathogens. The primary response of *Arabidopsis* includes the perception of pathogens by cell surface pattern recognition receptors (PRRs) and is referred to as PAMP-triggered immunity (PTI). Activation of FLS2 and EFR triggers MAPK signaling pathway that activates defense genes for synthesis of antimicrobial compounds. *Arabidopsis* possess specific intracellular surveillance proteins (R proteins) to monitor the presence of pathogen virulence proteins. This ETI occurs with localized programmed cell death to arrest pathogen growth, resulting in cultivar-specific disease resistance.

1.5.1 *Arabidopsis*-Virus Interactions

Viral infections and their spread throughout a plant require numerous interactions between the host and the virus. Systemic viral infections in plants are complex processes that require compatible virus-host interactions in multiple tissues. These interactions include viral genome replication in the cytoplasm of the initially infected cells, cell-to-cell movement toward neighboring tissues, long-distance movement through the vascular tissue, phloem unloading, and cell-to-cell movement in non-inoculated systemic tissues (Carrington et al. 1996). Incompatibilities between virus and host factors at any of these stages could therefore lead to restrictions and delay establishment of a systemic infection. The utility of *Arabidopsis* as a model system has not gone unnoticed, and several viruses previously found to be pathogenic on crucifers have also been found to infect *Arabidopsis*. This model organism has proven to be useful to understand the relationship between the host plant and the virus replication and movement processes (Kunkel 1996; Yoshii et al. 1998). Susceptible interactions between plants and viruses can result in a variety of visible symptoms ranging from mild stunting to overall necrosis.

Although plant viruses are among the least genetically complex pathogens, they use a variety of strategies to suppress or bypass host defense and infect susceptible hosts. In plants, these strategies may be an enhancement of infection by manipulating host resources, such as the formation of replication complexes (Hills et al. 1987), enlargement of the plasmodesmata size-exclusion limit (Wolf et al. 1989; Waigmann et al. 1994), evolution of viral suppressors of RNA silencing to counteract antiviral silencing (Burgyan and Havelda 2011), interference with plant cell cycle regulation (Lai et al. 2009), and using host components for its own replication. In turn, plants have evolved intricate mechanisms to fight viral infection, such as pathways mediated by gene silencing, hormone-mediated signaling pathways, and regulation of metabolism (Gao et al. 2016).

A number of viruses such as *turnip crinkle virus*, *cucumber mosaic virus*, *tobacco ringspot virus*, *tobacco etch virus*, *plum pox virus*, *lettuce mosaic virus*, *plantago asiatica mosaic virus*, and *watermelon mosaic virus* are reported to infect *A. thaliana*. In *Arabidopsis-turnip crinkle virus* pathosystems, *turnip crinkle virus* (TCV) is the smallest and simplest plant RNA virus belonging to *Tombusviridae* family *Carmoviruses*. It infects many *Arabidopsis* ecotypes (Dempsey et al. 1993). The ecotype Dijon (Di-O) of *A. thaliana* has been reported to be resistant to TCV-M infection. This resistance is not expressed at the cellular level but rather at the level of whole plant. In *Arabidopsis*, resistance to most viral pathogens does not involve an HR (Ishikawa et al. 1991; Leisner et al. 1993; Lee et al. 1994; Callaway et al. 1996). However, inoculation of turnip crinkle virus (TCV) on plants from the resistant ecotype Dijon (Di-0 or Di-17) shows both an HR and the induction of PR gene expression (Simon et al. 1992; Dempsey et al. 1993, 1997; Uknes et al. 1993). Furthermore, it was found that the HR and resistance are dependent on SA but independent of NPR1, ethylene, and JA-mediated defense signaling (Kachroo et al. 2000). Genetic analyses revealed that HR development is conferred by a single dominant gene termed HRT (for HR to TCV) (Dempsey et al. 1997). HRT gene has been reported to trigger defense gene expression and SA accumulation (Cooley et al. 2000; Ren et al. 2000; Jeong et al. 2008). Thus, HRT also appears to be required for resistance to TCV infection along with a second locus, named RRT (which regulates resistance to TCV). In contrast, TCV-susceptible ecotypes, including Columbia (Col-0), fail to mount an HR, exhibit delayed and weak PR gene expression, and develop systemic disease symptoms (Li and Simon 1990; Dempsey et al. 1993). TCV infection triggers the production of ROS as well as alteration of cellular redox in *Arabidopsis* (Pu et al. 2016).

1.5.2 *Arabidopsis*-Bacterium Interactions

Perhaps, bacterial pathogens are the more facile pathogens to work with in a laboratory setting as bacteria have several advantages over the other classes of pathogens for pathological studies. They can be cultured in vitro and have relatively rapid generation times (only minutes not even days). In addition, most bacterial plant pathogens elicit rapid host responses (hours to days) and have pathogenicity and avirulence factors that have been documented in other plant species. Thus, they provide a foundation to begin work in *Arabidopsis*. Only a small number of bacterial species are pathogenic on *Arabidopsis*. The predominant bacterial pathogen utilized in *Arabidopsis* studies is *Pseudomonas syringae* on which many scientists have reported various important findings. Additional bacterial pathogens utilized in *Arabidopsis* research include *Erwinia* species (causal agents of soft rots), *Ralstonia* species (causal agents of vascular wilts), and *Xanthomonas campestris* pathovars (causal agents of blights and rots). A brief review of *Arabidopsis-Pseudomonas* pathosystems is presented in this chapter.

The interaction between *A. thaliana* and various phytopathogenic *Pseudomonas* pathovars presents an outstanding model to genetically define plant and bacterial loci necessary for generation of a hypersensitive response (HR) (Dangl et al. 1991). Certain isolates of *Pseudomonas syringae pv. maculicola* are virulent on *Arabidopsis*, while others are not. *Pseudomonas syringae* is a Gram-negative, rod-shaped bacterium with polar flagella (Agrios 1997). Different strains of *P. syringae*, however, are known for their diverse and host-specific interactions with plants (Hirano and Upper 2000). Understanding the molecular basis of this high level of host specificity has been a driving force in using *P. syringae* as a model for the study of host-pathogen interactions. In crop fields, infected seeds are often an important source of primary inoculum in *P. syringae* diseases, and epiphytic bacterial growth on leaf surfaces often precedes disease development (Hirano and Upper 2000). *P. syringae* enters the host tissues (usually leaves) through wounds or natural openings such as stomata, and in a susceptible plant, it multiplies to high population levels in intercellular spaces. Infected leaves show water-soaked patches, which eventually become necrotic. Depending on *P. syringae* strains, necrotic lesions may be surrounded by diffuse chlorosis. Some strains of *P. syringae* also cause cankers and galls (Agrios 1997). In resistant plants, on the other hand, *P. syringae* triggers the hypersensitive response (HR), a rapid, defense-associated death of plant cells in contact with the pathogen (Klement et al. 1964). In this situation, *P. syringae* fails to multiply to high population levels and causes no disease symptoms. Several strains belonging to pathovars *tomato*, *maculicola*, *pisi*, and *atropurpurea* of *Pseudomonas syringae* may infect the model plant *A. thaliana* (Crute et al. 1994). The establishment of the *Arabidopsis-P. syringae* pathosystem triggered a period of highly productive research that has contributed to the elucidation of the fascinating mechanisms underlying plant recognition of pathogens, signal transduction pathways controlling plant defense responses, host susceptibility, and pathogen virulence and avirulence determinants.

Upon infection with *P. syringae*, PTI and the subsequent SA-dependent and independent defenses are activated in *Arabidopsis* (Tsuda et al. 2008). About 36 protein effectors (Cunnac et al. 2009; Lindeberg et al. 2009) are secreted into the host cell by a remarkable system conserved in Gram-negative bacteria called the type III secretion system (TTSS). These effectors suppress both SA-independent and SA-mediated basal defenses (Nomura et al. 2005; Kim et al. 2008). Furthermore, *P. syringae* pathovars secrete the phytotoxin coronatine (COR) involved in bacterial virulence (Kloek et al. 2001). COR has been reported to target JA, ET, AUX, and ABA pathways, which play important roles to antagonize the SA-mediated defenses (Thilmony et al. 2006). COR also suppresses PTI and induces the reopening of

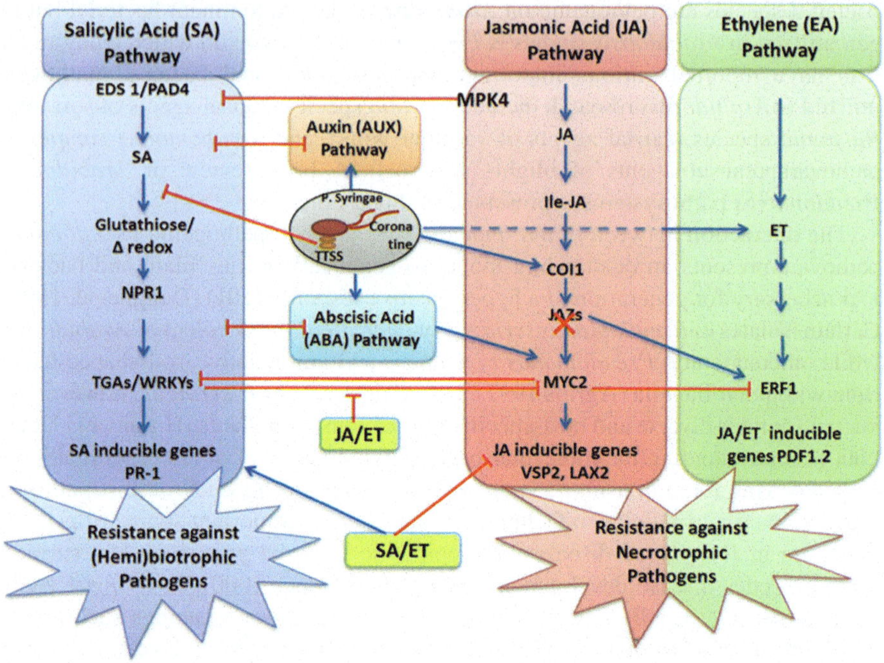

Fig. 1.2 Networking between the principal plant defense hormones during *Arabidopsis-Pseudomonas* interaction. The signaling cascades of the major plant defense pathways (salicylic acid (SA), jasmonic acid (JA), ethylene (ET), which are the backbone of defense signaling network) along with other hormonal signaling pathways feeding into it are shown here. The cross-communications between these signaling pathways that lead to protection against different pathogens may have synergistic (shown in blue color) as well as antagonistic (shown in red color) effects. In addition, the hemibiotrophic bacteria *Pseudomonas syringae* can suppress host defenses by manipulating plant hormone pathways through secretion of the phytotoxin coronatine and effectors delivered by the type III secretion system (TTSS) (Adapted from Pieterse et al. 2009)

stomata that have been closed upon recognition of the bacterial flagellin and lipopolysaccharides by *Arabidopsis* (Melotto et al. 2006). Both COR and TTSS effectors display common and distinct virulence effects in *Arabidopsis* such as targeting the JA pathway, thereby repressing the SA-inducible defenses (Zhao et al. 2003; He et al. 2004; Uppalapati et al. 2007). Thus, during *P. syringae* infections, TTSS effectors and COR plays important role in suppression of SA-inducible responses and promote full susceptibility (Laurie-Berry et al. 2006). Some TTSS effectors of *P. syringae* also interfere with the AUX (Navarro et al. 2006; Chen et al. 2007; Zhang et al. 2007) and the ABA pathways (de Torres-Zabala et al. 2007), both promoting the repression of SA-inducible defenses. These effectors interfere with the hormonal network of plant defenses, thereby leading to successful infection through the suppression of basal defenses including the SA pathway (Fig. 1.2).

Importantly, and in contrast to most fungal pathogens, *Pseudomonas syringae* is amenable to molecular genetic manipulations such as gene introduction, transposon mutagenesis, and targeted gene replacement. Some *P. syringae* pv. tomato and pv. maculicola strains are virulent on some *Arabidopsis* ecotypes (Quirino and Bent 2003). Among its various strains, pathovar tomato strain DC3000 (PstDC3000) can usually infect the plant host *A. thaliana*. This *A. thaliana-Pseudomonas syringae* pv. tomato DC3000 (DC3000) pathosystem has been reported to be an ideal system to understand both microbial-associated molecular pattern (MAMP)-triggered immunity (MTI) and effector-triggered susceptibility (ETS) processes at the transcriptional level (Xin and He 2013). DC3000 is highly virulent on *Arabidopsis* as it directly delivers 28 effector proteins (Cunnac et al. 2009) into the host cell through type III secretion system (TTSS) as well as small molecules such as coronatine (COR). These virulence factors collectively suppress MAMP-triggered immunity (MTI) and enhance nutrient availability, therefore allowing bacterial multiplication. A key structural component of the TTSS pilus is the HrpA protein. DC3000*hrpA*-mutants activate MTI but cannot form a TTSS to deliver the suite of effectors that can suppress MTI. They also produce minimal amounts of coronatine. Thus, DC3000*hrpA*-infection triggers MTI in the host (Lewis et al. 2015).

1.5.3 *Arabidopsis*-Fungus Interactions

A large number of fungal and oomycete pathogens have been reported to infect *A. thaliana*, either naturally or in the laboratory. These include *Ustilaginoidea*, *Botrytis*, *Fusarium*, *Colletotrichum*, obligate biotrophs (e.g., *Peronospora*, *Albugo*, and *Erysiphe*), hemibiotrophs (i.e., facultative biotrophs; e.g., *Phytophthora*), and necrotrophs (e.g., *Alternaria*, *Botrytis*, and *Rhizoctonia*). On infection, *Arabidopsis* respond to each pathogen in a specific manner. A brief review of *Arabidopsis-Ustilaginoidea* pathosystems is described here.

The biotrophic ascomycete fungus, *Ustilaginoidea virens* (Cooke) Tak (teleomorph: *Villosiclava virens*), causes one of the most severe false smut diseases in rice (*Oryza sativa*) (Ford et al. 1994; Talbot and Foster 2001). This pathogenic fungus has been reported to interact compatibly with the model plant *A. thaliana* also (Andargie and Li 2016). In rice, *U. virens* infection converts infected rice grains into smut balls, which results in sterility of the florets, whereas in *Arabidopsis* sterility of the flowers is evident without smut ball formation. A 39.4 Mb draft of *U. virens* genome that encodes 8426 predicted genes have been sequenced (Zhang et al. 2014). In addition to this, the pathogen has showed a decreased gene inventories for different metabolisms including nutrient uptake and polysaccharide degradation. This could arise possibly due to the adaptation of the pathogen to the specific floret infection and biotrophic lifestyles. *Once U. virens* colonize and attack the leaves, flowers, pods, and roots of a dicotyledonous plant, *A. thaliana*, different defense response proteins become induced in the infected plants as witnessed through gene expression analysis. *A. thaliana* has been reported to activate a hypersensitive response (HR) mechanism which precedes a slower systemic response that ultimately leads to systemic defense response. This mechanism reduces the damage and destruction caused by this biotrophic fungus. The progression of the fungal hyphae on the surface of the leaves is through the attachment to the trichomes. Leaves, along with retention of fungal spores, also determine the survival, attachment, and penetration of the hyphae. The trichomes present on the leaf surface provide physical adhesion point for the hyphae in addition to retaining water on the plant surface and provide nutrients and protected environment for microbial growth (Lindow and Brandl 2003; Monier and Lindow 2003; Calo et al. 2006).

Basically both *Arabidopsis* and rice share similar defense response mechanisms. Chao et al. (2014) reported that during the first stage of *U. virens* infection in the infected rice spikelet, proteins that are involved in protein modification, protein degradation, and receptor phosphorylation become activated to a great extent. During this time, some receptor protein kinases can activate corresponding substrates to facilitate downstream signal transduction, and this is shown by MPK3 and MPK6, which phosphorylate WRKY33 to initiate phytoalexin biosynthesis in *Arabidopsis*. In addition, a protein kinase APK1B which is involved in stamen development and its repression can prevent pollen tube germination causing self-incompatibility in *Arabidopsis, is also involved in defense against U. virens causing rice spikelet infection*. The presence of protein kinase APK1B indicates that the pathogen manipulates host development signaling by prohibiting protein phosphorylation, hence allowing further infection of the plant with *U. virens* to occur.

Andargie and Li (2016) performed semiquantitative RT-PCR analysis with the RNA isolated from leaves, roots, flowers, and siliques of *Arabidopsis* plants following infection with the *U. virens* spores in order to test whether the pathogenesis-related genes or plant defensin genes were inducible by pathogen infection. Their findings indicate that *U. virens* isolates infect *Arabidopsis*, and the plant subsequently activates different defense response mechanisms, witnessed by the expression of pathogenesis-related genes, *PR-1*, *PR-2*, *PR-5*, *PDF1.1*, and *PDF1.2*. Thus *Arabidopsis* plants activate different defense strategies in order to limit the damage and destruction which are caused by this biotrophic fungus *U. virens* and may serve as a good model host species to study the interaction between infected plants and the rice false smut fungus *U. virens*.

Thus it is a novel pathosystem based on *U. virens* and *Arabidopsis as* rice pathogenic *U. virens* transformed colonies may infect and colonize endophytically on the different parts of the *Arabidopsis* plant. Since the processes that determine the outcome of an interaction between a microbial pathogen and a host plant are complex, understanding the molecular details of these interactions, like the pathogen genes required for infection, effective host defense responses, as well as mechanisms by which host and pathogen signaling networks are regulated, might be utilized to design new plant protection strategies. Generally, the established *A. thaliana-U. virens* pathosystem could be able to expand the model systems investigating fungi-plant interactions and will facilitate a full understanding of *U. virens* biology and pathology. It can advance our knowledge in order to describe the plant immune system. This pathosystem will now permit various follow-up molecular genetics and gene expression experiments to be performed to identify the defense signals and responses that restrict fungal hyphae colonization in plants and also provide initial evidence for tissue-adapted fungal infection strategies.

1.6 Conclusion

This chapter has focused on the well-suited model interactions of *A. thaliana with various plant pathogens (fungi, bacteria, and virus).* Till now, various studies have been done pertaining to the interaction of *A. thaliana* with a large number of virulent and avirulent bacterial, fungal, and viral pathogens. This model species exhibits all of the major kinds of defense responses described in other plants. Regardless of the setting, nature, or the laboratory, *Arabidopsis* responds to the pathogen attack in a similar fashion as other higher plant species when exposed to pathogens. *A. thaliana* has been used extensively as a model plant not for any particular pathogen but to understand diverse range of plant-pathogen interactions with a wide variety of pathogens. The exhaustive analysis of *Arabidopsis* genes, whose expression is modulated during disease development, paves the way for dissecting plant networks activated by recognition of *pathogen* effectors in susceptible plants.

The lessons learned and information gathered from the model plant, *Arabidopsis*, can be applied to agronomically important crops. The study of this model plant interaction with various pathogens will contribute to our appreciation of how plants and pathogens have evolved to survive each other's attacks and counterattacks, which will in turn help us to develop sustained control measures by guided interception of pathogen virulence and/or by selective activation of plant defense. Thus it may address the key challenges of understanding how both plant defense and pathogen attacks are integrated and translating knowledge from *Arabidopsis* to crop plants. It would enable researchers to limit disease spread due to a better knowledge of the pathogen as well as provide them with a better understanding of the mechanisms involved in plant defense. Interfamily transfer of R genes from *Arabidopsis* can develop pathogen-resistant crops. It has been reported when RRS1/RPS4 pair of R genes from *Arabidopsis thaliana* was transferred to tomato, it caused resistance against *Ralstonia solanacearum*.

Glossary

Avirulence (Avr) protein Small proteins produced by pathogens and recognized by the host cell resistance proteins. These proteins trigger defense responses in plants. Avr proteins are often type III secretion system effectors, involved in pathogenicity.

Basal defense Plant defense that occurs early in the host-pathogen interaction in response to the perception by plant pattern recognition receptors of microbial-associated molecular patterns (MAMPs). Basal defense is MAMP-triggered immunity (MTI) plus weak effector-triggered immunity (ETI), minus effector-triggered susceptibility (ETS).

Biotroph A pathogen that colonizes living tissues for its growth and reproduction.

Effector molecules Pathogen-produced molecules that interfere with and suppress plant defense mechanisms, e.g., bacterial proteins, delivered by the bacterial type III secretion system (TTSS) to the plant cell interior.

Effector-triggered immunity (ETI) Immune responses triggered by recognition of specific pathogen effectors. The ETI response relies on R genes. Plant ETI often causes an apoptotic hypersensitive response.

Effector-triggered susceptibility (ETS) The state of a plant in which the plant's defense mechanism become suppressed by pathogen effector molecules.

Elicitor Any metabolite isolated from pathogens that at a very low concentration induces a hypersensitive response in host plants.

Hypersensitive response (HR) A complex defense response that is often associated with resistance (R) protein-mediated immunity. HR culminates in programmed cell death in cells in the vicinity of the pathogen, which may inhibit pathogen spreading.

MAMP-triggered immunity (MTI) Immunity raised after recognition of MAMPs by pattern recognition receptors (PRRs) localized on the surface of plant cells.

Microbial-associated molecular patterns (MAMPs) More recent term used for PAMPs. A series of essential and conserved molecular motifs of both pathogenic and nonpathogenic microbes that can be recognized by pattern recognition receptors in plants.

Necrotroph A pathogen that rapidly kills the host tissue and feeds on the dead tissue.

PAMP-triggered immunity (PTI) Immunity raised from the interaction of pattern recognition receptors (PRRs) in plant cells with elicitor molecules. It is a part of the first line of defense and results in a basal level of resistance.

Pathogen-associated molecular patterns (PAMPs) A set of molecular structures (epitopes) not shared with the host but shared by related pathogens, relatively invariant.

Pathogenesis-related (PR) genes Plant genes that activates after infection by pathogens.

Pathogenesis-related (PR) proteins Plant proteins that are synthesized in response to microbial attack and that serve to limit the growth pathogens. They are induced as a part of systemic acquired resistance.

Pattern recognition receptors (PRRs) Germ line-encoded proteins that can recognize microbe-associated molecular patterns and induce signaling cascade in innate immunity responses.

Plant defensin (PDF) Small, highly stable, cysteine-rich peptides that constitute a part of the innate immune system, mostly involved in defense against a broad range of fungal pathogens.

R genes and R proteins Plants have R genes (resistance genes) whose products mediate resistance to specific microbes, e.g., virus, bacteria, fungus, oomycete, nematode, etc. The product of R gene is R protein that allows recognition of specific pathogen effectors, either through direct binding or by recognition of the effector's alteration of a host protein.

Systemic acquired resistance (SAR) Inducible whole-body resistance. The development of a general immune capacity throughout the entire plant following an initial invasion by a pathogen.

References

Agrios GN (1997) Plant pathology. Academic, San Diego

Alonso-Blanco C, Aarts MG, Bentsink L et al (2009) What has natural variation taught us about plant development, physiology, and adaptation? Plant Cell 21:1877–1896

Andargie M, Li J (2016) *Arabidopsis thaliana*: a model host plant to study plant–pathogen interaction using rice false smut isolates of *Ustilaginoidea virens*. Front Plant Sci 7:192. https://doi.org/10.3389/fpls.2016.00192

Arabidopsis-Genome-Initiative (2000) Analysis of the genome sequence of the flowering plant *Arabidopsis thaliana*. Nature 408:796–815

Asano Y, Hiramoto T, Nishino R, Aiba Y, Kimura T, Yoshihara K et al (2012) Critical role of gut microbiota in the production of biologically active, free catecholamines in the gut lumen of mice. Am J Physiol Gastrointest Liver Physiol 303:G1288–G1295

Atwell S, Huang YS, Vilhjálmsson BJ et al (2010) Genome-wide association study of 107 phenotypes in *Arabidopsis thaliana* inbred lines. Nature 465:627–631

Berger S, Sinha AK, Roitsch T (2007) Plant physiology meets phytopathology: plant primary metabolism and plant-pathogen interactions. J Exp Bot 58:4019–4026

Buell CR (1998) *Arabidopsis*: a weed leading the field of plant-pathogen interactions. Plant Physiol Biochem 36(1):177–186

Burgyan J, Havelda Z (2011) Viral suppressors of RNA silencing. Trends Plant Sci 16:265–272

Callaway A, Liu W, Andrianov V et al (1996) Characterization of cauliflower mosaic virus (CaMV) resistance in virus-resistant ecotypes of *Arabidopsis*. Mol Plant-Microbe Interact 9:810–818

Calo L, García I, Gotor C, Romero LC (2006) Leaf hairs influence phytopathogenic fungus infection and confer an increased resistance when expressing a *Trichoderma* α-1,3-glucanase. J Exp Bot 57:3911–3920

Carrington J, Kasschau K, Mahajan S, Schaad M (1996) Cell-to-cell and long-distance movement of viruses in plants. Plant Cell 8:1669–1681

Chao J, Jin J, Wang D et al (2014) Cytological and transcriptional dynamics analysis of host plant revealed stage-specific biological processes related to compatible rice-*Ustilaginoidea virens* interaction. PLoS One 9:e91391. https://doi.org/10.1371/journal.pone.0091391

Chen ZY, Agnew JL, Cohen JD et al (2007) *Pseudomonas syringae* type III effector AvrRpt2 alters *Arabidopsis thaliana* auxin physiology. Proc Natl Acad Sci U S A 104:20131–20136

Clarke JD, Volko SM, Ledford H et al (2000) Roles of salicylic acid, jasmonic acid, and ethylene in cpr-induced resistance in *Arabidopsis*. Plant Cell 12:2175–2190

Cooley MB, Pathirana S, Wu HJ et al (2000) Members of the *Arabidopsis* HRT/RPP8 family of resistance genes confer resistance to both viral and oomycete pathogens. Plant Cell 12:663–676

Crute IJ, Beynon J, Dangl E et al (1994) Microbial pathogenesis of *Arabidopsis*. In: Meyerowitz EM, Somerville CR (eds) Arabidopsis. Cold Spring Harbor Laboratory Press, Cold Spring Harbor, pp 11705–11747

Cunnac S, Lindeberg M, Collmer A (2009) *Pseudomonas syringae* type III secretion system effectors: repertoires in search of functions. Curr Opin Microbiol 12:53–60

Dangl JL, Jones JD (2001) Plant pathogens and integrated defence responses to infection. Nature 411:826–833

Dangl J, Lehnackers H, Kiedrowski S et al (1991) Interactions between *Arabidopsis thaliana* and phytopathogenic *Pseudomonas* pathovars: a model for the genetics of disease resistance. In: Hennecke H, Verma DPS (eds) Advances in molecular genetics of plant-microbe interactions, vol 1. Springer, Dordrecht, pp 78–83

de Torres-Zabala M, Truman W, Bennett MH et al (2007) *Pseudomonas syringae* pv. *tomato* hijacks the *Arabidopsis* abscisic acid signalling pathway to cause disease. EMBO J 26:1434–1443
Dempsey DA, Wobbe KK, Klessig DF (1993) Resistance and susceptible responses of *Arabidopsis thaliana* to Turnip Crinkle Virus. Mol Plant Pathol 83:1021–1029
Dempsey DA, Pathirana MS, Wobbe KK, Klessig DF (1997) Identification of an *Arabidopsis* locus required for resistance to turnip crinkle virus. Plant J 2:301–311
Ford TL, Cooley JT, Christou P (1994) Current status for gene transfer into rice utilizing variety-independent delivery systems. In: Ziegler RS, Leong SA, Teng PS (eds) Rice blast disease. CAB International, Wallingford, pp 195–208
Francois O, Blum MG, Jakobsson M, Rosenberg NA (2008) Demographic history of European populations of *Arabidopsis thaliana*. PLoS Genet 4:e1000075
Gao R, Liu P, Yong Y, Wong SM (2016) Genome-wide transcriptomic analysis reveals correlation between higher WRKY61 expression and reduced symptom severity in Turnip crinkle virus infected Arabidopsis thaliana. Sci Rep 6:24604
Garcia-Brugger A, Lamotte O, Vandelle E et al (2006) Early signaling events induced by elicitors of plant defenses. Mol Plant-Microbe Interact 19:711–724
Glazebrook J, Rogers EE, Ausubel FM (1997) Use of *Arabidopsis* for genetic dissection of plant defense responses. Annu Rev Genet 31(1):547–569
He P, Chintamanani S, Chen Z et al (2004) Activation of a COI1-dependent pathway in *Arabidopsis* by *Pseudomonas syringae* type III effectors and coronatine. Plant J 37:589–602
Heath MC (2000) Hypersensitive response-related death. Plant Mol Biol 44:321–334
Hills GJ, Plaskitt KA, Young ND et al (1987) Immunogold localization of the intracellular sites of structural and nonstructural tobacco mosaic virus proteins. Virology 161:488–496
Hirano SS, Upper CD (2000) Bacteria in the leaf ecosystem with emphasis on *Pseudomonas syringae*-a pathogen, ice nucleus, and epiphyte. Microbiol Mol Biol Rev 64:11624–11653
Holub EB, Beynon JL, Crute IR (1994) Phenotypic and genotypic characterizations of interactions between isolates of Peronospora parasitica and accessions of Arabidopsis thaliana. Mol Plant-Microbe Interact 7:223–239
Holub EB, Brose E, Tor M et al (1995) Phenotypic and genotypic variation in the interaction between *Arabidopsis thaliana* and *Albugo candida*. Mol Plant Microbe Interact 81(1):916–928
Hossain MM, Sultana F, Kubota M et al (2007) The plant growth-promoting fungus *Penicillium simplicissimum* GP17-2 induces resistance in *Arabidopsis thaliana* by activation of multiple defense signals. Plant Cell Physiol 48:1724–1736
Ishikawa M, Obata F, Kumagai T, Ohno T (1991) Isolation of mutants of *Arabidopsis thaliana* in which accumulation of tobacco mosaic virus coat protein is reduced to low levels. Mol Gen Genet 230:33–38
Jeong RD, Chandra-Shekara AC, Kachroo A et al (2008) HRT-mediated hypersensitive response and resistance to Turnip crinkle virus in *Arabidopsis* does not require the function of TIP, the presumed guardee protein. Mol Plant-Microbe Interact 21:1316–1324
Jones JD, Dangl JL (2006) The plant immune system. Nature 444:323–329
Jones DA, Takemoto D (2004) Plant innate immunity – direct and indirect recognition of general and specific pathogen-associated molecules. Curr Opin Immunol 16:48–62
Kachroo P, Yoshioka K, Shah J et al (2000) Resistance to turnip crinkle virus in *Arabidopsis* is regulated by two host genes and is salicylic acid dependent but NPR1, ethylene, and jasmonate independent. Plant Cell 12:677–690
Keen NT (1990) Gene-for-gene complementarity in plant-pathogen interactions. Annu Rev Genet 24:447–463
Kim MG, Kim WY, Lee JR et al (2008) Host immunity-suppressive molecular weapons of phytopathogenic bacteria. J Plant Biol 51:235–239
Klement ZG, Farkas L, Lovrekovich L (1964) Hypersensitive reaction induced by phytopathogenic bacteria in the tobacco leaf. Phytopathology 54:11474–11477
Kloek AP, Verbsky ML, Sharma SB et al (2001) Resistance to *Pseudomonas syringae* conferred by an *Arabidopsis thaliana* coronatine-insensitive (coi1) mutation occurs through two distinct mechanisms. Plant J 26:509–522

Koornneef M, Alonso-Blanco C, Vreugdenhil D (2004) Naturally occurring genetic variation in *Arabidopsis thaliana*. Annu Rev Plant Biol 55:141–172

Kunkel BN (1996) A useful weed put to work: genetic analysis of disease resistance in *Arabidopsis thaliana*. Trends Genet 12:63–69

Kunkel BN, Brooks DM (2002) Cross talk between signaling pathways in pathogen defense. Curr Opin Plant Biol 5:325–331

Lai J, Chen H, Teng K et al (2009) RKP, a RING finger E3 ligase induced by BSCTV C4 protein, affects geminivirus infection by regulation of the plant cell cycle. Plant J 57:905–917

Laurie-Berry N, Joardar V, Street IH, Kunkel BN (2006) The *Arabidopsis thaliana* JASMONATE INSENSITIVE 1 gene is required infection by *Pseudomonas syringae*. Mol Plant-Microbe Interact 19:789–800

Lee S, Stenger DC, Bisaro DM, Davis KR (1994) Identification of loci in *Arabidopsis* that confer resistance to geminivirus infection. Plant J 6:525–535

Leisner S, Turgeon R, Howell SH (1993) Effects of host plant development and genetic determinants on the long-distance movement of cauliflower mosaic virus in *Arabidopsis*. Plant Cell 5:191–202

Lewis LA, Polanski K, de Torres-Zabala M et al (2015) Transcriptional dynamics driving MAMP-triggered immunity and pathogen effector-mediated immunosuppression in *Arabidopsis* leaves following infection with *Pseudomonas syringae pv tomato* DC3000. Plant Cell 27(11):3038–3064

Li XH, Simon AE (1990) Symptom intensification on cruciferous hosts by the virulent satellite RNA of turnip crinkle virus. Phytopathology 80:238–242

Lindeberg M, Cunnac S, Collmer A (2009) The evolution of *Pseudomonas syringae* host specificity and type III effector repertoires. Mol Plant Pathol 10:767–775

Lindow SE, Brandl MT (2003) Microbiology of the phyllosphere. Appl Environ Microbiol 69:1875–1883

Matthews RE (1991) Plant virology, 3rd edn. Academic Press, San Diego

McDowell JM, Dangl JL (2000) Signal transduction in the plant immune response. Trends Biochem Sci 25:79–82

Melotto M, Underwood W, Koczan J et al (2006) Plant stomata function in innate immunity against bacterial invasion. Cell 126:969–980

Monier JM, Lindow SE (2003) Differential survival of solitary and aggregated bacteria cells promotes aggregate formation on leaf surfaces. Proc Natl Acad Sci U S A 100:15977–15982

Narasimhan ML, Damsz B, Coca MA et al (2001) A plant defense response effector induces microbial apoptosis. Mol Cell 8:921–930

Navarro L, Dunoyer P, Jay F et al (2006) A plant miRNA contributes to antibacterial resistance by repressing auxin signaling. Science 312:436–439

Nimchuk Z, Eulgem T, Holt BF 3rd, Dangl JL (2003) Recognition and response in the plant immune system. Annu Rev Genet 37:579–609

Nomura K, Melotto M, He SY (2005) Suppression of host defense in compatible plant-*Pseudomonas syringae* interactions. Curr Opin Plant Biol 8:361–368

Oliver RP, Ipcho SV (2004) *Arabidopsis* pathology breathes new life into the necrotrophs-vs.-biotrophs classification of fungal pathogens. Mol Plant Pathol 5:347–352

Pieterse CM, Leon-Reyes A, Van Der Ent S, Van Wees SC (2009) Networking by small-molecule hormones in plant immunity. Nat Chem Biol 5:308–316

Pu XJ, Li YN, Wei LJ et al (2016) Mitochondrial energy-dissipation pathway and cellular redox disruption compromises *Arabidopsis* resistance to turnip crinkle virus infection. Biochem Biophys Res Commun 473(2):421–427

Quirino BF, Bent AF (2003) Deciphering host resistance and pathogen virulence: the *Arabidopsis*/ Pseudomonas interaction as a model. Mol Plant Pathol 4(6):517–530

Ren T, Qu F, Morris TJ (2000) HRT gene function requires interaction between a NAC protein and viral capsid protein to confer resistance to turnip crinkle virus. Plant Cell 12:1917–1926

Roux F, Bergelson J (2016) Chapter four-the genetics underlying natural variation in the biotic interactions of *Arabidopsis thaliana*: the challenges of linking evolutionary genetics and community ecology. Curr Top Dev Biol 119:111–156

Scofield SR, Tobias CM, Rathjen JP et al (1996) Molecular basis of gene-for-gene specificity in bacterial speck disease of tomato. Science 274:2063–2065

Simon AE, Li XH, Lew JE et al (1992) Susceptibility and resistance of *Arabidopsis thaliana* to turnip crinkle virus. Mol Plant-Microbe Interact 5:496–503

Stahl EA, Bishop JG (2000) Plant-pathogen arms races at the molecular level. Curr Opin Plant Biol 3:299–304

Talbot N, Foster A (2001) Genetics and genomics of the rice blast fungus Magnaporthe grisea: developing an experimental model for understanding fungal diseases of cereals. Adv Bot Res 34:263–287

Tang X, Frederick RD, Zhou J et al (1996) Initiation of plant disease resistance by physical interaction of AvrPto and Pto Kinase. Science 274:2060–2063

Thakur M, Sohal BS (2013) Role of elicitors in inducing resistance in plants against pathogen infection: a review. ISRN Biochem 2013:762412

Thilmony R, Underwood W, He SY (2006) Genome-wide transcriptional analysis of the *Arabidopsis thaliana* interaction with the plant pathogen *Pseudomonas syringae* pv. *tomato* DC3000 and the human pathogen *Escherichia coli* O157: H7. Plant J 46:34–53

Thomma BP, Penninckx IA, Broekaert WF, Cammue BP (2001) The complexity of disease signaling in *Arabidopsis*. Curr Opin Immunol 13:63–68

Torres MA, Dangl JL, Jones JD (2002) *Arabidopsis* gp91phox homologues AtrbohD and AtrbohF are required for accumulation of reactive oxygen intermediates in the plant defense response. Proc Natl Acad Sci U S A 99:517–522

Tsuda K, Sato M, Glazebrook J et al (2008) Interplay between MAMP-triggered and SA-mediated defense responses. Plant J 53:763–775

Tsuji J, Somerville SC (1992) First report of the natural infection of *Arabidopsis thaliana* by *Xanthomonas campestris pv. campestris*. Plant Dis 761(1):539

Uknes S, Winter AM, Delaney T et al (1993) Biological induction of systemic acquired resistance in *Arabidopsis*. Mol Plant-Microbe Interact 6:692–698

Uppalapati SR, Ishiga Y, Wangdi T et al (2007) The phytotoxin coronatine contributes to pathogen fitness and is required for suppression of salicylic acid accumulation in tomato inoculated with *Pseudomonas syringae* pv. *tomato* DC3000. Mol Plant-Microbe Interact 20:955–965

Uren AG, O'Rourke K, Aravind LA et al (2000) Identification of paracaspases and metacaspases: two ancient families of caspase-like proteins, one of which plays a key role in MALT lymphoma. Mol Cell 6:961–967

Waigmann E, Lucas WJ, Citovsky V, Zambryski P (1994) Direct functional assay for tobacco mosaic virus cell-to-cell movement protein and identification of a domain involved in increasing plasmodesmal permeability. Proc Natl Acad Sci U S A 91:1433–1437

Wolf S, Deom CM, Beachy RN, Lucas WJ (1989) Movement protein of tobacco mosaic virus modifies plasmodesmatal size exclusion limit. Science 246:377–379

Xin XF, He SY (2013) *Pseudomonas syringae pv. tomato* DC3000: a model pathogen for probing disease susceptibility and hormone signaling in plants. Annu Rev Phytopathol 51:473–498

Yoshii M, Yoshioka N, Ishikawa M, Naito S (1998) Isolation of an *Arabidopsis thaliana* mutant in which accumulation of *Cucumber Mosaic Virus* coat protein is delayed. Plant J 13:211–219

Yu L, Quinn MT, Cross AR, Dinauer MC (1998) Gp91(phox) is the heme binding subunit of the superoxide-generating NADPH oxidase. Proc Natl Acad Sci U S A 95:7993–7998

Zhang Z, Li Q, Li Z et al (2007) Dual regulation role of GH3.5 in salicylic acid and auxin signaling during *Arabidopsis-Pseudomonas syringae* interaction. Plant Physiol 145:450–464

Zhang Y, Zhang K, Fang A et al (2014) Specific adaptation of *Ustilaginoidea virens* in occupying host florets revealed by comparative and functional genomics. Nat Commun 5:3849

Zhao Y, Thilmony R, Bender CL et al (2003) Virulence systems of *Pseudomonas syringae* pv. *tomato* promote bacterial speck disease in tomato by targeting the jasmonate signaling pathway. Plant J 36:485–499

2 Fungal and Bacterial Biotrophy and Necrotrophy

Geeta and Reema Mishra

Abstract

Plant pathogens have been divided into two classes, namely, biotrophs and necrotrophs. These pathogens lead to significant economic losses by infecting various crops. Biotrophs complete their life cycle by using the living host cell machinery, while necrotrophs feed on the host cell after killing them. Hemibiotrophs, a third group, show both the forms for obtaining nutrition i.e., early biotrophic stage to later necrotrophic phase. After infecting the plants, both the groups of plant pathogens can trigger and suppress plant immune responses by synthesizing and secreting effector proteins. In case of biotrophic pathogens, effector proteins were found to be Avr proteins (identified by resistance proteins), *hrp* genes, and cell wall-degrading enzymes, while necrotrophic pathogen has additional effectors called as host-selective toxins. Significant differences have been observed between these two groups in the disease symptoms they cause, their host range, morphogenesis of the infection, production of secondary metabolites and hormones, and nature of plant resistance. Biotrophs possess a sophisticated way of infection, i.e., it enters the host cell using the haustoria, colonizes the intercellular space, and overpowers the host defenses. Necrotrophs have been further grouped into host-specific and broad host range necrotrophs depending on the toxins they secrete. In case of necrotrophic infections, host cell death has been shown to trigger production of hormones like ethylene, abscisic acid, salicylic acid, and jasmonic acid. Both bacterial and fungal plant pathogens belonging to the above mentioned category have been identified. In this chapter we are going to discuss the current state of knowledge about bacterial and fungal biotrophs and necrotrophs.

Geeta and R. Mishra have contributed equally to this chapter.

Geeta (✉) · R. Mishra
Department of Botany, Gargi College, University of Delhi, New Delhi, Delhi, India
e-mail: geeta.biotech@gmail.com

A. Singh, I. K. Singh (eds.), *Molecular Aspects of Plant-Pathogen Interaction*,
https://doi.org/10.1007/978-981-10-7371-7_2

Keywords
Bacteria · Biotrophs · Effector proteins · Fungal · Necrotrophs · Pathogen

2.1 Introduction

During the course of evolution, there has been an arms race between plant pathogens and their hosts or virulence and resistance. To date, many phytopathogens having varied infection strategies have been identified. These phytopathogens are broadly divided into two major categories on their modes of nutrition, namely, necrotrophs and biotrophs. These pathogens can be bacterial and fungal and cause various diseases in plants. Necrotrophic pathogens kill the host tissue rapidly and survive on the contents (Stone 2001), while biotrophic pathogens thrive on the nutrients of the living host. Biotrophs have a sophisticated or stealthy mechanism of invasion, and they develop specialized structures such as haustoria, appressoria, and hyphae for gaining entry into the host plant for absorbing nutrients and for blocking the immune responses of plants (Schulze-Lefert and Panstruga 2003; Mendgen and Hahn 2004). Biotroph sustains the host viability, causes comparatively little injury to the host, and suppresses hypersensitive response (HR; localized programmed cell death of host plant) as it restricts the nutrient supply. Necrotrophs are less sophisticated, rather are notorious and have various virulence strategies for killing and absorbing the nutrient from the host plant cells for their growth and reproduction. They macerate the tissues or can cause soft rots and stimulate HR-like host cell death. Based on the secretion of toxins, necrotrophs have been further divided into two subcategories, namely, host-specific necrotrophs (HSNs) and broad host range necrotrophs (BHNs) (Wolpert et al. 2002). As the name suggests, HSNs secrete host-specific toxins for the development of virulence. The fungal necrotrophs *Cochliobolus carbonum* and *Alternaria* spp. have been shown to secrete host-specific toxins for the pathogenicity (Walton 1996). The examples of necrotrophs belonging to BHN category include the bacterial pathogen *Erwinia carotovora* and fungal pathogens *Plectosphaerella cucumerina*, *Botrytis cinerea*, *Alternaria brassicicola*, and *Sclerotinia sclerotiorum* (Mengiste 2012). All these reports suggest the fact that there are significant differences between the two major classes of pathogens in terms of infection symptoms, host range, and effector proteins they secrete (Laluk and Mengiste 2010).

Another category of phytopathogens has been named as hemibiotrophs, which exhibit two phases of nutrient acquisition: an early biotrophic phase and later manifests necrotrophic phase. Hemibiotrophic plant pathogens include bacterium *Pseudomonas syringae*, fungi *Colletotrichum graminicola* and *Magnaporthe oryzae*, and oomycete *Phytophthora infestans* (Lee and Rose 2010). Irrespective of the category of pathogens, the basic events after the invasion include production of effector proteins by the pathogen for their colonization, while, in case of host plant, the defense mechanism is activated for limiting the spread of pathogen. In case of necrotrophic invasion, host cell death, and secretion of various secondary metabolites and hormones named such as jasmonic acid, salicylic acid, abscisic acid, and ethylene, accumulation of reactive oxygen species and callose takes place (Mengiste 2012).

Plant fungal pathogens cause various diseases on economically important crops. They exhibit different infection strategies and develop specialized infection structures for deriving the nutrition. All the three modes of pathogenesis (necrotrophy, biotrophy, and hemibiotrophy) have been identified in fungi. Different fungi secrete different pathogen effector molecules during the infection process for their invasion and colonization. Pathogen effectors have been shown to be involved in pathogenicity, as silencing or disrupting pathogen effector coding genes showed reduced virulence (Stergiopoulos et al. 2013). Fungi also secrete certain toxins, cell wall-degrading enzymes, and toxic secondary metabolites to aid the infection process (Horbach et al. 2011). It is believed that infection strategies of biotrophic fungi are more complex when compared to necrotrophic ones.

Bacterial pathogens belonging to both the categories secrete virulence proteins, cell wall-degrading enzymes, toxins, and extracellular polysaccharides during the infection for their colonization, growth, and replication (Alfano and Collmer 1996). Bacteria possess a multi-protein secretory system for translocating the virulence factors to the host plant cell (Alfano and Collmer 2004) and also employ the quorum sensing to suppress the host immune system. The most well-characterized bacterial pathogens include *Pseudomonas*, *Erwinia*, *Xanthomonas*, and *Ralstonia.* All the genera have three common features, i.e., they kill the cells, inhabit the intercellular spaces, and have *hrp* genes (hypersensitive response and pathogenicity; Alfano and Collmer 1997). The Hrp protein secretion system is very important for pathogenesis.

These pathogens cause various pre- and post-harvest diseases in economically important plants incurring significant economic losses. The plant responds to these pathogens by activating the immune system called as pathogen-associated molecular pattern (PAMP)-triggered immunity (PTI), effector-triggered immunity (ETI), and phytohormone signaling (Bent and Mackey 2007; Willment and Brown 2007).

Whole genome sequencing and comparative genomics studies have helped in better understanding of the various effector proteins, strategies of pathogenesis, and resistance mechanism in plants. Different effector proteins have also been used as an imperative tool in disease resistance breeding (Vleeshouwers and Oliver 2014). This chapter provides an overview on the different lifestyles and infection strategies of fungal and bacterial phytopathogens characterized.

2.2 Fungal Biotrophy and Necrotrophy

Of the 1.5 million species of fungi projected to exist (Hawksworth 1991), very few of them have been functionally characterized. The lifestyle of pathogenic fungi can be saprophytic, biotrophic, or necrotrophic (Kahmann and Basse 2001; Mendgen and Hahn 2002; Glazebrook 2005; Howlett 2006; Ferreira et al. 2007). It has been suggested that most of the fungi are saprotrophic in nature, i.e., they derive their nutrition from decaying organic matter (Kahmann and Basse 2001; Ahmad et al.

2006). Fungal necrotrophs first infect the host cells, and this requires various stages of conidial attachment followed by germination, formation of lesion, and finally the softening of tissue and sporulation (Prins et al. 2000). After the initial phase of infection, the penetration is facilitated by production of toxins, appressoria, haustoria, hyphae formation, and secretion of cell wall-degrading enzymes (CWDEs) (Prell and Day 2001; Idnurm and Howlett 2001; Mendgen and Hahn 2002; Oliver and Ipcho 2004). All through the infection process, fungi also dynamically suppress the host cell defense by manipulating various processes, thereby aiding their own proliferation (Prins et al. 2000). Biotrophic pathogens obtain the nutrients from living cells by forming complex infection structures like hyphae or haustoria for exchanging nutrients from the host plant cell (Heath 1997, 2002; Schulze-Lefert 2004). The secretory activity of CWDEs is limited in case of fungal biotrophs (Idnurm and Howlett 2001; Tudzynski and Sharon 2003; Oliver and Ipcho 2004; Schulze-Lefert 2004). The definitions are categorical, but the relationship between necrotrophs and biotrophs is probably best represented as continuum with intermediates, i.e., hemibiotrophs, pathogens that initiate infection as biotrophs but later switch to necrotrophs (Glazebrook 2005). The economically important biotrophic, hemibiotrophic, and necrotrophic plant fungal pathogens have been recorded in Tables 2.1 and 2.2.

2.2.1 Effector Proteins Secreted by Fungal Plant Pathogens

Plants recognize fungal pathogens on the basis of pathogen-associated molecular patterns (PAMPs). PAMPs are conserved molecules important for pathogen and characterize a class of microbes (Bent and Mackey 2007; Willment and Brown 2007). Examples of PAMPs include fungal xylanase (Ron and Avni 2004) and chitin (Kaku et al. 2006; Miya et al. 2007). Detection of these PAMPs by pattern recognition receptors (PRRs) leads to a PAMP-triggered immune response (Bent and Mackey 2007; Hückelhoven 2007; Zipfel 2008, 2009). To counter the PTI, fungal pathogens secrete effector molecules that can directly suppress the host defense response and also alter the host cell physiology for their proliferation (Boller and He 2009; Göhre and Robatzek 2009; Stergiopoulos and de Wit 2009; Stergiopoulos et al. 2010; Koeck et al. 2011).

Chitin (the major structural component of filamentous fungus cell wall) is released by the action of plant chitinases and triggers the immune response of plants (Kaku et al. 2006; Miya et al. 2007). The secreted effector Ecp6 from *Cladosporium fulvum* competes with the plant receptors CEBiP, by sequestering chitin oligosaccharides, and suppresses chitin-triggered immunity (de Jonge et al. 2010).

As discussed earlier *Avr* genes are another category of pathogen effectors that are specifically recognized by plant resistance (R). R genes consist of leucine-rich repeat (LRR) domains and nucleotide-binding sites (NBS) (Jones and Dangl 2006). It is not clear how fungal pathogens translocate the effector proteins into the host cell and whether they have type III secretion system or not as it is present in bacterial pathogens (Jin et al. 2003; Büttner and Bonas 2006). Most of the fungal *Avr*

Table 2.1 List of important biotrophic, hemibiotrophic, and necrotrophic fungal plant pathogens

Name of the pathogen	Host plant of the pathogen	Name of the disease
Biotrophic pathogens		
Cladosporium fulvum	Tomato	Tomato leaf mold
Ustilago maydis	Maize	Maize smut
Blumeria graminis	Barley; wheat	Powdery mildew
Plasmopara halstedii	Sunflower	Downy mildew
Synchytrium endobioticum	Potato	Potato wart
Plasmopara viticola	Grapevine	Downy mildew
Hemibiotrophic pathogens		
Magnaporthe oryzae	Rice	Rice blast
Mycosphaerella graminicola	Wheat	*Septoria tritici* leaf blotch
Bipolaris sorokiniana	Barley; wheat	Spot blotch disease
Colletotrichum graminicola	Maize	Anthracnose stalk rot disease
Moniliophthora perniciosa	*Theobroma cacao*	Witches' broom disease
Necrotrophic pathogens		
Cochliobolus heterostrophus	Maize	Southern leaf blight
Ascochyta rabiei	Chickpea	Blight disease
Cochliobolus carbonum	Maize	Northern leaf spot and ear rot
Cochliobolus victoriae	Oat	Victoria blight
Alternaria alternata	Pear; strawberry; tangerine; apple; tomato; tobacco; citrus	Black/dark leaf spot
Alternaria solani	Tomato and potato	Tomato early blight; collar and fruit rot
Alternaria brassicicola	*Brassica* species (broccoli, cabbage, canola, mustard; cauliflower; turnip)	Black spot (leaf, stem, or pod spots)
Periconia circinata	Sorghum	Milo
Pyrenophora tritici-repentis (*Drechslera tritici-repentis*)	Wheat	Tan spot
Bipolaris sacchari	Sugarcane	Eyespot
Phyllosticta maydis (*Mycosphaerella zeae-maydis*)	Maize	Yellow corn leaf blight
Stagonospora nodorum (*Phaeosphaeria nodorum*)	Wheat	*Stagonospora nodorum* blotch
Stemphylium vesicarium	European pear	Brown spot
Botrytis fabae	Bell bean (*Vicia faba*)	Chocolate spot
Name of the pathogen	Host plant of the pathogen	Name of the disease
Botrytis elliptica	Lilly	Gray mold
Botrytis cinerea	Dicots; some monocots	Gray mold "*Botrytis* blight"

(continued)

Table 2.1 (continued)

Name of the pathogen	Host plant of the pathogen	Name of the disease
Sclerotinia sclerotiorum	Cabbage; bean; citrus; celery; coriander; melon; squash; soybean; tomato; lettuce; cucumber	White mold
Monilinia fructicola	*Prunus* species (apples; pears; and other pome fruits in Rosaceae)	Brown fruit rots
Fusarium graminearum/ Gibberella zeae	Cereals	*Fusarium* head blight
Cercospora zeae-maydis	Maize	Gray leaf spot
Exserohilum turcicum	Maize	Northern leaf blight
Leptosphaeria maculans	Oilseed rape (or canola) (*Brassica napus*)	Blackleg or stem canker disease
Diaporthe toxica	Lupin	*Phomopsis* stem blight
Phoma medicaginis	Pea	Leaf spot and spring black stem
Colletotrichum gloeosporioides	Lupin and mango	Anthracnose
Fusarium oxysporum	Tomato, banana, cotton, and many others	*Fusarium* wilt
Rhizoctonia solani	Lucerne, clovers, pasture grasses, grain legumes, cereals, and oilseed crops	Rhizoctonia canker/ root rot
Pythium spp.	Very broad	Seedling damping off
Leptosphaerulina trifolii	*Medicago* spp.	Lepto leaf spot
Pleiochaeta setosa	Lupin	Brown leaf spot
Stagonospora meliloti	Lucerne and medics	*Stagonospora* crown rot
Stemphylium botryosum	Tomato; alfalfa; lettuce	Leaf spot and foliage blight
Pseudopeziza medicaginis	Lucerne	Common leaf spot

Adapted from Laluk and Mengiste (2010) and Wang et al. (2014)

genes identified have been shown to contain two important motifs, namely, dEER (aspartate, glutamate, glutamate, arginine) and RxLR (arginine, any amino acid, leucine, arginine) (Dodds et al. 2009; Tyler 2009). It is believed that these two motifs are essential for the entry into host cell. Mutation in these two motifs prevented the delivery of Avr1b from *Phytophthora sojae* and Avr3a from *P. infestans* into the host cell (Whisson et al. 2007; Dou et al. 2008). N-terminal "RxLR-like" motifs have been identified in the *Avr* genes of *Fusarium oxysporum* f. sp. *lycopersici*, *Melampsora lini*, and *Leptosphaeria maculans* (Kale et al. 2010). Based on the observation that oomycete and fungal RxLR-like motifs interact with the phosphatidylinositol-3-phosphate (an abundant phospholipid present in the outer surface of plasma membrane of plants), Kale et al. (2010) proposed that lipid raft-mediated endocytosis allows the pathogen effectors to enter the plant cell. Approximately 1500 effectors with RxLR and dEER motifs have been identified in

Table 2.2 List of important bacterial plant pathogens

Name of the pathogen	Host plant of the pathogen	Name of the disease
Dickeya	Potato tubers, bulbs of vegetables, and ornamental crops	Necrosis, blight, and soft rot
Erwinia amylovora	Apple, pear	Fire blight
Erwinia carotovora	Carrots, potatoes, cucumbers, onions, tomatoes, lettuce	Soft rot
Erwinia chrysanthemi	*Chrysanthemums*, maize, *Dieffenbachia*, *Euphorbia pulcherrima*, bananas, *Philodendron*	Soft rots and wilts
Pectobacterium atrosepticum	Potato	Potato blackleg disease
Phytophthora infestans	Potato, tomato	Late blight of potato
		Late blight of tomato
Pseudomonas syringae pv. *tomato*	*Tomato, Arabidopsis thaliana, Brassica oleracea* var. *botrytis*	Bacterial speck
Pseudomonas viridiflava	Tomato, soybean, pepper	Bacterial leaf blight
Pseudomonas syringae pv. *syringae*	*Prunus* sp., tomato, cereals, citrus, and kiwi	Bacterial canker and blast
Ralstonia solanacearum	Potato, tomato, banana, groundnut, tobacco	Bacterial wilt
Xanthomonas campestris pv. *Vesicatoria*	Tomato, pepper	Bacterial leaf spot

Phytophthora species (Jiang et al. 2008; Tyler 2009). After entering the host cell, these effector molecules suppress the host defense and elicit cell death (Bos et al. 2006; Dou et al. 2008).

Another category of fungal effectors identified to suppress plant basal resistance and cause plant cell death is crinkler and necrosis (CRN) effectors (Torto et al. 2003; Win et al. 2007; Haas et al. 2009). The CRN effectors also contain a conserved FLAK (phenylalanine, leucine, alanine, lysine) domain followed by a signal peptide (Win et al. 2007; Haas et al. 2009). *Phytophthora* genome has been shown to contain 61–451 CRN genes (Haas et al. 2009: Liu et al. 2011). Transient expression of CRN C- terminal domain has been shown to elicit host cell death (Liu et al. 2011). CRN8 of *Phytophthora infestans* has also been shown to regulate host signaling during infection by kinase activity (van Damme et al. 2012). Stam et al. (2013) identified 84 CRN proteins in *P. capsici*.

In fairly recent reports, small RNAs (sRNA) produced by the fungal pathogens have been shown to suppress the host immunity and are termed as non-proteinaceous sRNA effectors (Weiberg et al. 2013, 2014; Wang et al. 2015). Fungal phytopathogens also produce various small secretory proteins (SSPs) important for virulence. A family of such SSPs implicated as effector proteins required for pathogen host association has been identified in different fungal species, and it is believed that biotrophs are likely to secrete more SSPs as compared to necrotrophs (Cheng et al. 2014; Kim et al. 2016).

2.2.2 Effectors Produced by Pathogens Disturb the Phytohormonal Signaling

Plant hormones like auxin, gibberellins (GAs), abscisic acid (ABA), jasmonic acid (JA), ethylene (ET), and salicylic acid (SA) have been implicated in regulating plant defense responses. The three main players extensively characterized for the defensive roles till date are SA, JA, and ET. It has been reported that SA signaling is involved in eliciting resistance to biotrophic and hemibiotrophic phytopathogens, while ET and JA provide resistance to plants against necrotrophs (Glazebrook 2005; Di et al. 2016). Many fungal pathogens produce phytohormone and mimic during the infection for altering host cell response (Inomata et al. 2004). Many fungal pathogens themselves also produce plant hormones to disturb the host cell plant hormone signaling (Agrios 2005; Mobius and Hertweck 2009; Brodhun et al. 2013). Also many effector proteins synthesized by fungal pathogens have been shown to alter phytohormone biosynthesis and signaling pathways (Kazan and Lyons 2014).

In *Ustilago maydis* an increase in chorismate mutase 1 (Cmu1) that changes SA precursor chorismate into prephenate has been observed during infection process which alters SA biosynthesis pathway (Djamei et al. 2011). *Sclerotinia sclerotiorum* (a necrotrophic fungal pathogen) has also been shown to have SA-degrading effector protein (Penn and Daniel 2013). A secretory effector protein, SSITL (*Sclerotinia sclerotiorum* integrin-like), secreted by *S. sclerotiorum* has been shown to inhibit JA-regulated defense responses (Zhu et al. 2013). PSE1 (penetration-specific effector 1), an effector secreted from thc pathogen *P. parasitica*, affected auxin efflux (Evangelisti et al. 2013). ABA confers susceptibility to fungal pathogens, namely, *B. cinerea* and *F. oxysporum* (Audenaert et al. 2002; Anderson et al. 2004).

2.2.3 Phytotoxins Released by Fungal Pathogens

Toxins released by fungal pathogens play an important role in pathogenicity (Mobius and Hertweck 2009; Stergiopoulos et al. 2013). These toxins damage the host cells thereby aiding in colonization of pathogen. These toxins are categorized into two classes: host-specific toxins (HSTs) and non-HSTs. HSTs are majorly synthesized by *Dothideomycetes* (Wolpert et al. 2002; Berestetskiy 2008). Various trichothecene phytotoxins (inhibits translation) are secreted by species of *Fusarium* for the proliferation inside the host plant (Desjardins et al. 2007). Fungal necrotrophs also synthesize phytotoxic proteins like NEP1 (necrosis and ethylene-inducing protein) and NEP1-like proteins involved in host cell death and ROS production (Pemberton and Salmond 2004). AAL (a toxin released by *Alternaria alternata*) and Fumonisin B1 toxins secreted by fungal pathogen inhibit the ceramide synthesis and result in buildup of free sphingoid bases which results in the production of ROS and host cell death (Abbas et al. 1994; Shi et al. 2007). Fumonisin B1, secreted by *Fusarium verticillioides*, lessens extracellular ATP and causes PCD.

2.2.4 Significant Effector Proteins Secreted by Few Biotrophic, Hemibiotrophic, and Necrotrophic Fungi

2.2.4.1 *Cladosporium fulvum*

C. fulvum is a fungal pathogen which is biotrophic in nature and causes leaf mold disease of tomato (Thomma et al. 2005). Avr9, a 28 amino acid cysteine-rich effector protein of this genus, elicits HR in tomatoes. At the time of infection, *C. fulvum* secretes Ecp6 effector protein containing LysM chitin-binding domain. This domain and another effector, Avr4, bind to the chitin and prevent the host recognition (van Esse et al. 2007; Sánchez-Vallet et al. 2013). It also produces Avr2 effector molecule which binds the cysteine proteases of plants and prevents the defense responses of plants (van Esse et al. 2008).

2.2.4.2 *Ustilago maydis*

U. maydis is a biotrophic fungus which is the causal agent of maize smut. It infects the aerial organs of maize and enters the cell wall with the help of special structures named as hyphae for deriving nutrition from the host plant (Brefort et al. 2009). During infection process 20% of its genes were found to be upregulated. Effector gene cluster deletion analysis was performed to delineate the function of different effector proteins in the infection process (Kamper et al. 2006). Pep1 effector of this genus codes for a 178 amino acid secretory protein and plays an important role during the invasion process. It also inhibits the activity of apoplastic plant peroxidases thereby suppressing the immune response of plants (Doehlemann et al. 2009; Hemetsberger et al. 2015). Hum3 (a hydrophobin domain and a repetitive repellent protein-like repetitive domain-containing protein) and Rsp1 (repetitive secreted protein 1, a protein with internal hydrophilic repeats) effector proteins from *U. maydis* have been shown to be significant for cell adhesion during the infection process (Muller et al. 2008). Another effector See1 (Seedling efficient effector 1) has been shown to be involved in reactivating DNA synthesis important for tumor proliferation in leaf cells (Redkar et al. 2015).

2.2.4.3 *Blumeria graminis*

It is an obligate biotrophic fungus and causes powdery mildew disease on barley and wheat. It forms filamentous hyphae for infecting the leaves and haustoria for deriving nutrition from its host. More than 500 candidate secreted effector proteins (CSEPs) also called as Blumeria effector candidates (BECs) have been identified in this fungus. Eight such BECs including BEC1054 (RNase-like effector) are required for the formation of feeding structure, i.e., haustoria (Pennington et al. 2016).

2.2.4.4 *Botrytis cinerea*

B. cinerea is a broad host range necrotrophic fungus and is the causal agent of gray mold disease. It penetrates the host epidermal cells by secreting various cell wall-degrading enzymes, cutinases, proteases, lipases, oxalic acid, and a nonhost-selective toxin named as botrydial (Prins et al. 2000; Kars et al. 2005).

2.2.4.5 *Sclerotinia sclerotiorum*

S. sclerotiorum is a necrotrophic fungus that infects more than 400 plant species. It secretes various enzymes like glucanases, pectinases, cellulases, and proteases for disrupting the tissue (Bolton et al. 2006). It also synthesizes oxalic acid during the infection process to suppress the host immune responses (Guimaraes and Stotz 2004). It causes necrosis of crucifers by the secretion of polyketides (Pedras and Ahiahonu 2004). A secretory protein of *S. sclerotiorum*, SSITL (SS1G_14133), has also been shown to suppress the host immune response (Zhu et al. 2013).

2.2.4.6 *Alternaria brassicicola*

A. brassicicola is a broad host range necrotrophic pathogen known to synthesize various toxins important for pathogenesis. Host-specific toxins are secreted by *Alternaria alternata*, another necrotrophic fungus, to suppress host resistance (Tsuge et al. 2013). Cell wall-degrading enzymes and lipases produced by this genus affect the various cellular process of the host (Cho 2015).

2.2.4.7 *Magnaporthe oryzae*

It is a hemibiotrophic fungus, which causes rice blast disease. It forms appressorial pegs for the penetration of leaf cuticle and epidermal cells of the host. The effector proteins of this genus accumulate in the biotrophic interfacial complex (BIC, a lobed structure formed at the tip of invasive hypha) (Khang et al. 2010). BAS1, PWL2, BAS2, BAS3, BAS4, and AvrPita1 are the different effector proteins found to be upregulated during the infection process (Mosquera et al. 2009; Khang et al. 2010). Another effector protein of this genus named as AVR-Pii has been shown to suppress the plant immune response by inhibiting the activity of host Os-NADP-ME (a rice NADP-malic enzyme2 protein), which is important for the production of ROS (Singh et al. 2016).

2.2.5 Bacterial Biotrophy and Necrotrophy

Phytopathogenic bacteria are accountable for the devastating damage to the agriculture. The interaction between bacteria and its plant host is very dynamic. Bacteria cause many diseases in higher plants by entering the intercellular spaces through the natural openings, stomata, and wounds. Fire blight, halo blights cankers, galls, and leaf spots are the various diseases elicited by the bacteria in higher plants. They have varied infection strategies and belong to both the biotrophic and necrotrophic categories of pathogens. Necrotrophic bacteria follows the brute force strategy in which the bacteria rapidly kill the parenchymatous tissues of the host plants, while biotrophic bacteria reproduces within the host cell and slowly kills the host (Collmer and Bauer 1994). Necrotrophic bacteria trigger hypersensitive response (HR), i.e., a localized programmed cell death of host plant that restricts the pathogen by cutting off the nutrient supply; cell death occurs that enhances the colonization of the bacteria and acts as a sign of successful infection (Van Kan 2006). For a successful infection to occur, phytopathogenic bacteria have many genes that are induced when the contact between bacteria and plant is established. Using an in vivo expression technology, Boch et al. (2002) identified several genes that are induced when *P. syringae* pv. *tomato* attacks *Arabidopsis thaliana*. Few bacteria employ quorum-sensing mechanism to counter the host defense (Alfano and Collmer 1996).

Sometimes it becomes difficult to place the bacterial pathogens in a particular class as we have examples like *Pseudomonas syringae* which can be placed in biotrophic, hemibiotrophic, and partly in necrotrophic category; *Ralstonia solanacearum* is both a necrotrophic and biotrophic bacteria, *Xanthomonas* species are biotrophic, and *Erwinia amylovora* is often mentioned under necrotrophic category (Kraepiel and Barny 2016).

2.2.6 Basis of Bacterial Pathogenicity: The HRP System

The necrotrophic bacteria elicit the HR response using *hrp* genes (hypersensitive response and pathogenicity). These genes were first identified in *Pseudomonas syringae* pv. *syringae* (brown spot of bean) and were found to be present in clusters and are essential for the transport of the *Avr* gene-derived signal from the bacteria into host plant cells (Niepold et al. 1985; Lindgren et al. 1986). *hrp* genes have also been identified in Gram-negative bacterial pathogens, namely, *Xanthomonas campestris* pv. *vesicatoria*, *Erwinia amylovora*, and *Ralstonia solanacearum*. Mutation in any one of the *hrp* genes leads to failure of pathogen invasion (Van Gijsegem et al. 1993). *Hrp* and *Hrc* (hypersensitive response and conserved) genes encode the type III secretion system (TTSS) that helps in the injection of bacterial virulence "effector" proteins into host cells (Alfano and Collmer 2004; Cornelis and Van Gijsegem 2000). The HRP system delivers various effector proteins which promote the induction of disease by suppressing the basal defense mechanism of host plant and also triggers HR response (Chang et al. 2005; Collmer et al. 2002; Zwiesler-Vollick et al. 2002). These effector proteins are designated as hairpins, which are cysteine-lacking, glycine-rich proteins (Bauer et al. 1995), for example, Avr proteins (Keen 1990), Hop (Hrp outer protein, Alfano and Collmer 1997) of *Pseudomonas*, Xop (*Xanthomonas* outer protein; Noel et al. 2001), or Pop (*Pseudomonas* outer protein; Arlat et al. 1994) of *Ralstonia*. NLPs or Nep1-like proteins are a new family of proteins identified in pathogenic bacteria that induce HR response in plants (Pemberton and Salmond 2004). Various reports on the localization of effector proteins suggested nucleus and plasma membrane to be their target site (Yang and Gabriel 1995; Deslandes et al. 2003).

2.2.7 Effector Proteins Synthesized by the Bacteria

Phytopathogenic bacteria synthesize effector proteins also named as double-edged swords during the course of infection. These proteins enter the host cell cytoplasm and helps in suppressing the plant immune system, thereby playing an important part in virulence and pathogen survival (Mattoo et al. 2007; Shames and Finlay

2012). These effector proteins target various processes of host cell like RNA metabolism, secretion of proteins, and activation of kinases (Kaffarnik et al. 2009). These effector proteins can be recognized by the plant R proteins (resistance proteins) leading to development of plant resistance against the pathogen. Such proteins are called as avirulence proteins (Avr). The pathogen has an *Avr* (avirulence) gene that interacts with the corresponding R (resistance) gene of the host plant leading to establishment of HR and avirulence preventing development of disease (Flor 1971). Reports also suggest that pathogens deploy the secretion system of host plant invasion (Kaffarnik et al. 2009). Pathogenic bacteria *Pseudomonas syringae* have been shown to secrete about 30 effector proteins into the cytosol of plant (Chang et al. 2005). One of the effector proteins of *Pseudomonas syringae* named as AvrPto has been shown to bind the receptor kinases of *Arabidopsis* and blocks the immune response (Zong et al. 2008; Xiang et al. 2008).

2.2.8 Extracellular Polysaccharide and Toxins Produced by Bacteria

Extracellular polysaccharides (EPSs) produced by the bacteria alter the defense-activating signal, block the xylem, and protect the bacteria from the various environmental stresses (Denny 1995). EPSs secreted by *P. syringae* lead to the development of chlorotic and necrotic symptoms (Corsaro et al. 2001). The toxins mainly the secondary metabolites or peptides produced by various necrogenic bacteria are important for pathogenesis or have an antimicrobial activity that reduces the microbial competition (Gross 1991). Coronatine, a toxin produced by *P. syringae*, causes chlorosis in plants and also aids the entry of bacteria by triggering the opening of stomata and by defeating the host defenses (Brooks et al. 2005; Uppalapati et al. 2007). Some pathovars of *P. syringae* produces antimetabolite toxins (mangotoxin, phaseolotoxin, and tabtoxin) that inhibit the biosynthesis of aromatic amino acids thereby interfering with the nitrogen metabolism of host plant (Snoeijers et al. 2000).

2.2.9 Plant Cell Wall-Degrading Enzymes (PCWDE)

Soft rot pectinolytic bacteria under necrotrophic category, namely, *Erwinia carotovora*, *E. chrysanthemi*, *Pseudomonas viridiflava*, *Pectobacterium*, and *Dickeya*, using type II secretion system secrete plant cell wall-degrading enzymes (pectic enzymes, cellulases, proteases, and hemicellulases) that soften the host tissue and kill the cells and also help in nutrient uptake (Perombelon and Kelman 1980; Barras et al. 1994). A single pectate lyase (Pel) is secreted by *P. viridiflava* (Liao et al. 1988), while *E. carotovora* and *E. chrysanthemi* produce complex of pectic enzymes (Barras et al. 1994). Investigation also suggests that PCWDE are produced in a population density-dependent mechanism called quorum sensing (QS). It plays a very important role in regulating bacterial pathogenesis (Whitehead et al. 2001). In case of soft rot-causing bacteria *Pectobacterium atrosepticum*, QS control the production of cell wall-degrading enzymes (Liu et al. 2008).

2.3 Plant Defense Responses Against Pathogen Attack

Necrotrophs and biotrophs confront almost the same basal plant defenses, with the major component being the waxy cuticle layer and rigid cell walls (Heath 2000a, 2002). Plants recognize pathogens on the basis of pathogen-associated molecular patterns (PAMPs). PAMPs are conserved molecules important for pathogen and characterize a class of microbes (Bent and Mackey 2007; Willment and Brown 2007). Examples of PAMPs include bacterial flagellin (Zipfel et al. 2004), fungal xylanase (Ron and Avni 2004), and chitin (Kaku et al. 2006; Miya et al. 2007). Detection of these PAMPs by pattern recognition receptors (PRRs) leads to a PAMP-triggered immune response (PTI) (Bent and Mackey 2007; Hückelhoven 2007; Zipfel 2008, 2009). This comprises the activation of a mitogen-activated protein (MAP) kinases followed by WRKY transcription factor phosphorylation leading to induction of plant defense responses (Heath 2000a; Dodds and Rathjen 2010). Plant responses associated with PTI include callose deposition in the cell wall, production of reactive oxygen species (ROS), and induction of pathogenesis-related proteins and defensins (Silverstein et al. 2005; van Loon et al. 2006). In addition to PAMPs, plant cells can recognize degraded damaged host cells mainly polysaccharides called damage-associated molecular patterns (DAMPs) (Lotze et al. 2007; Matzinger 2007; Hückelhoven 2007). Effector-triggered immunity response happens when an *Avr* gene is recognized by a plant *R* gene (Heath 1998, 2000b; Tudzynski and Sharon 2003; Schulze-Lefert 2004; Ferreira et al. 2007). ETI is associated with the production of reactive oxygen species (ROS) which results in HR response at the point of infection (Heath 1998, 2000b; Lamb and Dixon 1997; Torres and Dangl 2005).

2.4 Use of Effector Proteins in Plant Breeding

Effector molecules are the pathogen-induced molecules that help the bacteria or fungi to invade plants. "Effectoromics" is a high-throughput functional genomics approach used to identify the R genes, and it has played a very important role in modern resistance breeding. For the first time, this strategy was deployed for *Phytophthora infestans* and potato, leading to identification of many *R* and *avr* genes (Vleeshouwers et al. 2008, 2011). Detailed characterization of the effector molecules will help in hastening the R gene cloning and determining the redundancy and can help in deploying R gene in agriculture (Vleeshouwers and Oliver 2014).

2.5 Prospects

Plant-pathogen interactions are highly coevolved and dynamic at molecular and cellular levels. The model illustrating the interaction between plants and fungal/bacterial phytopathogens (necrotrophs/biotrophs) has been depicted in Fig. 2.1. Despite this substantial progress in this field, a critical challenge still remains in understanding the mechanism by which fungal/bacterial pathogens cause the disease and how plants respond to the elicitors secreted by these pathogens during invasion. Also it is not known whether fungi also have their own transport machinery, analogous to the bacterial type III secretion system. Research focused on the identification of diverse virulence factors important for pathogenesis will provide information about the genes present in the pathogenicity islands, which will definitely catapult our understanding of fungal and bacterial pathogenesis. In the coming years, amalgamation of genomics, transcriptomics, proteomics, and metabolomics data will help us in clarifying the various genes encoding different proteins important in plant-pathogen (biotroph/necrotroph) interactions. The study of interplay between the effector proteins secreted by pathogen and defense-promoting signaling in host cells will help us in elucidating the mechanism of disease development. It is very important to develop new methods to manage the biotrophic and necrotrophic plant pathogens. Plant biologists studying these plant-pathogen interactions will undoubtedly help in saving our crops from the devastating attacks.

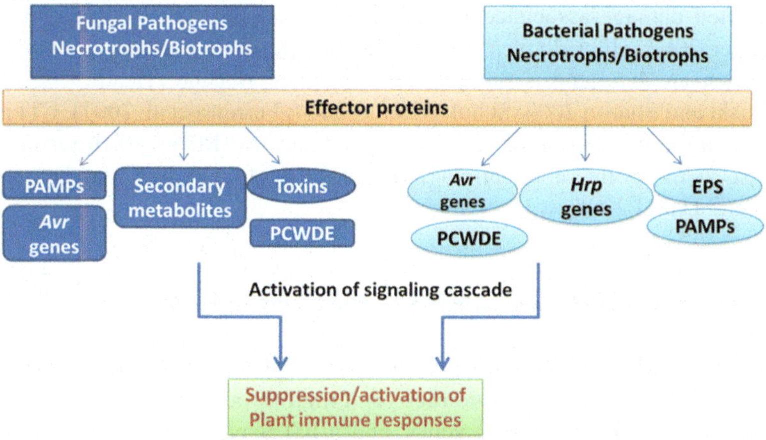

Fig. 2.1 Model for plant/pathogen (necrotroph/biotroph) interaction. *PAMPs* pathogen-associated molecular patterns, *PCWDE* (plant cell wall-degrading enzyme), *EPSs* extracellular polysaccharides

Glossary

Appressorial pegs It is a specialized cell characteristic of fungal plant pathogens and is used during infection process.

Effector proteins Proteins secreted by bacterial pathogens during the infection process and help in suppressing the immune system of host.

Extracellular polysaccharides High molecular weight sugar polymers synthesized by microorganisms. They play important roles in protecting the microorganism and also mediate their pathogenicity.

Hypersensitive response It is a defense mechanism evoked by pathogens and involves localized cell death to stop the spread of infection.

Phytohormone Chemicals or signal molecules synthesized by plants and play an important role in their growth and development.

Phytopathogens Pathogenic bacteria, viruses, or fungi which infect plants and cause many plant diseases.

Quorum sensing It is a phenomenon of cell-cell communication which helps bacteria to sense the cell density and coordinate their behavior accordingly.

Small RNAs Noncoding RNA molecules which are less than 200 nucleotide in length and have a role in RNA silencing and regulation of gene expression.

Virulence The extent of injury caused by pathogen to its host.

References

Abbas HK, Tanaka T, Duke SO, Porter JK, Wray EM, Hodges L, Sessions AE, Wang E, Merrill AH, Riley RT (1994) Fumonisin- and AAL-toxin-induced disruption of sphingolipid metabolism with accumulation of free sphingoid bases. Plant Physiol 106:1085–1093

Agrios GN (2005) Plant pathology (5). Elsevier Academic Press, San Francisco

Ahmad S, Soanes DM, Barooah MC, Talbot NJ (2006) Investigating the evolution of fungal virulence by functional genomics. In: Brown AJP (ed) The mycota vol XII. Fungal genomics. Springer, Berlin, pp 35–49

Alfano JR, Collmer A (1996) Bacterial pathogens in plants: life up against the wall. Plant Cell 8:1683–1698

Alfano JR, Collmer A (1997) The type III (hrp) secretion pathway of plant pathogenic bacteria: trafficking harpins, Avr proteins, and death. J Bacteriol 179:5655–5662

Alfano JR, Collmer A (2004) Type III secretion system effector proteins: double agents in bacterial disease and plant defense. Annu Rev Phytopathol 42:385–414

Anderson JP, Badruzsaufari E, Schenk PM, Manners JM, Desmond OJ, Ehlert C, Maclean DJ, Ebert PR, Kazan K (2004) Antagonistic interaction between abscisic acid and jasmonate-ethylene signaling pathways modulates defense gene expression and disease resistance in *Arabidopsis*. Plant Cell 16:3460–3479

Arlat M, Van Gijsegem F, Huet JC, Pernollet JC, Boucher CA (1994) PopA1, a protein which induces a hypersensitive like response on specific *Petunia* genotypes, is secreted via the Hrp pathway of *Pseudomonas solanacearum*. EMBO J 13:543–553

Audenaert K, De Meyer GB, Höfte MM (2002) Abscisic acid determines basal susceptibility of tomato to *Botrytis cinerea* and suppresses salicylic acid-dependent signaling mechanisms. Plant Physiol 128:491–501

Barras F, Van Gijsegem F, Chatterjee AK (1994) Extracellular enzymes and pathogenesis of soft-rot *Erwinia*. Annu Rev Phytopathol 32:201–234

Bauer DW, Wei Z-M, Beer SV, Collmer A (1995) *Erwinia chrysanthemi* harpin$_{\mathrm{EChA}}$: an elicitor of the hypersensitive response that contributes to soft-rot pathogenesis. Mol Plant-Microbe Interact 8:484–491

Bent AF, Mackey D (2007) Elicitors, effectors and R genes: the new paradigm and a lifetime supply of questions. Annu Rev Phytopathol 45:399–436

Berestetskiy AO (2008) A review of fungal phytotoxins: from basic studies to practical use. Appl Biochem Microbiol 44:453–465

Boch J, Joardar V, Gao L, Robertson TL, Lim M, Kunkel BN (2002) Identification of *Pseudomonas syringae* genes induced during infection of *Arabidopsis thaliana*. Mol Microbiol 44:73–88

Boller T, He SY (2009) Innate immunity in plants: an arms race between pattern recognition receptors in plants and effectors in microbial pathogens. Science 324:742–744

Bolton MD, Thomma BP, Nelson BD (2006) *Sclerotinia sclerotiorum* (Lib.) de Bary: biology and molecular traits of a cosmopolitan pathogen. Mol Plant Pathol 7:1–16

Bos JI, Kanneganti TD, Young C, Cakir C, Huitema E, Win J, Armstrong MR, Birch PR, Kamoun S (2006) The C-terminal half of *Phytophthora infestans* RXLR effector AVR3a is sufficient to trigger R3a-mediated hypersensitivity and suppress INF1-induced cell death in *Nicotiana benthamiana*. Plant J 48:165–176

Brefort T, Doehlemann G, Mendoza-Mendoza A, Reissmann S, Djamei A, Kahmann R (2009) *Ustilago maydis* as a pathogen. Annu Rev Phytopathol 47:423–445

Brodhun F, Cristobal-Sarramian A, Zabel S, Newie J, Hamberg M, Feussner I (2013) An iron 13S-lipoxygenase with an a-linolenic acid specific hydroperoxidase activity from *Fusarium oxysporum*. PLoS One 8:e64919

Brooks DM, Bender CL, Kunkel BN (2005) The *Pseudomonas syringae* phytotoxin coronatine promotes virulence by overcoming salicylic acid-dependent defenses in *Arabidopsis thaliana*. Mol Plant Pathol 6:629–639

Büttner D, Bonas U (2006) Who comes first? How plant pathogenic bacteria orchestrate type III secretion. Curr Opin Microbiol 9:193–200

Chang JH, Urbach JM, Law TF, Arnold LW, Hu A, Gombar S et al (2005) A high-throughput, near saturating screen for type III effector genes from *Pseudomonas syringae*. Proc Natl Acad Sci U S A 102:2549–2554

Cheng Q, Wang H, Xu B, Zhu S, Hu L, Huang M (2014) Discovery of a novel small secreted protein family with conserved N-terminal IGY motif in Dikarya fungi. BMC Genomics 15:1151

Cho Y (2015) How the necrotrophic fungus *Alternaria brassicicola* kills plant cells remains an enigma. Eukaryot Cell 14:335–344

Collmer A, Bauer DW (1994) *Erwinia chrysanthemi* and *Pseudomonas syringae*: plant pathogens trafficking in virulence proteins. Curr Top Microbiol Immunol 192:43–78

Collmer A, Lindeberg M, Petnicki-Ocwieja T, Schneider DJ, Alfano JR (2002) Genomic mining type III secretion system effectors in *Pseudomonas syringae* yields new picks for all TTSS prospectors. Trends Microbiol 10:462–469

Cornelis GR, Van Gijsegem F (2000) Assembly and function of type III secretory systems. Annu Rev Microbiol 54:735–774

Corsaro MM, Evidente A, Lanzetta R, Lavermicocca P, Molinaro A (2001) Structure determination of the phytotoxic mannan exopysaccharide from *Pseudomonas syringae* pv. *ciccaronei*. Carbohydr Res 330:6208–6215

de Jonge R, van Esse HP, Kombrink A, Shinya T, Desaki Y et al (2010) Conserved fungal LysM effector Ecp6 prevents chitin-triggered immunity in plants. Science 329:953–955

Denny TP (1995) Involvement of bacterial polysaccharide in plant pathogenesis. Annu Rev Phytopathol 33:173–197

Desjardins AE, McCormick SP, Appell M (2007) Structure−activity relationships of trichothecene toxins in an *Arabidopsis thaliana* leaf assay. J Agric Food Chem 55:6487–6492

Deslandes L, Olivier J, Peeters N, Feng DX, Khounlotham M et al (2003) Physical interaction between RRS1-R, a protein conferring resistance to bacterial wilt, and PopP2, a type III effector targeted to the plant nucleus. Proc Natl Acad Sci U S A 100:8024–8029

Di X, Takken FLW, Tintor N (2016) How phytohormones shape interactions between plants and the soil-borne fungus *Fusarium oxysporum*. Front Plant Sci 7:170. https://doi.org/10.3389/fpls.2016.00170

Djamei A, Schipper K, Rabe F, Ghosh A, Vincon V et al (2011) Metabolic priming by a secreted fungal effector. Nature 478:395–398

Dodds PN, Rathjen JP (2010) Plant immunity: towards an integrated view of plant pathogen interactions. Nat Rev Genet 11:539–548

Dodds PN, Rafiqi M, Gan PHP, Hardham AR, Jones DA, Ellis JG (2009) Effectors of biotrophic fungi and oomycetes: pathogenicity factors and triggers of host resistance. New Phytol 183:993–1000

Doehlemann G, van der Linde K, Amann D, Schwammbach D, Hof A, Mohanty A et al (2009) Pep1, a secreted effector protein of *Ustilago maydis*, is required for successful invasion of plant cells. PLoS Pathog 5:e1000290

Dou D, Kale SD, Wang X, Chen Y, Wang Q, Wang X, Jiang RHY, Arredondo FD, Anderson R, Thakur P, McDowell J, Wang Y, Tyler BM (2008) Carboxy-terminal motifs common to many oomycete RXLR effectors are required for avirulence and suppression of BAX-mediated programmed cell death by *Phytophthora sojae* effector Avr1b. Plant Cell 20:118–1133

Evangelisti E, Govetto B, Minet-Kebdani N, Kuhn ML, Attard A, Ponchet M, Panabières F, Gourgues M (2013) The *Phytophthora parasitica* RXLR effector penetration-specific effector 1 favours *Arabidopsis thaliana* infection by interfering with auxin physiology. New Phytol 199:476–489

Ferreira RB, Monteiro S, Freitas R, Santos CN, Chen Z, Batista LM, Duarte J, Borges A, Teixeira AR (2007) The role of plant defence proteins in fungal pathogenesis. Mol Plant Pathol 8:677–700

Flor HH (1971) Current status of the gene-for-gene concept. Annu Rev Phytopathol 9:275–296

Glazebrook J (2005) Contrasting mechanisms of defence against biotrophic and necrotrophic pathogens. Annu Rev Phytopathol 43:205–227

Göhre V, Robatzek S (2009) Breaking the barriers: microbial effector molecules subvert plant immunity. Annu Rev Phytopathol 46:189–215

Gross DC (1991) Molecular and genetic analysis of toxin production by pathovars of *Pseudomonas syringae*. Annu Rev Phytopathol 29:247–278

Guimarães RL, Stotz HU (2004) Oxalate production by *Sclerotinia sclerotiorum* deregulates guard cells during infection. Plant Physiol 136:3703–3711

Haas BJ, Kamoun S, Zody MC, Jiang RH et al (2009) Genome sequence and analysis of the Irish potato famine pathogen *Phytophthora infestans*. Nature 461:393–398

Hawksworth DL (1991) The fungal dimension of biodiversity: magnitude, significance, and conservation. Mycol Res 95:641–655

Heath MC (1997) Signalling between pathogenic rust fungi and resistant or susceptible host plants. Ann Bot 80:713–720

Heath MC (1998) Apoptosis, programmed cell death and the hypersensitive response. Eur J Plant Pathol 104:117–124

Heath MC (2000a) Nonhost resistance and non-specific plant defences. Curr Opin Plant Biol 3:315–319

Heath MC (2000b) Hypersensitive response-related death. Plant Mol Biol 44:321–334

Heath MC (2002) Cellular interactions between biotrophic fungal pathogens and host or nonhost plants. Can J Plant Pathol 24:259–264

Hemetsberger C, Mueller AN, Matei A, Herrberger C, Hensel G, Kumlehn J, Mishra B, Sharma R, Thines M, Hückelhoven R, Doehlemann G (2015) The fungal core effector Pep1 is conserved across smuts of dicots and monocots. New Phytol 206:1116–1126

Horbach R, Navarro-Quesada AR, Knogge W et al (2011) When and how to kill a plant cell: infection strategies of plant pathogenic fungi. J Plant Physiol 168:51–62

Howlett BJ (2006) Secondary metabolite toxins and nutrition of plant pathogenic fungi. Curr Opin Plant Biol 9:371–375
Hückelhoven R (2007) Cell wall-associated mechanisms of disease resistance and susceptibility. Annu Rev Phytopathol 45:101–127
Idnurm A, Howlett BJ (2001) Pathogenicity genes of phytopathogenic fungi. Mol Plant Pathol 2:241–255
Inomata M, Hirai N, Yoshida R, Ohigashi H (2004) The biosynthetic pathway to abscisic acid via ionylideneethane in the fungus *Botrytis cinerea*. Phytochemistry 65:2667–2678
Jiang RHY, Tripathy S, Govers F, Tyler BM (2008) RXLR effector reservoir in two *Phytophthora* species is dominated by a single rapidly evolving superfamily with more than 700 members. Proc Natl Acad Sci U S A 105:4874–4879
Jin Q, Thilmony R, Zwiesler-Vollick J, He SY (2003) Type III protein secretion in *Pseudomonas syringae*. Microbes Infect 5:301–310
Jones JDG, Dangl JL (2006) The plant immune system. Nature 444:323–329
Kaffarnik FA, Jones AM, Rathjen JP, Peck SC (2009) Effector proteins of the bacterial pathogen *Pseudomonas syringae* alter the extracellular proteome of the host plant, *Arabidopsis thaliana*. Mol Cell Proteomics 8:145–156
Kahmann R, Basse C (2001) Fungal gene expression during pathogenesis-related development and host plant colonization. Curr Opin Microbiol 4:374–380
Kaku H, Nishizawa Y, Ishii-Minami N, Akimoto-Tomiyama C, Dohmae N et al (2006) Plant cells recognize chitin fragments for defense signaling through a plasma membrane receptor. Proc Natl Acad Sci U S A 103:11086–11091
Kale SD, Gu B, Capelluto DGS, Dou D, Feldman E et al (2010) External lipid PI3P mediates entry of eukaryotic pathogen effectors into plant and animal host cells. Cell 142:284–295
Kamper J, Kahmann R, Bolker M, Ma LJ, Brefort T, Saville BJ et al (2006) Insights from the genome of the biotrophic fungal plant pathogen *Ustilago maydis*. Nature 444:97–101
Kars I, McCalman M, Wagemakers L, van Kan JAL (2005) Functional analysis of *Botrytis cinerea* pectin methylesterase genes by PCR-based targeted mutagenesis: *Bcpme1* and *Bcpme2* are dispensable for virulence of strain B05.10. Mol Plant Pathol 6:641–652
Kazan K, Lyons R (2014) Intervention of phytohormone pathways by pathogen effectors. Plant Cell 26:2285–2309
Keen NT (1990) Gene-for-gene complementarity in plant-pathogen interactions. Annu Rev Genet 24:447–463
Khang CH, Berruyer R, Giraldo MC, Kankanala P, Park SY, Czymmek K et al (2010) Translocation of *Magnaporthe oryzae* effectors into rice cells and their subsequent cell-to-cell movement. Plant Cell 22:1388–1403
Kim K, Jeon J, Choi J, Cheong K, Song H, Choi G, Kang S, Lee Y (2016) Kingdom-wide analysis of fungal small secreted proteins (SSPs) reveals their potential role in host association. Front Plant Sci 7:186. https://doi.org/10.3389/fpls.2016.00186
Koeck M, Hardham AR, Dodds PN (2011) The role of effectors of biotrophic and hemibiotrophic fungi in infection. Cell Microbiol 13:1849–1857
Kraepiel Y, Barny M (2016) Gram-negative phytopathogenic bacteria, all hemibiotrophs after all? Mol Plant Pathol 17:313–316
Laluk K, Mengiste T (2010) Necrotroph attacks on plants: wanton destruction or covert extortion? In: The *Arabidopsis* book. The American Society of Plant Biologists, Rockville, pp 1–34
Lamb C, Dixon RA (1997) The oxidative burst in plant disease resistance. Annu Rev Plant Physiol Plant Mol Biol 48:251–275
Lee S-J, Rose JK (2010) Mediation of the transition from biotrophy to necrotrophy in hemibiotrophic plant pathogens by secreted effector proteins. Plant Signal Behav 5:769–772
Liao C-H, Hung HY, Chatterjee AK (1988) An extracellular pectate lyase is the pathogenicity factor of the soft-rotting bacterium *Pseudomonas viridiflava*. Mol Plant-Microbe Interact 1:199–206

Lindgren PB, Peet RC, Panopoulos NJ (1986) Gene cluster of *Pseudomonas syringae pv. "phaseolicola"* controls pathogenicity of bean plants and hypersensitivity on nonhost plants. J Bacteriol 168:512–522

Liu H, Coulthurst SJ, Pritchard L, Hedley PE, Ravensdale M, Humphris S, Burr T, Takle G, Brurberg MB, Birch PR et al (2008) Quorum sensing coordinates brute force and stealth modes of infection in the plant pathogen *Pectobacterium atrosepticum*. PLoS Pathog 4:e1000093–e1000010

Liu T, Ye W, Ru Y, Yang X, Gu B, Tao K et al (2011) Two host cytoplasmic effectors are required for pathogenesis of *Phytophthora sojae* by suppression of host defenses. Plant Physiol 155:490–501

Lotze MT, Zeh HJ, Rubartelli A, Sparvero LJ, Amoscato AA, Washburn NR, Devera ME, Liang X, Tör M, Billiar T (2007) The grateful dead: damage-associated molecular pattern molecules and reduction/oxidation regulate immunity. Immunol Rev 220:60–81

Mattoo S, Lee YM, Dixon JE (2007) Interactions of bacterial effector proteins with host proteins. Curr Opin Immunol 19:392–401

Matzinger P (2007) Friendly and dangerous signals: is the tissue in control? Nat Immunol 8:11–13

Mendgen K, Hahn M (2002) Plant infection and the establishment of fungal biotrophy. Trends Plant Sci 7:352–356

Mendgen K, Hahn M (2004) Plant infection and the establishment of fungal biotrophy. Curr Opin Plant Biol 7:356–364

Mengiste T (2012) Plant immunity to necrotrophs. Annu Rev Phytopathol 50:267–294

Miya A, Albert P, Shinya T, Desaki Y, Ichimura K et al (2007) CERK1, a LysM receptor kinase, is essential for chitin elicitor signaling in *Arabidopsis*. Proc Natl Acad Sci U S A 104:19613–19618

Mobius N, Hertweck C (2009) Fungal phytotoxins as mediators of virulence. Curr Opin Plant Biol 12:390–398

Mosquera G, Giraldo MC, Khang CH, Coughlan S, Valent B (2009) Interaction transcriptome analysis identifies *Magnaporthe oryzae* BAS1–4 as biotrophy-associated secreted proteins in rice blast disease. Plant Cell 21:1273–1290

Muller O, Schreier PH, Uhrig JF (2008) Identification and characterization of secreted and pathogenesis related proteins in *Ustilago maydis*. Mol Gen Genomics 279:27–39

Niepold F, Anderson D, Mills D (1985) Cloning determinants of pathogenesis from *Pseudomonas syringae pathovar syringae*. Proc Natl Acad Sci U S A 82:406–410

Noel L, Thieme F, Nennstie D, Bonas U (2001) cDNA-AFLP analysis unravels a genome-wide hrpG-regulon in the plant pathogen *Xanthomonas campestris* pv. *vesicatoria*. Mol Microbiol 41:1271–1281

Oliver RP, Ipcho SVS (2004) *Arabidopsis* pathology breathes new life into necrotrophs vs.- biotrophs classification of fungal pathogens. Mol Plant Pathol 5:347–352

Pedras MS, Ahiahonu PW (2004) Phytotoxin production and phytoalexin elicitation by the phytopathogenic fungus *Sclerotinia sclerotiorum*. J Chem Ecol 30:2163–2179

Pemberton CL, Salmond GPC (2004) The Nep1-like proteins – a growing family of microbial elicitors of plant necrosis. Mol Plant Pathol 5:353–359

Penn CD, Daniel SL (2013) Salicylate degradation by the fungal plant pathogen *Sclerotinia sclerotiorum*. Curr Microbiol 67:218–225

Pennington HG, Gheorghe DM, Damerum A, Pliego C, Spanu PD, Cramer R, Bindschedler LV (2016) Interactions between the powdery mildew effector BEC1054 and barley proteins identify candidate host targets. J Proteome Res 15:826–839

Perombelon MCM, Kelman A (1980) Ecology of the soft rot Erwinias. Annu Rev Phytopathol 18:361–387

Prell HH, Day PR (2001) Plant–fungal pathogen interaction: a classical and molecular view. Springer, Berlin

Prins TW, Tudzynski P, Tiedemann AV, Tudzynski B, ten Have A, Hansen ME, Tenberge K, van Kan JAL (2000) Infection strategies of *Botrytis cinerea* and related necrotrophic pathogens. In: Kronstad JW (ed) Fungal pathology. Kluwer Academic Publishers, Dordrecht, pp 33–64

Redkar A, Hoser R, Schilling L, Zechmann B, Krzymowska M, Walbot V, Doehlemann G (2015) A secreted effector protein of *Ustilago maydis* guides maize leaf cells to form tumors. Plant Cell 27:1332–1351

Ron M, Avni A (2004) The receptor for the fungal elicitor ethylene-inducing xylanase is a member of a resistance-like gene family in tomato. Plant Cell 16:1604–1615

Sánchez-Vallet A, Saleem-Batcha R, Kombrink A, Hansen G, Valkenburg DJ, Thomma BP, Mesters JR (2013) Fungal effector Ecp6 outcompetes host immune receptor for chitin binding through intrachain LysM dimerization. elife 2:e00790

Schulze-Lefert P (2004) Knocking in the heaven's wall: pathogenesis of and resistance to biotrophic fungi at the cell wall. Curr Opin Plant Biol 7:377–383

Schulze-Lefert P, Panstruga R (2003) Establishment of biotrophy by parasitic fungi and reprogramming of host cells for disease resistance. Annu Rev Phytopathol 41:641–667

Shames SR, Finlay BB (2012) Bacterial effector interplay: a new way to view effector function. Trends Microbiol 20:214–219

Shi L, Bielawski J, Mu J, Dong H, Teng C, Zhang J, Yang X, Tomishige N, Hanada K, Hannun YA, Zuo J (2007) Involvement of sphingoid bases in mediating reactive oxygen intermediate production and programmed cell death in *Arabidopsis*. Cell Res 17:1030–1040

Silverstein KA, Graham MA, Paape TD, VandenBosch KA (2005) Genome organization of more than 300 defensin-like genes in *Arabidopsis*. Plant Physiol 138:600–610

Singh R, Dangol S, Chen Y, Choi J, Cho YS, Lee JE, Choi MO, Jwa NS (2016) *Magnaporthe oryzae* effector AVR-Pii helps to establish compatibility by inhibition of the rice NADP-malic enzyme resulting in disruption of oxidative burst and host innate immunity. Mol Cell 39:426–438

Snoeijers SS, Pérez-García A, Joosten MHAJ, De Wit PJGM (2000) The effect of nitrogen on disease development and gene expression in bacterial and fungal plant pathogens. Eur J Plant Pathol 106:493–506

Stam R, Jupe J, Howden AJ, Morris JA, Boevink PC, Hedley PE, Huitema E (2013) Identification and characterization CRN effectors in *Phytophthora capsici* shows modularity and functional diversity. PLoS One 8:e59517

Stergiopoulos I, de Wit PJGM (2009) Fungal effector proteins. Annu Rev Phytopathol 47:233–263

Stergiopoulos I, van den Burg HA, Ökmen B, Beenen HG, van Liere S, Kema GHJ, de Wit PJGM (2010) Tomato Cf resistance proteins mediate recognition of cognate homologous effectors from fungi pathogenic on dicots and monocots. Proc Natl Acad Sci U S A 107:7610–7615

Stergiopoulos I, Collemare J, Mehrabi R, De Wit PJ (2013) Phytotoxic secondary metabolites and peptides produced by plant pathogenic Dothideomycete fungi. FEMS Microbiol Rev 37:67–93

Stone JK (2001) Necrotroph. In: Maloy OC, Murray TD (eds) Encyclopedia of plant pathology. Wiley, New York, pp 676–677

Thomma B, Van Esse HP, Crous PW, De Wit P (2005) *Cladosporium fulvum* (syn. *Passalora fulva*), a highly specialized plant pathogen as a model for functional studies on plant pathogenic Mycosphaerellaceae. Mol Plant Pathol 6:379–393

Torres MA, Dangl JL (2005) Functions of the respiratory burst oxidase in biotic interactions, abiotic stress and development. Curr Opin Plant Biol 8:397–403

Torto TA, Li S, Styer A, Huitema E, Testa A, Gow NA, van West P, Kamoun S (2003) EST mining and functional expression assays identify extracellular effector proteins from the plant pathogen *Phytophthora*. Genome Res 13:1675–1685

Tsuge T, Harimoto Y, Akimitsu K, Ohtani K, Kodama M, Akagi Y, Egusa M, Yamamoto M, Otani H (2013) Host selective toxins produced by the plant pathogenic fungus *Alternaria alternata*. FEMS Microbiol Rev 37:44–66

Tudzynski P, Sharon A (2003) Fungal pathogenicity genes. In: Arora DDK, Khachatourians GG (eds) Applied mycology and biotechnology vol. 3: fungal genomics. Elsevier, Berlin, pp 187–212

Tyler BM (2009) Entering and breaking: virulence effector proteins of oomycete plant pathogens. Cell Microbiol 11:13–20

Uppalapati SR, Ishiga Y, Wangdi T, Kunkel BN, Anand A, Mysore KS, Bender CL (2007) The phytotoxin coronatine contribute to pathogen fitness and is required for suppression of salicylic

acid accumulation in tomato inoculated with *Pseudomonas syringae* pv. tomato DC3000. Mol Plant-Microbe Interact 20:955–965
van Damme M, Bozkurt TO, Cakir C, Schornack S, Sklenar J et al (2012) The Irish potato famine pathogen *Phytophthora infestans* translocates the CRN8 kinase into host plant cells. PLoS Pathog 8:e1002875
van Esse HP, Bolton MD, Stergiopoulos L, de Wit P, Thomma B (2007) The chitin-binding *Cladosporium fulvum* effector protein Avr4 is a virulence factor. Mol Plant Microbe Interac 20:1092–1101
van Esse HP, van't Klooster JW, Bolton MD, Yadeta KA, van Baarlen P, Boeren S et al (2008) The *Cladosporium fulvum* virulence protein Avr2 inhibits host proteases required for basal defense. Plant Cell 20:1948–1963
Van Gijsegem F, Genin S, Boucher C (1993) Evolutionary conservation of pathogenicity determinants among plant and animal pathogenic bacteria. Trends Microbiol 1:175–180
Van Kan JA (2006) Licensed to kill: the lifestyle of a necrotrophic plant pathogen. Trends Plant Sci 11:247–253
van Loon LC, Rep M, Pieterse CMJ (2006) Significance of inducible defense-related proteins in infected plants. Annu Rev Phytopathol 44:135–162
Vleeshouwers VGAA, Oliver RP (2014) Effectors as tools in disease resistance breeding against biotrophic, hemibiotrophic, and necrotrophic plant pathogens. Mol Plant-Microbe Interact 27:196–206
Vleeshouwers VGAA, Rietman H, Krenek P, Champouret N, Young C, Oh S-K, Wang M, Bouwmeester K, Vosman B, Visser R (2008) Effector genomics accelerates discovery and functional profiling of potato disease resistance and *Phytophthora infestans* avirulence genes. PLoS One 3:e2875
Vleeshouwers VGAA, Finkers R, Budding DJ, Visser M, Jacobs MMJ, van Berloo R, Pel M, Champouret N, Bakker E, Krenek P, Rietman H, Huigen DJ, Hoekstra R, Goverse A, Vosman B, Jacobsen E, Visser RGF (2011) SolRgene: an online database to explore disease resistance genes in tuber-bearing *Solanum* species. BMC Plant Biol 11:116
Walton JD (1996) Host-selective toxins: agents of compatibility. Plant Cell 8:1723–1733
Wang X, Jiang N, Liu J, Liu W, Wang G-L (2014) The role of effectors and host immunity in plant–necrotrophic fungal interactions. Virulence 5:722–732
Wang M, Weiberg A, Jin H (2015) Pathogen small RNAs: a new class of effectors for pathogen attacks. Mol Plant Pathol 16:219–223
Weiberg A, Wang M, Lin FM, Zhao H, Zhang Z, Kaloshian I, Huang HD, Jin H (2013) Fungal small RNAs suppress plant immunity by hijacking host RNA interference pathways. Science 342:118–123
Weiberg A, Wang M, Bellinger M, Jin H (2014) Small RNAs: a new paradigm in plant–microbe interactions. Annu Rev Phytopathol 52:495–516
Whisson SC, Boevink PC, Moleleki L, Avrova AO, Morales JG et al (2007) A translocation signal for delivery of oomycete effector proteins into host plant cells. Nature 450:115–119
Whitehead NA, Barnard AM, Slater H, Simpson NJ, Salmond GP (2001) Quorum-sensing in Gram-negative bacteria. FEMS Microbiol Rev 25:365–404
Willment JA, Brown GD (2007) C-type lectin receptors in antifungal immunity. Trends Microbiol 16:27–32
Win J, Morgan W, Bos J, Krasileva KV, Cano LM, Chaparro-Garcia A, Ammar R, Staskawicz BJ, Kamoun S (2007) Adaptive evolution has targeted the c-terminal domain of the RXLR effectors of plant pathogenic oomycetes. Plant Cell 19:2349–2369
Wolpert TJ, Dunkle LD, Ciuffetti LM (2002) Host-selective toxins and avirulence determinants: what's in a name? Annu Rev Phytopathol 40:251–285
Xiang T, Zong N, Zou Y, Wu Y, Zhang J, Xing W et al (2008) *Pseudomonas syringae* effector AvrPto blocks innate immunity by targeting receptor kinases. Curr Biol 18:74–80
Yang Y, Gabriel DW (1995) *Xanthomonas* avirulence/pathogenicity gene family encodes functional plant nuclear targeting signals. Mol Plant-Microbe Interact 8:627–631

Zhu W, Wei W, Fu Y, Cheng J, Xie J, Li G, Yi X, Kang Z, Dickman MB, Jiang D (2013) A secretory protein of necrotrophic fungus *Sclerotinia sclerotiorum* that suppresses host resistance. PLoS One 8:e53901

Zipfel C (2008) Pattern recognition receptors in plant innate immunity. Curr Opin Immunol 20:10–16

Zipfel C (2009) Early molecular events in PAMP-triggered immunity. Curr Opin Plant Biol 12:414–420

Zipfel C, Robatzek S, Navarro L, Oakeley EJ, Jones JD, Felix G, Boller T (2004) Bacterial disease resistance in *Arabidopsis* through flagellin perception. Nature 428:764–767

Zong N, Xiang T, Zou Y, Chai J, Zhou J-M (2008) Blocking and triggering of plant immunity by *Pseudomonas syringae* effector AvrPto. Plant Signal Behav 3:583–585

Zwiesler-Vollick J, Plovanich-Jones AE, Nomura K, Brandyopadhyay S, Joardar V et al (2002) Identification of novel *hrp*-regulated genes through functional genomic analysis of the *Pseudomonas syringae* pv. *tomato* DC3000 genome. Mol Microbiol 45:1207–1218

3 Plant–Virus Interactions

Sunita Yadav and Anju K. Chhibbar

Abstract

Viruses are small pathogens not visible under light microscope and are causal agents for many common plant diseases. They lead to heavy economic losses in crop production and quality in different parts of the world. The simplest viruses are composed of nucleic acid and protein coat. Plant viruses mostly have single-stranded ribonucleic acid (ssRNA), but in few cases single- or double-stranded DNA may also be present. They are obligate parasites and require host machinery for their reproduction. They make their passive entry into plant cells through the wounds caused by either physical injuries, through environmental factors, or by the vectors which could be insects, nematodes, fungi, and even mites. Viral RNA disassembles, replicates, and converts its mRNA to proteins in the host cytoplasm using energy and proteins from the host cell. Once viruses enter the host, they move from infected cells to healthy neighboring cells locally. Long-distance transport via the vascular system for systemic infection is also the key feature of plant viruses. In response to the infection by viruses, plants also develop certain defense mechanisms. In this chapter the aspects related to movement of viruses in plant system, general response of plants to viruses, defense mechanisms developed by the plant like RNA silencing, virus-encoded suppressor proteins, development of disease-free tissues, and future aspects are considered.

Keywords

Pathogen · ssRNA · RNA silencing · Suppressor proteins

S. Yadav
Department of Botany, HansRaj College, University of Delhi, New Delhi, India
e-mail: response2sunita@gmail.com

A. K. Chhibbar (✉)
Department of Botany, Deshbandhu College, University of Delhi, New Delhi, India
e-mail: anjuchhibbar@gmail.com

A. Singh, I. K. Singh (eds.), *Molecular Aspects of Plant-Pathogen Interaction*,
https://doi.org/10.1007/978-981-10-7371-7_3

3.1 Introduction

About 15% of the global crop production is lost due to plant diseases highlighting the fact that there is a continuous threat to the plants throughout the world. And more than one third of these plant diseases are thought to be caused by phytopathogenic viruses (Boualem et al. 2016). Viral infections pose serious threat to agricultural and horticultural crops and cause huge economic loss. Plant virus infections cannot be directly controlled by the use of various chemicals and insect vector management, although virus population and distribution can be restricted to some extent through the use of these chemical treatments. Thus the development of disease-resistant cultivars is a major challenge in plant breeding research because the usage of chemicals has more negative effect on humans and surrounding environment (Islam et al. 2017). Viruses being the most abundant life forms on earth can be found in all kingdoms of life. About one third of all the viruses are represented by plant viruses. Plant viruses are simple, obligate, intracellular parasites. One of the important characteristics shared by all viruses is their relatively small genome size, which has a very limited coding capacity. Viruses are nucleic acid-based pathogens and packed with a protein called capsid. They contain single-stranded (ss) or double-stranded (ds) RNA or DNA genome. For instance, a vast majority of known plant viruses are positive ssRNA viruses that typically encode not more than a dozen proteins. They lack necessary components for their independent survival, so they rely entirely on host machinery for their life cycle. They use protein and energy from the host cell to perform different processes inside/outside the host cytoplasm like entry into plant cell or uncoating of nucleic acid, translating viral proteins, viral nucleic acid replication, intracellular and systemic movement, encapsulation and suppression of host defenses on their accumulation, progeny virions assembly, and further transmission (Nelson and Citovsky 2005). They enter their host cells passively either through wounds caused by environmental physical injuries or by different vectors, which may be insects, nematodes, soil fungi, or mites. The obligate intracellular nature of viruses provides a platform for host–virus interaction (Pallas and Garcia 2011).

Viruses can replicate within the living cells of their respective hosts using their own enzymes like RNA-dependent RNA polymerase, DNA replicase, or reverse transcriptase, whereas subviral agents called viroids use RNA polymerase of their hosts for replication. Viroids till now are considered to be the smallest plant pathogens. They are nonprotein-encoding and highly structured, single-stranded RNA molecules (Navarro et al. 2012). The absence of a protein coat distinguishes viroids from viruses. Viroids completely have different ways of transmission as compared to plant viruses because they are not enveloped and do not encode any (movement) proteins. The most important transmission route is believed to be mechanical transmission through physical contact with insect parts and/or plant products (e.g., pollen).

During each stage of viral cycle, variant interactions are generated between the plant and the virus, which might lead to either a compatible or non-compatible relationship between virus and host plant. The development of disease in the plant is an exception and not an outcome of the viral infection as plants are capable of counteracting the harmful effects of viruses. The resistance is offered either through passive

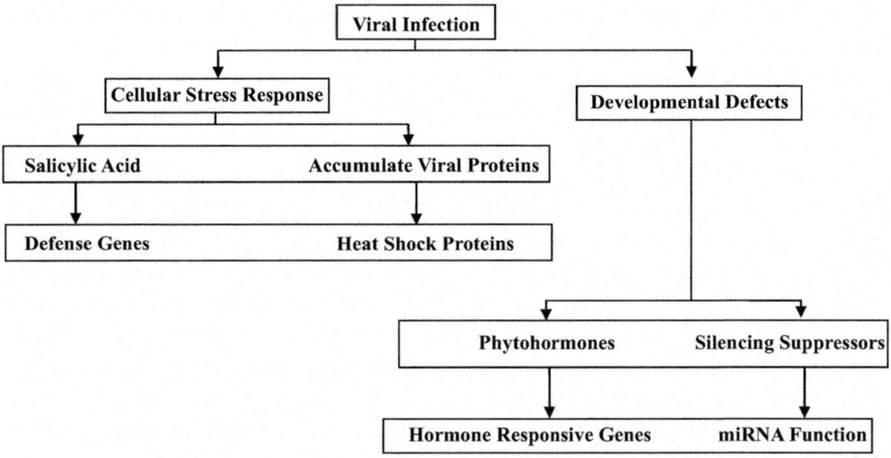

Fig. 3.1 Host plant and virus interactions leading to the expression of defense-related and heat shock genes under the control of signal transduction pathways. Viruses also cause developmental defects as they disrupt the phytohormone signaling and regulatory microRNAs

means where essential host susceptibility factors are absent or through the presence of defensive physical and chemical barriers which the virus has to cross. Once the virus has crossed these barriers, then it is exposed to certain nonspecific defense reactions. The virus causes the infection only if it produces virulence factors against these defense mechanisms, which may or may not be challenged by the plants depending on the defense response to virulence factors of virus (Jones and Dangl 2006).

Various plant proteins (like RNA polymerases, RNA helicases, Dicer-type dsRNA RNases, ssRNA RNases) induced by viral RNA participate in antiviral defense response via RNA silencing mechanism. Viruses have to escape this universal mechanism for its successful infection in the plant. The strategy that is generally adopted by the virus is the production of silencing suppressors (Valli et al. 2009). Viruses also have to complete their reproductive cycle apart from overcoming defense barriers. The positive polarity viruses uncoat and translate the genomic RNA once they enter the plant cell followed by their movement to neighboring cells and finally spread all over the plant (Fig. 3.1). Therefore undertaking plant–virus interaction mechanism is a better way to develop novel tools for plant protection against virus attack. In this chapter, we have highlighted the present state of knowledge about plant–virus interactions.

3.2 Environment and Evolution Modulate Plant Pathogenesis

Since the beginning of virology, a primary motivation for the study of plant viruses is that these are important pathogens of plants. And a major area of research has focused on viral plant pathogenesis (study of the capacity of viruses to cause diseases in plants). Since the identification of first virus genes in mid-1980s that had the properties of avirulence (Avr) factors or that encoded functions associated with the development of specific symptoms, the knowledge on plant virus pathogenesis has increased (Baughman et al. 1988). An obvious antagonistic plant–virus interaction has focused on those viruses that cause disease in crops. However, recently, the concept of viruses being commensals or even mutualists has been emphasized, although based on limited experimental evidence. Still, plant pathogenesis is an ongoing subject of debate, and it is highly important to understand which factors favor the evolution of viruses toward pathogenicity and under which conditions viruses will act as mutualists or commensals. Although much studies on plant–virus interactions in wild ecosystems have not been undertaken (Roossinck and García-Arenal 2015), evidence point toward asymptomatic virus infection in wild plants (Stobbe and Roossinck 2014). For viruses that infect wild plants, the interactions affecting the composition and dynamics of wild plant ecosystems have been well documented (Prendeville et al. 2014).

However, the universal outcome of plant–virus interactions is not always a disease. Viruses that lead to manifestation of disease in some host plants can be asymptomatic in others emphasizing the role of plant–virus interactions in plant pathogenicity. The interaction between viral infection and environmental conditions may change the host–virus interaction from detrimental to beneficial to the host. For example, the effects of CMV (*Cucumber mosaic virus*) infection on *Arabidopsis thaliana* plants over a range of temperature and light intensity conditions varied according to plant genotype and environment, infection being detrimental, neutral, or beneficial to the host plant in terms of viable seed production (Roossinck and García-Arenal 2015).

Environmental modulation during pathogenesis does not involve any genetic change either in the virus or the host. However, it has been observed that change in the environment can result in virus evolution towards more severely pathogenic genotypes. For several plant–virus interactions, symptom expression has often been considered as an alternative to virulence (Doumayrou et al. 2013). Although evidences fail to show any correlation between virus multiplication and virulence, the trade-off hypothesis (there is evolution of highly virulent isolates of virus in environmental conditions that favor high rates of horizontal transmission) can be useful for the analyses of virulence/pathogenicity evolution in plant viruses (Stukenbrock and McDonald 2008). Diversity of habitat also modulates the evolution of pathogenesis. Viruses with a narrow host range have higher host availability in agricultural habitats than in wild resulting in higher rates of horizontal transmission. An increase in the transmission rate may result in the evolution toward high pathogenicity. Factors favoring or hindering the evolution of high pathogenicity in specific environments or across environments await future research.

3.3 Plant–Virus Relationships

The studies conducted on extremely complex plant–virus interactions for more than half a century have led to a partial elucidation of the mechanisms linked to the accumulation of viruses inside host cells, virus movement within the plant, and the plant defense mechanisms. Most of the plant virus combinations do not necessarily lead to effective infections because virus must be able to find appropriate supportive host factors and should be able to escape the host defense responses. A virus may not be always pathogenic even though it is able to replicate. Numerous viral and host components interact within the context of viral needs, including host proteins, cell membranes, lipids, and metabolites (Stapleford and Miller 2010). Developmental abnormalities and other phenotypic manifestations described as disease symptoms appear only when virus infection disturbs host physiology. It is only at this stage that pathogenesis takes place (Pallas and García 2011). Numerous interactions between virus and host factors involved in the replication of plant viruses having either positive or negative effects result in an outcome of the infection. Both the factors, i.e., defense responses raised by the host against the infectious agent and effects of virus replication in the host, can lead to pathogenesis. Thus a single virus factor can even be the main responsible agent of the pathogenic process (Nagy and Pogany 2012).

Viruses do not normally cause the death of the host because they need living tissue for their multiplication, although there are exceptions. Plant viruses need to counter plant defense mechanisms and to usurp the functions of different host factors in order to complete their life cycle. Plants can defend themselves against various parasites and pathogens like insects, animals, other parasitic plants, viruses, and bacteria as they have well-developed recognition mechanisms which act as barriers preventing infection in the case of pathogens. Therefore, an interaction and/or interference is needed between viral and host components which, in some instances, would lead to symptom development through an alteration in the plant physiology. According to recent discoveries, plant–virus interactions are also known to be involved in plant development such as hormonal regulation, cell-cycle control, endogenous transport of macromolecules, etc. Nevertheless, the identities of all host factors involved in the viral cycle are still unknown (Hull 2009). However, the interactions differ widely in the mechanisms involved in the display of symptoms. When there is viral infection in the plant, both compatible and incompatible host–virus interactions exist.

3.3.1 Compatible Host–Virus Relationships

If the host is not able to recognize the virus, the two develop a compatible interaction, which may be favorable for the virus. In such a relationship, virus infects the host cell, and, depending on where the virus symptoms appear, local and systemic hosts are distinguished. Sometimes both the symptoms appear simultaneously. Local and systemic symptoms are called *external symptoms*.

A variety of disease symptoms can be observed due to viral infections in plants, like leaf rolling, wilting, yellowing, stunting necrosis, and mosaic pattern formation. These symptoms were found to be of great commercial importance even before viruses were discovered, for example, the appearance of flame-like streaks in tulip flowers called "tulip mania" which were sold at very high prices in Netherlands in

Table 3.1 The most frequently observed viruses, their hosts, and symptoms in compatible host–virus interaction

Virus	Host	Symptoms
Potato virus (PVY)	*Solanum tuberosum*	Leaf drop (LED)
	Nicotiana tabacum	Necrosis (N)
	Chenopodium quinoa	
Tobacco mosaic virus (TMV)	*N. tabacum*	Necrotic lesions (N)
	N. glutinosa	
	N. sylvestris	
Tomato mosaic virus (ToMV)	*Lycopersicon esculentum*	Leaf deformation (Ldef)
Cucumber mosaic virus	*Cucumis sativus*	Mosaic (M)
Alfalfa mosaic virus (AMV)	*Chenopodium album*	Yellow mosaic

the seventeenth century. It was later found to be a result of infection by *Tulip breaking virus* (Dekker et al. 1993).

Apart from external symptoms, internal symptoms or microsymptoms caused by virus infection are not uncommon. Some plants can react by forming special inclusion bodies in the cytoplasm of infected cell. For example, light microscopic studies have revealed the formation of pinwheel inclusion bodies in the cytoplasm of *Potyvirus*-infected cells. Symptoms of viral infection can occur on all parts of the plant like the root, stem, leaf, and flower, but most commonly they occur on the leaves of the susceptible hosts (Table 3.1).

Hypersensitive reaction (HR) is also a special type of host–virus relationship to avoid spreading of virus particles or nucleic acids. This mechanism is a survival strategy in which the infected cell dies before the virus can be translocated to other cells. Thus, virus infection might not lead to symptoms although the plants are susceptible to virus infection. Plants are thus able to confine virus particles at the site of infection. This trait of producing HR could be used in breeding procedures for resistance against viruses.

Viral symptoms can be greatly influenced through environmental factors. For example, TMV at lower temperature produces local chlorotic and necrotic lesions, but at higher temperature virus can be translocated in *Nicotiana tabacum* Xanthi plants. There may be absence of external symptoms although viral movement from cell to cell can be perceived in a special type of compatible host–virus relationship as illustrated in *Medicago sativa* infected by AMV (*Alfalfa mosaic virus*). The diagnosis of viruses in symptomless host can be detected by other specialized methods (Stange 2006).

3.3.2 Non-compatible Host–Virus Relationships

A non-comatible host-virus relationship develops when the host plant recognizes the virus. In this case, a series of defense reactions are induced in the host which obstructs the replication of virus and also movement within the host. When a host plant possesses extreme immunity, viruses cannot infect a host, external symptoms do not appear, and virus cannot be detected. The resistance of host plants to viruses may be qualitative or quantitative. In the case of qualitative resistance, there is a special relationship between the host's resistance genes and viral genes which can be expressed as hypersensitivity or resistance to the spreading of viruses (Kiraly et al. 2008), whereas in quantitative resistance, specific relationship between their genes cannot be observed. This type of resistance is in the form of resistance to viral replication and spreading, field resistance, and tolerance to plant disease.

3.4 Methods of Diagnosis of Viruses

To establish a control program for any plant disease specifically viral disease, it must always be preceded with an approved and accurate diagnosis. Lack of correct information on the causal agents of viruses, their means of spreading to distances, and their survival strategies could lead to a total failure in an attempt to control plant disease, including viral diseases. For a correct and definitive diagnosis of a viral disease, several methods can now be used. But at the beginning of the study of plant virology, the causal viruses were identified and characterized through the external symptoms on plants (Purcifull et al. 2001; Lima et al. 2012). Because different viruses due to their synergistic effect may cause similar symptoms and same virus can cause different symptoms, it is now believed that the symptoms on plants may be an unreliable source of identifying a plant disease. The symptoms caused by viruses may vary according to the plant variety, environment, viral strain, and temperature. Some plant species show different symptoms of virus infections, but there are some which do not show any symptoms at all. For example, *Datura stramonium* hosts PVX (*Potato virus X*) but is resistant to PVY. It is therefore almost impossible to diagnose plant virus infection by just observing host symptoms. Still, for the denominations of plant virus, the original symptoms are of great importance, although several other methods are additionally used for a correct and definite diagnosis of a viral plant disease. For the identification of viruses, several laboratory methods have now been developed and adapted (Fig. 3.1). The important ones are discussed below.

3.4.1 Serological Assays: Traditional Molecular Method of Disease Detection

One of the easiest and specific methods involving a rapid and precise identification of plant viruses is serology (Purcifull et al. 2001). Because the viruses cannot be cultivated as other pathogens, bacteria, and fungi can be, therefore serological assays were developed for the identification and characterization of plant viruses. Pathogens can be detected by using polyclonal and monoclonal antisera and techniques such as enzyme-linked immunosorbent assay (ELISA), Western blots, immunostrip assays, dot–blot immune-binding assays, and serologically specific electron microscopy (SSEM). The first and most widely used immunodiagnostic technique was ELISA in the 1970s. Because of its high-throughput potential, specificity, adaptability, economic advantage in the use of reagents, and sensitivity (Clark and Adams 1977), the virus particles at very low concentrations could be detected, and a large number of samples can be indexed in a relatively short period of time. These methods are relatively simple involving antigen-antibody reactions in vitro, and any sophisticated and expensive equipment is not required. The advent of ELISA has facilitated the use of serology for virus identification on a large scale, and, thus, ELISA can be used in a wide range of situations (Purcifull et al. 2001; Lima et al. 2012). Direct and indirect ELISA are the most frequently used methods for the diagnosis of plant virus diseases although different variations of this serologic technique have further been developed (Lima et al. 2012).

The following variations of the ELISA technique were successfully used for the detection: indirect ELISA or the plate-trapped antigen technique (PTA-ELISA), immune virus particle precipitation followed by ELISA (IP-ELISA), and a simple kit for plate-trapped antigen ELISA (Dorokhov and Komarova 2016).

Polyclonal antisera for many viruses and bacteria have been developed for commercial use or research labs and have been used in numerous protocols, but their frequent cross-reactivity inspired the development of more effective monoclonal antisera using hybridoma technology (with cell lines with specificity to single epitopes). Because of the difficulty in producing a good virus-specific antiserum, serology for plant virus identification and detection becomes a serious limitation. Purified plant viruses or different types of viral protein used for immunizing warm-blooded animals can be used for preparing most of the antisera used in plant virus identification and detection. For example, most plant viruses like PLYV can serve as good and effective antigens stimulating the production of specific antibodies that can be used in different serologic tests. ELISA procedures (using both monoclonal and polyclonal antibodies) and rapid detection kits are commercially available for numerous taxa.

3.4.2 Molecular Techniques or Nucleic Acid-Based Methods for Virus Detection

Usage of molecular techniques for plant virus identification and characterization is increasing throughout the world although serology has been used extensively for the same on a large scale. Several molecular methods have now been developed, out of which some are DNA based like fluorescence in situ hybridization (FISH), many PCR variants (PCR), real-time PCR (q-PCR), and DNA fingerprinting, while others are RNA based like reverse transcriptase PCR (RT-PCR) and nucleic acid sequence-based amplification (NASBA). Reverse transcriptase polymerase chain reaction has been shown to be a suitable method of research with RNA plant viruses (Lima et al. 2012). These enable a rapid and accurate detection and quantification of pathogens and therefore can overcome uncertain diagnosis or pathogen taxonomy. However, reproducible and efficient protocols are needed for critical preparation of samples required during this method. To avoid the presence of inhibitory compounds that compromise detection, many published protocols for RNA and DNA isolation were developed. The primary compounds interfering with these methods include polysaccharides, phenolic compounds, or humic substances from plants or other substrates. Different genomes such as ssRNA, ssDNA, or dsDNA have been used for several different protocols developed for PCR-based methods. To extract nucleic acids from different types of plant material, many commercial kits have been specifically designed and widely used. However, these methods are not always effective with all types of plant material. Each combination of pathogen and plant needs to be evaluated before these can be adopted for regular detection. For example, loss of production may vary from 20% to 100% depending on which begomovirus is causing tomato yellow leaf curl disease (TYLCD), and molecular analysis seems to be the only way by which different begomoviruses can be distinguished (Davino et al. 2006). However, PCR and RT-PCR have become attractive and efficient methods for the diagnosis of plant virus diseases, because of their power to amplify a target nucleic acid present at an extremely low level in a complex mixture of heterologous sequences. Mumford et al. (2000) developed a novel real-time quantitative PCR assay for the detection and quantification of plant viruses.

3.4.3 Innovative Detection Methods

Because many pathogens like nematodes, fungi, bacteria, viruses, and viroids can simultaneously affect cultivated plants, therefore relatively novel approaches detecting different infections in the same plant and detecting pathogen at presymptomatic to early spread stages are highly desirable. Novel methods allow detection of pathogen infections when symptoms are still unclear and limited to a few plants only. Traditional methods, on the other hand, can detect pathogens only at a later symptomatic stage.

3.4.3.1 Lateral Flow Microarrays (LFM)

This method allows rapid, hybridization-based nucleic acid detection and utilizes strong and reliable host plus pathogen biomarkers discovered through transcriptomic approaches to easily visualize the colorimetric signal. These arrays are built on miniaturized lateral flow chromatography nitrocellulose membrane. This method minimizes the need for expensive laboratory instruments, has detection limits similar to microarrays, and needs less time for hybridization. Key plant metabolites of primary and secondary metabolism can be used as biomarkers for different environmental stresses or pathogen infections, and they have been widely identified through metabolomics. For example, highly interactive proteins (acting as possible indicators of plant health status) such as dehydrins or heat shock proteins upregulated by different environmental factors can be identified by an OMIC approach (Dandekar et al. 2010).

3.4.3.2 Methods Based on the Analysis of Volatile Compounds as Biomarkers

Many volatile organic compounds (VOCs) are emitted from leaf surfaces into their immediate surroundings that serve vital functions in growth, communication, defense, and survival (Baldwin et al. 2006). VOCs are low molecular weight biomolecules and terminal metabolites of the host plant. VOCs have a high vapor pressure and low boiling point which can indicate physiological health status of the plant. They generally exist in the gaseous phase under standard temperature and pressure. A new avenue of research is opened by VOC profiling which may detect mechanisms for "plant-to-plant" and "plant-to-insect" communication. It provides with an ability to frequently and noninvasively monitor the health status of high-value commodity crops. This method offers potential for immediate applications within the plant sciences gaining new insights into host responses to pathogens and abiotic stressors. Development of hardware and software tools leading to novel analytical methods and instrumentation is required to make and interpret these data sets, which is critical to bring these concepts into the field. Several studies have been done involving VOC profiling of plants using gas chromatography mass spectrometry (GC-MS) like investigations into *Cucumber mosaic virus* (CMV) through VOC profiling. CMV-infected cultivated squash (*Cucurbita pepo* cv. Dixie) plants showed an overall net increase in VOCs, but no major qualitative difference in VOC profiles could be identified in infected plants (Mauck et al. 2010). The altered VOC profile emitted by CMV-infected plants demonstrates that the plant is inducing a change in VOC profile in response to viral infection, a mechanism known as "supernormal stimulus."

A variety of techniques (Fig. 3.2) have been used to identify the changes occurring in host genetic expression due to virus infection. The most common ones are in situ hybridization for individual genes and global profiling of host transcripts using cDNA microarrays or oligonucleotides (Zhu et al. 2001). The microarray-based studies have been done on model host systems, e.g., *Nicotiana benthamiana* and *Arabidopsis thaliana*. *Nicotiana benthamiana* experiments have utilized a heterologous array from potato (*Solanum tuberosum*, also a member of Solanaceae). Heterologous arrays are microarrays containing cDNA sequences or oligonucleotides from plants providing the RNA for transcription profiles. A large number of ESTs from six solanaceous species potato, tobacco, pepper, tomato, petunia, and *N. benthamiana* were analyzed, and it was found that 51–81% of ESTs had significant sequence similarity. Only 16–19% transcripts did not match among Solanaceae, rice, or *Arabidopsis* (Rensink et al. 2005).

Engineered viruses different from wild type in terms of virulence can also be used to study plant responses to viral strains and help to analyze the specific viral protein functions. Another strategy to discover common or specific responses of host to diverse viruses is by comparative analysis of expression data from viral infections of plants with different pathogens or stress.

To determine the control of host responses and to study the effects on viral pathogenicity, the different plant mutants and genotypes are analyzed. Another approach is to reveal the expression of individual viral components under the control of promoters either inducible or constitutive for studying the gene expression in host cells or tissues.

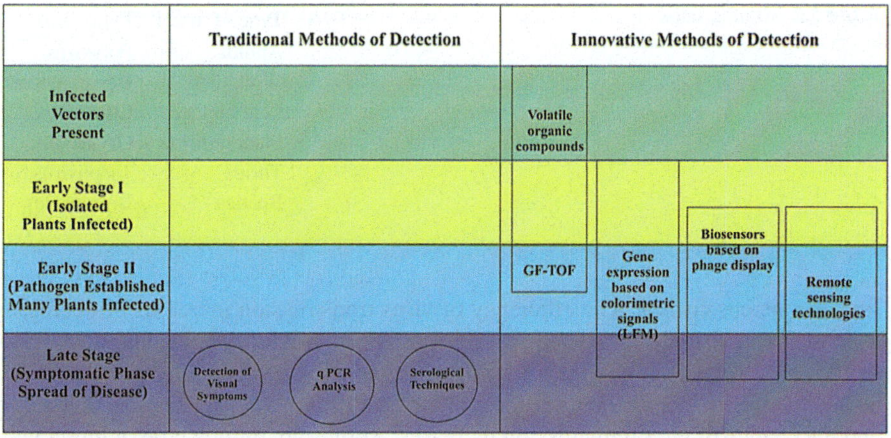

Fig. 3.2 Traditional and innovative methods of detection used for diagnosis of diseases in plants caused by viruses. Four disease stages are indicated along with their timing of use during plant disease progression

3.5 Transmission of Viruses

Agriculture is essential for continuous supply of food and other important products for human being. Therefore developing efficient strategies for protecting plants against pests and pathogens has been an important concern. Viruses are different from other plant pathogens because they require other organisms for their transmission. For instance, winged insects with sucking mouthparts help in transmission of most of the viruses from one place to another even at a distance (Vuorinen et al. 2011). Other insects like chewing insects are also responsible for the transmission of plant diseases especially those caused by plant viruses. The major insect vectors are aphids, whiteflies, and leafhoppers (Table 3.2). The viral transmission is completed in four steps: First is the acquisition of viruses from an infected plant. Second is the retention of virus in the vector by binding of virion to receptor-like elements in the digestive tract or circulation from the gut to the salivary gland. In the next step, virions are delivered from the retention sites to new site, and finally virions are deposited in a susceptible cell of the host plant. The transmission mode could be of two types depending on the length of the period during which the acquired virus can be transmitted to an uninfected plant by the vector. For persistent transmission few hours are required for acquisition of virus, and a retention time, varying from few hours to days, is required before the insect vector becomes viruliferous. Sometimes an intermediate mode called semi-persistent transmission occurs for viruses which require intermediate time period between acquisition and retention.

The transmission can also be classified as *non-circulative* or *circulative* depending on the route followed by the virus in the insect vector. In non-circulative transmission, the virus is retained in the anterior tract, i.e., the mouthparts or foregut of the digestive system. Viruses following this mode are generally nonpersistent or

Table 3.2 Main groups of insect vectors of plant viruses

Type of insect	Plant viruses
Aphids	Potyvirus
Whiteflies	Begomovirus
Leaf hoppers	Curtovirus
Plant hoppers	Oryzavirus
Thrips	Tospovirus
Beetles	Bromovirus

semi-persistent viruses. The efficiency of virus transmission increases or decreases as in semi-persistent viruses (where the acquisition time is longer) and nonpersistent viruses, respectively. The main difference is associated with the virus stability in the vector, so that virions get accumulated until the retention sites are saturated and the chance of later transmission increases. Aphids are unique type of insect vectors which transmit mainly nonpersistent viruses where the virions may rapidly be lost if the feeding period is extended beyond the limit. It is, therefore, difficult to control the spread of such viruses by insecticide treatments because both acquisition and inoculation occur during feeding only. In circulative transmission, the acquired virus must pass across the gut to reach the hemolymph after feeding on the infected plant. Finally it reaches the salivary glands and inoculated through saliva. This process might take few days due to latency period during which the virus crosses cellular barriers within the insect.

An important molecular interface which determines the acquisition from infected host plants and transmission to new hosts is the protein–protein interaction between plant viruses and insect vectors. These interactions are highly specific and open avenues for the control of insect vectors and viral transmission. Highly potential OMICS technology can be used to identify and validate the virus-interacting proteins in the vector and understand the innate immune responses to viral infection (Dietzgen et al. 2016).

3.6 Virus Accumulation and Movement in the Host

The virus accumulation in host plant requires replication and translation of viral gene sequences. The host further enhances their accumulation by providing factors for accumulation. The NAC (from the first word of these three genes: no apical meristem gene (NAM), Arabidopsis transcription activation factor gene (ATAF), cup-shaped cotyledon gene (CUC)) domain protein from tomato is found to interact with replication enhancer (REn) of geminivirus and hence participate in virus replication (Selth et al. 2005). NAC proteins also interact with Rep A (another viral protein) inhibiting the specialization of viruses (Xie et al. 1999). Interaction of NAC protein with viral coat protein is necessary in *Arabidopsis* during a resistance response (Ren et al. 2000).

Using yeast as an alternative host, screening for host factors affecting virus accumulation has been done which shows that the host gene involved in viral accumulation may vary from virus to virus (Panavas et al. 2005).

3.6.1 Viral Factors Involved in Plant Pathogenesis

The different virus factors, mainly proteins and nucleic acids, contributing to the pathogenesis of plant virus infections are discussed below:

1. **RNA replicase-related proteins/RNA-dependent RNA polymerase (RdRp):**

Some enzymes like viral RNA-dependent RNA polymerase or RNA replicase can affect viral genome replication and therefore viral accumulation and thus contribute to pathogenesis in an indirect way. For example, serial passages of singly mutated RNA replicase sequence of *plum pox virus* (PPV) in an experimental host, *Pisum sativum*, causes a drastic enhancement of disease symptoms which is associated with a notable increase in viral accumulation (Wallis et al. 2007). Like most viral proteins, RNA replicases can also lead to HR by functioning as elicitors of R-gene-driven effector-triggered immunity (ETI). For example, the helicase domain of the TMV RNA replicase (an avirulent replicase) induces HR (Erickson et al. 1999).

2. **Coat proteins/capsid proteins (CPs):**

Viral coat proteins or capsid proteins are the multifunctional proteins that, in addition to having a structural role (except some virus groups notably umbraviruses where the genomic RNA is not encapsidated by the coat protein), are involved in virus entry, disassembly, replication and translation of viral RNAs, movement or transmission of viruses, activation or control of antiviral resistance, symptom development, and host defense responses (Ni and Cheng Kao 2013). There has been a great advancement in the understanding of contribution of CPs to plant–virus interaction (Peng and Kao 2013).

Pathogen-derived resistance (PDR) can be engineered through mechanisms underlying the interaction of CP with host genes. RNA silencing of CP gene or inhibition of CP activity can mediate such resistance (Hafrén et al. 2010). It is not only the coat protein sequence but other sequences like those of MP and viral replicase that can also be used to engineer PDR. Although a tiny proportion of the transgenic crops carries the disease resistance traits, these engineered crops are of great interest in developing countries. Plant viral CPs like RdRps can also act as elicitors of R-gene-mediated HR. For example, *Tobacco mosaic tobamovirus* CP can function as an elicitor of hypersensitive response (HR) genes (Taraporewala and Culver 1997). Similarly the CPs of TMV and CMV act as avirulence factors that elicit resistance controlled by a dominant R-gene (Moffett 2009). Also a single amino acid substitution in the CP has been demonstrated to be responsible for conferring the ability of TMV to elicit HR in *Nicotiana sylvestris* or to alter the symptom phenotype and host range in CMV-infected tobacco plants. Likewise antiviral defense generated upon viral infection can be enhanced through metabolic genes via CP production. For example, sulfur metabolism genes in the infected kenaf plants are upregulated by CP of *Hibiscus chlorotic ringspot carmovirus* (HCRSV).

Some kinds of CP interactions might target a redirection of host resources toward viral replication. And some CP can serve as an efficient suppressor of RNA silencing after binding to a host chaperone, thereby sequestering the silencing signal.

3. **Movement proteins (MPs):**

The viral pathogenesis is strongly influenced by the translocation of viruses from cell to cell throughout the plant body. An early event in the infective process is the movement of viruses from one cell to another. Viruses must cross the cell wall barrier to infect the adjacent cells and thus establish a systemic infection in plants. Derrick et al. (1992) described that TMV in *Nicotiana tabacum* and *Tobacco rattle virus* in *Nicotiana clevelandii* takes 4–5 h to move from one cell to another. Viruses first move to plasmodesmata (PD) from their replication sites on the periphery of cell and then pass through these channels entering the adjoining cell. This intercellular movement of virus occurs mainly through plasmodesmata in mesophyll and epidermal cells. Plasmodesmata are specialized intercellular organelles establishing cytoplasmic and endomembrane continuity between adjacent cells after crossing the cell walls. Although PD regulate the intercellular movement of macromolecules or macromolecular complexes, such as viral particles and ribonucleoprotein (RNP) complexes, they allow small molecules to diffuse between cells. This movement is mediated by viral encoding factors named movement proteins (MPs) (Waigmann et al. 2004). These facilitate the translocation of plant viruses among cells and through plant. During the first stage, these MPs bind to viral genome forming ribonucleoprotein complexes or to tubular structures which hold viruses allowing them to cross plasmodesmata. These then transport the virus through plasmodesmata from the epidermis to mesophyll and then to vascular bundles. There is still a poor understanding on how MP opens PD and how viral RNPs or virions pass through MP-gated or MP-formed tubule at PD. PD have a very small aperture with the cytoplasmic sleeve not more than 10 nm in diameter. Factors restricting the PD aperture include callose and pectic polysaccharide. Callose is biosynthesized at the cell wall by callose synthases, while it is degraded by β-1,3-glucanases. A positive correlation exists between β-1,3-glucanase expression and viral spread. And TMV has also been found to interact with an enzyme involved in de-esterification of pectic homogalacturonan, i.e., pectin methylesterase (PME) (Faulkner et al. 2008). MPs play a crucial role in host specificity for most of the plant viruses. Viral MPs have been shown to interact with products of R-genes to elicit HR response. During the hypersensitivity reaction (HR), callose (β 1–3 glycan) gets deposited in the plasmodesmata blocking the channels and restricting the viral movement from one cell to another. The *Potato virus X* (PVX) produces a protein interacting with

β-1,3-glucanase (callose-degrading enzyme) facilitating the movement of PVX through the plasmodesmata (Fridborg et al. 2003). Constitutive or tissue-specific expression of viral MPs may trigger typical viral infection symptoms such as abnormal sugar accumulation, diminished photosynthesis, chlorosis, and dwarfism. One of the most important strategies is blocking cell-to-cell or long-distance movement of the virus. It has been reported that a single-nucleotide substitution in MP gene which enhanced the viral movement efficiency of tymovirus led to greater viral accumulation and also increased severity of symptoms. Additionally, in TAV cucumovirus, the different levels of expression of MP can determine the difference in severity of symptoms between two virus strains (Moreno et al. 1997).

A coordinated action of host factors and virus-encoded movement proteins (MPs) is required for cell-to-cell movement of viruses through PD. More than ten host factors in addition to these cell wall enzymes have been found to be involved in intercellular movement of different viruses (Heinlein 2015).

Interaction of plant virus MPs has also been reported with the host protein promoting viral movement. These proteins can be localized in the nucleus, microtubules, and also plasma membrane which may be required for intra- or intercellular virus movement and affect the symptom development (Table 3.3).

Transgenic plants overexpressing these proteins have been used to understand the involvement of MPs in viral pathogenesis. These proteins interfere with the cytoplasmic communication channels as transgenically expressed MPs are located in the plasmodesmata, hence increasing the size exclusion limit of plasmodesmata further triggering alteration in metabolism and distribution of carbohydrates (Pallas et al. 2011).

Another such host apparatus is the cytoskeleton and its components, which further facilitate viral transport through plasmodesmata. It may act together with the endomembrane system of the host cells. To deliver viral MPs, vRNPs, and virions

Table 3.3 Plant viral movement proteins

Virus-interacting protein	Host protein	Role in viral movement	References
Pvx-tgb2	Tip I	Cell-to-cell movement	Fridborg et al. (2003)
TMV-MP	Calreticulin	Regulate cell-to-cell movement	Chen et al. (2005)
	ANK		Ueki et al. (2010)
	PME		Dorokhov et al. (1999)
PMTV-TGB-2	RME-8 family	Intracellular and intercellular movement	Haupt et al. (2005)
GRV-MP	Fibrillarin	Vascular transport intracellular movement – nucleus	Kim et al. (2007a, b)
Geminivirus – NSP	Protein kinase		
CaMV-MP	MPI-7	Susceptibility – cytoplasm	Huang et al. (2001)

to PD, viruses must hijack the endomembrane system, early secretory pathways, and/or the cytoskeleton network along with its motor proteins. The various viral MPs are transported via the ER to plasmodesmata, and actin/myosin filaments control the flow of proteins in ER membrane. In TMV, microfilaments assist in the cell-to-cell movement of virus, and microtubule-associated proteins degrade the viral movement proteins (Kragler et al. 2003). Calreticulin is a chaperone protein localized in the lumen of ER which helps in protein degradation via the proteasome. The overexpression of calreticulin in transgenic plants redirects *Tobacco mosaic virus* to microtubules from plasmodesmata, and cell-to-cell transport of virus is compromised (Chen et al. 2005).

Viral intercellular movement may be summarized into four major types on the basis of the characteristics of CP involvement and MP behavior:

1. The first type is illustrated through tobamovirus, bromovirus, and cucumovirus. The movement of viruses is in the form of the viral RNA-MP complex (vRNP). In the case of bromovirus and cucumovirus, the viral movement requires CP but is independent of the virion. Single dedicated MP of TMV however increases the size exclusion limit of PD for CP-independent viral RNP movement between cells.
2. Rod-shaped viruses such as hordeivirus-like viruses and potexviruses illustrate the second type. The three TGB (triple gene block) proteins encoded by these viruses plus CP coordinate the intercellular transport of vRNPs (for hordeivirus-like viruses) and virions or vRNPs (for potexviruses).
3. The third type including comovirus and nepovirus transverses and modifies the PD and is in the form of a virion guided by the MP-assembled tubules.
4. The last type illustrated by potyviruses assists the passage of virion by the formation of cone-shaped and not tubular structures at the PD.

3.6.2 Viral Long-Distance Movement

The virus moves from the mesophyll via bundle sheath cells, phloem parenchyma, and companion cells into phloem sieve elements (SEs) for long-distance or phloem-dependent movement. Following the source-to-sink flow of photoassimilates, the virus is then passively transported and unloaded from SEs to sink tissues at distant sites. It then results in further infection. PME seems to affect long-distance movement as well. In a PME-silenced tobacco line, there is a substantial delay in viral systemic transport. This happens because TMV movement out of the vascular system is blocked because of the significant degree of pectin methylesterase (PME) suppression in the vascular tissues (Chen MH and Citovsky 2003). However, the exact roles and underlying mechanisms of the host factors in viral long-distance transport are still far from understanding.

3.7 Responses of Plants to Viruses

Plant–virus interplay where the virus tries to take over plant resources for its multiplication and then counteracts the antiviral defenses raised by the plant often leads to manifestation of disease symptoms. The cross talk among different hormones governing the various pathways in antiviral defense is also disturbed through virus infection (Pallas and García 2011; Nagy and Pogany 2012). Thus the most common symptom of plant virus infections appears to the developmental abnormalities resulting from such disturbances. Different subcellular organelles, cells, and organs when adapting to the new requirements of virus replication and movement can interfere with normal plant homeostasis and therefore lead to disease symptoms. However, sometimes interactions among definite viral and host factors can lead to specific symptoms through viral infection as illustrated in the case of HR-related necrosis elicited by discrete viral elements or misregulation of auxin response factor 8 caused by several viral silencing suppressors leading to developmental defects. Still the interactions between numerous viral and host factors conditioned by multiple environmental conditions can lead to much more complex manifestations of many plant virus infections. Unique experimental designs based on systems biology and the development of powerful technological tools in the coming years are considered to open new chapters in understanding the molecular basis of these plant virus pathogenicity relationships. On viral attack, plant responses can be broadly classified into two categories: (a) cellular stress and (b) developmental defects. The changes in plant gene expression profiles due to viral infection are very much like the defense and stress responses. There may be induction of heat shock proteins (HSPs) during stress-like responses and pathogenesis-related (PR) genes under defense like responses.

3.7.1 Cellular Stress

PR genes are the defense-related genes like PR-1, β-1 glucanase (PR-2), chitinase (PR-3) PR-4, thaumatin-like protein (PR-5), superoxide dismutase (SOD), and glutathione S-transferases (GT). Some other genes are also co-induced with the PR genes. For example, it has been observed that in *Arabidopsis thaliana* and *Nicotiana benthamiana*, few genes of WRKY family of transcription factors also get induced during viral infection.

Under stress conditions and also in normal growth and development, HSPs get induced. Their expressions also get induced by DNA and RNA viruses in the host plants. The first report of HSP70 induction in response to viral pathogen in pea embryos was demonstrated by in situ hybridization (Aranda et al. 1996). Further evidence was provided in *Arabidopsis* in response to different RNA viruses (Whitham et al. 2006) (Fig. 3.1).

3.7.2 Development Defects

Some of the genes of the host plants found to express themselves on viral infection may have a connection with the developmental defects and symptoms of disease. Viruses interfere in the signaling pathways and hence affect plant growth and development. The abnormal growth forms of virus-infected plants have led to the experiments revealing effects of viruses on hormone levels in host plants. Depending on host–virus combination, auxin, cytokinin, gibberellin, ethylene, and abscisic acid levels get altered on viral attack. For TMV and RDV (*Rice dwarf virus*), a link between auxin–gibberellin levels has been well established (Jameson 2000). The interaction of helicase domain of TMV 126 and 183 kDa replicase protein with IAA transcription factor (IAA26) was established (Padmanabhan et al. 2005). Silencing of IAA 26 resulted in TMV-infected plant phenotypes directing the loss of function of IAA 26 by TMV replicase. The P2 protein of RDV also affects gibberellic acid signaling (Zhu et al. 2005).

The effects of viral RNA silencing suppressors (RSS) on host gene expression cannot be neglected as they promote viral infections by interfering with host defense mechanisms and also regulatory miRNAs and trans-acting small interfering RNAs ultimately leading to developmental defects in host plants (Dunoyer and Voinnet 2005). To counter the defense system of plants, viruses have acquired a strategy by disruption of host antiviral silencing.

3.7.3 Abnormalities in Chloroplast

After studying viruses from around 12 families covering major genera and those responsible for devastating disease (having either sense ssRNA, antisense ssRNA, or ssDNA genomes), it has been confirmed that chloroplast abnormality is a common event across diverse plant–virus interactions. Chloroplast has been implicated as a common target of plant viruses for a long time. Typical photosynthesis-related symptoms, such as chlorosis and mosaic pattern, cause disruption in normal chloroplast function although the causes leading to development of viral symptoms are different (Rahoutei et al. 2000). For example, infection by CMV in *Nicotiana tabacum* cv. Xanthi associated with fewer grana and reduced chloroplasts led to severe chlorosis on systemic leaves (Roberts and Wood 1982). In accordance with the prevailing studies, the two main causes of virus-induced chloroplast symptomatology include ultrastructural alteration of chloroplast and the reduced abundance of proteins involved in photosynthesis. Typical chloroplast malformations include:

1. Overall decrease in the number and clustering of chloroplasts
2. Change in the appearance of chloroplast, such as swollen or globule or amoeboid-shaped chloroplast, chloroplast with membrane-bound extrusions, or chloroplast with a dynamic tubular extensions from chloroplast known as stromule
3. Broken envelope and irregular out-membrane structures such as peripheral vesicle, cytoplasmic invagination, and membrane proliferations

4. Any change in the content of the chloroplast such as large intermembranous sac, increase in the number and size of electron-dense granules/plastoglobules/bodies, enlarged or numerous starch grains, and small vesicles or vacuoles in stroma
5. Absence of stroma or grana stacks and distorted, loosened, or dilated thylakoid
6. Completely destroyed chloroplasts and disorganized grana scattered in the cytoplasm (Zhao et al. 2016)

Recently carried out studies reveal that chloroplast ultrastructure and symptom development are affected by viral factors, especially CPs, as shown in the work of Neeleman et al. (1991).

On comparison with healthy plants, virus-infected plant cells were seen to contain reduced amounts of chlorophyll–protein complex. No significant differences were observed in the photosystem I (PSI) reaction center upon TMV infection, but an inhibition of photosystem II (PS II) activity was observed by selectively decomposing the light-harvesting antenna complex of photosystem II. Likewise, among the stroma proteins, chlorotic tissues of *Cucumber mosaic virus* (CMV)-infected plants observed a loss in the activity of the small subunit of the enzyme ribulose-1,5-biphosphate-carboxylase-oxygenase (RuBisCO). A significant increase was measured in the enzyme activity of chlorophyllase and catalase vis-a-vis decrease in the chlorophyll content. This could be due to release of the enzyme chlorophyllase bound to the chloroplast inner membrane after the disorganization in the chloroplast. Rarely does virus replication occur on chloroplast, but still the products of viral infection like coat proteins are found to inhibit the photosynthetic activity. The interaction of CP with chloroplast might affect chloroplast function and stability, which gives rise to chlorosis. Similar to replicases and silencing suppressors, viral CPs have been recently shown to interfere with/modulate hormone signaling pathways.

Transcriptomic and proteomic analysis of expression of chloroplast proteins upon virus infection shall provide an insight into the molecular events during symptom expression. For example, in response to virus infection in susceptible plants, a majority of significantly changed proteins are identified to be located in chloroplasts or associated with chloroplast membranes. Most of them correlate with severity of chlorosis and are seen to be downregulated (Dardick 2007).

3.8 Viral Infection and Physiological Functioning of Host Plants

A successful infection by plant virus results from molecular network of interactions during viral infection process. Therefore understanding this complex molecular interplay between the host plant and the invading virus may assist in the development of novel antiviral strategies. Theoretically, viral infection in plants can be divided into several major steps including viral particle disassembly, viral genome translation (in case of + ssRNA viruses), cellular membrane modifications coupled with formation of VRC (viral replication complex), viral genome encapsidation, cell-to-cell movement, and long-distance transport. Viruses are thus evolutionarily empowered with an ability to take over the host cellular pathways and manipulate the cellular components. Additionally, diverse host-encoded proteins or host factors are also present at each of these steps (Wang and Krishnaswamy 2012).

3.8.1 Virion Disassembly and Viral Genome Translation

3.8.1.1 Host Factors in Virion Decoating

Positive-sense RNA viruses do not encapsidate the virally encoded RNA-dependent RNA polymerase (RdRp; an absolute requirement for viral genome replication) in their virions. Physical barrier is a major bottleneck that plant viruses have to overcome to spread from cell to cell for the subsequent systemic invasion of their host. Thus, plant wounding is an obligatory condition for virus entry. Mechanical damage to the CW provides the virus with an opportunity to enter the cell. As soon as the virus gains entry into the plant cell, the first step toward a successful infection is the removal of viral shell or capsid (made of capsid protein subunits) so that the viral genome is exposed to cellular translation machinery to begin viral genome translation. Although viral particles are highly stable and can survive for extended periods under the harsh extracellular environment, there remains a poor understanding on how the stable virions disassemble in the environmentally friendly symplast of their host. Decoating is a passive process triggered by the change of pH and positively charged cation concentration in the cell as suggested in some of the early studies on TMV. This thus leads to removal of few CP subunits from 5′ end of the encapsidated RNA (Culver 2002) and thus leading to initiation of virion disassembly. Thus, an important, if not universal, strategy for plant positive-sense RNA viruses is to disassemble the cotranslational mechanism. In order to prevent the activation of host antiviral mechanisms (like RNA silencing triggered by ds RNA intermediates), to provide a scaffold for tethering the VRC, and to restrict the process of viral replication to a specific safe cytoplasmic site, viral replication is associated with the virus-induced intracellular membranous structures. The origins of such membranous structures are diverse and dependent on the type of virus (Laliberte and Sanfacon 2010). Example, tobamoviruses and *Tomato mosaic virus* (ToMV) recruit and modify the ER membrane for their genome replication in plants.

3.8.2 Viral Replication Complex Composition

All VRC components (in addition to remodeling of cellular membranes by viral proteins and co-opted host factors) must be present together at the site of replication to form a functional unit for catalysis of viral replication. The following constitutes the VRC present in the modified cellular membranes:

- Viral replicase proteins, such as RdRp and viral helicase
- A viral RNA template
- Diverse host proteins like heat shock proteins (HSPs) and RNA-binding proteins (RBPs) (Verchot 2012)

3.8.2.1 Host Factors in Viral Genome Translation

Viruses have evolved diverse strategies to recruit the host translation apparatus for quick and efficient translation of the viral genome because they lack functional ribosomes and their genomes do not encode other translation-required components. Translation is known to be a tightly regulated process including several phases, i.e., initiation, elongation, termination, and ribosome recycling. Wang and Krishnaswamy (2012) have discussed the possible mechanistic roles of eIF4E and its isoforms (best characterized host factors for plant potyviruses) in potyviral infection. Because the viral genome translation and replication of positive-sense RNA viruses are intimately linked by sharing the same site and the same template (often considered to be a coupled process), it is highly possible that these translation factors may contribute to viral genome multiplication (Kawamura-Nagaya et al. 2014).

3.8.2.2 Regulation of Viral Replication by Viral Protein Modifications

The functional diversity of the plants can be increased by posttranslational modifications (PTMs) of viral proteins in the infected cells. Various PTMs have been documented till date to affect viral replication in plants. Examples include phosphorylation (the most common and reversible protein phosphorylation which seems to negatively regulate viral replication), ubiquitination, SUMOylation, and other PTMS (like methylation which modifies eEF1A and thus affects tombusvirus replication) which have also been shown to affect viral replication (Li et al. 2014).

Viruses not only infect and damage the crops, but weed plants are also seriously infected. Keeping in view the considerable losses and destruction of crop plants, earlier investigations were conducted only on crop plants. But later it was observed that weeds also play an important role in infecting and spreading viral diseases and not just compete with crop plants for nutritional benefits. Thus, biological decline of weeds is considered to be an important strategy for promoting crop production, and the level of weeds is therefore maintained under economic threshold to sustain agriculture. Viruses can directly contribute to weed control by reducing their competitive ability. The investigations done earlier were focusing more on physiology and biochemistry of virus-infected crops although viruses can damage both crops & weeds and weed–virus interaction was almost not focused upon. The various physiological effects of viruses to crop plants and weeds reduce their growth and development. There are certain reports such as in *Datura stramonium*, where a reduction in photosynthetic pigment content (Chl-b) was detected in the leaves infected with CMV (*Cucumber mosaic virus*) and HeMV (*Henbane mosaic virus*). Depending upon the host–virus relationships, there was 30–80% reduction in shoot dry weight of test plants. Virus infections do have an important effect on germination characteristics of weeds. Viral infection influenced seed dormancy rather than viability of seeds in *Solanum nigrum* plants infected with Obpv (*Obuda pepper virus*) and PepMV (*Pepino mosaic virus*), whereas it was reverse in the case of chenopodium plants infected with SoMV. Considerable effect of viral infection on the nutrient uptake of plants was observed in *S. nigrum* plants infected with TMV. Seed production is also reduced by TMV infection in *S. nigrum* plants (Takacs et al. 2014). Weeds can even function as sources of primary infection in the spreading of plant diseases and overwintering of plant viruses apart from its role as direct competitors for nutrients, light, and water. Weeds can also function as host plants of virus vectors. Therefore, in sustainable agricultural practice, the level of weeds is kept under economic threshold, and considerable efforts are taken to maintain biological diversity.

3.8.2.3 Interference of Viral and Plant Factors with Hormone Regulation

Hormones are those signal molecules which move around the plant to stimulate responses to different environmental stresses. Hormones like salicylic acid (SA), jasmonic acid (JA), and ethylene (Et) have long been known for their roles in tuning plant responses to biotic stresses. However, the other hormones known for their roles in plant growth and development like auxins (Auxs), brassinosteroids (BRs), cytokinins (CKs), and abscisic acid (ABA) are also known to be involved in plant–pathogen interactions (Denance et al. 2013). Certain hormones can prevail over others under specific circumstances by showing antagonistic or synergistic interrelations. For example, SA, JA, and Et, regulating defense pathways, exhibit antagonistic interactions.

On recognition of viral effectors by *R*-gene products, SA biosynthesis and signaling are activated which conditions incompatible interaction. It activates SAR in distal tissues. To limit viral propagation at the infection site, activation of the incompatible reaction occurs resulting in several responses like callose deposition, tissue disorganization, nuclear and nucleolar degradation, alterations in the shape and size of chloroplasts, accumulation of reactive oxygen species (ROS) and pathogenesis-related (PR) proteins, induction of the hypersensitive response (HR), and programmed cell death (PCD). Several examples of incompatible plant–virus interactions exist. For example, a parallel and significant increase in the SA and expression of PR genes is observed in the inoculated and systemic leaves of resistant tobacco plants after infection with *Tobacco mosaic virus* (TMV) (Baebler et al. 2014).

ABA appears to have multiple roles against the pathogens depending on the stage of infection in addition to its antagonistic roles in defense hormone pathways such as SA and JA/Et. In cases where pathogen overcomes the first line of defense, ABA helps in regulating plant defense at the early stages of infection by the mediation of stomatal closure against invaders or induction of callose deposition. But, if activated at later stages, ABA can suppress ROS induction and SA or JA signaling transduction, thereby negating defenses controlled by these two pathways (Ton et al. 2009). Although the involvement of ABA in biotic stress has been studied thoroughly, the roles of ABA in virus replication and movement are not well characterized. The involvement of ABA in virus interaction was first studied in the context of the effect of TMV on ABA accumulation in *N. tabacum* and tomato, which revealed that ABA increases callose deposition and limits virus movement (Fraser and Whenham 1989). The defensive role of ABA against viruses is mediated through inhibition of the basic β-1,3-glucanase which is responsible for the degradation of

β-1,3-glucan (callose). Subsequently β-1,3-glucan (callose) is deposited on plasmodesmata which strengthens them against virus movement (Mauch-Mani and Mauch 2005). ABA also affects plant defenses at the level of the RNA silencing machinery, which is considered to be a broader defense system against viruses. For example, ABA seems to partially control *ARGONAUTE1* (*AGO1*) levels and thus have direct and indirect links with this system.

3.8.2.4 Interference of Viral and Plant Factors with Cell Cycle and Gene Expression

Few viruses infect the host cells, which are dividing actively. As most of the plant cells do not divide actively, viruses develop mechanisms involved in alteration of host cell cycle. For example, Rep protein of geminiviruses interacts with a retinoblastoma-related protein family pRBR, involved in negative regulation of cell cycle. This interaction inhibits the activity of pRBR protein, cell enters S-phase, and host DNA replication machinery is produced which is required to reproduce the virus (Kong et al. 2000).

Plant viruses can also reprogram host gene expression just like the animal viruses. The evidence came from the reports of Havelda et al. (2008), where a connection between reprogramming of host gene expression and pathogenicity of viruses was provided. A correlation between the interruption of cellular gene expression (switch off mechanism) and viral symptom development was provided.

3.8.2.5 Effect of 5′ and 3′ NCRs (Noncoding Regions) on Viral Pathogen

The viral translation and replication process are influenced by NCR sequences (5′ NCR and 3′ NCR) of viral RNA. There was reduction in viral RNA replication along with effects on symptom development where 5′ and 3′ NCRs were altered. A correlation was observed between the viral movement and alteration in 5′ NCR conditioning the viral pathogenicity (Petty et al. 1990). When a single nucleotide from ORF near the 5′ end of RNA in hordeivirus barley stripe mosaic virus was changed, there was prevention of viral movement encoded by immediately adjacent gene via negative regulation of synthesis of viral replicase.

The presence of 4 repetitions of 14 nucleotide sequence in 3′ NCR of TVMV (polyvirus *Tobacco vein mottling virus*) reduces the gravity of symptoms of disease without affecting viral accumulation.

3.9 Defense Responses in Plants and Viruses

Although viruses are relatively simple genetic entities, resistance molecular mechanisms and viral diseases susceptibility are still not fully comprehended and understood. Several mechanisms exist in plants for disease resistance against virus infections, but it is very difficult to explain them for various pathosystems separately (Brown 2015). An understanding of plant–virus interactions and molecular mechanisms of these interactions has been achieved through the unveiling of several model bacteria–plant systems. Gene theory proposed in the early 1970s has served as a model for many years, explaining how disease resistance is turned on against pathogens (Keen 1990). According to this theory, a single resistance gene (R-gene) encoded by the host recognizes the presence of avirulence (Avr) proteins and triggers a hypersensitive response of resistance (HR) leading toward rapid cell death. The most general R-gene types can be classified either as genes encoding protein nucleotide-binding leucine-rich repeat (NB-LRR) or genes encoding receptor-like kinase/receptor-like proteins (Rathjen and Moffett 2003). About a decade later, another model was proposed known as zigzag model (Jones and Dangl 2006). There are two distinct defense responses for the plant defense system in the zigzag model. The primary defense level is called PAMP/MAMP-triggered immunity (PTI), and the secondary defense level is called effector-triggered immunity (ETI). A basic defense mechanism presented by PTI is preventing invasion of the pathogen through cell wall thickening in response to specific structures or pathogen-associated proteins so-called pathogen-associated molecular patterns (PAMPs) or microbe-associated molecular patterns (MAMPs). Plants show susceptibility only when a pathogen successfully establishes both PTI response suppression and its pathogenic effector's facilitation. ETI, the second defense response level, is triggered when the products of R-gene directly or indirectly sense the presence of specific effectors called as Avr factors. Consequently, an effective ETI will keep the plants resistant; however, an insufficient ETI will lead to the establishment of disease, i.e., the susceptibility of the plant. Models of general resistance of most pathogens don't fit well with viral resistance because of intracellular parasitic nature of virus. For example, receptors of pattern recognition which serve as a component of major defense by triggering the first layer of resistance when a receptor of plasma membrane perceives a fungal or bacterial MAMP or PAMP (Tena et al. 2011) cannot play a role in plant viruses fighting because viruses do not express extracellular PAMPs. Pathogens, however, have evolved counter measures, such as delivering effector proteins into the plant cell to suppress host PTI. The defense and counter-defense between host and pathogen never end. Plants, in turn, have acquired resistance (R) proteins to recognize these pathogen effectors and trigger the ETI, a more robust and specific response. Plant viruses are, thus, both trigger and target of RNA silencing. This reinforces the concept that RNA silencing evolved as an antiviral defense mechanism and highlights the arms race between plants and plant viruses (Chen 2010). The major strategies developed by plants to counteract virus infections are:

Resistance (R) gene mediated
RNA silencing-based defense
Recessive gene-mediated defense

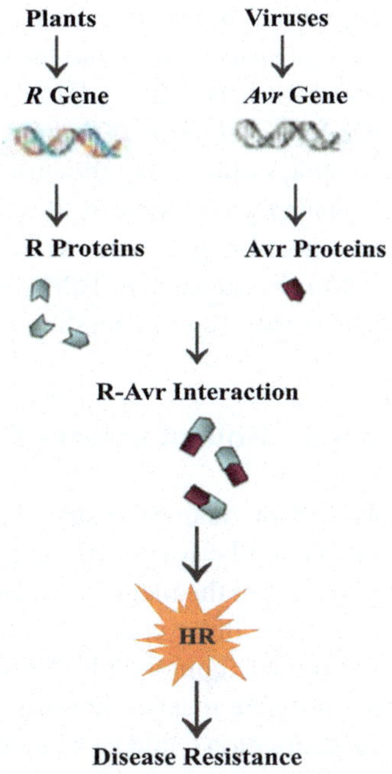

Fig. 3.3 A simplified representation of R-gene-mediated resistance in plants. R-genes (receptors) are expressed in plants which can bind to Avr gene products (elicitor protein) of the pathogen resulting in a cascade of transduction signals developed by receptor–ligand complex

3.9.1 Resistance (R) Gene-Mediated Response

More often it has been observed that viral disease symptoms are caused by toxic effect of some viral components. R-gene-mediated resistance which is the most intensively explored form of resistance toward the diverse bacteria, fungi, and viruses is frequently HR responsible and is an effective way to gain resistance against plant viruses. HR is the most common defense response to the viruses. R-genes (receptors) are expressed in the plant which can bind to the Avr gene products (elicitor protein) of the pathogen. HR response is an outcome of a cascade of transduction signals developed by receptor–ligand complex (Fig. 3.3). As the R-genes restrict the pathogen to the inoculated area, its spread to the entire plant is obstructed. The first viral R-gene to be cloned and characterized was from *Nicotiana* and *Tobacco mosaic virus* (TMV) is the first virus discovered and isolated from *Nicotiana glutinosa*, and its counterpart R-gene served as a model for studying HR-based resistance, systemic acquired resistance (SAR), and gene-for-gene theory (Holmes 1929). R-gene product RX_1 interacts with the capsid protein of *Potato virus X* to cause programmed cell death at the infection site. At the same time, R-gene can block virus replication before the generation of sufficient amount of capsid protein to cause necrosis. In some instances, it doesn't impede the viral spread throughout the plant, but systemic necrotic spots are generated. HR may be induced by any viral gene other than CP also which may or may not be capable of impeding viral propagation. The cytoplasmic inclusion RNA helicase from *Soybean mosaic virus* (SMV) and RSV3 gene from *Glycine max* (Zhang et al. 2009) and the interaction between p50 helicase domain of TMV replicase and N gene from *Nicotiana glutinosa* (Padgett et al. 1997) are the examples of HR inducers being capable of hindering the virus propagation.

There are reports of the other type also where viral gene product inducing HR may not be capable of restricting viral propagation. Systemic necrosis of *Arabidopsis thaliana* is caused by TuMV (*Turnip mosaic virus*), which is the result of gene interaction between P_3 encoding region of virus and TuNI resistance gene.

3.9.2 Role of Volatile Organic Compounds in Plant Defense

The plant defense system includes the organization of communication between plants and between different parts of the same plant. VOCs emitted by a damaged plant act on the plant's own leaves and on the organs of neighboring plants, modifying intercellular communication. In the absence of physical contact, they use volatile organic compounds (VOCs) for communication. Theoretically, three responses are possible in a neighboring plant upon the release of VOCs as a signal of damage. First situation could be when there is no response by the neighboring plant. Secondly, plant could respond by creating conditions that promote plant–virus interaction (increasing sensitivity of plant, its sensitization). Finally, the plant can create conditions that prevent virus infection by increasing its antiviral resistance. Thus, we cannot define a negative or a neutral influence of VOCs in plant–virus relationships. Plants release terpenes, fatty acid derivatives, benzenoids, phenylpropanoids, and amino acid-derived metabolites. Many of these products are made more lipophilic before their release into the air by the removal or masking of hydrophilic functional groups through reduction, methylation, or acylation reactions (Pichersky et al. 2006). VOC-mediated signals are transmitted to neighboring plants promoting the survival of entire community. Such VOC-mediated signals protect plants not only against insects, bacteria, fungi, and nematodes but also viral pathogens.

We currently have a fairly complete understanding of the processes and metabolic pathways involved in the production of many VOCs, but we have an extremely limited understanding of how VOCs affect intercellular traffic and thus what impact VOCs have on the plant virus–host interaction. This modification of intercellular transport has a significant effect on pathogenic infections. The VOC-mediated process of preparing plants for a putative attack can be referred to as "priming" (i.e., the initiation of reactions before the impact of the pathogen). In general, the process of priming is related to plant immunity, whereby plants trigger their defenses in response to a signal or previous challenge so that they can react with increased severity (Holopainen and Blande 2012). However, when we consider the role of priming in the relationship of the virus and the plant host, the picture is ambiguous.

Because the life cycle of viruses include the intercellular transport, cell-to-cell movement of viral genome, and also long-distance spread throughout the plant, therefore, unlike bacteria, fungi, oomycetes, and nematodes, viruses depend more on the state of intercellular transport, and their life cycle is completed in the symplast (Xu and Jackson 2010).

Volatile communication plays an important role in mediating the interactions between plants, aphids, and viruses in the environment. Thus, VOCs can also influence the virus–host–vector combination apart from regulating the life cycle of viruses. Because the plant cell is bounded by a rigid cell wall (preventing direct contact between adjacent cells and virus particles), there is no evidence of the participation of constitutively emitted VOCs in the innate immunity of plants to viruses.

Transgenic plants with increased VOC emission may be used as disinfectants to trigger and enhance a protective response against pathogens and plant-eating insects (Holopainen and Blande 2012).

3.9.3 RNA Silencing-Based Defense

In 1990, a defense mechanism was described where silencing of viral RNA was done as viruses are capable of overcoming complex defense barriers developed by host plants. This genetic surveillance system seems to be conserved in most eukaryotic organisms ranging from fission yeast to human beings. It is also known as RNA interference (RNAi) in animals and posttranscriptional gene silencing (PTGS) in plants (Carrington 2000). Intracellular presence of dsRNA triggers this system which ultimately leads to downregulation of the expression of genes that share substantial sequence homologies with this dsRNA trigger.

The silencing-induced dsRNAs are perceived by the RNA silencing machinery and then acted upon by a dsRNA-specific nuclease designated Dicer or Dicer-like (DCL) nuclease in plants (Fig. 3.4). It results into the formation of short RNA duplexes of 21–25 nucleotides known as small interfering RNAs or siRNAs (Hamilton and Baulcombe 1999). They function as key specific determinants of RNA silencing. Another nuclease, referred to as argonaute (AGO), recruits one strand of the siRNA duplex and guides it to ssRNAs containing complementary sequences to the siRNAs. siRNAs also help in amplifying the silencing process with the help of RNA-dependent RNA polymerases (RDRps). siRNA-complementary ssRNAs act as templates to synthesize more of dsRNAs, which are again processed by DCLs, and thus leading to an amplification in silencing process. Additionally a family of dsRNA-binding proteins (DRBs) having abundant differences in their structures and functions have been reported as partners of DCLs at various steps of RNA silencing cascade. As the function of RNA silencing is defense of host cells against virus invasion, few DRBs play an important role in antiviral defense not only in animals but also in plants. DRB_4 has been involved in defense against a large number of viruses in *Arabidopsis* (Qu et al. 2008). Thus four different families of proteins primarily constitute the plant RNA silencing machinery: DCLs, DRBs, AGOs, and RDRps.

The best understood RNA silencing pathway in plants is probably the *microRNA* (*miRNA*) pathway. In plants, an *MIR* gene is first transcribed to primary miRNAs

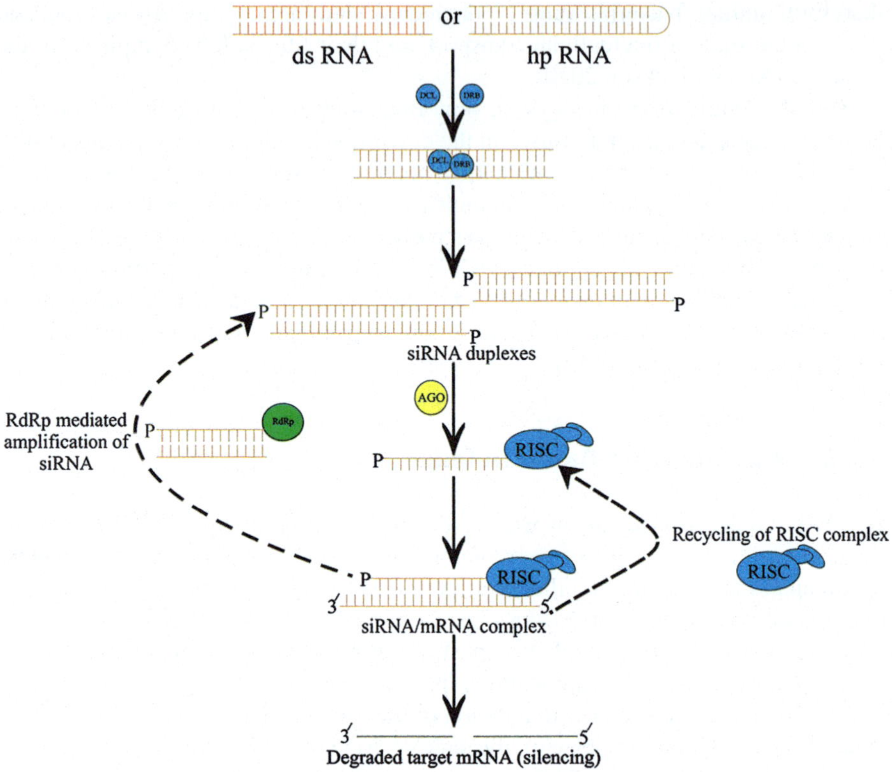

Fig. 3.4 An overview of RNA silencing cascade in plants. Both long dsRNA and hairpin RNA with a significant length of double-stranded region (hpRNA) can be processed by DCL/DRB complex into siRNA duplexes. This in turn mediates degradation of ssRNAs with the help of AGO and RISC complex. Additionally, siRNA could also serve as primers to prime the synthesis of new dsRNAs

(pri-miRNA) by DNA-dependent RNA polymerase II; pri-miRNAs form partially double-stranded hairpins through extensive intramolecular base pairs. These pri-miRNAs are processed by RNase III-like Dicer-like I endonuclease (DCL1) to generate miRNA precursors (pre-miRNAs). These are then sequentially processed by DCL1 to produce miRNA/miRNA duplex. The duplex moves to the cytoplasm, and mature miRNA (20–22 nucleotides) is selectively introduced into RISCs associated with AGO1. miRNA-programmed RISCs are directed to mRNAs with complementary sequences to mediate cleavage as well as translational repression and thus inhibit gene expression. RNA silencing serves as a major component in the antiviral defense mechanism; however, the strategy of R-gene-mediated resistance is effective against viruses as well as other phytopathogens (Nakahara and Masuta 2014).

3.9.4 Recessive Gene-Mediated Defense

Viruses being intracellular parasites are exclusively dependent on host mechanisms for their life cycle. After entering a plant cell, viral genome is released, and early viral proteins are translated. The virus also takes over some of the host functions. Because of limited number of genes coded by viral genome, a number of host factors are required to continue the viral cycle. Any alteration in the host factor or its absence can be an efficient strategy for plant defense and can be considered as a passive form of resistance. Such resistance mechanism has been frequently shown to be recessively inherited. More than half of plant virus resistances are recessively inherited, and many are still to be characterized. In the case of recessive inheritance resistance, the majority of genes have been identified in plant–virus pathosystems, and several recessive resistance genes have been characterized in bacterial and fungal pathogens research including *xa5* (a *Xanthomonas*-resistant gene in rice) and *mlo* (resistance gene for powdery mildew in barley). Much of the identified R-genes confer resistance to various potyviruses. Recessive r-genes conferring resistance of potyvirus have been identified and deployed for decades in various crops. For example, an essential host factor required for virus infection is translation factor of eukaryotes 4E (eIF4E) which plays a major role in host translation initiation by recruiting the ribosomal complex. Thus, natural variations in eIF4E provide an effective resistance against potyvirus infection in multiple crop species by preventing sequestration of virus. Thus, alteration in translation initiation factors (host factors) is a common strategy for developing viral resistance in plants (e.g., sbm1 in pea seed-borne mosaic virus of *Pisum sativum* and pot1 in PVY of *Solanum lycopersicum*). Another example of recessive resistance in tomato is the recently characterized ty5 which encodes messenger RNA surveillance factor pelo and confers resistance in *Tomato yellow leaf curl virus* (TYLCV). Pelo leads to impaired functionalities in protein synthesis and ribosome recycling phases which triggers viral infection suppression in resistant ty5 genotypes.

3.10 Virus Counter Defense Mechanism

Virus pathogenesis not only induces a resistance mechanism in the host but also suppresses it. Almost all plant viruses have developed a viral-encoded silencing suppressor-mediated counterdefense mechanism inhibiting RNA silencing at different steps. HC-Pro protein of potyviruses was the first identified silencing suppressor interfering with RNA silencing mechanism. The inactive forms of miRNAs (microRNAs) get accumulated on ectopic expression of HC-Pro protein which have negative regulatory function. HC-Pro also performs long-distance movement of viruses and also enhances pathogenicity of viruses (Saenz et al. 2002). To counteract the antiviral RNA silencing, most plant viruses have evolved silencing suppressor proteins that block one or more steps in the RNA silencing pathway.

3.11 Engineering Crops for Resistance Against Viruses

Geminiviruses are spread worldwide due to climate change, increasing insect vector population changes in crop cultivation, and changes in agricultural practices causing significant crop loss globally. Many of these crops are staple foods in tropical and subtropical regions; therefore developing resistance against geminiviruses is important socially and economically. The strategies used against geminiviruses include eradicating the vectors by using insecticides, but most of the times, the virus has already been transmitted to the plant even before the vector is killed. Also insecticides are extremely toxic and have adverse effects on the environment. Therefore, environment-friendly agricultural practices are to be raised. These aspects necessitate the development of natural resistance through genetically engineered crops providing more durable resistance to geminiviruses. The objective is to develop varieties by introducing selected nucleic acid sequences into plants either from the pathogen itself (pathogen-derived resistance, PDR) or not from the pathogen (non-pathogen-derived resistance) (Khan et al. 2014). Transgenic papaya cultivars were developed in 1998 against *Papaya ring spot virus* (PRSV) which is an RNA virus saving papaya industry from severe destruction in Hawaii. After herbicide- and insect-resistant crops, the third important transgenic crop having resistance against a virus is the RNA resistance papaya which is grown commercially. Currently, transgenic resistance is the most important research area to provide protection to plants against a large number of viruses.

3.12 Future Prospects

Viruses are obligate parasites infecting only living cells. The use of resistant cultivars is an effective strategy for reducing crop loss due to viral infection. Counter defense mechanism by virus encoding silencing suppressors gives rise to a battle between host and virus; ultimately variable symptoms are produced due to evolution of viral strains. The global transcriptomic and proteomic analyses have helped us to understand the plant–virus interaction at molecular level and also to acquire a broad vision of viral pathogens. Identification of host factors has long been one of the few major goals of virology research. The driving force comes from the obvious practical applications for the development of novel antiviral strategies and for beneficial biotechnological uses of viruses. Host factors may be targeted for the development of inheritable recessive genetic resistance in plants through advanced biotechnology, i.e., mutation, silencing, or downregulation of the host factor gene(s) (Wang and Krishnaswamy 2012).The challenge is to learn the influence of environmental factors, plant phenology, mode of viral entry into host, evolutionary adaptations of host and virus, and development of pathological process in the host for producing virus-resistant crops in the forthcoming years.

References

Aranda MA, Escaler M, Wang D, Maule AJ (1996) Induction of HSP70 and polyubiquitin expression associated with plant virus replication. Proc Natl Acad Sci U S A 93:15289–11529

Baebler S, Witek K, Petek M, Stare K, TusekZnidaric M, Pompe Novak M (2014) Salicylic acid is an indispensable component of the Ny-1 resistance gene-mediated response against potato virus Y infection in potato. J Exp Bot 65:1095–1109

Baldwin IT, Halitschke R, Paschold A, von Dahl CC, Preston CA (2006) Volatile signaling in plant-plant interactions: 'talking trees' in the genomics era. Science 311:812–814

Baughman GA, Jacobs JD, Howell SH (1988) Cauliflower mosaic-virus gene VI produces a symptomatic phenotype in transgenic tobacco plants. Proc Natl Acad Sci 85:733–737

Boualem A, Dogimont C, Bendahmane A (2016) The battle for survival between viruses and their host plants. Curr Opin Virolol 17:32–38

Brown J (2015) Durable resistance of crops to disease: a Darwinian perspective. Annu Rev Phytopathol l53:513–539

Carrington JC (2000) RNA silencing – moving targets. Nature 408:150–151

Chen X (2010) Small RNAs-secrets and surprise of the genome. Plant J 61:941–958

Chen MH, Citovsky V (2003) Systemic movement of a tobamovirus requires host cell pectin methylesterase. Plant J 35:386–392

Chen MH, Tian GW, Gafni Y, Citovsky (2005) Effects of calreticulin on viral cell to cell movement. Plant Physiol 138:1866–1876

Clark MF, Adams AN (1977) Characteristics of the microplate method of enzyme-linked immunosorbent assay for the detection of plant viruses. J Gen Virol 34:475–483

Culver JN (2002) *Tobacco mosaic virus* assembly: determinants in pathogenicity and resistance. Annu Rev Phytopathol 40:287–308

Dandekar AM, Martinelli F, Davis CE, Bhushan A, Zhao W, Fiehn O, Skogerson K, Wohlgemuth G, D'Souza R, Roy S, Reagan RL, Lin D, Bruce Cary R, Pardington P, Gupta G (2010) Analysis of early host responses for asymptomatic disease detection and management of specialty crops. Crit Rev Immunol 30:277–289

Dardick C (2007) Comparative expression profiling of *Nicotiana benthamiana* leaves systemically infected with three fruit tree viruses. Mol Plant Microbe Interact 20:1004–1017

Davino S, Napoli C, Davino M, Accotto GP (2006) Spread of tomato yellow leaf curl virus: partial displacement of another geminivirus originally present. Eur J Plant Pathol 114:293–299

Dekker EL, Derkes FLM, Asjes CJ, Lemmers MEC, Bol JF, Langeveld SA (1993) Characterization of potyviruses from tulip and lily which cause flower breaking. J Gen Virol 79:881–887

Denance N, Sanchez-Vallet A, Goffner D, Molina A (2013) Disease resistance or growth: the role of plant hormones in balancing immune responses and fitness costs. Front Plant Sci 4:1–12. Article155

Derrick PM, Barker H, Oparka KJ (1992) Increase in plasmodesmatal permeability during cell-to-cell spread of tobacco rattle virus from individually inoculated cells. Plant Cell 4:1405–1412

Dietzgen RG, Mann KS, Johnson KN (2016) Plant virus insect vector interaction: current and potential future research directions. Virus 8:303

Dorokhov YL, Komarova TV (2016) Volatile organic compounds and plant virus–host interaction. In: Gaur RK, Hohn T, Sharma P (eds) Plant virus host interaction: molecular approaches and viral evolution. Academic (Elsevier Inc), Amsterdam/Heidelberg, pp 1–16

Dorokhov YL, Mäkinen K, Frolova OY, Merits A, Saarinen J, Kalkkinen N, Atabekov JG, Saarma M (1999) Anovel function for a ubiquitous plant enzyme pectin methylesterase: the host-cell receptor for the tobacco mosaicvirus movement protein. FEBS Lett 461:223–228

Doumayrou J, Leblaye S, Froissart R, Michalakis Y (2013) Reduction of leaf area and symptom severity as proxies of disease-induced plant mortality the example of the cauliflower mosaic virus infecting two Brassicaceae hosts. Virus Res 176:91–100

Dunoyer P, Voinnet (2005) The complex interplay between plant viruses and host RNA-silencing pathways. Curr Opin Plant Biol 8:415–423

Erickson FL, Holzberg S, Calderon-Urrea A, Handley V, Axtell M, Corr C, Baker B (1999) The helicase domain of the TMV replicase proteins induces the N-mediated defence response in tobacco. Plant J 18:67–75

Faulkner C, Akman OE, Bell K, Jeffree C, Oparka K (2008) Peeking into pit fields: a multiple twinning model of secondary plasmodesmata formation in tobacco. Plant Cell 20:1504–1518

Fraser RSS, Whenham RJ (1989) Abscisic-acid metabolism in tomato plants infected with tobacco mosaic virus relationships with growth, symptoms and the Tm-1 gene for Tmv resistance. Physiol Mol Plant Pathol 34:215–226

Fridborg I, Grainger J, Page A, Coleman M, Findlay K, Angell S (2003) TIP, a novel host factor linking callose degradation with the cell-to-cell movement of potato virus X. MPMI 16:132–140

Hafren A, Hofius D, Ronnholm G, Sonnewald U, Makinen K (2010) HSP70 and its cochaperone CPIP promote potyvirus infection in *Nicotiana benthamiana* by regulating viral coat protein functions. Plant Cell 22:523–535

Hamilton AJ, Baulcombe DC (1999) A species of small antisense RNA in posttranscriptional gene silencing in plants. Science 286:950–952

Haupt S, Cowan GH, Ziegler A, Roberts AG, Oparka KJ, Torrance L (2005) Two plant-viralmovement proteinstraffic in the endocytic recycling pathway. Plant Cell 17:164–181

Havelda Z, Várallyay É, Válóczi A, Burgyán J (2008) Plant virus infection-induced persisitent host genedownregulation in systemically infected leaves. Plant J 55:278–288

Heinlein M (2015) Plasmodesmata: channels for viruses on the move. Methods Mol Biol 1217:25–52

Holmes FO (1929) Local lesions in tobacco mosaic. Bot Gazzet 87:39–55

Holopainen JK, Blande JD (2012) Molecular plant volatile communication. Adv Exp Med Biol 739:17–31

Huang Z, Andrianov VM, Han Y, Howell SH (2001) Identification of Arabidopsis proteins that interact with thecauliflower mosaic virus (CaMV) movement protein. Plant Mol Biol 47:663–675

Hull R (2009) Overview of plant viruses. In: Hull (ed) Comparative plant virology, 2 nd edn. Elsevier, Oxford

Islam W, Zhang J, Adnan M, Noman A, Zaynab M, Wu Z (2017) Plant virus ecology: a glimpse of recent accomplishments. Appl Ecol Environ Res 15(1):691–705

Jameson P (2000) Cytokinins and auxins in plant-pathogen interactions-an overview. Plant Growth Regul 32:369–380

Jones JDG, Dangl JL (2006) The plant immune system. Nature 444:323–329

Kawamura-Nagaya K, Ishibashi K, Huang YP, Miyashita S, Ishikawa M (2014) Replication protein of *Tobacco mosaic virus* cotranslationally binds 5 untranslated region of genomic RNA to enable viral replication. Proc Natl Acad Sci U S A 111:e1620–e1628

Keen NT (1990) Gene-for-gene complementarity in plant-pathogen interactions. Annu Rev Genet 24:447–463

Khan AJ, Akhtar S, Mansoor S, Amin I (2014) Engineering crops for resistance to geminiviruses. In: Gaur RK, Hohn T, Sharma P (eds) Plant virus host interaction: molecular approaches and viral evolution. Academic (Elsevier Inc), Oxford, pp 1–16

Kim SH, Macfarlane S, Kalinina NO, Rakitina DV, Ryabo EV, Gillespie T, Haupt S, JWS B, Taliansky M (2007a) Interaction of a plant virus-encoded protein with the major nucleolar protein fibrillarin is required forsystematic virus infection. Proc Na Acad Sci U S A 104:11115–11120

Kim SH, Ryabov EV, Kalinina NO, Rakitina DV, Gillespie T, MacFarlane S, Haupt S, Brown JWS, Taliansky M (2007b) Cajal bodies and the nucleous are required for a plant virus systemic infection. EMBO J 26:2169–2179

Kiraly Z, Hafez MY, Fodor J, Kiraly Z (2008) Suppression of tobacco mosaic virus-induced hypersensitive-type necrotization in tobacco at high temperature is associated with downregulation of NADPH oxidase and superoxidase and stimulation of dehydroascorbate reductase. J Gen Virol 89:799–808

Kong LJ, Orozco BM, Roe JL, Nagar S, Ou S, Feiler HS, Durfee T, Miller AB, Gruissem W et al (2000) A geminivirus replication protein interacts with the retinoblastoma protein through a novel domain to determine symptoms and tissue specificity of infection in plants. EMBO J 19:3485–3495

Kragler F, Curin M, Trutnyeva K, Gansch A, Waigmann E (2003) Mpb2c, a microtubule-associated plant protein binds to and interferes with cell to cell transport of tobacco mosaic virus movement protein. Plant Physiol 132:1870–1883

Laliberte JF, Sanfac (2010) Cellular remodeling during plant virus infection. Annu Rev Phytopathol 48:69–91

Li Z, Gonzalez PA, Sasvari Z, Kinzy TG, Nagy PD (2014) Methylation of translation elongation factor1A by the METTL10-like See1 methyltransferase facilitates tombusvirus replication in yeast and plants. Virology 448:43–54

Lima JAA, Nascimento AKQ, Radaelli P, Purcifull DE (2012) Serology applied to plant virology. In: Molish-Al-Moslih M (ed) Serological diagnosis of certain human, animal and plant diseases. InTech, Rijekax, pp 71–94

Mauch-Mani B, Mauch F (2005) The role of abscisic acid in plant–pathogen interactions. Curr Opin Plant Biol 8:409–414

Mauck KE, De Moraes CM, Mescher MC (2010) Deceptive chemical signals induced by a plant virus attract insect vectors to inferior hosts. PNAS 23:3600–3605

Moffett P (2009) Mechanisms of recognition in dominant R gene mediated resistance. Adv Virus Res 75:1–33

Moreno IM, Bernal JJ, de Garcia B, Rodriguez-Cerezo E, Garcia-Arenal F (1997) The expression level of the 3a movement of protein determines differences in severity of symptoms between two strains of tomato aspermy cucumovirus. Mol Plant-Microbe Interact 10:171–179

Mumford RA, Walsh K, Barker I, Boonham N (2000) Detection of *Potato mop top virus* and *Tobacco rattle virus* using multiplex real-time X fluorescent reverse-transcription polymerase chain reaction assay. Phytopathology 90:448–453

Nagy PD, Pogany J (2012) The dependence of viral RNA replication on co-opted host factors. Nat Rev Microbiol 10:137–149

Nakahara KS, Masuta C (2014) Interaction between viral RNA silencing suppressors and host factors in plant immunity. Curr Opin Plant Biol 20:88–95

Navarro B, Gisel A, Rodio ME, Delgado S, Flores R, Di Serio F (2012) Viroids: how to infect a host and cause disease without encoding proteins. Biochimie 94:1474–1480

Neeleman L, VanDerKuyl AC, Bol JF (1991) Role of alfalfa mosaic virus coat protein gene in symptom formation. Virology 181:687–693

Nelson RS, Citovsky V (2005) Plant viruses invaders of cells and pirates of cellular pathways. Plant Physiol 138:1809–1814

Ni P, Cheng Kao C (2013) Non-encapsidation activities of the capsid proteins of positive-strand RNA viruses. Virology 446:123–132

Padgett HS, Watanabe Y, Beachy RN (1997) Identification of the TMV replicase sequence that activates the N gene-mediated hypersensitive response. Mol Plant-Microbe Interact 10:709–715

Padmanabhan MS, Goregaoker SP, Golem S, Shiferaw H, Culver JN (2005) Interaction of the tobacco mosaic virus replicase protein with AUX/IAA protein PAP1/IAA 26 is associated with disease development. J Virol 79:2549–2558

Pallas V, Garc'ıa JA (2011) How do plant viruses induce disease? Interactions and interference with host components. J Gen Virol 92:2691–2705

Pallas V, Genoves A, Sánchez-Pina MA, Navarro JA (2011) Systemic movement of viruses via the plant phloem. In: Caranta C, Aranda MA, Tepfer M, López-Moya JJ (eds) Advances in plant virology. Caister Academic Press, Norwich, p 470

Panavas, Serviene E, Brasher J, Nagy PD (2005) Yeast genome-wide screen reveals dissimilar sets of host genes affecting replication of RNA viruses. Proc Natl Acad Sci U S A 102:7326–7331

Peng Ni and Kao CC (2013) Non-encapsidation activities of the capsid proteins of positive-strand RNA viruses.Virology 446: 123–132

Petty ITD, Edwards MC, Jackson AO (1990) Systemic movement of an RNA plant virus determined by a point substitution in a 59 leader sequence. Proc Natl Acad Sci U S A 87:8894–8897
Pichersky E, Noil JP, Dudareva N (2006) Biosynthesis of plant volatiles; nature's diversity and ingenuity. Science 311(5762):801–811
Prendeville HR, Tenhumberg B, Pilson D (2014) Effects of virus on plant fecundity and population dynamics. New Phytol 202:1346–1356
Purcifull DE, Hiebert E, Petersen M, Webb S (2001) Virus detection serology. In: Maloy OC, Murray TD (eds) Encyclopedia of plant pathology. Wiley, New York, pp 1100–1109
Qu F, Ye X, Mortis TJ (2008) *Arabidopsis* DRB4, AGO1, AGO7 and RDR6 participate in a DCL4-intiated antiviral RNA silencing pathway negatively regulated by DCL1. Proc Nat Acad Sci U S A 105:14732–14737
Rahoutei J, García Luque, Barón M (2000) Inhibition of photosynthesis by viral infection: effect on PSII structure and function. Physiol Plant 110:286–292
Rathjen JP, Moffett P (2003) Early signal transduction events in specific plant disease resistance. Curr Opin Plant Biol 6:300–306
Ren T, Qu F, Morris TJ (2000) HRT gene function requires interaction between a NAC protein and viral capsid protein to confer resistance to turnip crinkle virus. Plant Cell 12:1917–1925
Rensink WA, Lee Y, Liu J, Iobst S, Ouyan S, Buell CR (2005) Comparative analyses of six solanaceous transcriptomes reveal a high degree of sequence conservation and species-specific transcripts. BMC Genomics 6:124
Roberts PL, Wood KR (1982) Effects of a severe (P6) and a mild (W) strain of cucumber mosaic virus on tobacco leaf chlorophyll, starch and cell ultrastructure. Physiol Plant Pathol 21:31–37
Roossinck MJ, Garcia Arenal F (2015) Ecosystem simplification, biodiversity loss and plant virus emergence. Curr Opin Virol 10:56–62
Saenz P, Salvador B, Simon-Mateo C, Kasschau KD, Carrington JC, Garcia JA (2002) Host-specific involvement of the HC protein in the long-distance movement of potyviruses. J Virol 76:1922–1931
Selth LA, Dogra SC, Rasheed MS, Healy H, Randles JW, Rezaian MA (2005) A NAC domain protein interacts with tomato leaf curl virus replication accessory protein and enhances viral replication. Plant Cell 17:311–325
Stange C (2006) Plant-virus interactions during the infective process. Cien Inv Agr 33(1):1–18
Stapleford KA, Miller DJ (2010) Role of cellular lipids in positive-sense RNA virus replication complex assembly and function. Viruses 2:1055–1068
Stobbe AH, Roossinck MJ (2014) Plant virus metagenomics what we know and why we need to know more. Front Plant Sci 5:150
Stukenbrock EH, McDonald BA (2008) The origins of plant pathogens in agro-ecosystems. Annu Rev Phytopathol 46:75–100
Takacs A, Horvath J, Gaborjanyi R, Mikulas J (2014) Hosts and non-hosts in plant virology and the effects of plant viruses on host plants. In: Gaur RK, Hohn T, Sharma P (eds) Plant virus host interaction: molecular approaches and viral evolution. Academic (Elsevier Inc), Amsterdam/Heidelberg, pp 1–16
Taraporewala ZF, Culver JN (1997) Structural and functional conservation of the tobamovirus coat protein elicitor active site. Mol Plant-Microbe Interact 10:597–604
Tena G, Boudsocq M, Sheen J (2011) Protein kinase signaling networks in plant innate immunity. Curr Opin Plant Biol 14:519–529
Ton J, Flors V, Mauch-Mani B (2009) The multifaceted role of ABA in disease resistance. Trends Plant Sci 14:310–317
Ueki S, Spektor R, Natale DM, Citovsky V (2010) ANK, a host cytoplasmic receptor for the tobacco mosaicvirus cell-to-cell movement protein, facilitates intercellular transport through plasmodesmata. PLoS Pathog 6:e1001201
Valli A, López-Moya JJ, Garcıa JA (2009) RNA silencing and its suppressors in the plant-virus interplay. Wiley, Chichester

Verchot J (2012) Cellular chaperones and folding enzymes are vital contributors to membrane bound replication and movement complexes during plant RNA virus infection. Front Plant Sci 3:1–12. Article 275

Vuorinen AL, Kelloniemi J, Valkonen JP (2011) Why do viruses need phloem for systemic invasion of plants? Plant Sci 181:355–363

Waigmann E, Ueki S, Trutnyeva K, Citovsky V (2004) The ins and outs of non-destructive cell-to-cell and systemic movement of plant viruses. Crit Rev Plant Sci 23:195–250

Wallis CM, Stone AL, Sherman DJ, Damsteegt VD, Gildow FE, Schneider WL (2007) Adaptation of plum pox virus to a herbaceous host (*Pisum sativum*) following serial passages. J Gen Virol 88:2839–2845

Wang A, Krishnaswamy S (2012) Eukaryotic translation initiation factor 4E-mediated recessive resistance to plant viruses and its utility in crop improvement. Mol Plant Pathol 13:795–803

Whitham SA, Yang C, Gooding MM (2006) Global impact: elucidating plant responses to viral infection. MPMI 19(11):1207–1215

Xie Q, Sanz-Burgos AP, Guo H, Garcia JA, Gutierrez C (1999) GRAB proteins, novel members of the NAC domain family, isolated by their interaction with a geminivirus protein. Plant Mol Biol 39:647–656

Xu XM, Jackson D (2010) Lights at the end of the tunnel: new views of plasmodesmatal structure and function. Curr Opin Plant Biol 13(6):684–692

Zhang C, Hajimorad MR, Eggenberger AL, Tsang S, Whitham SA, Hill JH (2009) Cytoplasmic inclusion cistron of soybean mosaic virus serves as a virulence determinant on Rsv3-genotype soybean and a symptom determinant. Virology 391:240–248

Zhao J, Zhang X, Hong Y, Liu Y (2016) Chloroplast in plant virus interaction. Front Microbiol 7:1–20

Zhu T, Chang HS, Schmeitz J, Gil P, Shi L, Budworth PR, Zhou H, Chen C, Wang X (2001) Gene expression microarrays: improvements and applications towards agricultural gene discovery. JALA 6:95–98

Zhu S, Gao F, Cao X, Chen M, Ye G, Wei C, Li Y (2005) The Rice dwarf virus P2 protein interacts with ent-kaurene oxidases in vivo, leading to reduced biosynthesis of gibberellins and rice dwarf symptoms. Plant Physiol 139:1935–1945

Pathogen-Associated Molecular Patterns and Their Perception in Plants

4

Jitendra Singh Rathore and Chaitali Ghosh

Abstract

In plants, innate immunity, the first line of microbial recognition leading to active defense responses, relies on the perception of pathogen-associated molecular patterns (PAMPs) by pattern recognition receptors (PRRs). Pattern recognition receptors (PRRs) enable plants to sense non-self molecules exhibited by microbes and raise proper defense responses or establish symbiosis. This recognition leads to PAMP-triggered immunity (PTI). Despite the numerous PAMPs recognized by plants, only a handful of PRRs are characterized. Most of them correspond to the transmembrane proteins with a ligand-binding ectodomain. PRRs interact with additional transmembrane proteins that act as signaling adapters or amplifiers to achieve full functionality. The crucial role of PRRs in antimicrobial immunity is demonstrated by the direct targeting of PRRs and their associated proteins by pathogenic virulence effectors. In recent years the importance of PRR subcellular trafficking to plant immunity has become apparent. PRRs traffic through the endoplasmic reticulum (ER) and the Golgi apparatus to the plasma membrane, where they recognize their cognate ligands. At the plasma membrane, PRRs can be recycled or internalized via endocytic pathways. By using genetic and biochemical tools in combination with bio-imaging, the trafficking pathways and their role in PRR perception of microbial molecules are now being revealed.

J. S. Rathore
School of Biotechnology, Gautam Buddha University, Greater Noida, Uttar Pradesh, India

C. Ghosh (✉)
Department of Zoology, Gargi College, University of Delhi, New Delhi, India
e-mail: ghoshchaitali@gmail.com

A. Singh, I. K. Singh (eds.), *Molecular Aspects of Plant-Pathogen Interaction*,
https://doi.org/10.1007/978-981-10-7371-7_4

Keywords
Pattern recognition receptor (PRR) · Pathogen associated molecular pattern (MAMPs/PAMPs) · Receptor like kinase (RLK) · Receptor like protein (RLP) · Effector triggered immunity (ETI) · PRR triggered immunity (PTI)

4.1 Introduction

Plants are sessile organisms and are in continuous exposure to various environmental signals. The perception of environmental signals and the ability to respond accordingly are essential for plants to survive. To defend themselves against a plethora of pathogenic microbes or pests, plants relies only on innate immunity as they lack specialized immune cells. The plant innate immunity has a two-tier perception system (Dodds and Rathjen 2010). The first layer of defense is mediated by surface-localized pattern recognition receptors (PRRs), leading to PRR-triggered immunity (PTI). PTI relies on the perception of the specific molecular patterns, which includes microbe- or pathogen-associated molecular patterns (MAMPs/PAMPs) or self molecules (damage-associated molecular pattern – DAMPs), that are released upon pathogen-induced cell damage (Boller and Felix 2009) (Table 4.1). MAMPs/PAMPs are conserved among pathogen and nonpathogenic bacteria; hence they are essential structures. MAMPs/PAMPs are recognized in plants by pattern recognition receptors (PRRs). These receptors are localized on the surface of plant cells, and their induction provides the first line of defense. PRRs are membrane localized and broadly classified as receptor-like kinase (RLK) or receptor-like protein (RLP). RLK and RLP have multidomain architecture, including extracellular ectodomain, transmembrane domain, and cytoplasmic domain. The major difference between RLK and RLP lies in the cytoplasmic domain which is essential for downstream signaling. As kinase is the key function of RLKs, therefore they have a long cytoplasmic domain, responsible for signaling. However, on the other hand, RLPs have a small cytoplasmic domain, and hence they depend upon other cytoplasmic proteins for downstream signaling. To cross the first line of defense, microbes like bacteria have developed a mechanism by which they directly inject effector proteins into the host cells. Type three secretion system (TTSS) in bacteria is the best studied example of this type. To overcome such conditions, plants have developed a second line of defense, which recognizes the effector proteins directly or indirectly through plant-resistant (R) gene product called effector-triggered immunity (ETI) (Jones and Dangl 2006). Most of the R proteins are intracellular immune receptor proteins and interact with effector proteins indirectly.

PTI's early cellular response includes the rapid generation of reactive oxygen species (ROS) and nitrogen oxide (NO), activation of mitogen-activated kinases, expression of immune-related genes, alteration in cell wall architecture and synthesis of pathogenesis-related (PR) protein and antimicrobial compounds. ROS and NO have antimicrobial effect, whereas NO is responsible for the cross-linking of the

Table 4.1 Pathogen-associated molecular patterns (PAMPs), damage-associated molecular patterns (DAMPs), and nematode-associated molecular patterns (NAMPs)

Name	Corresponding plant receptor (PRR)	References
PAMPs		
Beta-glycan (GE)	GEBP (putative receptor soybean)	Umemoto et al. (1997)
Flagellin	FLS2 (*Arabidopsis*)	Felix et al. (1999) and Gómez-gómez et al. (2001)
Lipopolysaccharide (LPS)	Not identified	Newman et al. (1995)
Chitin	CeBip and CERK1 (rice): AtCERK1 (*Arabidopsis*)	Kaku et al. (2006), Miya et al. (2007), and Shimizu et al. (2010)
Xylanase (ElX)	ElX (tomato)	Ron and Avni (2004) and Bailey et al. (1990)
Elongation factor TU (EF-Tu, elf18/elf26)	EFR (*Arabidopsis*)	Kunze et al. (2004)
Pep-13 (an oligopeptide of 13 amino acids from *P. megasperma*)	Not identified	Nurnberger et al. (1994)
Cellulose-binding elicitor lectin (CBEL) from *Phytophthora*	Not identified	Mateos et al. (1997), Sejalon-Delmas et al. (1997), and Gaulin et al. (2006)
Peptidoglycan (PGN)	Lym1 and Lym3 (*Arabidopsis*)	Gust et al. (2007), Erbs et al. (2008), and Willmann et al. (2011)
Bacterial cold shock proteins (RNP1 motif)	Not identified	Felix and Boller (2003)
Bacterial superoxide dismutase (Sod)	Not identified	Watt et al. (2006)
Activator of XA21 (Ax21)	XA21 and XA21D (*Oryza sativa*)	Song et al. (1995), Wang et al. (1998), and Gust et al. (2007)
Avirulence on Ve1 tomato (Ave 1)	Ve1 putative tomato receptor (*Solanum lycopersicum*)	de Jonge et al. (2012), Kawchuk et al. (2001), and Thomma et al. (2011)
DAMPs		
Systemin	Not identified	Narvaez-Vasquez and Ryan (2004)
Pep1 (23 aa part of a cytosolic protein from *Arabidopsis*)	PEPR1 (*Arabidopsis*)	Huffaker et al. (2006), Yamaguchi et al. (2006)
Oligogalacturonides (OGs)	WAK1 (*Arabidopsis*)	Brutus et al. (2010) and Nothnagel et al. (1983)
Cutin	Not identified	Schweizer et al. (1996) and Kauss et al. (1999)
Hydroxyproline-rich systemin	Not identified	Pearce and Ryan (2003) and Heiling et al. (2010)
Extracellular ATP (eATP)	Does not respond to nucleotide 1	Choi et al. (2014) and Jeter et al. (2004)

(continued)

Table 4.1 (continued)

Name	Corresponding plant receptor (PRR)	References
Plant elicitor peptides (Peps)	PEPR1/PEPR2	Pearce et al. (1991, 2001), Huffaker et al. (2006), Yamaguchi et al. (2010), Huffaker et al. (2011), Yamaguchi and Huffaker (2011), Huffaker and Ryan (2007), and Huffaker et al. (2013)
AtHMGB3	Not identified	Choi et al. (2016)
NAMPs		
Ascr#18	Not identified	Choi and Klessig (2016)

polymer of plant cell wall and hence provides strength against degradation caused by pathogen. To inhibit the pathogen multiplication, plants also produce PR proteins such as β-1-3 glucanase and chitinase. All these responses are sufficient to overcome most of the microbes or pathogens (Dou and Zhou 2012; Dodds and Rathjen 2010). Plants, defective in PRR or PTI signaling components, are often more susceptible to pathogens.

The general elicitors for PTI in plants were oligosaccharides or glycoproteins (Boller and Felix 2009). The increasing amount of genomic information leads to the identification of novel protein elicitors and their corresponding epitopes. Bacterial peptides like Flg22 and elf18 are derived from surface-associated flagellin protein (Flg) and translocation elongation factor Tu (Ef-Tu), respectively (Felix and Boller 2003; Zipfel et al. 2006). Apart from these two well-studied elicitors, bacterial glycoconjugates including peptidoglycan (PGN) which provides strength to the cell envelope of Gram-positive and Gram-negative bacteria also work as elicitor of plant innate immunity (Erbs et al. 2008; Willmann et al. 2011). Lipopolysaccharide (LPS) carrying lipid A moiety from the Gram-negative bacteria is a potent MAMP for mammals. It is responsible for activating pro-inflammatory responses via toll-like receptor 4 (TLR4) in mammals. Similarly, LPS from the outer membrane of Gram-negative and Gram-positive bacteria also evokes innate immune response in plants (Silipo et al. 2005; Erbs and Newman 2012). Recently, it has been discovered that *Arabidopsis thaliana* can sense LPS, specifically from *Pseudomonas* and *Xanthomonas*. Gene knockout studies showed that bulb-type lectin S-domain-1 receptor-like kinase LORE (SD1-29) is essential for the perception of LPS in plants. These results were further confirmed when LORE mutants were found hypersusceptible to *Pseudomonas syringae* infection. Further studies with chemically degraded LPS, isolated from *Pseudomonas* species, proved that LORE detected mainly the lipid A moiety of LPS. The heterologous expression of LORE in tobacco conferred sensitivity to LPS, which confirmed its key function in LPS sensing (Ranf et al. 2015). Oligosaccharide derived from fungus and oomycetes also act as MAMP/PAMP. Fungal chitin and its derivatives N-acetylglucosamine are responsible for induction of innate immunity in monocot and dicot plants.

Apart from bacteria and fungus, viruses and nematodes also target plants and cause pathogenicity. So far not a single conserved viral MAMP/PAMP has been reported. However, plants generally defend themselves from viruses via RNA silencing mechanism (RNAi). To counter such defense mechanism, plant viruses developed RNAi silencing suppressors, many of which bind with double-stranded RNA (dsRNA) (which generally acts as PAMP) and attenuate RNAi mechanism (Csorba et al. 2009; Ruiz-Ferrer and Voinnet 2009). Nematodes are also reported to parasitize plants, but detailed perceived signal is not known. Recently, defense signaling molecules from different plant-parasitic nematodes have been identified. They belong to conserved nematode pheromone family called ascarosides. Ascr#18 is among one of them which induces innate immunity in different plants via activating defense genes, MAPKs, enhancing resistance against bacteria, viruses, fungi, and oomycetes.

Due to their sessile lifestyle, plants are subjected to biotic stresses as well as a multitude of abiotic stress factors (e.g., cold, excess water, increased salt concentrations). In animal systems, the C-type lectins represent a key player in the recognition of pathogens and the induction of the immune response, whereas C-type lectins in plants are rather rare. To recognize lectins in plants, the membrane-bound PRRs carry extracellular lectin domains, that are coupled to intracellular Ser/Thr kinase domains. These lectin receptor kinases (LecRKs) are further categorized into four types: G-, C-, L-, and LysM-type (Bouwmeester and Govers 2009; Cambi et al. 2005; Vaid et al. 2012, 2013; Singh and Zimmerli 2013). In plants the G-type LecRKs carry a lectin domain belonging to the *Galanthus nivalis* agglutinin (GNA) family. There are 32 G-type LecRKs identified from *Arabidopsis* and about 100 from rice. However it is not clear whether these lectin domains play a role in the interaction with pathogen or not (Vaid et al. 2012).

Currently 11 PRRs with known ligands have been characterized from different plant species. Flagellin sensing 2 (FLS2) and EF-TU receptor (EFR), resistant to *Xanthomonas oryzae* pv. *oryzae* (Xa21), detect the bacterial MAMP flagellin (flg22), elongation factor Tu (elf18), and the Ax21 sulfated protein (AxYS22), respectively. The LysM proteins LYM1, LYM2, and chitin elicitor receptor kinase 1 (CERK1) are responsible for the perception of bacterial peptidoglycans (PGN), whereas ethylene-inducing xylanase receptors (LeEIX1/2) sense fungal xylanase, and both chitin elicitor-binding protein (CEBiP) and CERK1 recognize fungal chitin (Borner et al. 2003; Kaku et al. 2006; Miya et al. 2007; Shimizu et al. 2010). The PRRs differ in their dependence on BRI1-associated kinase 1/somatic embryo receptor kinase 3 (BAK1/SERK3), a LRR-RLK identified as a co-receptor for the brassinosteroid insensitive 1 (BRI1), but now it is known to regulate multiple signaling pathways (Chinchilla et al. 2009). PRRs not only mediate MAMP-/PAMP-triggered immunity (PTI), the first layer of active defense in plants, but are also required for non-self-discrimination during symbiotic plant-microbe interactions (Boller and Felix 2009). This includes LysM receptor-like kinase 3 (LYK3) and NOD factor perception (NFP) perceiving bacterial nodulation factors (NF) and symbiosis receptor kinases (SYMRK) necessary for bacterial and fungal symbiosis (Gherbi et al. 2008; Limpens et al. 2003; Stracke et al. 2002; Madsen et al. 2011).

4.2 Bacterial PAMPs

4.2.1 Flagellin (Flg)

Flagellin is an essential structure of most of the bacteria as it provides motility and in some cases assists in binding to the host cells. It is an oligomeric proteinaceous structure made up of monomeric multidomain subunit of flagellin (Flg). In animals, flagellin is recognized by surface-localized toll-like receptors (TLR5) (Ramos et al. 2004). In plants such as *Arabidopsis* and tomato, it is conserved in the N-terminal of flagellin which elicits innate immune response. Studies in plants showed that only 22 amino acids of flagellin (Flg22) are responsible to activate immune response in plants (Felix et al. 1999). To decipher the receptor of Flg22, various genetic approaches have been applied in *Arabidopsis*. It has been observed that among various mutants of *Arabidopsis*, it is FLS2 mutant which is susceptible to pathogen. The gene responsible for coding FLS2 is located on chromosome number 5 in *Arabidopsis* and belongs to the RLK family. Further studies proved that it is LRR domain of FLS2 which is responsible for binding with Flg22 (Gomez-Gomez and Boller 2000; Bauer et al. 2001). However, it is not the same conserved domain of Flg22, which is recognized by all plants. In case of tomato (*Solanum lycopersicum*), it is Flg15, a shorter version of Flg22 from N-terminal, which elicits the immune response, whereas in rice it is the full-length flagellin which gives better immune response as compared to Flg22. Heterologous expression studies of FLS2 from *Arabidopsis* in tomato cells further proved its specificity for Flg22. Different plants perceived different epitopes of flagellin. For example, 15-amino-acid-long peptide from *E.coli* Flg has shown high response in tomato, whereas the same peptide is not able to elicit immune responses in tobacco (Newman et al. 2013). Differences in the perception of flagellin are not limited to different plants, but it is also observed in different species of the same family. In other studies, when FLS2 from tomato is expressed in *Nicotiana benthamiana* cells, they gained flagellin perceptions, specific for tomato (Robatzek et al. 2007).

4.2.2 Elongation Factor TU (EF-Tu)

Protein biosynthesis is essential for the survival of microorganisms where mRNA and ribosomes play a crucial role. During protein synthesis, elongation factors are associated with ribosomes. EF-Tu is among one of them and is present in abundance in the bacterial cells (Jeppesen et al. 2005). N-terminal of elongation factor is highly conserved and has elicitor property in plants. Either 28-amino-acid- or 16-amino-acid-long protein form N-terminus named elf28 or elf16, respectively, is responsible for elicitor property. Each elicitor is perceived by specific or common receptors in plants. EF-Tu is also perceived by EF-Tu receptor in plants (EFR) and is independent from the Flg, as FLS2 mutants are active against EF-Tu and elicit innate immune responses in plants (Kunze et al. 2004). Cross-linking assays in *Arabidopsis* cells confirmed that elf18 and flg22 recognized different receptors, whereas they induced a small pool of genes. Further combined treatments of elf26/elf18 and flg22 did not show any additive effect of immune responses (Zipfel et al. 2006). Studies with EFR mutant further showed no response-like oxidative burst and increased ethylene synthesis, and resistance against *Pseudomonas syringe* pv. *tomato* (*Pst*) DC3000 was observed when challenged with EF-Tu derivatives. On the other hand *Arabidopsis* Col-0 and *fls2* mutant respond perfectly against EF-Tu elicitor. *N. benthamiana* does not have a perception system for Elf-Tu. Heterologous expression of EFR in *N. benthamiana* makes it active against EF-Tu elicitor and has been confirmed that it acts as the functional receptor (Zipfel et al. 2006). Further, the efficiency of *Agrobacterium tumefaciens (At)*-mediated transformation in EFR mutant has been increased as compared to the wild type. This indicates that EF-Tu recognition reduced the *Agrobacterium*-mediated transformation (Zipfel et al. 2006).

4.2.3 Peptidoglycan (PGN)

Gram-negative and Gram-positive bacteria cell envelopes have peptidoglycan which provides rigidity to their membranes. Since eukaryotes do not have PGN, therefore PGN becomes excellent target to host immune system (McDonald et al. 2005; Dziarski and Gupta 2006). PGNs are complex glycan structures which are hold together with oligopeptides. Chemically, they are composed of alternative N-acetylglucosamine (NAG) and N-acetylmuramic (NAM) with short peptides in between (Newman et al. 2013). Studies in tomato showed that pre-inoculation with *Staphylococcus aureus's* PGN reduced the bacterial infection in PGN treated tissues. *S. aureus* PGN was considered as an active elicitor as it induces extracellular alkalinization in cultured tobacco cells. No such response was observed in cultured tomato cells, indicating that PGN is perceived differently in *Solanaceae* (Felix and Boller 2003). Studies in *Arabidopsis* confirmed that PGN from Gram-negative and Gram-positive bacteria acts as an elicitor and evokes innate immune responses in plants (Gust et al. 2007; Erbs et al. 2008). Further, it has been shown that sugar backbone of Gram-positive PGN is responsible for the immune response in plants. Derivatives of PGN like muramyl dipeptide (MDP) or the muropeptide dimer are not responsible for triggering immune responses in plants, whereas these are easily perceived in insects and vertebrates (Traub et al. 2006). On other hand, PGN from two Gram-negative bacterial plant pathogens *Xcc* and *At* and their derivatives induces immune responses including ROS generation, PR1 gene expression, extracellular pH increase, and callous deposition (Erbs et al. 2008). Further it has been observed that derivatives like muropeptides are more effective as compared to the total PGN structures. Hence, we can say that PGN from Gram-negative and Gram-positive bacteria is perceived by different mechanisms. It could be due to different structure and recognition sites of muropeptides of human pathogen vs. plant pathogen. Studies with *Arabidopsis cerk1* mutant plant enhanced ROS response after treatment with PGN from Gram-negative *Pst* DC3000 and indicated its independent perception system for CERK1 (Newman et al. 2013).

4.2.4 Lipopolysaccharide (LPS)

Gram-negative bacteria have two-membrane architecture. It has an outer membrane followed by periplasmic space and an inner membrane. Each structure has its own function. Outer membrane contributes to the permeability as well as supports the growth of bacteria in unfavorable environments. LPS, a major component of the outer membrane, interacts with plants and plays a major role in inducing innate immune response (Newman et al. 1995; Bedini et al. 2005; Dow et al. 2000; Silipo et al. 2005). LPS consists of core oligosaccharides and a lipid O-polysaccharide part. Since like dissolves like, it is the lipid part which is embedded in the outer phospholipid bilayer of bacterial membrane and referred as lipid A. Lipid A is linked to the core oligosaccharide via 3-deoxy-D-manno-2-octulosonate (KDO). The core oligosaccharide is made up of short sugar and O-antigen. The composition of O-antigen is repeating oligosaccharide units (Raetz and Whitfield 2002). LPS of various phytopathogenic bacteria has O-antigen, composed of oligorhamnans (Bedini et al. 2002). To decipher the structure of LPS responsible to trigger immune response in plant, different O-antigen polysaccharides and different length of oligorhamnans were tested in *Arabidopsis*. It has been observed that PR genes including *PR1* and *PR2* were induced by tri-hexa- and nanosaccharides. Further, the increasing length of sugar and their coiled structure of O-antigen (synthetic) evoke plant innate immune responses ((Bedini et al. 2002). *Xcc* lipooligosaccaride (LOS) and its derivatives have shown the induction of *PR* genes in *Arabidopsis*. Therefore we can say that *Xcc* LOC and *Xcc* core oligosaccharides induce innate immunity in plants (Newman et al. 2013). However, in the case of tobacco, neither lipid A nor O-chain of LPS (*Xcc*) induces any innate immunity. Interestingly the inner core of the LPS was responsible for the induction of oxidative bursts (Braun et al. 2005). It has been observed that phosphorylation of lipid A affects its biological activity in mammalian system (Gutsmann et al. 2007). The same has been determined in *Arabidopsis* too. Studies with dephosphorylated *Xcc* LOS in *Arabidopsis* showed no localized induced responses (LIR) (Silipo et al. 2005). This result indicates that the phosphate group has its important function in the binding with specific receptor in plants. LPS also plays an important role in prime expression of defense response in plants, post bacterial infection. This includes synthesis of antimicrobial compound including feruloyl tyramine (FT) and p-coumaroyl tyramine (CT) (Newman et al. 2002, 2007; Conrath et al. 2006). Induced systemic response (ISR) was also observed in *Arabidopsis*, and it was the O-antigen of LPS responsible for such kind of response. This observation was further confirmed in an experiment where bacterial mutant lacking O-antigen was not able to induce ISR (van Loon et al. 1998). Apart from ISR, systemic acquired response is also observed in plants and is responsible for increase in salicyclic acid (SA) and related signaling (Schneider et al. 1996; Ryals et al. 1996). Recent study showed that it is not the necrotic lesion formation (which induces SAR in plants) but MAMPs, Flg, or LPS that is responsible for bacterial induction of SAR. Studies with *Pseudomonas aeruginosa* LPS, Flg, or nonhost bacteria in *Arabidopsis* showed induction of immune response such as accumulation of SA and *PR* gene expression as well as SAR marker, flavin-dependent monooxygenase 1 gene in treated as well as in nontreated distal leaves (Mishina and Zeier 2006, 2007). The exact downstream signaling system acquired response is not known clearly.

LPS is known to induce immune response in dicots. Recent studies in monocots like rice with LPS from pathogenic or nonpathogenic bacterium induced defense-related gene as well as ROS. These results indicate that monocot and dicot may have a common mechanism (Desaki et al. 2006, 2012). However, LPS from some bacteria showed programmed cell death, which is not elicited in the case of dicot (Desaki et al. 2006). The exact mechanism for LPS perception in plant is not known. However, fluorescein-labeled *Xcc* LPS showed that they are internalized in *N. tabacum* cells, similar to the endocytosis process in mammals (Gross et al. 2005). Still there is no specific PRR identified for perception of LPS in plant.

Different groups are attempted to identify PRR for LPS in plants. Transcriptomic analysis of *Arabidopsis thaliana* cells post *B. cepacia* LPS treatment showed that there was no induction of callous synthesis genes (Livaja et al. 2008). Moreover, genes involved in ROS production were also not upregulated. However, LPS from *B. cepacia* induced *PR3* and *PR4* gene expression in *Arabidopsis.* Other sources of LPS like *E.coli* and *P. aeruginosa* induced *PR1* and *PR5* expression in *Arabidopsis* leaves (Mishina and Zeier 2007). Differences in these results could be due to different plant system as well as different sources of LPS (Newman et al. 2013).

Plants also have the ability to modify LPS. This property is observed when lipid A or structures within LPS are altered in symbiotic interaction with plants (Kannenberg and Carlson 2001). These changes are responsible for the increase in the resistance against bacteria or to inhibit activity of lipid A in inducing host defense (Newman et al. 2013). Acylation and phosphorylation of lipid A are responsible for the induction of immune response in plants (Silipo et al. 2008). Lipid A from *Halomonas magadiensis*, an extremophilic and alkalinophilic Gram-negative bacterium, has low degree of acylation and is known to inhibit *E. coli* lipid A immune response in mammalian cells (Ialenti et al. 2006). Studies showed that differences in lipid A structure like acylation made them act as agonist or antagonist (Munford and Varley 2006). *H. magadiensis's* lipid A acts as antagonist for the action of *E. coli* lipid A in *Arabidopsis*. LPS for the experimental studies in plants is prepared from bacterial culture and grown in artificial medium. Alteration in LPS structure could occur when they multiply in plants. These changes could be important for the recognition and downstream signaling point of view. Mass spectroscopy and transcriptomic analysis of bacteria isolated from plants may provide fruitful information in this direction. Lipid A-like molecules are absent in plants like *Arabidopsis*; however, the six orthologous genes for the synthesis of lipid A, out of nine in *E. coli*, are present in *Arabidopsis*. Knockout studies of these genes showed that mutations are viable; however they are able to produce lipid A-like precursors (Li et al. 2011). It may be possible that higher plants may acquire lipid A biosynthetic genes from Gram-negative bacterium via endosymbiosis, and hence lipid A may play an important role in the structure of mitochondria or chloroplast membrane (Newman et al. 2013). Lipid A-like molecules may be involved in signaling in *Arabidopsis*, but the mechanism of LPS perception in plant is still unknown.

4.2.5 XA21-Mediated Immunity (Ax21)

XA21 receptor is identified in rice; however its ligand Ax21 has been discovered recently. Ax21 is present in *Xanthomonas* spp., *Xylella fastidiosa*, and *Stenotrophomonas maltophilia* (a human pathogen), and hence it is conserved in nature and has an important biological role. Ax21 is composed of 194 amino acids, and only 17-amino acid (sulfated) peptide is responsible for its biological activity (Bogdanove et al. 2011). *Xa21* gene is located in chromosome number 11 and encodes receptor kinase-like protein, having LRR transmembrane (TM), juxtamembrane (JM), and cytosolic kinase domain. *XA21* gene is responsible for providing resistance to a large number of *Xanthomonas oryzae* pv. *oryzae* (*Xoo*) species (Song et al. 1995; Wang et al. 1996). *XA21* is among the seven members of gene family. Among the seven, *XA21D* is closest to *XA21* and provides resistance to plants. However, they differ in the degree of resistance as *XA21D* provides mild resistance. Structural LRR domain of *XA21* and *XA21D* is 99% identical. However, *XA21D* lacks transmembrane and kinase domain (Song et al. 1997; Wang et al. 1998). Different proteins are known to interact with *XA21*. For example, ATPase XB24 binds and promotes XA21 phosphorylation and keeps it in inactive state. Later binding of Ax21 to XA21 dissociates XB24 from XB21/XA21 complex and activates XA21 (Chen et al. 2010).

4.3 Fungal and Oomycete MAMPs

4.3.1 Chitin

Fungal chitin and β-glucan from the *P. megasperma* act as MAMPs. In fungus, branched β-glucan is cross-linked with chitin, whereas in oomycetes cross-linking is done with cellulose. Chitin and its fragments are responsible for inducing immune response in plants. Fungal chitin is recognized by CEBiP and CERK1 (Kaku et al. 2006; Shimizu et al. 2010). Studies in rice showed that RLP CEBiP binds with chitin fragments at cell surface and interacts with LysM-RLK OsCERK1 for downstream signaling (Miya et al. 2007). In *Arabidopsis*, three CEBiP-like proteins have been reported, and they are designated as LYS1, LYS2, and LYSM. Heterologous expression of each in tobacco BY cells, followed by test for binding ability with chitin oligosaccharide, showed that LysM2 (AtCEBiP) have the highest affinity (Shinya et al. 2012). Biotinylated $(GlcNAc)_8$ binding studies with CEBiP prove its cell surface localization. However, single or triplet knockout of LysM1, AtCEBiP, and LysM3 as well as overexpression of AtCEBiP suggests that they are not responsible for signaling in *Arabidopsis* (Shinya et al. 2012). Therefore, we may conclude that rice and *Arabidopsis* have different receptors for chitin perception. In *Arabidopsis*, five LysM RLKs1-5 have been reported (Lyk1, Lyk2, Lyk3, Lyk4, and Lyk5) (Wan et al. 2012). Lyk1 is also known as CERK1. Knockout studies have been performed to determine their role in chitin signaling. It has been observed that in Lyk4, mutant immune response has been reduced. Moreover, Lyk mutant plants were more susceptible to *Alternaria brassicicola* and bacterial pathogen *Pst*

DC3000 (Wan et al. 2012). Apart from this, lysine motif-containing proteins, LYP4 and LYP6, are responsible for binding with chitin as well as PGN and may function as dual PRR in rice. Therefore, in rice, overlapping perception system does exist for fungal and PGN. However, LYM1 and LYM3, which are orthologs of LYP4 and LYP6 in *Arabidopsis*, recognize only PGN and are not able to recognize chitin (Willmann et al. 2011).

4.3.2 Ethylene-Inducing Xylanase (EIX) and AVE1 Peptide

Tomato was resistant against *Verticillium dahliae* and *V. albo-atrum* (Kawchuk et al. 2001). Genes responsible for the resistance lie in Ve loci, *Ve1* and *Ve2*. Both of them encode the surface receptors which belonged to receptor-like protein (RLP) class. Studies showed that among Ve1 and Ve2, Ve1 was found responsible for developing resistance. Further knockout studies were done with BAK1, which confer more susceptibility to tomato against *Verticillium* infection (Fradin et al. 2009). A putative ligand for Ve1 was AVe1 (avirulence on Ve1 tomato) peptide. It was observed that Ave1 was conserved among many fungus and plant pathogenic bacteria like *Xanthomonas axonopodis* pv. *citri*. Ave1 was also found to have homology with plant natriuretic peptides (PNPs) which is responsible for maintaining homeostasis in stress conditions (Wang et al. 2011). Ave1 acts as an elicitor and induces immune response via Ve1-mediated downstream signaling. Hence, we can say that Ave1 peptide acts as MAMP, and Ve1 acts as PRR (Thomma et al. 2011). There are other PRRs in tomato including RLPs S1Eix1 and S1Eix2. They also have homology with tomato Ve1 and Cf PRRs (Ron and Avni 2004). They are known to recognize fungal ethylene-inducing xylanase. EIX1 (β-1-4-endoxylanase) is proteinaceous in nature having molecular weight of 22 kDa. It is isolated from *Trichoderma viride* and is responsible for inducing innate immunity in tomato and tobacco. The primary sequence of S1Eix1 and S1Eix2 is 81.4% identical and has the ability to bind with EIX. However, they have different functions, such as S1Ex2 when binds with Ex1 induces innate immune response, while binding of Ex1 with S1Ex1 inhibits plant defense (Bar et al. 2010; Ron and Avni 2004).

4.3.3 Damage-Associated Molecular Pattern (DAMP)

Plant-derived molecules in certain cases are responsible to induce immune response. This system is similar to mammals, where immune system can detect danger through DAMPS (Seong and Matzinger 2004; Boller and Felix 2009) and is responsible for inflammatory response. Systemin, which is an 18-amino-acid peptide, induces immune response in plants (tomato) (Pearce et al. 1991; McGurl et al. 1992). Systemin is derived from its cytoplasmic protein precursor, prosystemin (Narvaez-Vasquez and Ryan 2004). Upon cell damage, systemin is released, which then acts as DAMP. Earlier studies showed RLK SR160, which is an ortholog of BRI1 of tomato, acts as a receptor for systemin (Scheer and Ryan 1999, 2002).

Putative cytoplasmic peptide (Pep1) in *Arabidopsis* is responsible for activating defense genes and alkalization in cell culture. Pep1 is a 23-amino-acid-long peptide and has seven homologues, which are derived from PROPEP1-7 (Huffaker et al. 2006). Pep1 is perceived by PEPR1 receptor which belongs to LRR X1 subfamily (Yamaguchi et al. 2006). On the basis of primary sequence, a second receptor for Pep peptide was found and designated as PEPR2. *PEPR1* and *PEPR2* transcription occurred under different conditions including wounding Pep peptide and specific MAMP. Binding studies with peptide and receptor showed redundancy, as Pep1 and Pep2 both bind with PEPR1 and PEPR2, whereas PEPR1 binds with Pep3-6 (Yamaguchi et al. 2010).

Cutin and oligogalacturonides (OGs) released from plant cell walls also act as DAMPs (Denoux et al. 2008; Brutus et al. 2010; Schweizer et al. 1996). Wall-associated kinase1 (WAK1) acts as a receptor for OG (Brutus et al. 2010). However, receptor for cutin is still not known. OG is known to evoke immune responses in plants by activation of MAPK, ROS production, callous deposition, and calcium release in cytoplasm as well as activation of defense genes (Chandra et al. 1997; Denoux et al. 2008). Another elicitor which comes in DAMP category in plants is the extracellular ATP (eATP). Surface-associated receptor for eATP perception does not respond to nucleotide 1 (DORN1) (Choi et al. 2014). It has been observed that a mutant of DORN1 suppressed transcriptional response to wounding and upregulated genes post eATP application, which was wound-inducible (Tanaka et al. 2014). eATP induces innate immune response as it is able to activate MAPK, Ca^{2+} influx, synthesis of JA, and ethylene as well as induce defense-related genes (Jeter et al. 2004; Song et al. 2006; Tanaka et al. 2014).

In *Arabidopsis*, HMGB protein AtHMGB3 is reported to act as DAMP. In *Arabidopsis* there are 15 genes, responsible to encode HMG-box domain-containing protein. They are further divided in four groups: (1) HMGB-type protein, (2) 3X HMGB which contains three HMG boxes, (3) A-/T-rich interaction domain (ARID-HMG protein), and (4) SSRP1, structural-specific recognition protein 1 (Merkle and Grasser 2011). On the basis of domain structure and nuclear location, there are eight different types including HMGB1/2/3/4/5/6/12/14. Among these HMGB2/3/4 are cytoplasmic as well as nuclear localized (Launholt et al. 2006; Pedersen and Grasser 2010; Merkle and Grasser 2011). Recombinant AtHMGB3 when infiltrated in leaves induced MAPK activation and deposition of callus, induced defense gene, and developed resistance against *Botrytis cinerea* (Choi et al. 2016).

SA binds to AtHMGB3 through its conserved Arg and Lys residues and abolishes its DAMP activity (Choi et al. 2016). However, SA is a positive regulator of immune response in plants and provides resistance to biotrophic and hemi-biotrophic pathogen. JA is responsible for providing resistance against necrotrophic pathogen as well as insects (Vlot et al. 2009; Dempsey et al. 2011). Signaling pathway of JA and SA is antagonistic (Thaler et al. 2012). Necrotrophic pathogen caused infection and cellular damage, which leads to the release of AtHMGB3 into extracellular space, which activated JA-/ethylene-associated defense genes. However, biopathogenic pathogen infection increased SA levels which suppressed AtHMGB3

DAMP-mediated immune defense and activated assay-associated defense gene expression (Choi et al. 2016).

4.3.4 Nematode-Associated Molecular Pattern (NAMP)

Nematodes are known to parasitize both plants and animals (Lambert et al. 1999; Vercauteren et al. 2001; Kyndt et al. 2012). However, elicitor for plants is not known. Recently a group of defense signaling molecules from root-knot and cyst nematodes have been identified (Manosalva et al. 2015). They are conserved nematode pheromones called ascarosides. The most abundant ascaroside is Ascr#18. It is responsible to induce innate immune response in the plant including activation of MAPKs, defense genes, SA, and JA defense signaling pathways. It is responsible for enhanced resistance against viral, bacterial, fungal, and oomycete pathogen and root-knot nematodes in different monocot and dicot plants (Choi and Klessig 2016).

4.4 Binding of Bacterial PAMPs with Plant PRRs

Perception of PAMPs requires either homodimerization, heterodimerization, or heteromultimerization of PRRs. Recent advances in biochemical structure and genetic studies open the new window to know about the molecular mechanisms underlying PAMP binding to plant PRR.

4.5 Heterodimerization: Flagellin Perception in Arabidopsis

In *Arabidopsis*, bacterial flagellin is recognized by PRR, FLS2, which is LRR-RLK. Among full-length bacterial flagellin, the conserved 22 amino acids behave like epitope flg22 (Gomez-Gomez and Boller 2000). The FLS2 ectodomain has 28 LRRs and directly binds to flg22 (Chinchilla et al. 2006). FLS2 exists in dimer form even in the absence of elicitor, but its relevance is not clear yet (Sun et al. 2012, 2013a). Although FLS2 is conserved in most of the plant species, still there are lots of differences in specificity of recognition (Boller and Felix 2009). This has been proved in a study where a shorter peptide flg15 behaves as an agonist in tomato, whereas it acts adversely in *Arabidopsis* (Mueller et al. 2012; Bauer et al. 2001; Felix et al. 1999; Robatzek et al. 2007). Comparative and mutagenic studies of both FLS2 LRRs and flg peptides have identified possible LRRs involved in flg22 recognition (Dunning et al. 2007; Helft et al. 2011; Mueller et al. 2012). Further studies have shown that the N-terminal part of flg22 is responsible for receptor binding, whereas C-terminal part is a must for the activation of the immune response (Meindl et al. 2000; Sun et al. 2013a). Flg22 perception directs the FLS2 to form a complex with regulatory LRR/RLK and BAK1/SERK3 (Chinchilla et al. 2007; Schulze et al. 2010). This indicates the close proximity of FLS2 and BAK1 in the plasma membrane. It has been further proved by structural studies that the N-terminal of flg22

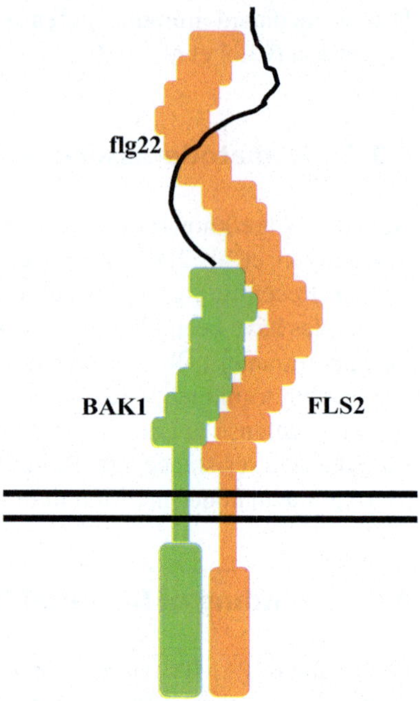

Fig. 4.1 The flg22 peptide perception in *Arabidopsis* directs the ectodomain of LRRs of FLS2 to form a stable heterodimer with the co-receptor BAK1

binds to the concave surface of FLS2 ectodomain and the C-terminal region of the FLS2-bound flg22 interacts with BAK1 ectodomain (Sun et al. 2013a). This interaction leads to the stabilization of FLS2-BAK1 dimerization, and thus FLS2-BAK1 heterodimerization is both ligand and receptor mediated as shown in Fig. 4.1. Hence, it is proven that flg22 binds to FLS2 first and later with BAK1; therefore BAK1 acts as a co-receptor for flg22 and is essential for signaling (Chinchilla et al. 2007; Sun et al. 2013a). Similar mechanism has been reported recently in *Arabidopsis* where LRR-RLK BRI1 (the receptor for the plant hormone brassinosteroid, which regulates growth and development) binds with BAK1 or the BAK1-related protein SERK1 (Santiago et al. 2013; Sun et al. 2013b). These results suggest that different LRR-containing RLKs (and RLPs) may be involved in similar heterodimeric complexes with BAK1 or related SERK proteins. The increasing number of newly identified LRR-RLKs and LRR-RLPs has shown their association with BAK1 and related SERK proteins. However some mechanistic differences exist between different plant species. For example, in rice (*Oryza sativa*) ortholog of BAK1, OsSERK2, forms a complex with the LRR-RLK XA21, and thus it confers resistance against *Xanthomonas oryzae* pv. *oryzae* (Chen et al. 2014). Due to the absence of confirmed ligand for XA21, it is not clear yet if the interaction between XA21-OsSERK2 is enhanced after PAMP perception or not (Bahar et al. 2014). FLS2-BAK1 dimerization is independent of kinase activity, whereas the association of OsSERK2 with the intracellular domains of ligand-binding receptors is kinase dependent (Schwessinger et al. 2011; Chen et al. 2014; Sun et al. 2013b).

4.6 Binding of Fungal PAMPs with Plant PRRs

Similar to bacterial PAMPs, fungal PAMPs also need homodimerization, and heteromultimerization.

4.6.1 Homodimerization: Chitin Perception in Arabidopsis

In *Arabidopsis thaliana* LysM-RLK CERK1/RLK1/LYK1 is required for chitin perception (Wan et al. 2008; Miya et al. 2007). Three extracellular LysM domains of CERK1 are responsible for the binding with oligomers of fungal chitin (Liu et al. 2012; Petutschnig et al. 2010; Miya et al. 2007). Seven to eight GlcNAc residues which constitute long chitin oligomers act as a bivalent ligand for CERK (Liu et al. 2012). This binding leads to the homodimerization of the CERK1 as shown in Fig. 4.2, which acts as an active receptor complex responsible for initiating chitin-induced immune signaling. However CERK1 is also able to bind with four to five GlcNac residues (shorter chitin oligomer), but their interaction does not induce CERK1 homodimerization, and hence does not trigger any immune responses (Liu et al. 2012). These studies highlighted that homodimerization of CERK1 is a must for signaling initiation, by bringing together CERK1 cytoplasmic domains, which contain an active kinase (Petutschnig et al. 2010), enabling intermolecular transphosphorylation. While CERK1 is the major chitin-binding protein in *Arabidopsis* and is strictly responsible for chitin-triggered immune responses, the other related LysM-RLK LYK4 is also known for chitin binding and is involved in chitin perception (Miya et al. 2007; Petutschnig et al. 2010; Wan et al. 2008, 2012).

4.6.2 Heteromultimerization: Chitin Perception in Rice

LysM-RLP CEBiP is the major chitin-binding protein in rice (Kouzai et al. 2014; Kaku et al. 2006). CEBiP is a GPI-anchored protein with three extracellular LysM domains along with a C-terminal tail (Hayafune et al. 2014; Kaku et al. 2006). Binding of long chitin oligomers with OsCEBiP leads to its homodimerization,

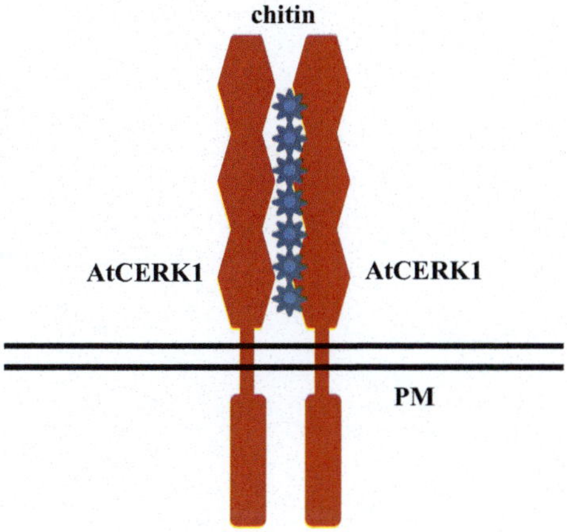

Fig. 4.2 Perception of seven to eight GlcNAc residues in *Arabidopsis* act as bivalent ligands and direct the formation of active receptor complex by homodimerization of AtCERK1

which resembles with that of chitin-binding mechanism shown for *Arabidopsis* chitin receptor AtCERK1 (Hayafune et al. 2014; Liu et al. 2012). As like other RLPs, CEBiP does not have cytoplasmic C-terminals responsible for signaling motifs, and therefore it requires assistance of additional proteins to initiate signaling (Kaku et al. 2006). In the presence of biologically active chitin, CEBiP is also known to form a hetero-oligomeric receptor complex with OsCERK1, which is the rice ortholog of AtCERK1 (Shimizu et al. 2010). OsCERK1 has only one single extracellular LysM domain and hence does not bind to chitin but is essential for chitin-mediated signaling (Shimizu et al. 2010). These results showed that the chitin perception system in rice requires hetero-oligomeric receptor complex formed by dimers of an elicitor-binding CEBiP (LysM-RLP) and a nonligand-binding signaling-active OsCERK1 (LysM-RLK) as shown in Fig. 4.3 (Kaku et al. 2006; Shinya et al. 2012). This confirmation of hetero-oligomer receptor form a sandwich type receptor for chitin oligomers (Hayafune et al. 2014). Apart from CEBiP (OsLYP4 and OsLYP6), the LysM-RLPs is also known for chitin binding and responsiveness (Liu et al. 2012). In *Arabidopsis*, CEBiP orthologs are not required for classical chitin immune responses, including ROS burst or immune gene expression (Shinya et al. 2012; Wan et al. 2012). However, the closest ortholog of CEBiP in *Arabidopsis*, AtLYM2, is known to bind chitin (Petutschnig et al. 2010; Shinya et al. 2012). The localization of AtLYM2 is in plasmodesmata and AtLYM2 is responsible for the closure of plasmodesmata in CERK1-independent manner but upon chitin perception chitin-induced. In addition AtLYMs also provides resistance to fungal pathogens (Faulkner et al. 2013; Narusaka et al. 2013). Therefore, in *Arabidopsis* some localized cellular responses which are initiated by chitin are also involved in hetero-oligomerization between a chitin-binding LysM-RLP (AtLYM2) and an unknown LysM-RLK

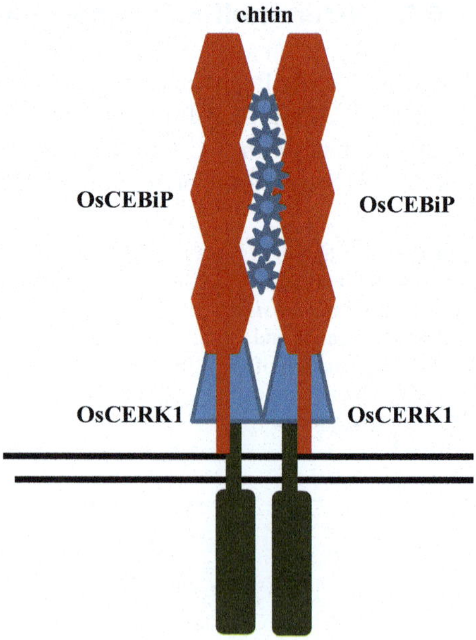

Fig. 4.3 Chitin perception in rice requires a multimeric receptor formed by dimers of OsCEBiP and OsCERK1

similar to CERK1. OsLYP4 and OsLYP6 (CEBiP paralog) are also known to bind with the bacterial cell wall component (peptidoglycan, PGN) (Liu et al. 2012). In *Arabidopsis*, AtLYM1 and AtLYM3 (orthologs of CEBiP) are known to bind with PGN and not with chitin (Willmann et al. 2011). Here CERK1 is also required for PGN-induced responses without binding with PGN itself (Willmann et al. 2011). These results show that CERK1 plays a multifaceted role, as it is able to function as a ligand-binding PRR for chitin and also as a positive regulator of PGN responses. In *Arabidopsis* the complex between AtLYM1, AtLYM3, and AtCERK1 has not been confirmed yet biochemically; however these results indicate that PGN perception system in *Arabidopsis* resembles with that of rice chitin receptor, which involves hetero-oligomeric complex of ligand-binding RLPs and RLKs.

4.6.3 Perception of Fungal Xylanase and Ave Peptide

In tomato the only RLP involved in PAMP perception receptor for the fungal ethylene-inducing xylanase (EIX) is LeEIX1/2, encoded by the genes *LeEIX1/2*, as shown in Fig. 4.4a (Ron and Avni 2004). Both LeEIX1 and LeEIX2 are reported to bind with EIX independently. However it is only LeEIX2 that can transduce the signal when expressed transiently in tobacco. In *Arabidopsis*, a few additional RLPs

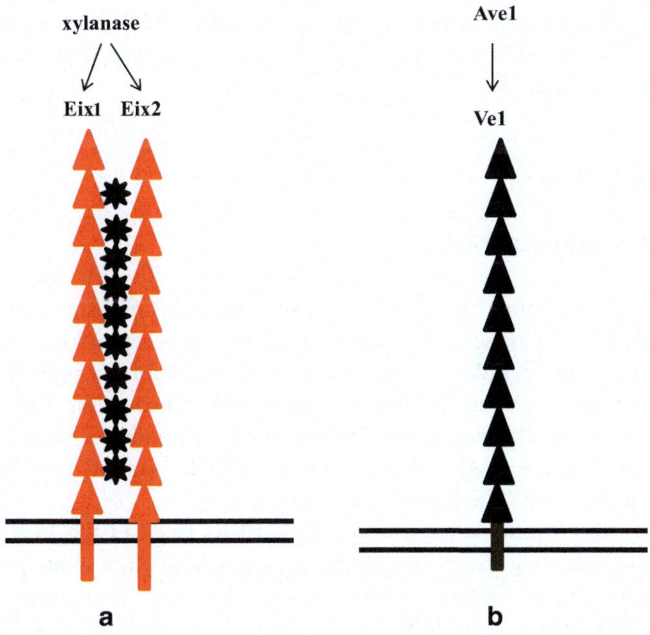

Fig. 4.4 (**a**) Perception of fungal xylanase by Eix1/2 PRR and (**b**) Ave peptide by Ve PRR in tomato

have been identified using reverse-genetic approaches, which linked to innate immunity. It has been reported that the chitin-inducible RLP52 gene mutant is more susceptible to adapted as well as non-adapted powdery mildews (Ramonell et al. 2005). Similarly, atrlp30 as well as atrlp18 mutants are found susceptible to the non-adapted bacterium *Pseudomonas syringae* pv. *phaseolicola 1448A* (Wang et al. 2008). In addition to this, tomato LRR-RLP Ve1 mediates resistance to fungus *Verticillium* by recognizing a peptide Ave1 which is conserved in several fungi and bacterium *Xanthomonas axonopodis* as shown in Fig. 4.4b. Ve1 is also reported to mediate resistance against *Fusarium oxysporum* (Monaghan and Zipfel 2012).

4.7 Plant Viruses and PRRs

In animal systems, viral patterns inducing PTI are well known, but nothing similar has been reported from plants so far. In plants, antiviral defense is mediated by well-studied posttranscriptional gene silencing of viral RNA as well as through effector-triggered immunity, which includes the recognition of virus-specific effectors by resistance proteins. In *Arabidopsis* the regulator BAK1 plays an important role for antiviral defense as *Arabidopsis* bak1 mutants show increased susceptibility to different RNA viruses during compatible interactions. It is also studied that crude viral extracts induce several PTI marker responses in a BAK1-dependent manner; however, purified virions don't. Hence, we may say that BAK1-dependent PTI contributes to antiviral resistance in plants; however specific PAMP for specific virus and its specific PRR are still not known, and further investigation is required in this direction (Korner et al. 2013).

4.8 Signaling

4.8.1 Activation of PRRs

The PRRs either contain a cytoplasmic kinase domain or are associated with RLKs. The binding of ligand to the extracellular domain of the receptor kinases leads to the activation of the intracellular kinase domain and which further phosphorylates the substrates, which contribute to intracellular signal transduction. Identified LRR-RLKs have an intracellular kinase domain with the potential for signaling, till they require dimerization with BAK1 (or a related SERK protein) post binding with ligand to transduce the signal as shown in Fig. 4.5.

BAK1 is a RD kinase, whereas most RLK PRRs (e.g., FLS2, EFR, XA21) are non-RD kinases, as they are having another amino acid in place of an otherwise conserved arginine residue in the catalytic loop of the kinase domain (Dardick et al. 2012). Non-RD kinases seem to be associated with immune functions across kingdoms, as the association between non-RD and RD kinases for the initiation of PTI is similar to *Drosophila* and humans (Dardick et al. 2012). It has been observed that

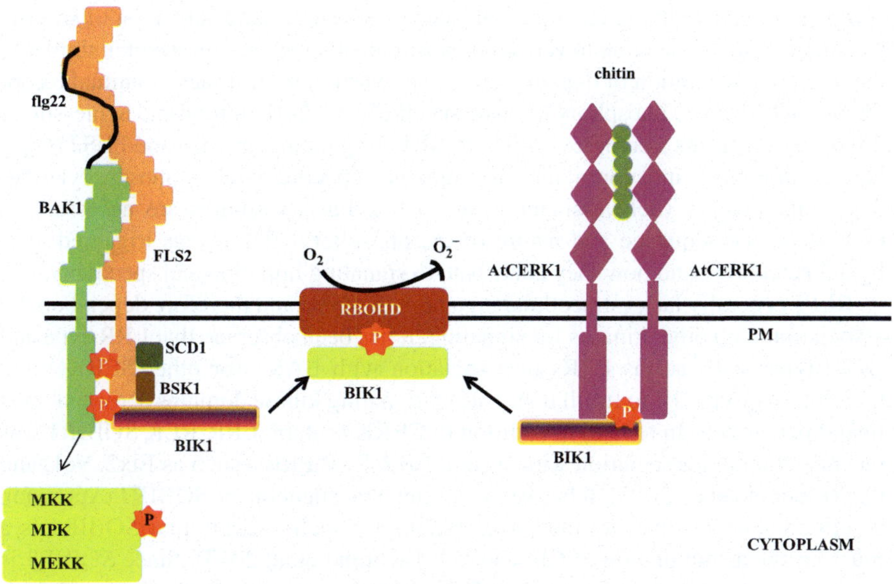

Fig. 4.5 The flg22 peptide perception in *Arabidopsis* phosphorylates the cytoplasmic domains of FLS2 and BAK1, as well as the receptor like cytoplasmic kinase BIK1. Phosphorylated BIK1 gets released from the receptor complex, and further phosphorylates and activates NADPH oxidase AtRBOHD. In addition to this, RLCK BSK1 and the endocytosis regulator SCD1 are also required for the Flg22-mediated ROS burst. Similarly, BIK1 also get phosphorylated followed by chitin perception and activation of AtCERK1 and is required for the ROS burst. The activation of MAPKs and other downstream substrates followed by flg22 perception is not clear yet

FLS2 or EFR kinase domains have weaker kinase activity as compared to other RD kinases (e.g., BAK1 or BRI1) (Schwessinger et al. 2011); therefore it may be possible that non-RD ligand-binding RLK PRRs require the association of a strong RD kinase (such as BAK1) to enhance the phosphorylation process and initiate signaling (Dardick et al. 2012). This has been further proved in a mutagenesis study where mutations that impair complex formation between FLS2 and BAK1 eliminate the phosphorylation of both proteins as well as initiation of downstream signaling (Sun et al. 2013b). Hence, it shows that kinase activities of FLS2 and EFR are essential for flg22- and elf18-mediated responses (Cao et al. 2013; Schwessinger et al. 2011). In *Arabidopsis*, CERK1 itself contains an RD kinase domain and therefore does not require BAK1 to initiate chitin-triggered signaling (Gimenez-Ibanez et al. 2009; Heese et al. 2007). However, signaling mediated by RD RLKs, including BRI1 or the PRRs PEPR1/PEPR2, still needs BAK1 (and other SERKs) (Gou et al. 2012; Krol et al. 2010; Li et al. 2002; Nam and Li 2002; Roux et al. 2011) indicating that SERK proteins are essential for the functioning of both RD and non-RD ligand-binding RLKs. SERKs act as kinase activity enhancers and are responsible for phosphorylation event in *cis* and *trans*, within PRR complexes. These phosphorylated PRR complexes regulate the interaction with specific substrates and activation of specific signaling branches, leading to various downstream responses. This event is similar to the observed generation of docking sites for downstream substrates of animal receptor kinases driven by phosphorylation on specific residues (Lemmon and Schlessinger 2010). BAK1 forms a complex with both BRI1 (regulating growth) and FLS2 (regulating immunity) and contributes to signaling specificity in a phosphorylation-dependent manner (Schwessinger et al. 2011). Further studies are necessary to understand the sequence and nature of the phosphorylation events triggered after ligand perception and how they contribute to signaling initiation and specificity.

RLPs generally lack intracellular signaling domains and therefore depend on the association with other kinases for signaling. It has been observed that LRR-type and LysM-type RLPs act as PRRs in association with BAK1 (or other SERKs) and CERK1, respectively, to fulfill their role of signaling kinase domains activated, post ligand perception. In tomato in addition to CERK protein, LRR-RLK SOBIR1 was recently found in association with several LRR-RLP PRRs, such as Eix2, Ve1, and Cf4 (Liebrand et al. 2013). It has been reported that silencing of SOBIR1 expression compromises Cf4- and Vel-mediated responses which indicate that SOBIR1 is a must for the accumulation of Cf4 and Vel (Liebrand et al. 2013). Since SOBIR1 is localized in the cytoplasmic vesicles, they may be necessary for adequate trafficking and providing stability to LRR-RLP-containing complexes (Liebrand et al. 2013). In *Arabidopsis* also, SOBIR1 is a must for proper functioning of several RLPs involved in innate immunity, which indicates that SOBIR1 is a common regulator of LRR-RLP PRRs in different plant species (Jehle et al. 2013; Zhang et al. 2013). Till date it is not clear how SOBIR1 regulates LRR-RLP accumulation. It has been observed that in rice, PRR XA21 (an LRR-RLK) autophosphorylation at several Ser and Thr amino acids subjected to phosphorylation-dependent mechanism is controlling protein stability (Xu et al. 2006).

Apart from this, for the accumulation of XA21, several non-kinase proteins are also required. It has been reported that ubiquitin ligase XB3 and XB25 (plant-specific ankyrin repeat) proteins interact with XA21 in plants as shown in Fig. 4.6. These are further transphosphorylated by XA21 kinase domain (Jiang et al. 2013; Wang et al. 1996). XB3 or XB25 is crucial for immune response as the reduced expression of XB3 or XB25 results in reduced accumulation of XA21 which consequently compromises XA21-mediated immunity (Wang et al. 2006; Jiang et al. 2013).

Recently it has been observed that the additional RLKs can be a part of PRR complexes. The LRR-RLK BIR2 is found to interact with BAK1 even in the absence of PAMP perception (Halter et al. 2014). The nature of BIR2 is a pseudo-kinase, and it negatively regulates BAK1 interaction with FLS2. Later during Flg22 perception, BIR2 dissociates from BAK1 which ultimately allows FLS2-BAK1 dimerization and downstream signaling (Halter et al. 2014).

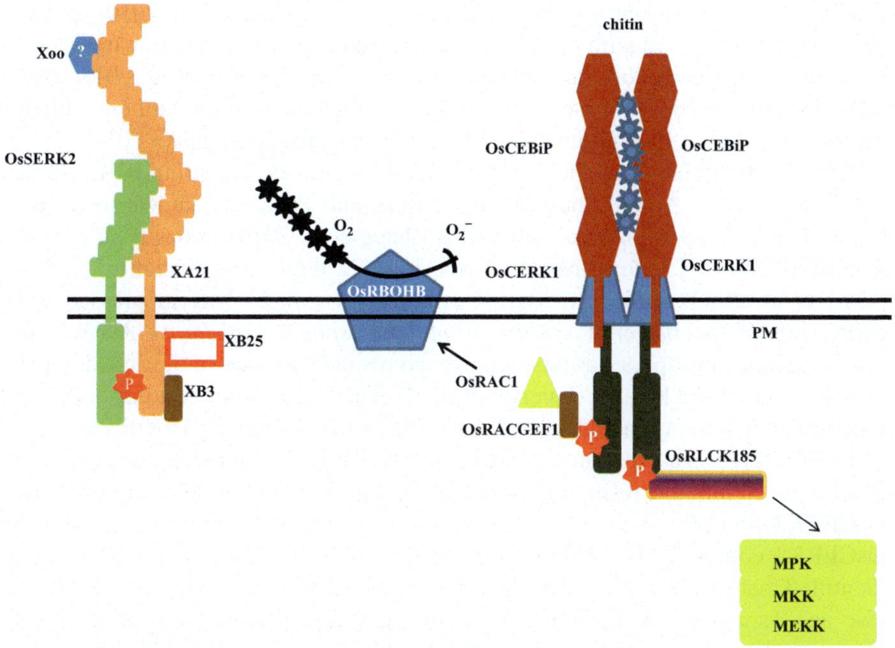

Fig. 4.6 For the stability of XA21 in rice, it requires the E3 ligase XB3 as well as the PANK protein XB25 and forms a constitutive complex with OsSERK2. However the exact mechanisms of signal transduction followed by perception of *Xanthomonas oryzae* pv. *oryzae* (Xoo) is not known. The heteromultimeric chitin receptor complex formed by dimeric OsCEBiP and OsCERK1 associates with various cytoplasmic proteins, which are required for signal transduction. Post chitin perception, OsCERK1 phosphorylates receptor like cytoplasmic kinase OsRLCK185, which gets dissociated from the receptor complex after phosphorylation and further responsible for the activation of MAPKs. In addition, OsCERK1 also activates the Os-RacGEF1/OsRac1 module, which is responsible for the activation of chitin-induced ROS burst

4.8.2 Downstream Events

For the downstream intracellular signaling, surface-associated PRRs and associated transmembrane proteins require cytoplasmic partners. Recently, receptor-like cytoplasmic kinases (RLCKs) have emerged as key substrates of PRR complexes and positive regulators of PTI signaling. In the absence of flg22 in *Arabidopsis*, RLCK BIK1 is associated with FLS2 and BAK1 (Lu et al. 2010; Zhang et al. 2010). Post flg22 perception, BIK1 gets phosphorylated by BAK1, which later phosphorylates both FLS2 and BAK1, and finally gets dissociated from the FLS2- BAK1 complex as shown in Fig. 4.5 (Lu et al. 2010; Zhang et al. 2010). BIK1 is also phosphorylated after elf18 perception and interacts with EFR and CERK1 and, therefore, plays a key role in downstream signaling (Lu et al. 2010; Zhang et al. 2010). This is further supported by mutagenesis studies, which showed that bik1 mutants are compromised in immune responses triggered by flg22, elf18, and chitin. Hence, they are more susceptible to the pathogenic fungus *Botrytis cinerea* and non-adapted bacterial pathogens

(Laluk et al. 2011; Veronese et al. 2006; Lu et al. 2010; Zhang et al. 2010). BIK1 is also reported to interact with PEPR1 and is involved in PTI amplification mechanism involving the gaseous hormone ethylene (Liu et al. 2013; Tintor et al. 2013; Zipfel 2013; Laluk et al. 2011). These results further highlighted the important role of BIK1 for the activation of PRR complexes in *Arabidopsis* innate immunity.

RLCK-VII subfamily contains more than 46 members, including BIK1, PBL1, PBL2, and PBL5. Among these, PBL1, PBL2, and PBL5 are known to regulate flg22-induced ROS burst (Liu et al. 2013; Zhang et al. 2010). It has been reported that BIK1 and PBL1 don't have any role in flg22-induced activation of MAPKs (Feng et al. 2012). However, MAPK activation via flg22 was suppressed upon expression of the bacterial *uridine 50-monophosphate transferase* AvrAC. The AvrAC inhibits phosphorylation in conserved residues located in the activation site of BIK1 and related kinases (Feng et al. 2012). These results suggest that additional BIK1-related proteins, including PBL2 and PBL5, may be required for this response.

OsRLCK185 is a substrate of OsCERK1 in rice and responsible for chitin- and PGN-induced immune responses (Yamaguchi et al. 2013). It has been observed that OsCERK1 phosphorylates OsRLCK185, which later partly dissociates from the OsCERK1 complex post chitin perception (Yamaguchi et al. 2013). Earlier it was identified that RLCK BSK1 acts as a substrate of brassinosteroid receptor BRI1 and also acts as a positive regulator of brassinosteroid responses (Tang et al. 2008). However, recently, it was found that BSK1was also associated with FLS2 and partially dissociated after flg22 perception. BSK1 was also required for a subset of flg22-triggered responses, but not for the activation of MAPK (Shi et al. 2013). BSK1 is a common positive regulator for both BRI1- and PRR-mediated signaling, but interaction of BIK1 with BRI1 acts as a negative regulator of brassinosteroid-triggered responses (Lin et al. 2013). After brassinosteroid perception, BRI1 phosphorylates BIK1 independently of BAK1, and this leads to the release of BIK1 from the BRI1 receptor complex (Lin et al. 2013). However antagonism exists between the BRI1 and FLS2 pathways, due to the indirect crosstalk by transcriptional regulator BZR1 and not caused by competition between common regulators at the plasma membrane (Malinovsky et al. 2014; Lozano-Duran et al. 2013; Albrecht et al. 2012; Belkhadir et al. 2012). Different PRRs recruit distinct RLCKs, such as BIK1, PBL1, PBL2, and PBL5, and are responsible for FLS2-mediated ROS burst, whereas only BKI1 and PBL1 are responsible for ROS burst mediated by PEPR1/2 activation (Liu et al. 2013; Zhang et al. 2010). Similarly, BSK1 is responsible for flg22-mediated ROS burst but not for elf18 (Shi et al. 2013). The involvement of different RLCKs for specific PRR responses indicates that the choice of specific RLCKs as PRR substrate constitutes another layer in the regulation of signaling, branching from PRR complexes. ROS burst and MAPK activation are the indicating responses to analyze signaling branching from PRR complexes, as both responses are traceable after PAMP treatment (<5 min) and constitute independent signaling (Ranf et al. 2011; Segonzac et al. 2011; Xu et al. 2014). The important role of RLCKs is in the initiation and specificity of PTI signaling and has initiated the search for different RLCK substrates. Recently NADPH oxidase, AtRBOHD, has been identified as direct target for BIK1 as shown in Fig. 4.5 (Kadota et al. 2014). It has been observed that AtRBOHD is the key enzyme responsible for the rapid production of apoplastic ROS upon PAMP perception (Nuhse et al. 2007; Zhang et al. 2007). For example, flg22 treatment leads to the activation of BIK1, which directly phosphorylates and activates AtRBOHD, leading to the further ROS burst and subsequent immunity (Kadota et al. 2014).

Apart from RLCKs, other substrates of PRR complexes have been identified to play an important role in PTI signaling. In rice, chitin treatment leads to OsCERK1-mediated phosphorylation of the Rac GDP/GTP exchange factor 1 (OsRacGEF1), which further activates OsRac1 as shown in Fig. 4.6 (Akamatsu et al. 2013). Therefore the activation of the OsCEBiP/OsCERK1 complex and OsRacGEF1/OsRac1 module together is essential for chitin-triggered responses and resistance against fungal pathogens in rice (Ono et al. 2001; Suharsono et al. 2002; Akamatsu et al. 2013).

4.8.3 Attenuation

Constitutive or hyperactivation of immune responses is always detrimental for plant growth. Therefore PRR and their immediate downstream signaling must be tightly regulated before and post ligand perception. The exact mechanism for the attenuation of PRR activation is not known exactly. Regulation mechanisms ensure that PRR complexes return to their steady state and are ready to get activated in case of further pathogen attack. As discussed earlier, phosphorylation is the key for the activation of many PRR complexes. Therefore negative regulation of PRR complexes is possible by protein phosphatases. Several phosphatases have been shown to have their association with PRR and/or with associated kinases, to keep the complex in inactive form through dephosphorylation in the absence of ligand binding. In rice, XA21 is kept inactive before ligand binding through the association with ATPase XB24, which promotes XA21 autophosphorylation at specific Ser-Thr sites. Binding of Ax21 to XA21 leads to the disassociation of XB24, thus further allowing XA21 activation and immune responses. Post activation, a phosphatase XB15 known as POL-type protein phosphatase 2C (PP2C) acts on XA21 and dephosphorylates it, further leading to its inactivation as shown in Fig. 4.7. Recently, novel phosphatases have been identified that regulate pre-ligand and post-ligand binding in FLS2/EFR complexes. A key regulatory aspect of surface-associated RLKs is their degradation after post-ligand binding and their subsequent replenishment via de novo synthesis at the plasma membrane. It has been reported that post flg22 perception, FLS2 is subjected to endocytosis and degradation (Lu et al. 2011; Robatzek et al. 2006; Smith et al. 2014). FLS2 degradation is controlled by two E3-ubiquitin ligases, PUB12 and PUB13. They exist in a constitutive complex with BAK1 and are therefore recruited into the FLS2 complex after flg22 binding takes place as shown in Fig. 4.8. It has been observed that phosphorylated PUB12/PUB13 by BAK1 is responsible for polyubiquitination of FLS2, and this polyubiquitination further leads to FLS2 degradation as shown in Fig. 4.8. However the exact role of PB12 and PB13 in FLS2 endocytosis is currently unknown. The negative regulation mediated by PUB12/PUB13 is highlighted in a study which proved that loss of PUB12/PUB13 results in heightened flg22-induced responses and enhanced resistance to PtoDC3000 (Lu et al. 2011; Robatzek et al. 2006). Additional PUBs,

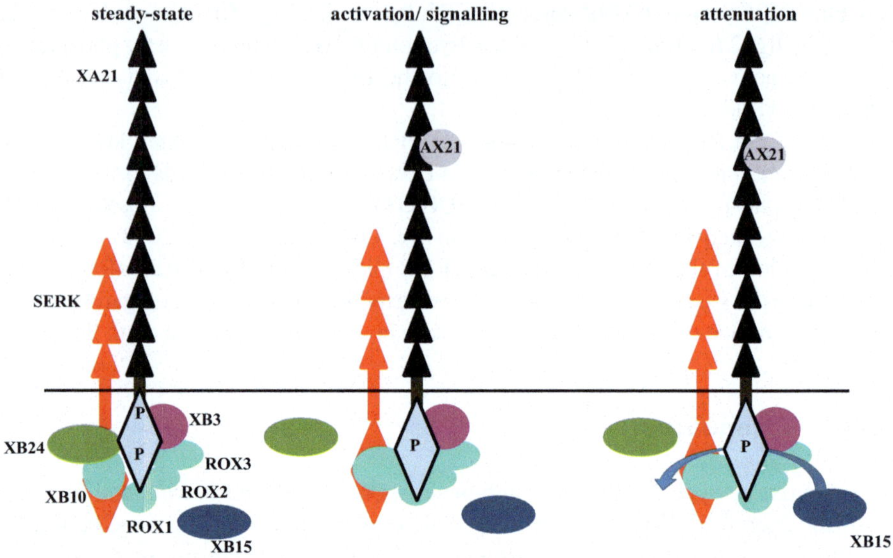

Fig. 4.7 The XA21 complex. In rice, constitutive interaction of XA21 has been observed with the E3-ligase XB3, the ATPase XB24, the PP2C XB15, the WRKY transcription factor XB10, the NOL1/NOL2/sun protein ROX2, the nudC-like nuclear migration factor ROX3 and the thiamine pyrophosphokinase ROX1. XB24 keeps XA21 in an inactive state by inducing auto-phosphorylation at specific Ser-Thr residues. Ligand binding is responsible for XB24 dissociation from XA21, followed by phosphorylation of XA21 at distinct residue(s), which further activates immune signaling. XB15, a phosphatase, further dephosphorylates XA21 and suppress the signaling. Phosphorylation is shown as labelled 'P', and arrows shows the direction of post-translational modifications

PUB22/PUB23/PUB24, were also reported as negative regulators of PTI responses, and their mechanism of regulation is still not known and therefore kept in separate clad forms, PUB12/PUB13 (Trujillo et al. 2008).

Degradation of FLS2, mediated by ligand-induced endocytosis, plays a key role to prevent continuous signaling from the activated surface-associated receptors (Smith et al. 2014). De novo synthesized FLS2 are incorporated into the plasma membrane to replenish the degraded FLS2, so that it restores the sensitivity of the cell for the upcoming pathogen attack (Smith et al. 2014). Hence appropriate PRR trafficking, after ligand perception, is essential for specific signal responses. During cytokinesis and cell expansion, DENN domain protein SCD1 is responsible for clathrin-mediated endocytosis (McMichael et al. 2013). It has been also reported that SCD1 also interacts with FLS2 and is essential for flg22-triggered immune responses (Korasick et al. 2010). Studies with chemical inhibitors of vesicular trafficking impaired flg22-triggered immune responses, suggesting that there is a potential link between endocytosis and flg22-mediated immune responses (Smith et al. 2014).

In *Arabidopsis* flagellin perception through FLAGELLIN-SENSITIVE 2 (FLS2) induces the activation of mitogen-activated protein kinases (MAPKs) and provides immunity. The precise molecular mechanism that connects activated FLS2 to

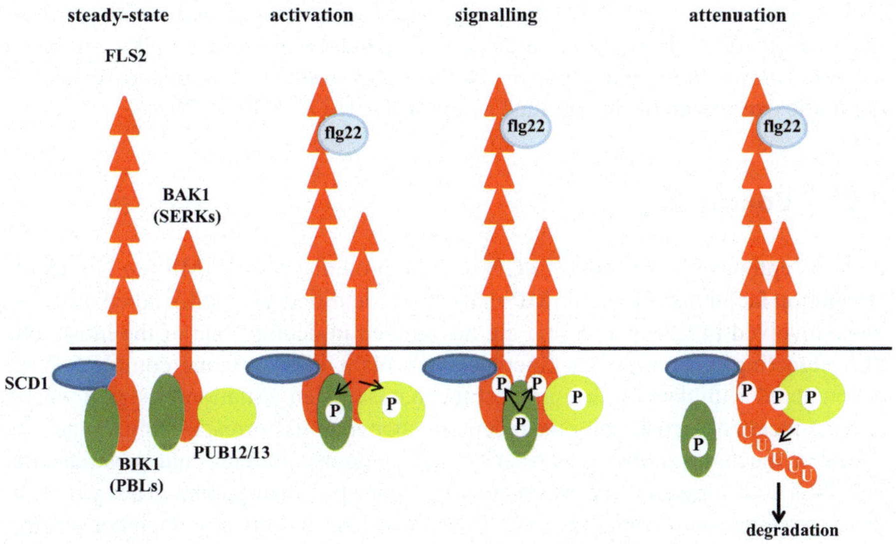

Fig. 4.8 The FLS2 heterodimeric complex. In *Arabidopsis*, constitutive interaction of FLS2 with BIK1 and related PBLs, as well as with SCD1 has been observed. BAK1 interacts with PUB12 and PUB13 constitutively and may associate with BIK1. Post flg22 binding, BAK1 interact with FLS2 and trans-phosphorylates BIK1 (and other PBLs) and PUB12/13. Phosphorylated BIK1 then trans-phosphorylates BAK1 and FLS2, resulting in a downstream signaling. Phosphorylated PUB12/13 then polyubiquitinate FLS2 resulting in its degradation and suppression of signaling. Phosphorylation is shown as labelled 'P', ubiquitination is as shown as labelled 'U', and arrows shows the direction of post-translational modifications

downstream MAPK cascades is still under investigation. Recently it has been identified that differentially phosphorylated MAP kinase also interacts with FLS2 (Mithoe et al. 2016). With the help of targeted proteomics and functional analysis, it has been observed that phosphorylated MKKK7 on specific serine residues negatively regulates flagellin-triggered signaling and basal immunity. MKKK7 is responsible for the attenuation of MPK6 activity and defense gene expression. Moreover, MKKK7 is also responsible for the suppression of the reactive oxygen species burst downstream of FLS2, indicating that MKKK7-mediated attenuation of FLS2 signaling occurs through direct modulation of the FLS2 complex (Mithoe et al. 2016).

In *Arabidopsis*, RLCKs have also been reported to play a key role in the regulation of plant innate immunity. The PBL13 kinase negatively regulates plant innate immunity to pathogenic bacteria *Pseudomonas syringae* and is also associated with RBOHD before pathogen perception (Lin et al. 2015). Hence these findings are consistent with the hypothesis that PBL13 acts to prevent inappropriate activation of defense responses in the absence of pathogen challenge. Similarly a recent study in rice related to functional characterization of four new rice RLCKs from subfamily VII – OsRLCK57, OsRLCK107, OsRLCK118, and OsRLCK176 – has been done. It has been observed that these OsRLCKs interact with the rice brassinosteroid receptor, OsBRI1, in yeast cell, but not the XA21 immune receptor. Silencing of these genes decreased XA21 gene expression and compromised XA21-mediated immunity to (*Xoo*). Hence, this study demonstrated that these OsRLCKs negatively regulate BR signaling, while positively regulating immune responses by contributing to the expression of the immune receptor XA21 (Zhou et al. 2016).

4.9 Conclusions

Plant receptors for the bacterial PGN and the proteinaceous PAMP Flg and EF-Tu elongation factor and fungal PAMPs have been identified and well-studied; however those involved in LPS perception are not studied in details. One of the most complex signaling molecular machineries is the plant PRR immune complex. When encountered with danger, the simplest mechanism is an initiation of signal when a surface-localized transmembrane receptor is bound to its corresponding ligand. The complete mechanism involves co-receptors, regulatory proteins, and substrates that link PRR activation and negative regulation. Since phosphorylation is a key event to drive the initiation of signaling from PRR, it will be important to decipher whether phosphorylation at specific sites is the key for recruitment or dissociation of the specific signaling component to the receptor complex or to determine the fate of activated PRR. Therefore specific phosphorylated sites would regulate directly or indirectly the branching of signaling from PRR complex. The expression pattern of different proteins, comprising of PRR complexes at the cell, tissue, or organ level, is still unknown. We need to address more PAMP/PRR perception systems to understand how plants integrate the different signals during natural infection. This is particularly important, as infections are often multi-tropic in nature. While PAMP induces overlapping responses, certain PAMP combinations seem to act synergistically or even antagonistically. Our knowledge of molecular mechanisms involved during downstream PRR signaling is very limited. Genetic screening, coupled with interaction studies complemented by the addition of more effector targets, is needed to fulfill this gap. The understanding of the basis of PRR complex formation, organization, activation, and subsequent connection of downstream networking leads to actual immunity and will be a key challenge for the future. In conclusion we say that with the assistance of expanding bioinformatics and molecular biology tools, we will be able to identify novel PAMPs, understanding their perception and signaling in plants.

References

Akamatsu A et al (2013) An OsCEBiP/OsCERK1-OsRacGEF1-OsRac1 module is an essential early component of chitin-induced rice immunity. Cell Host Microbe 13(4):465–476

Albrecht C et al (2012) Brassinosteroids inhibit pathogen-associated molecular pattern-triggered immune signaling independent of the receptor kinase BAK1. Proc Natl Acad Sci U S A 109(1):303–308

Bahar O et al (2014) The Xanthomonas Ax21 protein is processed by the general secretory system and is secreted in association with outer membrane vesicles. Peer J 2:e242

Bailey BA, Dean JF, Anderson JD (1990) An ethylene biosynthesis-inducing endoxylanase elicits electrolyte leakage and necrosis in Nicotiana tabacum cv Xanthi leaves. Plant Physiol 94(4):1849–1854

Bar M et al (2010) BAK1 is required for the attenuation of ethylene-inducing xylanase (Eix)-induced defense responses by the decoy receptor LeEix1. Plant J Cell Mol Biol 63(5):791–800

Bauer Z et al (2001) Sensitivity of different ecotypes and mutants of Arabidopsis thaliana toward the bacterial elicitor flagellin correlates with the presence of receptor-binding sites. J Biol Chem 276(49):45669–45676

Bedini E, Parrilli M, Unverzagt C (2002) Oligomerization of a rhamnanic trisaccharide repeating unit of O-chain polysaccharides from phytopathogenic bacteria

Bedini E et al (2005) Structure-dependent modulation of a pathogen response in plants by synthetic O-antigen polysaccharides. J Am Chem Soc 127(8):2414–2416

Belkhadir Y et al (2012) Brassinosteroids modulate the efficiency of plant immune responses to microbe-associated molecular patterns. Proc Natl Acad Sci U S A 109(1):297–302

Bogdanove AJ et al (2011) Two new complete genome sequences offer insight into host and tissue specificity of plant pathogenic Xanthomonas spp. J Bacteriol 193(19):5450–5464

Boller T, Felix G (2009) A renaissance of elicitors: perception of microbe-associated molecular patterns and danger signals by pattern-recognition receptors. Annu Rev Plant Biol 60:379–406

Borner GHH et al (2003) Identification of glycosylphosphatidylinositol-anchored proteins in Arabidopsis. A proteomic and genomic analysis. Plant Physiol 132(2):568–577

Bouwmeester K, Govers F (2009) Arabidopsis L-type lectin receptor kinases: phylogeny, classification, and expression profiles. J Exp Bot 60(15):4383–4396

Braun SG et al (2005) Characterization of the Xanthomonas campestris pv. campestris lipopolysaccharide substructures essential for elicitation of an oxidative burst in tobacco cells. Mol Plant-Microbe Interact MPMI 18(7):674–681

Brutus A et al (2010) A domain swap approach reveals a role of the plant wall-associated kinase 1 (WAK1) as a receptor of oligogalacturonides. Proc Natl Acad Sci U S A 107(20):9452–9457

Cambi A, Koopman M, Figdor CG (2005) How C-type lectins detect pathogens. Cell Microbiol 7(4):481–488

Cao Y et al (2013) Mutations in FLS2 Ser-938 dissect signaling activation in FLS2-mediated Arabidopsis immunity. PLoS Pathog 9(4):e1003313

Chandra S, Stennis M, Low PS (1997) Measurement of Ca2+ fluxes during elicitation of the oxida tive burst in aequorin-transformed tobacco cells. J Biol Chem 272(45):28274–28280

Chen X et al (2010) An ATPase promotes autophosphorylation of the pattern recognition receptor XA21 and inhibits XA21-mediated immunity. Proc Natl Acad Sci U S A 107(17):8029–8034

Chen X et al (2014) An XA21-associated kinase (OsSERK2) regulates immunity mediated by the XA21 and XA3 immune receptors. Mol Plant 7(5):874–892

Chinchilla D et al (2006) The Arabidopsis receptor kinase FLS2 binds flg22 and determines the specificity of flagellin perception. Plant Cell 18(2):465–476

Chinchilla D et al (2007) A flagellin-induced complex of the receptor FLS2 and BAK1 initiates plant defence. Nature 448(7152):497–500

Chinchilla D et al (2009) One for all: the receptor-associated kinase BAK1. Trends Plant Sci 14(10):535–541

Choi HW, Klessig DF (2016) DAMPs, MAMPs, and NAMPs in plant innate immunity. BMC Plant Biol 16(1):232

Choi J et al (2014) Identification of a plant receptor for extracellular ATP. Science (New York, NY) 343(6168):290–294

Choi HW et al (2016) Activation of plant innate immunity by extracellular high mobility group box 3 and its inhibition by salicylic acid. PLoS Pathog 12(3):e1005518

Conrath U et al (2006) Priming: getting ready for battle. Mol Plant-Microbe Interact MPMI 19(10):1062–1071

Csorba T, Pantaleo V, Burgyan J (2009) RNA silencing: an antiviral mechanism. Adv Virus Res 75:35–71

Dardick C, Schwessinger B, Ronald P (2012) Non-arginine-aspartate (non-RD) kinases are associated with innate immune receptors that recognize conserved microbial signatures. Curr Opin Plant Biol 15(4):358–366
de Jonge R et al (2012) Tomato immune receptor Ve1 recognizes effector of multiple fungal pathogens uncovered by genome and RNA sequencing. Proc Natl Acad Sci U S A 109(13):5110–5115
Dempsey DA et al (2011) Salicylic acid biosynthesis and metabolism. Arabidopsis Book 9:e0156
Denoux C et al (2008) Activation of defense response pathways by OGs and Flg22 elicitors in Arabidopsis seedlings. Mol Plant 1(3):423–445
Desaki Y et al (2006) Bacterial lipopolysaccharides induce defense responses associated with programmed cell death in rice cells. Plant Cell Physiol 47(11):1530–1540
Desaki Y et al (2012) Positive crosstalk of MAMP signaling pathways in rice cells. PLoS One 7(12):e51953
Dodds PN, Rathjen JP (2010) Plant immunity: towards an integrated view of plant-pathogen interactions. Nat Rev Genet 11(8):539–548
Dou D, Zhou J-M (2012) Phytopathogen effectors subverting host immunity: different foes, similar battleground. Cell Host Microbe 12(4):484–495
Dow M, Newman M-A, von Roepenack E (2000) The induction and modulation of plant defense responses by bacterial lipopolysaccharides. Annu Rev Phytopathol 38:241–261
Dunning FM et al (2007) Identification and mutational analysis of Arabidopsis FLS2 leucine-rich repeat domain residues that contribute to flagellin perception. Plant Cell 19(10):3297–3313
Dziarski R, Gupta D (2006) The peptidoglycan recognition proteins (PGRPs). Genome Biol 7(8):232
Erbs G, Newman M-A (2012) The role of lipopolysaccharide and peptidoglycan, two glycosylated bacterial microbe-associated molecular patterns (MAMPs), in plant innate immunity. Mol Plant Pathol 13(1):95–104
Erbs G et al (2008) Peptidoglycan and muropeptides from pathogens agrobacterium and Xanthomonas elicit plant innate immunity: structure and activity. Chem Biol 15(5):438–448
Faulkner C et al (2013) LYM2-dependent chitin perception limits molecular flux via plasmodesmata. Proc Natl Acad Sci U S A 110(22):9166–9170
Felix G, Boller T (2003) Molecular sensing of bacteria in plants. The highly conserved RNA-binding motif RNP-1 of bacterial cold shock proteins is recognized as an elicitor signal in tobacco. J Biol Chem 278(8):6201–6208
Felix G et al (1999) Plants have a sensitive perception system for the most conserved domain of bacterial flagellin. Plant J Cell Mol Biol 18(3):265–276
Feng F et al (2012) A Xanthomonas uridine 5′-monophosphate transferase inhibits plant immune kinases. Nature 485(7396):114–118
Fradin EF et al (2009) Genetic dissection of Verticillium wilt resistance mediated by tomato Ve1. Plant Physiol 150(1):320–332
Gaulin E et al (2006) Cellulose binding domains of a Phytophthora cell wall protein are novel pathogen-associated molecular patterns. Plant Cell 18(7):1766–1777
Gherbi H et al (2008) SymRK defines a common genetic basis for plant root endosymbioses with arbuscular mycorrhiza fungi, rhizobia, and Frankiabacteria. Proc Natl Acad Sci U S A 105(12):4928–4932
Gimenez-Ibanez S et al (2009) AvrPtoB targets the LysM receptor kinase CERK1 to promote bacterial virulence on plants. Curr Biol CB 19(5):423–429
Gomez-Gomez L, Boller T (2000) FLS2: an LRR receptor-like kinase involved in the perception of the bacterial elicitor flagellin in Arabidopsis. Mol Cell 5(6):1003–1011
Gómez-gómez L, Bauer Z, Boller T (2001) Both the extracellular leucine-rich repeat domain and the kinase activity of FLS2 are required for flagellin binding and signaling in arabidopsis. Plant Cell 13:1155–1163
Gou X et al (2012) Genetic evidence for an indispensable role of somatic embryogenesis receptor kinases in brassinosteroid signaling. PLoS Genet 8(1):e1002452
Gross A et al (2005) Endocytosis of Xanthomonas campestris pathovar campestris lipopolysaccharides in non-host plant cells of Nicotiana tabacum. New Phytol 165(1):215–226

Gust AA et al (2007) Bacteria-derived peptidoglycans constitute pathogen-associated molecular patterns triggering innate immunity in Arabidopsis. J Biol Chem 282(44):32338–32348
Gutsmann T, Schromm AB, Brandenburg K (2007) The physicochemistry of endotoxins in relation to bioactivity. Int J Med Microbiol IJMM 297(5):341–352
Halter T et al (2014) The leucine-rich repeat receptor kinase BIR2 is a negative regulator of BAK1 in plant immunity. Curr Biol CB 24(2):134–143
Hayafune M et al (2014) Chitin-induced activation of immune signaling by the rice receptor CEBiP relies on a unique sandwich-type dimerization. Proc Natl Acad Sci U S A 111(3):E404–E413
Heese A et al (2007) The receptor-like kinase SERK3/BAK1 is a central regulator of innate immunity in plants. Proc Natl Acad Sci U S A 104(29):12217–12222
Heiling S et al (2010) Jasmonate and ppHsystemin regulate key Malonylation steps in the biosynthesis of 17-Hydroxygeranyllinalool Diterpene glycosides, an abundant and effective direct defense against herbivores in Nicotiana attenuata. Plant Cell 22(1):273–292
Helft L et al (2011) LRR conservation mapping to predict functional sites within protein leucine-rich repeat domains. PLoS One 6(7):e21614
Huffaker A, Ryan CA (2007) Endogenous peptide defense signals in Arabidopsis differentially amplify signaling for the innate immune response. Proc Natl Acad Sci U S A 104(25):10732–10736
Huffaker A, Pearce G, Ryan CA (2006) An endogenous peptide signal in Arabidopsis activates components of the innate immune response. Proc Natl Acad Sci U S A 103(26):10098–10103
Huffaker A, Dafoe NJ, Schmelz EA (2011) ZmPep1, an ortholog of Arabidopsis elicitor peptide 1, regulates maize innate immunity and enhances disease resistance. Plant Physiol 155(3):1325–1338
Huffaker A et al (2013) Plant elicitor peptides are conserved signals regulating direct and indirect antiherbivore defense. Proc Natl Acad Sci U S A 110(14):5707–5712
Ialenti A et al (2006) A novel lipid A from Halomonas magadiensis inhibits enteric LPS-induced human monocyte activation. Eur J Immunol 36(2):354–360
Jehle AK et al (2013) Perception of the novel MAMP eMax from different Xanthomonas species requires the Arabidopsis receptor-like protein ReMAX and the receptor kinase SOBIR. Plant Signal Behav 8(12):e27408
Jeppesen MG et al (2005) Crystal structure of the bovine mitochondrial elongation factor Tu.Ts complex. J Biol Chem 280(6):5071–5081
Jeter CR et al (2004) Evidence of a novel cell signaling role for extracellular adenosine triphosphates and diphosphates in Arabidopsis. Plant Cell 16(10):2652–2664
Jiang Y et al (2013) The XA21 binding protein XB25 is required for maintaining XA21-mediated disease resistance. Plant J Cell Mol Biol 73(5):814–823
Jones JDG, Dangl JL (2006) The plant immune system. Nature 444(7117):323–329
Kadota Y et al (2014) Direct regulation of the NADPH oxidase RBOHD by the PRR-associated kinase BIK1 during plant immunity. Mol Cell 54(1):43–55
Kaku H et al (2006) Plant cells recognize chitin fragments for defense signaling through a plasma membrane receptor. Proc Natl Acad Sci U S A 103(29):11086–11091
Kannenberg EL, Carlson RW (2001) Lipid A and O-chain modifications cause rhizobium lipopolysaccharides to become hydrophobic during bacteroid development. Mol Microbiol 39(2):379–391
Kauss et al (1999) Cucumber hypocotyls respond to cutin monomers via both an inducible and a constitutive H(2)O(2)-generating system. Plant Physiol 120(4):1175–1182
Kawchuk LM et al (2001) Tomato Ve disease resistance genes encode cell surface-like receptors. Proc Natl Acad Sci U S A 98(11):6511–6515
Korasick DA et al (2010) Novel functions of stomatal cytokinesis-defective 1 (SCD1) in innate immune responses against bacteria. J Biol Chem 285(30):23342–23350
Korner CJ et al (2013) The immunity regulator BAK1 contributes to resistance against diverse RNA viruses. Mol Plant-Microbe Interact MPMI 26(11):1271–1280
Kouzai Y et al (2014) CEBiP is the major chitin oligomer-binding protein in rice and plays a main role in the perception of chitin oligomers. Plant Mol Biol 84(4–5):519–528

Krol E et al (2010) Perception of the Arabidopsis danger signal peptide 1 involves the pattern recognition receptor AtPEPR1 and its close homologue AtPEPR2. J Biol Chem 285(18):13471–13479
Kunze G et al (2004) The N terminus of bacterial elongation factor Tu elicits innate immunity in Arabidopsis plants. Plant Cell 16(12):3496–3507
Kyndt T et al (2012) Transcriptional reprogramming by root knot and migratory nematode infection in rice. New Phytol 196(3):887–900
Laluk K et al (2011) Biochemical and genetic requirements for function of the immune response regulator BOTRYTIS-INDUCED KINASE1 in plant growth, ethylene signaling, and PAMP-triggered immunity in Arabidopsis. Plant Cell 23(8):2831–2849
Lambert KN, Allen KD, Sussex IM (1999) Cloning and characterization of an esophageal-gland-specific chorismate mutase from the phytoparasitic nematode Meloidogyne javanica. Mol Plant-Microbe Interact MPMI 12(4):328–336
Launholt D et al (2006) Arabidopsis chromatin-associated HMGA and HMGB use different nuclear targeting signals and display highly dynamic localization within the nucleus. Plant Cell 18(11):2904–2918
Lemmon MA, Schlessinger J (2010) Cell signaling by receptor tyrosine kinases. Cell 141(7):1117–1134
Li J et al (2002) BAK1, an Arabidopsis LRR receptor-like protein kinase, interacts with BRI1 and modulates brassinosteroid signaling. Cell 110(2):213–222
Li C et al (2011) Pathway for lipid A biosynthesis in Arabidopsis thaliana resembling that of Escherichia coli. Proc Natl Acad Sci U S A 108(28):11387–11392
Liebrand TWH et al (2013) Receptor-like kinase SOBIR1/EVR interacts with receptor-like proteins in plant immunity against fungal infection. Proc Natl Acad Sci U S A 110(24):10010–10015
Limpens E et al (2003) LysM domain receptor kinases regulating rhizobial Nod factor-induced infection. Science (New York, NY) 302(5645):630–633
Lin W et al (2013) Inverse modulation of plant immune and brassinosteroid signaling pathways by the receptor-like cytoplasmic kinase BIK1. Proc Natl Acad Sci U S A 110(29):12114–12119
Lin Z-JD et al (2015) PBL13 is a serine/threonine protein kinase that negatively regulates Arabidopsis immune responses. Plant Physiol 169(4):2950–2962
Liu T et al (2012) Chitin-induced dimerization activates a plant immune receptor. Science (New York, NY) 336(6085):1160–1164
Liu Z et al (2013) BIK1 interacts with PEPRs to mediate ethylene-induced immunity. Proc Natl Acad Sci U S A 110(15):6205–6210
Livaja M et al (2008) Transcriptional responses of Arabidopsis thaliana to the bacteria-derived PAMPs harpin and lipopolysaccharide. Immunobiology 213(3–4):161–171
Lozano-Duran R et al (2013) The transcriptional regulator BZR1 mediates trade-off between plant innate immunity and growth. eLife 2:e00983
Lu D et al (2010) A receptor-like cytoplasmic kinase, BIK1, associates with a flagellin receptor complex to initiate plant innate immunity. Proc Natl Acad Sci U S A 107(1):496–501
Lu D et al (2011) Direct ubiquitination of pattern recognition receptor FLS2 attenuates plant innate immunity. Science (New York, NY) 332(6036):1439–1442
Madsen EB et al (2011) Autophosphorylation is essential for the in vivo function of the Lotus japonicus Nod factor receptor 1 and receptor-mediated signalling in cooperation with Nod factor receptor 5. Plant J Cell Mol Biol 65(3):404–417
Malinovsky FG et al (2014) Antagonistic regulation of growth and immunity by the Arabidopsis basic helix-loop-helix transcription factor homolog of brassinosteroid enhanced expression2 interacting with increased leaf inclination1 binding bHLH1. Plant Physiol 164(3):1443–1455
Manosalva P et al (2015) Conserved nematode signalling molecules elicit plant defenses and pathogen resistance. Nat Commun 6:7795
Mateos FV, Rickauer M, Esquerre-Tugaye MT (1997) Cloning and characterization of a cDNA encoding an elicitor of Phytophthora parasitica var. nicotianae that shows cellulose-binding and lectin-like activities. Mol Plant-Microbe Interact MPMI 10(9):1045–1053
McDonald C, Inohara N, Nunez G (2005) Peptidoglycan signaling in innate immunity and inflammatory disease. J Biol Chem 280(21):20177–20180

McGurl B et al (1992) Structure, expression, and antisense inhibition of the systemin precursor gene. Science (New York, NY) 255(5051):1570–1573
McMichael CM et al (2013) Mediation of clathrin-dependent trafficking during cytokinesis and cell expansion by Arabidopsis stomatal cytokinesis defective proteins. Plant Cell 25(10):3910–3925
Meindl T, Boller T, Felix G (2000) The bacterial elicitor flagellin activates its receptor in tomato cells according to the address-message concept. Plant Cell 12(9):1783–1794
Merkle T, Grasser KD (2011) Unexpected mobility of plant chromatin-associated HMGB proteins. Plant Signal Behav 6(6):878–880
Mishina TE, Zeier J (2006) The Arabidopsis flavin-dependent monooxygenase FMO1 is an essential component of biologically induced systemic acquired resistance. Plant Physiol 141(4):1666–1675
Mishina TE, Zeier J (2007) Pathogen-associated molecular pattern recognition rather than development of tissue necrosis contributes to bacterial induction of systemic acquired resistance in Arabidopsis. Plant J Cell Mol Biol 50(3):500–513
Mithoe SC et al (2016) Attenuation of pattern recognition receptor signaling is mediated by a MAP kinase kinase kinase. EMBO Rep 17(3):441–454
Miya A et al (2007) CERK1, a LysM receptor kinase, is essential for chitin elicitor signaling in Arabidopsis. Proc Natl Acad Sci U S A 104(49):19613–19618
Monaghan J, Zipfel C (2012) Plant pattern recognition receptor complexes at the plasma membrane. Curr Opin Plant Biol 15(4):349–357
Mueller K et al (2012) Chimeric FLS2 receptors reveal the basis for differential flagellin perception in Arabidopsis and tomato. Plant Cell 24(5):2213–2224
Munford RS, Varley AW (2006) Shield as signal: lipopolysaccharides and the evolution of immunity to gram-negative bacteria. PLoS Pathog 2(6):e67
Nam KH, Li J (2002) BRI1/BAK1, a receptor kinase pair mediating brassinosteroid signaling. Cell 110(2):203–212
Narusaka Y et al (2013) Presence of LYM2 dependent but CERK1 independent disease resistance in Arabidopsis. Plant Signal Behav 8(9)
Narvaez-Vasquez J, Ryan CA (2004) The cellular localization of prosystemin: a functional role for phloem parenchyma in systemic wound signaling. Planta 218(3):360–369
Newman MA, Daniels MJ, Dow JM (1995) Lipopolysaccharide from Xanthomonas campestris induces defense-related gene expression in Brassica campestris. Mol Plant-Microbe Interact MPMI 8(5):778–780
Newman M-A et al (2002) Prior exposure to lipopolysaccharide potentiates expression of plant defenses in response to bacteria. Plant J Cell Mol Biol 29(4):487–495
Newman M-A et al (2007) Priming, induction and modulation of plant defence responses by bacterial lipopolysaccharides. J Endotoxin Res 13(2):69–84
Newman M-A et al (2013) MAMP (microbe-associated molecular pattern) triggered immunity in plants. Front Plant Sci 4:139
Nothnagel EA et al (1983) Host-pathogen interactions: XXII. A Galacturonic acid oligosaccharide from plant cell walls elicits phytoalexins. Plant Physiol 71(4):916–926
Nuhse TS et al (2007) Quantitative phosphoproteomic analysis of plasma membrane proteins reveals regulatory mechanisms of plant innate immune responses. Plant J Cell Mol Biol 51(5):931–940
Nurnberger T et al (1994) High affinity binding of a fungal oligopeptide elicitor to parsley plasma membranes triggers multiple defense responses. Cell 78(3):449–460
Ono E et al (2001) Essential role of the small GTPase Rac in disease resistance of rice. Proc Natl Acad Sci U S A 98(2):759–764
Pearce G, Ryan CA (2003) Systemic signaling in tomato plants for defense against herbivores. Isolation and characterization of three novel defense-signaling glycopeptide hormones coded in a single precursor gene. J Biol Chem 278(32):30044–30050
Pearce G et al (1991) A polypeptide from tomato leaves induces wound-inducible proteinase inhibitor proteins. Science (New York, NY) 253(5022):895–897

Pearce G et al (2001) Production of multiple plant hormones from a single polyprotein precursor. Nature 411(6839):817–820
Pedersen DS, Grasser KD (2010) The role of chromosomal HMGB proteins in plants. Biochim Biophys Acta 1799(1–2):171–174
Petutschnig EK et al (2010) The lysin motif receptor-like kinase (LysM-RLK) CERK1 is a major chitin-binding protein in Arabidopsis thaliana and subject to chitin-induced phosphorylation. J Biol Chem 285(37):28902–28911
Raetz CRH, Whitfield C (2002) Lipopolysaccharide endotoxins. Annu Rev Biochem 71:635–700
Ramonell K et al (2005) Loss-of-function mutations in chitin responsive genes show increased susceptibility to the powdery mildew pathogen Erysiphe cichoracearum. Plant Physiol 138(2):1027–1036
Ranf S et al (2011) Interplay between calcium signalling and early signalling elements during defence responses to microbe- or damage-associated molecular patterns. Plant J Cell Mol Biol 68(1):100–113
Ranf S et al (2015) A lectin S-domain receptor kinase mediates lipopolysaccharide sensing in Arabidopsis thaliana. Nat Immunol 16(4):426–433
Robatzek S, Chinchilla D, Boller T (2006) Ligand-induced endocytosis of the pattern recognition receptor FLS2 in Arabidopsis. Genes Dev 20(5):537–542
Robatzek S et al (2007) Molecular identification and characterization of the tomato flagellin receptor LeFLS2, an orthologue of Arabidopsis FLS2 exhibiting characteristically different perception specificities. Plant Mol Biol 64(5):539–547
Ron M, Avni A (2004) The receptor for the fungal elicitor ethylene-inducing xylanase is a member of a resistance-like gene family in tomato. Plant Cell 16(6):1604–1615
Ramos HC, Rumbo M, Sirard J-C (2004) Bacterial flagellins: mediators of pathogenicity and host immune responses in mucosa. Trends Microbiol 12(11):509–517
Roux M et al (2011) The Arabidopsis leucine-rich repeat receptor-like kinases BAK1/SERK3 and BKK1/SERK4 are required for innate immunity to hemibiotrophic and biotrophic pathogens. Plant Cell 23(6):2440–2455
Ruiz-Ferrer V, Voinnet O (2009) Roles of plant small RNAs in biotic stress responses. Annu Rev Plant Biol 60:485–510
Ryals JA et al (1996) Systemic acquired resistance. Plant Cell 8(10):1809–1819
Santiago J, Henzler C, Hothorn M (2013) Molecular mechanism for plant steroid receptor activation by somatic embryogenesis co-receptor kinases. Science (New York, NY) 341(6148):889–892
Scheer, Ryan (1999) A 160-kD systemin receptor on the surface of lycopersicon peruvianum suspension-cultured cells. Plant Cell 11(8):1525–1536
Scheer JM, Ryan CAJ (2002) The systemin receptor SR160 from Lycopersicon peruvianum is a member of the LRR receptor kinase family. Proc Natl Acad Sci U S A 99(14):9585–9590
Schneider M et al (1996) Systemic acquired resistance in plants. Int Rev Cytol 168:303–340
Schulze B et al (2010) Rapid heteromerization and phosphorylation of ligand-activated plant transmembrane receptors and their associated kinase BAK1. J Biol Chem 285(13):9444–9451
Schweizer P et al (1996) Perception of free cutin monomers by plant cells. Plant J 10(2):331–341
Schwessinger B et al (2011) Phosphorylation-dependent differential regulation of plant growth, cell death, and innate immunity by the regulatory receptor-like kinase BAK1. PLoS Genet 7(4):e1002046
Segonzac C et al (2011) Hierarchy and roles of pathogen-associated molecular pattern-induced responses in Nicotiana benthamiana. Plant Physiol 156(2):687–699
Sejalon-Delmas N et al (1997) Purification, elicitor activity, and cell wall localization of a glycoprotein from Phytophthora parasitica var. nicotianae, a fungal pathogen of tobacco. Phytopathology 87(9):899–909
Seong S-Y, Matzinger P (2004) Hydrophobicity: an ancient damage-associated molecular pattern that initiates innate immune responses. Nat Rev Immunol 4(6):469–478
Shi H et al (2013) BR-signaling kinase1 physically associates with FLAGELLIN SENSING2 and regulates plant innate immunity in Arabidopsis. Plant Cell 25(3):1143–1157

Shimizu T et al (2010) Two LysM receptor molecules, CEBiP and OsCERK1, cooperatively regulate chitin elicitor signaling in rice. Plant J Cell Mol Biol 64(2):204–214

Shinya T et al (2012) Functional characterization of CEBiP and CERK1 homologs in arabidopsis and rice reveals the presence of different chitin receptor systems in plants. Plant Cell Physiol 53(10):1696–1706

Silipo A et al (2005) The elicitation of plant innate immunity by lipooligosaccharide of Xanthomonas campestris. J Biol Chem 280(39):33660–33668

Silipo A et al (2008) The acylation and phosphorylation pattern of lipid A from Xanthomonas campestris strongly influence its ability to trigger the innate immune response in Arabidopsis. Chembiochem Eur J Chem Biol 9(6):896–904

Singh P, Zimmerli L (2013) Lectin receptor kinases in plant innate immunity. Front Plant Sci 4:124

Smith JM et al (2014) Sensitivity to Flg22 is modulated by ligand-induced degradation and de novo synthesis of the endogenous flagellin-receptor FLAGELLIN-SENSING2. Plant Physiol 164(1):440–454

Song WY et al (1995) A receptor kinase-like protein encoded by the rice disease resistance gene, Xa21. Science (New York, NY) 270(5243):1804–1806

Song WY et al (1997) Evolution of the rice Xa21 disease resistance gene family. Plant Cell 9(8):1279–1287

Song CJ et al (2006) Extracellular ATP induces the accumulation of superoxide via NADPH oxidases in Arabidopsis. Plant Physiol 140(4):1222–1232

Stracke S et al (2002) A plant receptor-like kinase required for both bacterial and fungal symbiosis. Nature 417(6892):959–962

Suharsono U et al (2002) The heterotrimeric G protein alpha subunit acts upstream of the small GTPase Rac in disease resistance of rice. Proc Natl Acad Sci U S A 99(20):13307–13312

Sun W et al (2012) Probing the Arabidopsis flagellin receptor: FLS2-FLS2 association and the contributions of specific domains to signaling function. Plant Cell 24(3):1096–1113

Sun Y, Li L et al (2013a) Structural basis for flg22-induced activation of the Arabidopsis FLS2-BAK1 immune complex. Science (New York, NY) 342(6158):624–628

Sun Y, Han Z et al (2013b) Structure reveals that BAK1 as a co-receptor recognizes the BRI1-bound brassinolide. Cell Res 23(11):1326–1329

Tanaka K et al (2014) Extracellular ATP acts as a damage-associated molecular pattern (DAMP) signal in plants. Front Plant Sci 5:446

Tang W et al (2008) BSKs mediate signal transduction from the receptor kinase BRI1 in Arabidopsis. Science (New York, NY) 321(5888):557–560

Thaler JS, Humphrey PT, Whiteman NK (2012) Evolution of jasmonate and salicylate signal crosstalk. Trends Plant Sci 17(5):260–270

Thomma BPHJ, Nurnberger T, Joosten MHAJ (2011) Of PAMPs and effectors: the blurred PTI-ETI dichotomy. Plant Cell 23(1):4–15

Tintor N et al (2013) Layered pattern receptor signaling via ethylene and endogenous elicitor peptides during Arabidopsis immunity to bacterial infection. Proc Natl Acad Sci U S A 110(15):6211–6216

Traub S et al (2006) MDP and other muropeptides – direct and synergistic effects on the immune system. J Endotoxin Res 12(2):69–85

Trujillo M et al (2008) Negative regulation of PAMP-triggered immunity by an E3 ubiquitin ligase triplet in Arabidopsis. Curr Biol CB 18(18):1396–1401

Umemoto N et al (1997) The structure and function of a soybean beta-glucan-elicitor-binding protein. Proc Natl Acad Sci U S A 94(3):1029–1034

Vaid N, Pandey PK, Tuteja N (2012) Genome-wide analysis of lectin receptor-like kinase family from Arabidopsis and rice. Plant Mol Biol 80(4–5):365–388

Vaid N, Macovei A, Tuteja N (2013) Knights in action: lectin receptor-like kinases in plant development and stress responses. Mol Plant 6(5):1405–1418

van Loon LC, Bakker PA, Pieterse CM (1998) Systemic resistance induced by rhizosphere bacteria. Annu Rev Phytopathol 36:453–483

Vercauteren I et al (2001) Arabidopsis thaliana genes expressed in the early compatible interaction with root-knot nematodes. Mol Plant-Microbe Interact MPMI 14(3):288–299

Veronese P et al (2006) The membrane-anchored BOTRYTIS-INDUCED KINASE1 plays distinct roles in Arabidopsis resistance to necrotrophic and biotrophic pathogens. Plant Cell 18(1):257–273

Vlot AC, Dempsey DA, Klessig DF (2009) Salicylic acid, a multifaceted hormone to combat disease. Annu Rev Phytopathol 47:177–206

Wan J et al (2008) A LysM receptor-like kinase plays a critical role in chitin signaling and fungal resistance in Arabidopsis. Plant Cell 20(2):471–481

Wan J et al (2012) LYK4, a lysin motif receptor-like kinase, is important for chitin signaling and plant innate immunity in Arabidopsis. Plant Physiol 160(1):396–406

Wang GL et al (1996) The cloned gene, Xa21, confers resistance to multiple Xanthomonas oryzae pv. oryzae isolates in transgenic plants. Mol Plant-Microbe Interact MPMI 9(9):850–855

Wang GL et al (1998) Xa21D encodes a receptor-like molecule with a leucine-rich repeat domain that determines race-specific recognition and is subject to adaptive evolution. Plant Cell 10(5):765–779

Wang Y-S et al (2006) Rice XA21 binding protein 3 is a ubiquitin ligase required for full Xa21-mediated disease resistance. Plant Cell 18(12):3635–3646

Wang G et al (2008) A genome-wide functional investigation into the roles of receptor-like proteins in Arabidopsis. Plant Physiol 147(2):503–517

Wang YH, Gehring C, Irving HR (2011) Plant natriuretic peptides are apoplastic and paracrine stress response molecules. Plant Cell Physiol 52(5):837–850

Watt SA et al (2006) Identification of the bacterial superoxide dismutase (SodM) as plant-inducible elicitor of an oxidative burst reaction in tobacco cell suspension cultures. J Biotechnol 126(1):78–86

Willmann R et al (2011) Arabidopsis lysin-motif proteins LYM1 LYM3 CERK1 mediate bacterial peptidoglycan sensing and immunity to bacterial infection. Proc Natl Acad Sci U S A 108(49):19824–19829

Xu W-H et al (2006) The autophosphorylated Ser686, Thr688, and Ser689 residues in the intracellular juxtamembrane domain of XA21 are implicated in stability control of rice receptor-like kinase. Plant J Cell Mol Biol 45(5):740–751

Xu J et al (2014) A chemical genetic approach demonstrates that MPK3/MPK6 activation and NADPH oxidase-mediated oxidative burst are two independent signaling events in plant immunity. Plant J Cell Mol Biol 77(2):222–234

Yamaguchi Y, Huffaker A (2011) Endogenous peptide elicitors in higher plants. Curr Opin Plant Biol 14(4):351–357

Yamaguchi Y, Pearce G, Ryan CA (2006) The cell surface leucine-rich repeat receptor for AtPep1, an endogenous peptide elicitor in Arabidopsis, is functional in transgenic tobacco cells. Proc Natl Acad Sci U S A 103(26):10104–10109

Yamaguchi Y et al (2010) PEPR2 is a second receptor for the Pep1 and Pep2 peptides and contributes to defense responses in Arabidopsis. Plant Cell 22(2):508–522

Yamaguchi K et al (2013) A receptor-like cytoplasmic kinase targeted by a plant pathogen effector is directly phosphorylated by the chitin receptor and mediates rice immunity. Cell Host Microbe 13(3):347–357

Zhang J et al (2007) A Pseudomonas syringae effector inactivates MAPKs to suppress PAMP-induced immunity in plants. Cell Host Microbe 1(3):175–185

Zhang J et al (2010) Receptor-like cytoplasmic kinases integrate signaling from multiple plant immune receptors and are targeted by a Pseudomonas syringae effector. Cell Host Microbe 7(4):290–301

Zhang W et al (2013) Arabidopsis receptor-like protein30 and receptor-like kinase suppressor of BIR1-1/EVERSHED mediate innate immunity to necrotrophic fungi. Plant Cell 25(10):4227–4241

Zhou X et al (2016) Four receptor-like cytoplasmic kinases regulate development and immunity in rice. Plant Cell Environ 39(6):1381–1392

Zipfel C (2013) Combined roles of ethylene and endogenous peptides in regulating plant immunity and growth. Proc Natl Acad Sci U S A 110(15):5748–5749

Zipfel C et al (2006) Perception of the bacterial PAMP EF-Tu by the receptor EFR restricts Agrobacterium-mediated transformation. Cell 125(4):749–760

Role of NBS-LRR Proteins in Plant Defense

5

Namo Dubey and Kunal Singh

Abstract

The NBS-LRR proteins are encoded by one of the largest and most important gene family involved in disease resistance in plants. Many of these NBS-LRR proteins recognize effectors secreted by pathogens directly or indirectly that in turn activate downstream signaling pathways leading to activation of plant defense response against various classes of pathogens including bacterial, fungal, viral, nematode and insect. Defense response by NBS-LRR protein is a sophisticated strategy that induces effector-triggered immunity (ETI). The NBS-LRR proteins comprised of amino-terminal variable domain, a central nucleotide-binding site (NBS) and carboxy-terminal leucine-rich repeats (LRR) domain. The NBS domain binds and hydrolyzes ATP and primarily functions as a signal transduction switch following pathogen recognition. LRRs are highly adaptable structural domains that are involved in protein-protein interactions, and these LRRs can also evolve very different binding specificities. In the following chapter we have discussed in detail about the present knowledge pertaining to NBS-LRR class of proteins and their prospect in crop improvement against diseases.

Keywords

Plant Defense · NBS-LRR · R-Genes · Effectors · Phytopathogens

N. Dubey · K. Singh (✉)
Molecular Plant Pathology Laboratory, Division of Biotechnology, CSIR-Institute of Himalayan Bioresource Technology, Palampur, Himachal Pradesh, India
e-mail: kunal@ihbt.res.in

A. Singh, I. K. Singh (eds.), *Molecular Aspects of Plant-Pathogen Interaction*,
https://doi.org/10.1007/978-981-10-7371-7_5

5.1 Introduction

Plants are continuously attacked by a plethora of pathogens. To counter pathogen attack, plants have evolved an immune system, and the major part of this system is contributed by specificity determinants, which are the resistance (R) genes. R genes are set of genes that confer resistance to a wide variety of pathogens including bacteria, fungi, oomycetes, nematodes, and insects. Nucleotide-binding site-leucine-rich repeat (NBS-LRR) genes comprise the largest class in the category of all known R genes which encompasses more than 80% of characterized R genes (McHale et al. 2006; Meyers et al. 1999, 2003, 2005; Friedman and Baker 2007; Zhang et al. 2016; Song et al. 2017). The name NBS-LRR has been assigned due to a central nucleotide-binding domain, which is also known as the NB-ARC {nucleotide-binding adaptor shared by apoptotic protease activating factor 1 (APAF1), certain R genes and cell death protein 4 (CED4)} domain, and their C-terminal leucine-rich repeat (LRR) domain. NBS-LRR proteins form a subclass of the signal transduction ATPases with numerous domains (STAND) super family, a class of molecular switches that are involved in a variety of mechanisms, including immunity, apoptosis (e.g., APAF1 and CED4), and transcriptional regulation (Takken and Goverse 2012).

Plant NBS-LRR proteins show sequence resemblance with the members of the mammalian nucleotide-binding oligomerization domain (NOD)-LRR protein family{also called caspase recruitment domains (CARD), R (purine)-binding, transcription enhancer, pyrin, lots of leucine repeats (CATERPILLER) proteins}, which play crucial role in inflammatory and immune responses (Inohara et al. 2005; McHale et al. 2006). Evolutionary studies based on phylogeny to understand the origin and similarities between plant NBS-LRR and mammalian NOD proteins are inconclusive till date. Though a plain, convergent evolutionary origin was proposed initially (Ausubel 2005; McHale et al. 2006), the most recent study suggests that the plant NBS-LRR and mammalian NOD proteins may be a result of independent evolutionary events that occurred at least twice (Urbach and Ausubel 2017). Whether these events are due to horizontal gene transfer or due to some other mechanism, is yet to be determined.

NBS-LRR genes have been grouped into two classes, namely, TIR-NBS-LRR (TNL) and the non-TIR-NBS-LRR (nTNL). The last residue, D (aspartate) or W (tryptophan), of the conserved kinase-2 motif within the NBS domain helps in distinguishing TNL from nTNL by 95% accuracy (Meyers et al. 1999; Wan et al. 2012). Former class is different from the latter as the former class comprises a toll/interleukin-1 receptor-like (TIR) domain at the protein amino terminus (Bai et al. 2002; Meyers et al. 1999; Zhou et al. 2004; Zhang et al. 2016). In contrast, at the N-terminus, most nTNL genes encode a CC domain, and these nTNL genes are often called as CC-NBS-LRR (CNL) genes (Ameline et al. 2008; Meyers et al. 2003). Apart from CNL genes, a small group of nTNL genes that possesses a special N-terminal domain known as resistance to powdery mildew 8 (RPW8) domain has recently been discovered. RPW8 domain containing nTNL genes represents distinct class of NBS-LRR genes (RPW8-NBS-LRR, RNL) (Bonardi et al. 2011; Collier et al. 2011; Cannon et al. 2004; Shao et al. 2014; Xiao et al. 2001; Zhang et al. 2016).

Commonly, NBS-LRRs are intracellular multidomain proteins that recognize pathogen-derived effectors either directly or indirectly (Cesari et al. 2014; Dodds and

Rathjen 2010; Jones and Dangl 2006; Van der Hoorna and Kamoun 2008; Shao et al. 2016; Song et al. 2017). As per direct model, an NBS-LRR protein binds to the pathogen effector and serves as a substrate for the effector's enzymatic activity (Shao et al. 2016). According to indirect model, NBS-LRR recognizes the modifications of additional host protein(s). Further these modifications in the additional host proteins are targeted by the effector (Shao et al. 2016). Ultimately, the aim of NBS-LRR proteins is to sense and detect the respective pathogen effectors and virulence factors. This recognition leads to the activation of downstream signaling pathway, which results in various immune responses like hypersensitive response at the site of pathogen infection which provides immunity to the host plant (Bittle and Robatzek 2007).

5.2 Susceptibility and Resistance in Plants

Different pathogens such as viruses, bacteria, fungi, and nematodes are responsible to cause many destructive diseases in plants. As the diseases progress, visible symptoms can be seen, and in some cases severe infections can lead to the death of the plants. Often these symptoms include yellowing of leaves and stunted growth of plant, followed by necrosis at the infection site ultimately leading to cell death. To acquire resistance, plants need to suppress pathogen growth and replication at the site of infection.

Plant resistance in response to pathogen attack is associated with the rapid burst of reactive oxygen species (ROS), followed by a localized programmed cell death (PCD) called hypersensitive response (HR) at the infection sites, and increased expression of pathogenesis-related (PR) genes (Chisholm et al. 2006; Gururani et al. 2012; Hammond–Kosack and Jones 1996; Heath 2000; Morel and Dangl 1997; Shao et al. 2016; Takahashi et al. 2003). Other resistance responses include (but are not limited to) the activation of defense gene expression, induced biosynthesis, and accumulation of salicylic acid (SA) and jasmonic acid (JA) (Creelman and Mullet 1995). Hypersensitive response is an especially effective process in limiting pathogens (biotrophs) that require living host cells. Often HR is triggered when an appropriate R gene recognizes an effector or a pathogen elicitor gets recognized by suitable receptor (Minsavage et al. 1990; Nürnberger et al. 1994). The NBS-LRR class of protein play a very important role in both cases as major portion of R genes and immune receptors discovered till date belongs to NBS-LRR and LRR kinase (LRK) class of proteins in plant cell. A well-characterized example of HR mechanism through gene-for-gene interaction is provided by the tomato (*Solanum lycopersicon*) *Cf-9* gene, which confers resistance to races of the fungus *Cladosporium fulvum* expressing the *Avr9* gene (Van Kan and De Wit 1992).

5.3 NBS-LRR Gene-Mediated Resistance to Pathogens in Crop Plants

NBS-LRR gene-mediated resistance has several striking features for disease control. When the resistance is induced in a timely manner, then the signaling cascade can effectively halt pathogen growth with minimal collateral damage to the plant

(McDowell and Woffenden 2003). Most of the NBS-LRR genes exhibit high recognition specificity with elicitors. Specific amino acid changes in the NBS or ARC2 domain of various NBS-LRR R genes such as tomato *I-2*, potato *Rx*, and flax *L6* result in autoactivation and Avr-independent defense signaling (Tameling et al. 2006). Alternatively, a random in vitro mutagenesis of the domains in the R gene sequence can also result in pathogen recognition and subsequent recognition specificities of the selected gene variants. A subset of the mutant *Rx* gene sequences in potato was shown to confer resistance against the original viral strain as well as additional potato virus X strains and a second distantly related virus species (Farnham and Baulcombe 2006).

5.3.1 NBS-LRR Gene Family in Plant Genomes

Recent advancement in sequencing technologies holds the capacity to provide sequence information at a very high-throughput scale (Solexa, Roche; Illumina, single-molecule real-time sequencing). These technologies have opened new vistas to assess plant-microbe interactions. Their uses enable genome-wide analyses of NBS-LRR genes, and this analysis is based on the NB-ARC domain (Meyers et al. 2005). On the basis of the sequenced genomes of various plant species, hundreds of NBS-LRR genes have been identified (McHale et al. 2006). *Arabidopsis thaliana*, the first sequenced plant species, has 165 NBS-LRR genes including 52 CNLs, 106 TNLs, and 7 RNLs (Zhang et al. 2016). Plants that belong to Solanaceae family like *Solanum tuberosum* (potato) and *Solanum lycopersicum* (tomato) have more than twice the number of NBS-LRR genes than *Arabidopsis*, and plants of Solanaceae family possess more CNLs than TNLs (Consortium TG 2012; Guo et al. 2011; Kim et al. 2014; Jupe et al. 2012). It is considered that TNLs have evolved after the divergence of monocots and dicots (Goff et al. 2002; McHale et al. 2006; Meyers et al. 2003). The NBS-LRR class of R genes fit into rapidly evolving gene family, and the number of NBS-LRR genes differs among plant species (Clark et al. 2007; Rafiqi et al. 2009). For illustration, 198 NBS-LRR genes (138 CNLs, 55 TNLs, and 5 RNLs) were acknowledged in *Arabidopsis lyrata.* At the genus level, more differences in R gene composition have been observed. In recent times, identification of novel NBS-LRR genes has been possible with the help of *R* gene enrichment and sequencing (RenSeq) method (Andolfo et al. 2014; Jupe et al. 2013). Overexpression of these genes showed various phenotypes associated with basal defense response such as elevated SA level; therefore, the role of NBS-LRR proteins in basal defense has also been well characterized (Nandety et al. 2013). In this direction, genome-wide analysis of NBS-LRR genes may expand the scope to understand the role of these genes in plant disease resistance.

On the basis of the N-terminus, NBS-LRR proteins have been classified into two categories in plants which are TIR-NBS-LRR (TNL) and the non-TIR-NBS-LRR (nTNL). This classification is, however, not always precise. nTNL proteins comprise those NBS-LRR proteins which possess other domain at their N-terminus rather than TIR domain. NBS-LRR genes usually contain additional domains such as CC or RPW8 at the N-terminus and an unpredictable number of LRR domains at the carboxy-terminus (Shao et al. 2014; Zhang et al. 2016).

The functionality of NBS-LRR proteins is dependent on below described domains:

5.3.1.1 TIR Domain

TIR is most common domain in the group of NBS-LRR proteins (Zhang et al. 2016). Many factors which interact with animal TLRs (toll-like receptors) also possess TIR domain, for example, MyD88. During these interactions, TIR domains interact physically (Riedl et al. 2005). Thus TIR domains facilitate heterodimerization in plants like that of some animal TLR receptors (Nandety et al. 2013; Williams et al. 2014). In plants, TIR domain participates in the detection of Avr proteins, e.g., interaction between the TIR domain of tobacco N protein and the tobacco mosaic virus (TMV) p50 protein. This interaction leads to hypersensitive response (HR). Noticeably, NRIP1 (the additional host protein) is requisite for this interaction (Nandety et al. 2013). Recent studies suggest the role of TIR domain in pathogen recognition and in signaling cascade (Williams et al. 2014). However, TIR domain is notably absent from most monocotyledons (Tarr and Alexander 2009).

5.3.1.2 CC Domain

The coiled-coil (CC) domain usually ascribed a role analogous to the TIR domain as it acts as a mediator in interactions with other essentials of the signaling cascade. It has been observed that the EDVID motif is conservative motif in the coiled-coil domain of all CC-NBS-LRR proteins (Collier and Moffett 2009) but exceptions are also present in case of RPS2, RPS5, and Dm3 proteins. Mutations in this motif cause turbulence in the intramolecular interaction in between NBS and LRR domains; as a result decreased resistance response against pathogen attack has been observed (Zhang et al. 2016).

The CC structure is made up of two or more α-helices with a super helical twist. It represents heptad repeat sequence $(abcdefg)_n$ where 'a' and'd' are hydrobhobic amino acid residues, while the 'e' and 'g' represent polar amino acid residues. The CC structure is the most common in NBS-LRR proteins present in both monocots and dicots. It has been studied that CC domain alone is able to induce cell death, for instances, the *Arabidopsis RPS5*, *RPS2*, *RPM1* and *activated disease resistance 1* (*ADR1*); *Nicotiana benthamiana N requirement* gene 1 (*NRG1*); and barley *mildew-resistance locus A 10* (*MLA10*) genes (Jacob et al. 2013). But in contrast, central NBS domain of potato Rx protein is sufficient to trigger cell death (Moffett et al. 2002).

R proteins have also been known to contain a CC domain which binds to target proteins of pathogen effectors, e.g., the CC domain of *A. thaliana* RPS5 protein interacts with PBS1, which is a target of *Pseudomonas syringae* AvrPhB effector (Ade et al. 2007).

Fig. 5.1 Subdomains of NBS domain, showing comparison in between plants and animals

5.3.1.3 RPW8 Domain

Apart from CNL genes, a small group of nTNL genes that possess a special N-terminal domain known as RPW8 (resistance to powdery mildew8) domain has recently been discovered. RPW8 domain containing nTNL genes which represent a distinct class of NBS-LRR genes, i.e., RPW8-NBS-LRR, RNL (Bonardi et al. 2011; Cannon et al. 2004; Collier et al. 2011; Shao et al. 2014; Xiao et al. 2001; Zhang et al. 2016). Species-specific duplication studies on five *Rosaceae* species, namely, *Malus domestica* (apple), *Fragaria vesca* (strawberry), *Pyrus bretschneideri* (pear), *Prunus mume* (mei), and *Prunus persica* (peach), clearly described RNL as a distinct class of NBS-LRR genes (Zhong et al. 2015).

5.3.1.4 NBS Domain

The NBS domain was identified in NBS-LRR proteins, and it shows similarity to the homologous sequences in the animal APAF-1 and CED-4 proteins (Van der Biezen and Jones 1998). NTPase activity is a characterizing feature of NBS domain. This domain is suggested to play a crucial role as a molecular switch in activating signal transduction. In the signaling cascade, reversible nucleotide binding leads to changes in the conformation of the NBS domain. The conformational changes in NBS domain lead to the activation/deactivation of the whole receptor (Tameling et al. 2006).

After human APAF-1 studies, it has become possible to speculate about the structure and function of the NBS domain in plants. Four subdomains, NB, ARC1, ARC2 and ARC-3, are found in NBS domain of the human APAF-1 protein (Fig. 5.1) (Riedl et al. 2005). ARC3 domain is absent in the plant R proteins; nevertheless a short linker connecting ARC2 with LRR is found in the plant R proteins. ARC1 is made up of a bunch of α-helices and ARC2 of α-helices rolled up in a winged helix fold. The spatial structure of the NBS domain differs, and it depends on whether it is combined with ATP or ADP (Tameling et al. 2002). Many conservative motifs such as the P-loop (Walker A or kinase 1), the RNBS-A, kinase 2 (Walker B), kinase-3a, RNBS-B, RNBS-C, GLPL, and RNBS-D have been identified in this domain (Wan et al. 2012). P-loop, kinase-3a, kinase 2, and GLPL conserved motifs are common in both TNLs and nTNLs subfamilies (Wan et al. 2012).

NBS domain has the capability to bind and hydrolyse ATP, and thus it confers switching of R protein from active to inactive state (Riedl et al. 2005). Mutations in the NBS domain are attributed to the autoactivation of resistance response in the absence of an elicitor, and this is characterized by an increased susceptibility to pathogen. Constitutive active resistance response is because of disturbance in ATP hydrolysis. Therefore, the binding with ATP rather than ADP seems necessary to activate a receptor. It has been proven that the autoactivation mechanism of NBS-LRR protein is generally associated with the mutations in the NB and ARC2 subdomains (Takken et al. 2006). In spite of this, the ARC1 subdomain seems to be involved in binding to LRR domain; this has been proven after the evolution of loss of functional mutants. However, low stability of ARC1 domain is suggested for the cooperation of some additional factors which are necessary in the signaling cascade (Rairdan and Moffett 2006).

5.3.1.5 LRR Domain

The C-terminus of NBS-LRR protein is occupied by LRR domain. Tandem repeat of 20–30 amino acids containing a consensus sequence LxxLxLxxNxL is the essential structural element of LRR domain. In consensus sequence LxxLxLxxNxL, L is a leucine residue, N is asparagine/threonin/serine or cystein, and x is any amino acid (Bella et al. 2008; Kajava 1998; Matsushima et al. 2007; Stange et al. 2008). NBS-LRR protein with an LRR domain has to contain at least two LRR repeats. Horseshoe-shaped superhelix is the usual tertiary structure of LRR domain. Each repeat of amino acid forms other coils of the horseshoe-shaped super helix. It is proposed that the LRR domains constitute a platform for protein-protein interactions (Bella et al. 2008; Kobe and Kajava 2000). The tertiary structure of LRR usually has an inner surface composed of parallel β-strands. These parallel β-strands are composed of hydrophobic aliphatic residues of a consensus sequence and are assumed to be a site of interaction with other proteins, which in case of plant NBS-LRR proteins provides a crucial condition for the recognition of specific elicitors.

The outer part of the LRR domain is usually composed of α-helices which are connected with β-strands by β-turns (Bella et al. 2008). Consensus sequence LxxLxLxxNxL participates in interactions with microbe-associated molecular patterns (MAMPs) or microbe-induced molecular patterns (MIMPs), and this has been proven by the differences in the specificity of flax rust (*Melampsora lini*) effector recognition by the *P* and *P2* genes of flax (*Linum usitatissimum*).

After binding with a proper MIMP or MAMP, conformational changes within the LRR domain occur that lead to the dissociation of the LRR domain from the NBS domain (Wang et al. 2007). It is suggested that the dissociation of the LRR and NBS domains from NBS-LRR protein might not be required for the activation of the receptor (Rairdan and Moffett 2006; Van Ooijen et al. 2008). In contrast, rhythmic rounds of dissociation and reassociation could lead to the amplification of the signal originating from elicitor recognition (Rairdan and Moffett 2006).

5.3.1.6 Other Domains in NBS-LRR Protein

Despite the domains listed above, some additional domains are located at the N-terminus and more rarely at the C-terminus in many of the NBS-LRR proteins. One of the examples for such proteins is *Arabidopsis* RRS-1R that have WRKY domain at their N/C-terminal. In this protein plant WRKY transcription factors have been identified by the presence of WRKYGQK conserved motif situated at the N-terminus of the NBS-LRR protein. This WRKY transcription factor also has a typical domain similar to Zn (Zinc) finger motif and plays an essential role in regulating the expression of genes that are involved in plant resistance (Eulgem and Somssich 2007; Ulker and Somssich 2004).

In many instances, WRKY transcription factors are directly affected by some NBS-LRR proteins. Alleles of barley MLA proteins detect the related Avr proteins of *Blumeria graminis* and then disengage HvWRKY1/2 transcription factor, which has been found to repress resistance genes (Liu and Coaker 2008). It has been observed that some NBS-LRR proteins also contain a domain having similar structure as that of a WRKY transcription factor; thus, it became feasible to affect gene expression directly. An instance of such protein is the RRS-1R receptor of *A. thaliana* which is one of the TIR-NBS-LRR proteins. RRS-1R receptor possesses a C-terminal WRKY domain and recognizes the PopP2 effector of *Ralstonia solanacearum* (Deslandes et al. 2002). WRKY domain may act as a suppressor of signaling cascade responsible for bringing resistance against a pathogen (Noutoshi et al. 2005); this has been proven by the analysis of *Arabidopsis thaliana* mutants, known as SLH1. These mutants possess alteration in single amino acid in the WRKY domain of the TIR-NBS-LRR-WRKY protein (Noutoshi et al. 2005).

5.4 Intramolecular Interactions

Structural analyses of plant NBS-LRR are dependent on domain-swapping experiments, mutant analysis, and 3D modeling of NBS-LRR protein. Overall, it seems that plant NBS-LRR proteins have complex interactions among their domains; however, the function of those interactions in pathogen detection and activation of signal transductions seems to vary with some degree among plant NBS-LRR proteins. The first evidence for intramolecular interaction of plant NBS-LRR protein domains came after the work on Rx and Bs2 R proteins from pepper (Leister et al. 2005; Moffett et al. 2002). It has been known that plant NBS-LRR proteins have complex interactions among their domains; however, the role of intramolecular interactions in pathogen detection and activation of signal transductions seems to vary with some degree among plant NBS-LRR proteins. Co-immunoprecipitation experiments in Rx demonstrate physical interactions between the CC and NBS-LRR domains and between the CC-NBS and LRR domains.

It is considered that interaction of Rx domains with domains from the closely related NBS-LRR proteins Bs2 and HRT (an *Arabidopsis* R protein that confers resistance to turnip crinkle virus) has also been detected, suggesting some degree of 'promiscuity'; however, such physical interactions are not sufficient to reconstitute functional signaling molecules (Rairdan and Moffett 2006). Further, the intramolecular interactions of RPS5 have been studied, and it was found that the intramolecular interactions of RPS5 are similar but not identical to those reported for Rx and Bs2. Co-immunoprecipitation experiments have revealed that the CC and LRR domains individually interact with the NBS domain, but not with each other (Moffett et al. 2002).

5.5 NBS-LRR Formation of Oligomers

Several plant NBS-LRR proteins are known to form oligomers; however, the function of the NBS domain in that oligomerization is yet to be resolved. For example, several domains of Rx are able to interact when expressed in trans, as are domains of Bs2, although it has been found that some of those interactions are 'preferentially' intramolecular (Leister et al. 2005; Moffett et al. 2002). Oligomer formation of full-length Rx in the absence of pathogen effector has not been detected, and thus the direct interaction of NBS domains has not yet been detected for these proteins (Moffett et al. 2002). It has also been observed that in the presence of pathogen effector, NBS-LRR proteins form oligomers but the TIR domain is the only domain directly associated with that oligomerization, e.g., in N protein of tobacco (Mestre and Baulcombe 2006). But later on, it has been observed that in case of RPS5, the amino-terminal CC domain of RPS5 also forms oligomers. It suggests that in case of plant NBS-LRR proteins, the amino-terminal domain may be involved in the formation of oligomers, which differs from APAF1 and CED-4. Finally, study on tobacco N protein has revealed that the formation of oligomers is necessary but not sufficient for disease resistance. In addition, formation of oligomers is an early event in pathogen detection because downstream signal transduction mutants do not affect the formation of oligomers (Mestre and Baulcombe 2006).

5.6 Major Classes of R Proteins Containing at Least One NBS and/or LRR Domain

Plant resistance genes can be classified into nine groups. This classification is based on the organization of their amino acid motif and their membrane spanning domains (Fig. 5.2 and Table 5.1). In the majority of R proteins, LRR domain is present, and it plays important role in recognition specificity (Gururani et al. 2012; Jones 2001).

Class-I: CC-NBS-LRR genes are the first major class of R-genes. Proteins encoded by NBS-LRR genes are found in cytoplasm. First NBS-LRR protein encoded by NBS-LRR gene was discovered as a cytoplasmic protein. It possesses a putative CC at the N-terminus, a NBS, and a C-terminal LRR. *RPM1* resistance gene of *Arabidopsis* is the example of the first major class of R-genes (*I-2*).

Class-II: The second class of resistance genes includes the cytoplasmic proteins which possess N-terminal TIR domain, a NBS motif, and C-terminal LRR domain. The tobacco *N* gene, *RPP5* gene, and flax *L6* gene are the examples of the second class of R-genes (Gururani et al. 2012; Lawrence et al. 1995).

Class-III: The third class of R-genes lacks NBS motif and consists of extra cytoplasmic leucine-rich repeats (eLRR). This eLRR is attached to a transmembrane domain (TrD). eLRRs are supposed to play main role for certain defense proteins like polygalacturonase-inhibiting proteins PGIPs (Gururani et al. 2012; Jones and Jones 1997), but still no direct role of eLRRs in pathogen recognition and in activation of defense genes have been observed (Gururani et al. 2012; Jones

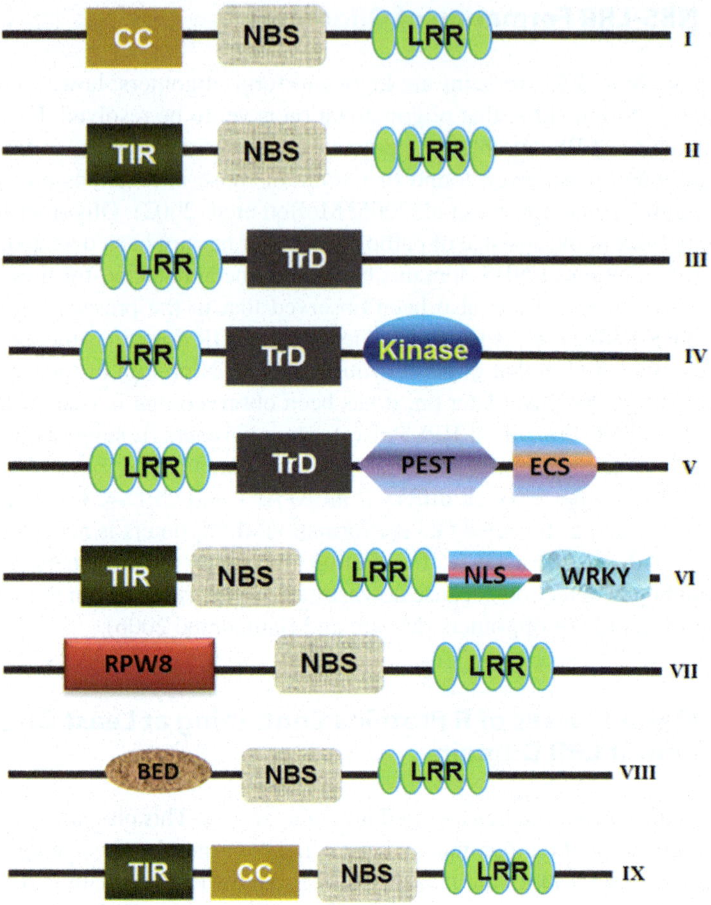

Fig. 5.2 Major classes of plant NBS-LRR genes with important domains: *LRR* leucine-rich repeats, *NBS* nucleotide-binding site, *ECS* endocytosis cell signaling domain, *TIR* toll/interleukin-1-receptors, *CC* coiled coil, *TrD* transmembrane domain, *PEST* amino acid domain, *NLS* nuclear localization signal, *WRKY* amino acid domain

2001). The *Cladosporium fulvum* resistance genes (*Cf-9*, *Cf-4*, and *Cf-2*) are the examples of this class.

Class-IV: The fourth major class of the resistance genes have an LRR extracellular domain, a transmembrane domain (TrD), and an intracellular KIN (serine-threonine kinase) domain (Gururani et al. 2012; Song et al. 1995). *Xa21* gene of rice which shows resistance against *Xanthomonas* is an example of this class.

Class-V: The fifth class encoding resistance genes contain the extracellular LRRs (putative), a Pro-Glu-Ser-Thr (PEST) domain, and an ECS (short proteins motifs). The PEST domain is involved in protein degradation, and it is found only in tomato *Ve2*, not in tomato *Ve1*. The ECS is supposed to target protein for receptor-mediated endocytosis.

Table 5.1 Major classes of plant NBS-LRR genes with important domains

S.No	Major R-gene classes	Domains											Examples
		LRR	NBS	RPW8	TIR	Kinase	CC	TrD	PEST	ECS	NLS	WRKY	
I	CC-NBS-LRR	+	+	–	–	–	+	–	–	–	–	–	*I2*, *RPS2*, *RPM1*
II	TIR-NBS-LRR	+	+	–	+	–	–	–	–	–	–	–	*N*, *L6*, *RRS1*, *RPP5*
III	LRR-TrD	+	–	–	–	–	–	+	–	–	–	–	*Cf-9*, *Cf-4*, *Cf-2*
IV	LRR-TrD-Kinase	+	–	–	–	+	–	+	–	–	–	–	*Xa21*
V	LRR-TrD-PEST-ECS	+	–	–	–	–	–	+	+	+	–	–	*Ve1*, *Ve2*
VI	TIR-NBS-LRR-NLS-WRKY	+	+	–	+	–	–	–	–	–	+	+	*RRS1*
VII	RPW8-NBS-LRR	+	+	+	–	–	–	–	–	–	–	–	*RPW8.1*, *RPW8.2*
VIII	BED-NBS-LRR	+	+	–	–	–	–	–	–	–	–	–	Poptr_1:787192
IX	TIR-CC-NBS-LRR	+	+	–	+	–	+	–	–	–	–	–	Reported in *Populus trichocarpa*

+ Present; – Absent

LRR leucine-rich repeats, *NBS* nucleotide-binding site, *ECS* endocytosis cell signaling domain, *TIR* toll/interleukin-1-receptors, *CC* coiled coil, *TrD* transmembrane domain, *PEST* amino acid domain, *NLS* nuclear localization signal, *WRKY* amino acid domain, *RPW8* resistance to powdery mildew8, "BED-type zinc finger domain"

Class-VI: *Arabidopsis RRS1-R* gene is an example of the seventh major class of R-genes. *RRS1-R* gene confers defense against bacterial phytopathogen *Ralstonia solanacearum*. Recently, *RRS1-R* gene has been included as a new member of the TIR-NBS-LRR class. TIR-NBS-LRR falls in R-protein class. RRS1-R has a C-terminal extension, along with a putative NLS (nuclear localization signal) and a WRKY domain (Deslandes et al. 2002, 2003). The WRKY domain comprises a 60 amino acid region.

Class-VII: The ninth class of R-genes contains RPW8 domain at the N-terminus, along with a NBS and LRR domain. *Arabidopsis RPW8.1* and *RPW8.2* genes are the examples of this class (Xiao et al. 2005).

Class VIII: This class of NBS-LRR genes contains a BED finger domain at their N-terminus (Kohler et al. 2007). This domain consists of about 50–60 amino acid residues which possess a shared pattern of cysteine and histidine amino acid residues which form a zinc-finger DNA-binding domain. This BED domain is present in rice *Xa1* gene and poplar's poptr_1:787192 (Bai et al. 2002; Kohler et al. 2008).

Class IX: Another class of NBS-LRR genes has been predicted through genome-wide study of NBS-LRR in *Populus trichocarpa* (Kohler et al. 2008). This class contains both TIR and CC domains at the N-terminus and was termed as TCNL (TIR-CC-NBS-LRR). Perhaps this may have emerged by gene recombination events as they contain both TIR and CC domains.

5.7 Instance of Plant NBS-LRR Genes Conferring Defense Against a Wide Range of Pathogens

Many plant R genes, which have been studied till now, confer resistance against a plethora of pathogens and for bulk of plant diseases; the genetics of susceptibility are less corporeal. Some of the examples of the plant NBS-LRR gene which confer resistance against pathogen attack have been discussed below.

5.7.1 Plant NBS-LRR Genes Conferring Resistance Against Bacterial Pathogens

It has been identified that both plant and animal bacterial pathogens transport virulence proteins into the host cytoplasm by means of the type-III secretion system (T3SS). T3SS is also known as injectisome (Gururani et al. 2012). This system enables Gram-negative bacteria to secrete and instill pathogenicity proteins into the cytosol of host cells (Galan and Collmer 1999). HR and pathogenicity (hrp) and HR and Conserved (hrc) genes encode T3SS. Mutations in hrp and hrc eliminate bacterial pathogenicity in susceptible host plants and the ability to bring out HR in non-host/cultivar-specific resistant plants. In rice, susceptible and resistant alleles of *Xa27* encode identical proteins. However, R allele expresses only when a rice plant faces bacterial AvrXa27 and the product of AvrXa27 is a nuclear localized T3SS effector. *Xa27* express only in the environs of infected tissue; however *Os8N3* (dominant rice gene) is an exception. It is upregulated by a bacterial type-III effector protein; thus it bestows gene-for-gene-specified disease susceptibility (Kaloshian 2004). Some plant R genes may confer resistance against unrelated or distantly related bacterial pathogens (Zhao et al. 2005) and demonstrate the transport feasibility of non-host *R* gene in between maize and rice. It was proposed that maize *Rxo1* detects *Xanthomonas oryzae* pv. oryzicola, and thus *Xanthomonas oryzae* is able to cause bacterial streak disease. In contrast, *Rxo1* has also been found to confer resistance to the distinct pathogen *Burkholderia andropogonis* (causes bacterial stripe of sorghum and maize).This study indicates that the same gene may control resistance toward both pathogens and nonpathogens of maize. This gives an idea that an NBS-LRR gene could be transferred between distinctly related cereals.

5.7.2 Plant NBS-LRR Genes Conferring Resistance Against Fungal Pathogens

Diseases which are caused by fungus contribute most devastatingly toward crop yield loses in approximately all major crops (Wani 2010). It has been studied that the variation in sequence and copy number within the central LRR domain of the gene plays an important role in identifying recognition specificity (Brande et al. 2001). For instance, sequence alterations in tomato *Cf*-4 and *Cf-9* genes play a key role in determination of recognition specificity. These genes confer resistance in tomato plants against *Cladosporium* (biotrophic leaf mold pathogen). HR is triggered upon detection of the fungus-encoded Avr4 and Avr9 peptides (Brande et al. 2001). Recently, a virus-induced gene silencing method for the validation of *Ve*-mediated signaling exposed that downstream signaling cascade of *Ve1* needs *EDS1* (enhanced disease susceptibility 1) and *NDR1* (non-race-specific disease resistance 1) gene. Another example of fungal R genes is the *RPW8.2* (*Arabidopsis thaliana* gene) which is induced by powdery mildew (Wang et al. 2009). It might also have been involved in reducing and enhancing of oxidative damage to the host cell and callosic encasement of the haustorial complex (EHC) formation, respectively. Marking of RPW8.2 to the extra haustorial membrane (EHM) needs normal functioning of the actin cytoskeleton.

5.7.3 Plant NBS-LRR Genes Conferring Resistance Against Oomycetes Pathogens

Many diseases like sudden oak death and late blight of potato are caused by phytopathogenic oomycetes. Cloning and functional analyses of four *Rpi* genes, *Rpi-blb3*, *R2*, *Rpi-abpt*, and *R2*-like, revealed that these genes contain all signature sequence characteristics of LZNBS-LRR (leucine zipper nucleotide-binding site-leucine-rich repeat). Also, several functional *R* genes which confer resistance to late blight have been cloned, and it has been observed that all of them belong to the NBS-LRR class of plant R genes (Bendahmane et al. 2000; Van der Vossen et al. 2003). Another instance of oomycetes R genes with NBS-LRR domain is Dm3 (downy mildew-resistance gene) (Shen et al. 2002). *Dm3* is present in *Bremia lactucae* and belongs to the large RGC2 (resistance gene candidate2) multigene family (McHale et al. 2006).

5.7.4 Plant NBS-LRR Genes Conferring Resistance Against Nematode Pathogens

Plant parasitic nematodes acquire nutrition from the cytoplasm of living cells. Basically, these are obligate parasites, and many ectoparasites and endoparasites come under the category of nematodes. Defense mechanism in plants against root-knot nematode was identified for the first time in *Lycopersicum peruvianum* Mill. (Watts 1947). *Lycopersicum peruvianum* Mill. is a wild variety of cultivated tomato (Watts 1947). Tomato *Mi* gene confers resistance to three root-knot nematodes that are *Meloidogyne arenaria*, *Meloidogyne incognita*, and *Meloidogyne javanica* (Gilbert and McGuire 1956). *Mi* gene-encoded protein having CC-NBS-LRR domains was transferred into cultivated variety of tomato. Positional cloning approach was used to isolate *Mi* gene. The defense mechanism triggered by *Mi* gene involves a HR response in the host (Dropkin 1969). Tomato *Mi* gene is the cloned root-knot nematode defense gene (Williamson and Kumar 2006). *Mi* gene participates in the resistance mechanism by following a gene-for-gene model. Similarly, the potato *Gpa2* gene shows resistance against *Globodera pallida* (potato cyst nematode). *Gpa2* gene is a member of the NBS-LRR gene family, and it also possesses a leucine zipper near its N-terminus. This *Gpa2* gene shows similarities to the *Rx1* gene in amino acid sequence. Studies indicate that *Rx1* gene confers defense in potato against potato virus X (Van der Voort et al. 1999).

5.7.5 Plant NBS-LRR Genes Conferring Resistance Against Viral Pathogens

Major characterized viral R genes from plants fall into the NBS-LRR class of R genes. For instance, resistance in tomato against TSWV (tomato spotted wilt virus) is associated with *Sw-5* gene. It has been studied that *Sw-5* gene is involved in broad and stable resistance (Rosello et al. 1998). *Sw-5* locus resistance allele encodes a CC-NBS-LRR resistance protein, and it has been studied that Sw-5 protein is similar to the Mi protein of tomato with the exception of 4 LZs at the amino terminus (Brommonschenkel et al. 2000). Another example is *RT4-4* gene which belongs to TIR-NBS-LRR class of NBS-LRR R gene and is generally involved in resistance response in common bean (*Phaseolus vulgari* cv. Othello) against viral attack (Seo et al. 2006). It functions across two plant families, which are Cucurbitaceae and Solanaceae.

5.7.6 Plant NBS-LRR Genes Conferring Resistance Against Insect

Resistance against insects has been studied since long in plants (Bent 1996; Dempsey et al. 1998; Panda and Khush 1995; Quisenberry and Clement 2002), and a number of single dominant resistance genes have been mapped (Venter and Botha 2000; Yencho et al. 2000). In staple crops such as wheat and rice, most of these mapped genes have been characterized.

To date, only some of the insect R genes belonging to NBS-LRR group of resistance genes were cloned. For instance, *Mi-1* confers defense to potato aphid (*Macrosiphum euphorbiae*) and whitefly (*Bemisia tabaci).* Lettuce *Nr*-gene confers resistance against only one single species of aphid (*Nasonovia ribisnigri*) (Reinink and Dieleman 1989). Other examples include apple *Sd1* gene, which shows resistance against *Dysaphis devecta* (rosy leaf curling aphid) and melon *Vat* gene which confers resistance against the cotton/melon aphid *Aphis gossypii* (Kaloshian 2004).

5.8 Sensing of Effector Protein by NBS-LRR Protein

It is well known that all LRR domains perform the same kind of functions. They sustain an NBS-LRR protein in an auto-inhibited state when pathogen is absent in host cell and this domain also confers specificity in pathogen recognition. The amino-terminal half of the LRR is supposed to act in intramolecular signal transduction, and the carboxy-terminal half plays a role in pathogen detection. The structural support for the function of LRR domain came from elucidation of the tertiary structure of the LRR domain of wild emmer wheat (*Triticum dicoccoides*) *Lr10* gene (Sela et al. 2012). In number of NBS-LRR proteins, such as L, Mla, and Rx, LRR domains possess a large number of hyper variable amino acid residues and plays a major role in pathogen recognition (Dodds et al. 2006; Seeholzer et al. 2010). Studies on domain swapping and individual mutations indicate that pathogen recognition specificity resides partly in the C-terminal of the LRR domain (Bendahmane et al. 2000; Sela et al. 2012).

5.9 Activation

Pathogen presence is perceived by the LRR; thus intramolecular signal is expanded to the remainder of the protein. In the resting state, the LRR and TIR-CC domain of NBS-LRR protein interact with the NB-ARC domain. These domains together form a closed conformation. It has been proven through biochemical studies carried out in tomato I-2, flax M, Mi-1, and L6 and barley MLA27. In auto-inhibited state, ADP binds the NB-ARC domain, or ADP bounded form on NB-ARC domain is an auto-inhibited state (Tameling et al. 2002; Williams et al. 2014). The crystal structure studies of CED4 show that in open configuration the NB-ARC domain binds to ATP. In activated state, LRR is suggested to trigger a conformational change in the NB-ARC domain. This conformational change allows exchange of ADP for ATP and helps in adopting more open conformation. Constitutive expression of NBS-LRR genes has been observed during disturbance in ATP hydrolysis activity. It has been proven by the studies on *I-2* mutants. Additional evidence comes from studies on flax rust NBS-LRR protein M. The wildtype M protein co-purified with ADP and an auto-active mutant (D555V) preferentially co-purified with ATP (Williams et al. 2014). It has earliar been studied that in the resting state of the NBS-LRR, part of the tobacco N protein was present in ATP-bound form, while interaction with the effector molecules triggered exchange of ATP for ADP (Ueda et al. 2006). Both events indicate that NBS-LRR protein activation takes place due to the changes in nucleotide-binding state.

Recently, an unexpected and additional enzymatic activity has been identified in the NB-ARC domain of NBS-LRR proteins: PSiP from corn, R1 from rice (NB), and RPM1 from *Arabidopsis*. It has been proven that ATP is hydrolyzed by this domain in anticipation of the nucleoside instead of diphosphate (Mayerhofer et al. 2016), thus indicating that phosphatase activity is associated with NBS domain. The phosphatase activity for NB is consistent and after activation of the NBS-LRR protein returns to an adenosine-bound state rather than an ADP bound conformation.

Once NBS-LRR protein gets activated, it acquires conformational changes in its structure. Conformational changes in NBS-LRR protein may expose different binding surfaces for the interactions with other proteins. Activation of an NB-LRR protein, and a subsequent change in conformation, may expose different potential binding surfaces allowing distinct interactions with other proteins. For instance, *I2* mutants having differences in their nucleotide-binding state (ADP, ATP, or vacant) illustrate diverse interaction patterns in a yeast two-hybrid assay (Lukasik-Shreepaathy et al. 2012).

5.10 NBS-LRR Signal Transduction

Gene-for-gene model is based on the physical interaction between R proteins and pathogen effectors. This physical interaction results in the plant defense responses and eventually leads to resistance (Keen 1990). The tomato non-NBS-LRR Pto protein kinase interacts directly with its cognate bacterial effector AvrPto within the serine-threonine kinase activation domain. Tomato Pto also interacts directly with second bacterial effector AvrPtoB. AvrPtoB has intrinsic E3 ubiquitin ligase activity. This direct interaction with effectors was demonstrated in NBS-LRR genes. This has led to the guard hypothesis, which predicts that an effector protein interacts with a host target, which is itself recognized by more than one NBS-LRR protein (Jones and Dangl 2006).

This is supported by *Arabidopsis* RIN4 protein. *Arabidopsis* RIN4 protein is an example of a host target for type-III bacterial effectors, which is recognized by at least two CNL R proteins (Mackey et al. 2002).Yeast two hybrid analyses of RPM1-RIN4 interactions proved that the amino-terminal domain seems to mediate the physical association between resistance proteins and pathogen effector targets (Young and Roger 2006). This is at least applicable for those resistance proteins that use an indirect recognition mechanism (Young and Roger 2006). Therefore, it has been suggested that the amino-terminal domain of plant NBS-LRR proteins may be involved in both detection of the pathogen signal and activation of the downstream response.

RIN4 is also targeted by structurally unrelated bacterial effectors AvrRpm1 and AvrB. Phosphorylation of RIN4 caused by AvrRpm1 and AvrB activates the R protein RPM1. A third effector, AvrRpt2 is recognized by RIN4 inside the plant cell. This effector cleaves RIN4 at two sites, and this cleavage activates the NBS-LRR protein, RPS2. Consequently, activation of HR on pathogen shot triggers a resistance response called SAR. SAR results in the accumulation of salicylic acid (SA) throughout the plant and the consequent expression of a characteristic set of defense genes. The SAR makes plants more resistant to subsequent attack by a range of other pathogens (Glazebrook 2001). Some defense responses require jasmonic acid (JA) and ethylene (ET) as signal molecules. The discovery of new genes and mutants allows dissection of local and systemic signaling cascade networks. These further highlight the complex interplay between defense molecules such as nitric oxide, SA, reactive oxygen intermediates, JA, and ET (Thomma et al. 2001).

Feys and Parker (2000) detected two *Arabidopsis* mutant *ndr1* and *eds1*. The mutant *ndr1* and *eds1* suppress race-specific resistance to strains of the bacterium *Pseudomonas syringae*. Generally, EDS1 and NDR1 are independently required for the function of different NBS-LRR genes. TIR-NBS-LRR R genes are suppressed by EDS1, whereas a subset of non-TIR-NBS-LRR R proteins is suppressed by NDR1 (Joshi and Nayak 2011).

RAR1/SGT1/HSP90 {RAR1 (required for Mla-dependent resistance 1), SGT1 (suppressor of G2 allele of SKP1), and HSP90 (heat shock protein 90)} are the cytoplasmically localized signaling complexes, and these proteins regulate R gene-mediated resistance (Austin et al. 2002; Azevedo et al. 2002; Muskett et al. 2002). All components work together to stabilize various NBS-LRR R protein complexes. SGT1 is the essential component for the function of SCF (Skp1-cullin-F-box protein) E3 ubiquitin ligase complex. SCF E3 ubiquitin ligase complex targets proteins for degradation and this degradation is caused the 26S proteasome. Therefore, R gene-triggered defense is also mediated by the ubiquitin-proteasome pathway. The components of the mitogen-activated protein kinase (MAPK) cascades are also other key regulators in the defense mechanism (Asai et al. 2002). Importantly, SA-inducible defenses are negatively regulated by the MAPK, EDR1 (Frye et al. 2001), whereas MAPK4 seems to differentially regulate JA and SA signals (Petersen et al. 2000). These studies strongly suggest that MAPK modules participate in different plant disease resistance pathways. Ankyrin repeat protein, NPR1, is also an important feature of the systemic signaling. Initially, it was identified as an SA response regulator which participates in both ISR (induced systemic resistance) and SAR. On treating *Arabidopsis* seedlings with SA, movement of NPR1 to the nucleus was observed. In the nucleus, NPR1 binds to many TGA (TGACG DNA motif) class transcription factors and confers a feasible route to defense gene induction (Fan and Dong 2002).

5.11 Conclusions

In plant innate immunity, NBS-LRR protein-based defense of plants to various pathogens is a foremost area of interest. Evolutionary studies on interactions between plants and pathogens provide a key idea to modify important crops for pathogen defense. However, only few such interactions have been successfully decoded regardless of the abundant NBS-LRR genes present in plant genomes. Further research should be focused on detecting new NBS-LRR genes and their related pathogen effectors and molding these NBS-LRR proteins and their interactions. Such studies would make it feasible to modify the plant defense against a range of plant pathogens.

Acknowledgment We are thankful to the Director of CSIR-IHBT, Palampur, for his kind support. We acknowledge the financial support from SERB-DST, New Delhi, under the Fast Track Scheme for Young Scientists (YSS/2015/001036). Namo Dubey acknowledges UGC for providing fellowship. The CSIR-IHBT communication number for this article is 4141.

References

Ade J, DeYoung BJ, Golstein C, Innes RW (2007) Indirect activation of a plant nucleotide binding site- leucine-rich repeat protein by a bacterial protease. Proc Natl Acad Sci U S A 104:2531–2536

Ameline TC, Wang BB, O'Bleness MS, Deshpande S, Zhu H, Roe B, Young ND, Cannon SB (2008) Identification and characterization of nucleotide-binding site-leucine-rich repeat genes in the model plant *Medicago truncatula*. Plant Physiol 146:5–21

Andolfo G, Jupe F, Witek K, Etherington GJ, Ercolano MR, Jones JDG (2014) Defining the full tomato NBLRR resistance gene repertoire using genomic and cDNA RenSeq. BMC Plant Biol 14:120

Asai T, Tena G, Plotnikova J, Willmann MR, Chiu WL, Gómez-Gómez L, Boller T, Ausubel FM, Sheen J (2002) MAP kinase signalling cascade in *Arabidopsis* innate immunity. Nature 415:977–983

Austin MJ, Muskett P, Kahn K, Feys BJ, Jones JD, Parker JE (2002) Regulatory role of SGT1 in early R-gene-mediated plant defenses. Science 295(5562):2032–2033

Ausubel FM (2005) Are innate immune signaling pathways in plants and animals conserved? Nat Immunol 6:973–979

Azevedo C, Sadanandom A, Kitagawa K, Freialdenhoven A, Shirasu K, Schulze-Lefert P (2002) The RAR1 interactor SGT1, an essential component of R gene-triggered disease resistance. Science 295:2073–2076

Bai J, Pennill LA, Ning J, Lee SW, Ramalingam J, Webb CA, Zhao B, Sun Q, Nelson JC, Leach JE, Hulbert SH (2002) Diversity in nucleotide binding site-leucine-rich repeat genes in cereals. Genome Res 12:1871–1884

Bella J, Hindle KL, McEwan PA, Lovell SC (2008) The leucine-rich repeat structure. Cell Mol Life Sci 65:2307–2333

Bendahmane A, Querci M, Kanyuka K, Baulcombe DC (2000) *Agrobacterium* transient expression system as a tool for disease resistance genes isolation: application to Rx2 locus in potato. Plant J 21(1):73–81

Bent AF (1996) Plant disease resistance: function meets structure. Plant Cell 8:1757–1771

Bittel P, Robatzek S (2007) Microbe-associated molecular patterns (MAMPs) probe plant immunity. Curr Opin Plant Biol 10:335–341

Bonardi V, Tang S, Stallmann A, Roberts M, Cherkis K, Dangl JL (2011) Expanded functions for a family of plant intracellular immune receptors beyond specific recognition of pathogen effectors. Proc Natl Acad Sci U S A 108:16463–16468

Brande BH, Wulff H, Thomas CM, Smoker M, Grant M, Jones JDG (2001) Domain swapping and gene shuffling identify sequences required for induction of an Avr-dependent hypersensitive response by the tomato Cf-4 and Cf-9 proteins. Plant Cell 13:255–272

Brommonschenkel SH, Frary A, Frary A, Tanksley SD (2000) The broad-spectrum tospovirus resistance gene Sw-5 of tomato is a homolog of root-knot nematode resistance gene Mi. Mol Plant-Microbe Interact 13:1130–1138

Cannon SB, Mitra A, Baumgarten A, Young ND, May G (2004) The roles of segmental and tandem gene duplication in the evolution of large gene families in *Arabidopsis thaliana*. BMC Plant Biol 4:10

Cesari S, Bernoux M, Moncuquet P, Kroj T, Dodds PN (2014) A novel conserved mechanism for plant NLR protein pairs: the"integrated decoy" hypothesis. Front Plant Sci 5:606

Chisholm ST, Coaker G, Day B, Staskawicz BJ (2006) Host-microbe interactions: shaping the evolution of the plant immune response. Cell 124:803e14

Clark RM, Schweikert G, Toomajian C, Ossowski S, Zeller G, Shinn P, Warthmann N, Hu TT, Fu G, Hinds DA, Chen H, Frazer KA, Huson DH, Schölkopf B, Nordborg M, Rätsch G, Ecker JR, Weigel D (2007) Common sequence polymorphisms shaping genetic diversity in *Arabidopsis thaliana*. Science 317:338–342

Collier SM, Moffett P (2009) NB-LRRs work a "bait and switch" on pathogens. Trends Plant Sci 14(10):521–529

Collier SM, Hamel LP, Moffett P (2011) Cell death mediated by the N-terminal domains of a unique and highly conserved class of NB-LRR protein. Mol Plant-Microbe Interact 24:918–931

Consortium TG (2012) The tomato genome sequence provides insights into fleshy fruit evolution. Nature 485:635–641

Creelman RA, Mullet JE (1995) Jasmonic acid distribution and action in plants: regulation during development and response to biotic and abiotic stress. Proc Natl Acad Sci U S A 92(10):4114–4119

Dempsey DA, Silva H, Klessig DF (1998) Engineering disease and pest resistance in plants. Trends Microbiol 6:54–61

Deslandes L, Olivier J, Theulieres F, Hirsch J, Feng DX, Bittner-Eddy PD, Beynon J, Marco Y (2002) Resistance to *Ralstonia solanacearum* in *Arabidopsis thaliana* is conferred by the recessive RRS1-R gene, a member of a novel family of resistance genes. Proc Natl Acad Sci U S A 99:2404–2409

Deslandes L, Olivier J, Peeters N, Feng DX, Khounlotham M, Boucher C, Somssich I, Genin S, Marco Y (2003) Physical interaction between RRS1-R, a protein conferring resistance to bacterial wilt, and PopP2, a type III effector targeted to the plant nucleus. Proc Natl Acad Sci U S A 100:8024–8029

Dodds PN, Rathjen JP (2010) Plant immunity: towards an integrated view of plant-pathogen interactions. Nat Rev Genet 11:539–548

Dodds PN, Lawrence GJ, Catanzariti AM, Teh T, Wang CI, Ayliffe MA, Kobe B, Ellis JG (2006) Direct protein interaction under lies gene-for-gene specificity and coevolution of the flax resistance genes and flax rust virulence genes. Proc Natl Acad Sci U S A 103:8888–8893

Dropkin VH (1969) The necrotic reaction tomato and other hosts resistant to *Meloidogyne*: reversal by temperature. Phytopathology 59:1632–1637

Eulgem T, Somssich IE (2007) Networks of WRKY transcription factors in defense signaling. Curr Opin Plant Biol 10:366–371

Fan W, Dong X (2002) In vivo interaction between NPR1 and transcription factor TGA2 leads to salicylic acid mediated gene activation in *Arabidopsis*. Plant Cell 14:1377–1389

Farnham G, Baulcombe DC (2006) Artificial evolution extends the spectrum of viruses that are targeted by a disease resistance gene from potato. Proc Natl Acad Sci U S A 103:18828–18833

Feys BJ, Parker JE (2000) Interplay of signaling pathways in plant disease resistance. Trends In Genet 16:449–455

Friedman AR, Baker BJ (2007) The evolution of resistance genes in multi-protein plant resistance systems. Curr Opin Genet Dev 17(6):493–499

Frye CA, Tang D, Innes RW (2001) Negative regulation of defense responses in plants by a conserved MAPKK kinase. Proc Natl Acad Sci U S A 98:373–378

Galan JE, Collmer A (1999) Type III secretion machines: bacterial devices for protein delivery into host cells. Science 284:1322–1328

Gilbert JC, McGuire DC (1956) Inheritance of resistance to severe root-knot from *Meloidogyne incognita* in commercial-type tomatoes. Proc Am Soc Hort Sci 68:437–442

Glazebrook J (2001) Genes controlling expression of defense responses in *Arabidopsis*-2001 status. Curr Opin Plant Biol 4:301–308

Goff SA, Ricke D, Lan TH, Presting G, Wang R, Dunn M, Glazebrook J, Sessions A, Oeller P, Varma H, Hadley D, Hutchison D, Martin C, Katagiri F, Lange BM, Moughamer T, Xia Y, Budworth P, Zhong J, Miguel T, Paszkowski U, Zhang S, Colbert M, Sun WL, Chen L, Cooper B, Park S, Wood TC, Mao L, Quail P, Wing R, Dean R, Yu Y, Zharkikh A, Shen R, Sahasrabudhe S, Thomas A, Cannings R, Gutin A, Pruss D, Reid J, Tavtigian S, Mitchell J, Eldredge G, Scholl T, Miller RM, Bhatnagar S, Adey N, Rubano T, Tusneem N, Robinson R, Feldhaus J, Macalma T, Oliphant A, Briggs S (2002) A draft sequence of the rice genome (*Oryza sativa* L. ssp. japonica). Science 296:92–100

Guo YL, Fitz J, Schneeberger K, Ossowski S, Cao J, Weigel D (2011) Genome-wide comparison of nucleotide-binding site-leucine-rich repeat-encoding genes in *Arabidopsis*. Plant Physiol 157:757–769

Gururani MA, Jelli V, Upadhyaya CP, Akula N, Pandey SK, Park SW (2012) Plant disease resistance genes: current status and future directions. Physiol Mol Plant Pathol 78:51–65

Hammond–Kosack KE, Jones JDG (1996) Inducible plant defense mechanisms and resistance gene function. Plant Cell 8:1773–1791

Heath MC (2000) Non host resistance and nonspecific plant defenses. Curr Opin Plant Biol 3:315–319

Inohara C, McDonald C, Nuñez G (2005) NOD-LRR proteins: role in host-microbial interactions and inflammatory disease. Annu Rev Biochem 74:355–383

Jacob F, Vernaldi S, Maekawa T (2013) Evolution and conservation of plant NLR functions. Front Immunol 4:297

Jones JD (2001) Putting knowledge of plant disease resistance genes to work. Curr Opin Plant Biol 4:281e7

Jones JDG, Dangl JL (2006) The plant immune system. Nature 444:323–329

Jones DA, Jones JDG (1997) The role of leucine rich repeat proteins in plant defenses. Adv Bot Res Inc Adv Plant Pathol 24:120e7

Joshi RK, Nayak S (2011) Functional characterization and signal transduction ability of nucleotide-binding site-leucine-rich repeat resistance genes in plants. Genet Mol Res 10(4):2637–2652

Jupe F, Pritchard L, Etherington GJ, MacKenzie K, Cock PJA, Wright F, Sharma SK, Bolser D, Bryan GJ, Jones JDG, Hein I (2012) Identification and localization of the NB-LRR gene family within the potato genome. BMC Genomics 13:75

Jupe F, Witek K, Verweij W, Sliwka J, Pritchard L, Etherington GJ, Maclean D, Cock PJ, Leggett RM, Bryan GJ, Cardle L, Hein I, Jones JD (2013) Resistance gene enrichment sequencing (RenSeq) enables reannotation of the NB-LRR gene family from sequenced plant genomes and rapid mapping of resistance loci in segregating populations. Plant J 76:530–544

Kajava AW (1998) Structural diversity of leucine-rich repeat proteins. J Mol Biol 277:519–527

Kaloshian I (2004) Gene of gene disease resistance: breeding insect pest and pathogen defense. J Chem Ecol 30:2419–2438

Keen NT (1990) Gene-for-gene complementarity in plant-pathogen interactions. Annu Rev Genet 24:447–463

Kim S, Park M, Yeom SI, Kim YM, Lee JM, Lee HA, Seo E, Choi J, Cheong K, Kim KT, Jung K, Lee GW, Oh SK, Bae C, Kim SB, Lee HY, Kim SY, Kim MS, Kang BC, Jo YD, Yang HB, Jeong HJ, Kang WH, Kwon JK, Shin C, Lim JY, Park JH, Huh JH, Kim JS, Kim BD, Cohen O, Paran I, Suh MC, Lee SB, Kim YK, Shin Y, Noh SJ, Park J, Seo YS, Kwon SY, Kim HA, Park JM, Kim HJ, Choi SB, Bosland PW, Reeves G, Jo SH, Lee BW, Cho HT, Choi HS, Lee MS, Yu Y, Do Choi Y, Park BS, Van Deynze A, Ashrafi H, Hill T, Kim WT, Pai HS, Ahn HK, Yeam I, Giovannoni JJ, Rose JK, Sørensen I, Lee SJ, Kim RW, Choi IY, Choi BS, Lim JS, Lee YH, Choi D (2014) Genome sequence of the hot pepper provides insights into the evolution of pungency in Capsicum species. Nat Genet 46:270–278

Kobe B, Kajava AV (2000) When protein folding is simplified to protein coiling: the continuum of solenoid protein structures. Trends Biochem Sci 25:509–515

Kohler A, Rinaldi C, Duplessis S, Baucher M, Geelen D, Duchaussoy F, Meyers BC, Boerjan W, Martin F (2008) Genome-wide identification of NBS resistance genes in *Populus trichocarpa*. Plant Mol Biol 66:619–636

Lawrence JG, Finnegan EJ, Ayliffe MA, Ellisai JG (1995) The L6 gene for flax rust resistance 1s related to the *Arabidopsis* bacterial resistance gene RPSP and the tobacco vira1 resistance gene N. Plant Cell 7:1195e206

Leister RT, Dahlbeck D, Day B, Li Y, Chesnokova O, Staskawicz BJ (2005) Molecular genetic evidence for the role of SGT1 in the intramolecular complementation of Bs2 protein activity in *Nicotiana benthamiana*. Plant Cell 17:1268–1278

Liu J, Coaker G (2008) Nuclear trafficking during plant innate immunity. Mol Plant 1:411–422

Lukasik-Shreepaathy E, Vossen JH, Tameling WIL, De Vroomen MJ, Cornelissen BJC, Takken FLW (2012) J Exp Bot 63(8):3047–3060

Mackey D, Holt BFI, Wiig A, Dangl JL (2002) RIN4 interacts with *Pseudomonas syringae* type III effector molecules and is required for RPM1-mediated resistance in *Arabidopsis*. Cell 108:743–754

Matsushima N, Tanaka T, Enkhbayar P, Mikami T, Taga M, Yamada K, Kuroki Y (2007) Comparative sequence analysis of leucine-rich repeats (LRRs) within vertebrate toll-like receptors. BMC Genomics 8(1):124

Mayerhofer H, Sautron E, Rolland N, Catty P, Seigneurin-Berny D, Pebay-Peyroula E, Stéphanie R (2016) Structural insights into the nucleotide-binding domains of the P1B-type ATPases HMA6 and HMA8 from *Arabidopsis thaliana*. PLoS One 11(11):e0165666

McDowell JM, Woffenden BJ (2003) Plant disease resistance genes: recent insights and potential applications. Trends Biotechnol 21:178–183

McHale L, Tan X, Koehl P, Michelmore RW (2006) Plant NBS-LRR proteins: adaptable guards. Genome Biol 7:212.1–212.11

Mestre P, Baulcombe DC (2006) Elicitor-mediated oligomerization of the tobacco N disease resistance protein. Plant Cell 18:491–501

Meyers BC, Dickerman AW, Michelmore RW, Sivaramakrishnan S, Sobral BW, Young ND (1999) Plant disease resistance genes encode members of an ancient and diverse protein family within the nucleotide binding superfamily. Plant J 20:317–332

Meyers BC, Kozik A, Griego A, Kuang H, Michelmore RW (2003) Genome-wide analysis of NB-LRR encoding genes in *Arabidopsis*. Plant Cell 15(4):809–834

Meyers BC, Kaushik S, Nandety RS (2005) Evolving disease resistance genes. Curr Opin Plant Biol 8:129–134

Minsavage GV, Dahlbeck D, Whalen MC, Kearney B, Bonas U, Staskawicz BJ, Stall RE (1990) Gene- for-gene relationships specifying disease resistance in *Xanthomonas campestris* pv vesicatoria – pepper interactions. Mol Plant-Microbe Int 3:41–47

Moffett P, Farnham G, Peart J, Baulcombe DC (2002) Interaction between domains of a plant NBS-LRR protein in disease resistance-related cell death. EMBO J 21:4511–4519

Morel J, Dangl J (1997) The hypersensitive response and the induction of cell death in plants. Cell Death Differ 4:671e83

Muskett PR, Kahn K, Austin MJ, Moisan LJ, Sadanandom A, Shirasu K, Jones JDG, Parker JE (2002) *Arabidopsis* RAR1 exerts rate-limiting control of R gene mediated defenses against multiple pathogens. Plant Cell 14:979–992

Nandety RS, Caplan JL, Cavanaugh K, Perroud B, Wroblewski T, Michelmore RW, Meyers BC (2013) The role of TIRNBS and TIR-X proteins in plant basal defense responses. Plant Physiol 162:1459–1472

Noutoshi Y, Ito T, Seki M, Nakashita H, Yoshida S, Marco Y, Shirasu K, Shinozaki K (2005) A single amino acid insertion in the WRKY domain of the *Arabidopsis* TIR-NBS-LRR-WRKY-type disease resistance protein SLH1 (sensitive to low humidity 1) causes activation of defense responses and hypersensitive cell death. Plant J 43:873–888

Nürnberger T, Nennstiel D, Jabs T, Sacks WR, Hahlbrock K, Scheel D (1994) High affinity binding of a fungal oligopeptide elicitor to parsley plasma membranes triggers multiple defense responses. Cell 78:449e60

Ooijen VG, Mayr G, Albrecht M, Cornelissen BJC, Takken FLW (2008) Transcomplementation, but not physical association of the CC-NBARC and LRR domains of tomato R protein Mi-1.2 is altered by mutations in the ARC2 subdomain. Mol Planta 1:401–410

Panda N, Khush GS (1995) Host plant resistance to insects. CAB Int, Wallingford, p 431

Petersen M, Brodersen P, Naested H, Andreasson E, Lindhart U, Johansen B, Nielsen HB, Lacy M, Austin MJ, Parker JE, Sharma SB, Klessig DF, Martienssen R, Mattsson O, Jensen AB, Mundy J (2000) *Arabidopsis* map kinase 4 negatively regulates systemic acquired resistance. Cell 103:1111–1120

Quisenberry SS, Clement SL (2002) Conservation and use of global plant genetic resources for insect resistance. Aust J Agric Res 53:865–872

Rafiqi M, Bernoux M, Ellis JG, Dodds PN (2009) In the trenches of plant pathogen recognition: role of NB-LRR proteins. Semin Cell Dev Biol 20:1017–1024

Rairdan GJ, Moffett P (2006) Distinct domains in the ARC region of the potato resistance protein Rx mediate LRR binding and inhibition of activation. Plant Cell 18:2082–2093

Reinink K, Dieleman FL (1989) Comparison of sources of resistance to leaf aphids in lettuce (*L. sativa* L.) Euphytica 40:21–29

Riedl SJ, Li W, Chao Y, Schwarzenbacher R, Shi Y (2005) Structure of the apoptotic protease-activating factor 1 bound to ADP. Nature 434:926–933

Rosello S, Diez MJ, Nuez F (1998) Genetics of tomato spotted wilt virus resistance coming from *Lycopersicon peruvianum*. Eur J Plant Pathol 104:499–509

Seeholzer S, Tsuchimatsu T, Jordan T, Bieri S, Pajonk S, Yang W, Jahoor A, Shimizu KK, Keller B, Schulze-Lefert P (2010) Diversity at the Mla powdery mildew resistance locus from cultivated barley reveals sites of positive selection. Mol Plant-Microbe Interact 23:497–509

Sela H, Spiridon LN, Petrescu AJ, Akerman M, Mandel-Gutfreund Y, Nevo E, Loutre C, Keller B, Schulman AH, Fahima T (2012) Ancient diversity of splicing motifs and protein surfaces in the wild emmer wheat (*Triticum dicoccoides*) LR10 coiled coil (CC) and leucine-rich repeat (LRR) domains. Mol Plant Pathol 13:276–287

Seo YS, Rojas MR, Lee JY, Lee SW, Jeon JS, Ronald P, Lucas WJ, Gilbertson RL (2006) A viral resistance gene from common bean functions across plant families and is up-regulated in a non-virus- specific manner. Proc Natl Acad Sci U S A 103:11856–11861

Shao ZQ, Zhang YM, Hang YY, Xue JY, Zhou GC, Wu P, Wu XY, Wu XZ, Wang Q, Wang B, Chen JQ (2014) Long-term evolution of nucleotide-binding site-leucine-rich repeat genes: understanding gained from and beyond the legume family. Plant Physiol 166:217–234

Shao ZQ, Xue JY, Wu P, Zhang YM, Wu Y, Hang YY, Wang B, Chen JQ (2016) Large-scale analyses of angiosperm nucleotide-binding site-leucine-rich repeat (NBS-LRR) genes reveal three anciently diverged classes with distinct evolutionary patterns. Plant Physiol 70:01487

Shen KA, Chin DB, Arroyo-Garcia R, Ochoa OE, Lavelle DO, Wroblewski T, Meyers BC, Michelmore RW (2002) Dm3 is one member of a large constitutively-expressed family of NBS-LRR encoding genes. Mol Plant-Microbe Int 15:251–261

Song WY, Wang GL, Chen LL, Kim HS, Pi LY, Holsten T, Gardner J, Wang B, Zhai WX, Zhu LH, Fauquet C, Ronald P (1995) A receptor kinase-like protein encoded by the rice disease resistance gene, Xa21. Science 180:4–6

Song H, Wang P, Li C, Han S, Zhao C, Xia H, Bi Y, Guo B, Zhang X, Wang X (2017) Comparative analysis of NBS-LRR genes and their response to *Aspergillus flavus* in Arachis. PLoS One 12(2):e0171181

Stange C, Matus JT, Domínguez C, Perez-Acle T, Arce-Johnson P (2008) The N-homologue LRR domain adopts a folding which explains the TMVCg- induced HR-like response in sensitive tobacco plants. J Mol Graph Model 26:850–860

Takahashi A, Casais C, Ichimura K, Shirasu K (2003) HSP90 interacts with RAR1 and SGT1 and is essential for RPS2-mediated disease resistance in Arabidopsis. Proc Natl Acad Sci U S A 100(117):77–82

Takken FW, Goverse A (2012) How to build a pathogen detector: structural basis of NB-LRR function. Curr Opin Plant Biol 15:1–10

Takken FLW, Albrecht M, Tameling WIL (2006) Resistance proteins: molecular switches of plant defence. Curr Opin Plant Biol 9:383–390

Tameling WIL, Elzinga SDP, Darmin PS, Vossen JH, Takken FLW, Haring MA, Cornelissen BJ (2002) The tomato R gene products I-2 and Mi-1 are functional ATP bindingproteins with ATPase activity. Plant Cell 14(11):2929–2939

Tameling WIL, Vossen JH, Albrecht M, Lengauer T, Berden JA, Haring MA, Cornelissen BJC, Takken FLW (2006) Mutations in the NBARC domain of I-2 that impair ATP hydrolysis cause auto activation. Plant Physiol 140:1233–1245

Tarr DEK, Alexander HM (2009) TIR-NBS-LRR genes are rare in monocots: evidence from diverse monocot orders. BMC Res Notes 2:197

Thomma BP, Penninckx IA, Broekaert WF, Cammue BP (2001) The complexity of disease signaling in Arabidopsis. Curr Opin Immunol 13:63–68

Ueda H, Yamaguchi Y, Sano H (2006) Direct interaction between the tobacco mosaic virus helicase domain and the ATP-bound resistance protein, N factor during the hypesensitive response in tobacco plants. Plant Mol Biol 61:31–45

Ulker B, Somssich IE (2004) WRKY transcription factors: from DNA binding towards biological function. Curr Opin Plant Biol 7:491–498

Urbach MJ, Ausubel MF (2017) The NBS-LRR architectures of plant R-proteins and metazoan NLRs evolved in independent events. Proc Natl Acad Sci U S A 114(5):1063–1068

Van den Ackerveken GF, Van Kan JA, De Wit PJGM (1992) Molecular analysis of the avirulence gene Avr9 of the fungal tomato pathogen *Cladosporium fulvum* fully supports the gene-for-gene hypothesis. Plant J 2:359–366

Van der Biezen EA, Jones JDG (1998) The NB-ARC domain: a novel signalling motif shared by plant resistance gene products and regulators of cell death in animals. Curr Biol 8:R226–R228

Van der Hoorna RAL, Kamoun S (2008) From guard to decoy: a new model for perception of plant pathogen effectors. Plant Cell 20(8):2009–2017

Van der Voort JR, Kanyuka K, Van der Vossen E, Bendahmane A, Mooijman P, Klein-Lankhorst R, Stiekema W, Baulcombe D, Bakker J (1999) Tight physical linkage of the nematode resistance gene Gpa2 and the virus resistance gene Rx on a single segment introgressed from the wild species *Solanum tuberosum* subsp. andigena CPC1673 into cultivated potato. Mol Plant-Microbe Int 12:197–206

Van der Vossen EA, Sikkema A, Hekkert BTL, Gros J, Stevens P, Muskens M, Wouters D, Pereira A, Stiekema WJ, Allefs S (2003) An ancient R-gene from the wild species *Solanum bulbocastanum* confers broad- spectrum resistance to *Phytophthora infestans* in cultivated potato and tomato. Plant J 36:867–882

Venter V, Botha AM (2000) Development of markers linked to *Diuraphis noxia* resistance in wheat using a novel PCR-RFLP approach. Theor Appl Genet 100:965–970

Wan H, Yuan W, Ye Q, Wang R, Ruan M, Li Z, Zhou G, Yao Z, Zhao J, Liu S, Yang Y (2012) Analysis of TIR- and non-TIR-NBS-LRR disease resistance gene analogous in pepper: characterization, genetic variation, functional divergence and expression patterns. BMC Genomics 13:502

Wang CIA, Gunčar G, Forwood JK, Teh T, Catanzariti AM, Lawrence GJ, Loughlin FE, Mackay JP, Schirra HJ, Anderson PA, Ellis JG, Dodds PN, Kobe B (2007) Crystal structures of flax rust avirulence proteins AvrL567-A and -D reveal details of the structural basis for flax disease resistance specificity. Plant Cell 19(9):2898–2912

Wang W, Wen Y, Berkey R, Xiao S (2009) Specific targeting of the *Arabidopsis* resistance protein RPW8.2 to the interfacial membrane encasing the fungal haustorium renders broad spectrum resistance to powdery mildew. Plant Cell 21(9):2898–2913

Wani SH (2010) Inducing fungus-resistance into plants through biotechnology. Not Sci Biol 2(2):14–21

Watts VM (1947) The use of *Lycopersicon peruvianum* as a source nematode resistance in tomatoes. Proc Am Soc Hort Sci 49:233–234

Williams SJ, Sohn KH, Wan L, Bernoux M, Sarris PF, Segonzac C, Ve T, Ma Y, Saucet SB, Ericsson DJ, Casey LW, Lonhienne T, Winzor DJ, Zhang X, Coerdt A, Parker JE, Dodds PN, Kobe B, Jones JD (2014) Structural basis for assembly and function of a heterodimeric plant immune receptor. Science 344:299–303

Williamson VM, Kumar A (2006) Nematode resistance in plants: the battle underground. Trends Genet 22:396–403

Xiao S, Ellwood S, Calis O, Patrick E, Li T, Coleman M, Turner JG (2001) Broad spectrum mildew resistance in *Arabidopsis thaliana* mediated by RPW8. Science 291:118–120

Xiao S, Calis O, Patrick E, Zhang G, Charoenwattana P, Muskett P, Parker JE, Turner JG (2005) The atypical resistance gene, RPW8, recruits components of basal defence for powdery mildew resistance in *Arabidopsis*. Plant J 42(1):95–110

Yencho GC, Cohen MB, Byrne PF (2000) Applications of tagging and mapping insect resistance loci in plants. Annu Rev Entomol 5:393–422

Young BJD, Roger WI (2006) Plant NBS-LRR proteins in pathogen sensing and host defense. Nat Immunol 7(12):1243–1249

Zhang YM, Shao ZQ, Wang Q, Hang YY, Xue JY, Wang B, Chen JQ (2016) Uncovering the dynamic evolution of nucleotide-binding site-leucine-rich repeat (NBS-LRR) genes in Brassicaceae. J Integr Plant Biol 58:165–177

Zhong Y, Yin H, Sargent DJ, Malnoy M, Cheng ZM (2015) Species-specific duplications driving the recent expansion of NBS-LRR genes in five Rosaceae species. BMC Genomics 16(1):77

Zhou T, Wang Y, Chen JQ, Araki H, Jing Z, Jiang K, Shen J, Tian D (2004) Genome-wide identification of NBS genes in japonica rice reveals significant expansion of divergent non-TIR NBS-LRR genes. Mol Gen Genomics 271:402–415

Zhao B, Lin X, Poland J, Trick H, Leach J, Hulbert S (2005) A maize resistance gene functions against bacterial streak disease in rice. Proc Natl Acad Sci 102(43):15383–15388

6 Recent Advancement on Map Kinase Cascade in Biotic Stress

Monika Jaggi

Abstract

Mitogen-activated protein kinases (MAPKs) are cell-signaling enzymes that govern an extraordinarily discrete range of biological processes in eukaryotes. MAPK cascade has come up as one of the most well-studied signaling pathways in recent years. It plays a vital role in transmitting extracellular signals to the nucleus in response to various environmental stresses. A MAPK cascade composes of a three strata system where each stratum is phosphorylated by upper stratum. It is depicted as a MAP3K-MAP2K-MAPK module that serves as a link between upstream receptors and downstream targets. MAP2K being the middle point of this cascade converge all the signals from upstream MAP3Ks and target genome through downstream MAPKs. Occasionally, MAP4Ks also get employed in coupling upstream signaling components to the core MAPK cascade. MAPKs then direct various genes involved in stress responses as well as cellular and developmental processes. Therefore, in this chapter, an endeavor has been made to compile the role of MAPK cascade in biotic stress in plants.

Keywords

Elicitor-triggered immunity · PAMP-triggered immunity · Plant defense · Signal transduction

M. Jaggi (✉)
National Institute of Science Communication and Information Resources (NISCAIR), New Delhi, Delhi, India
e-mail: monikajaggi@niscair.res.in; mj_molbio@rediffmail.com

A. Singh, I. K. Singh (eds.), *Molecular Aspects of Plant-Pathogen Interaction*,
https://doi.org/10.1007/978-981-10-7371-7_6

6.1 Introduction

Plants are one of the most essential components of our ecosystem. They photosynthesize and convert light energy into chemical energy, thus a rich source of energy/food for most life-forms on earth including microbes capable of infecting plants. These microbes or phytopathogens when infectious result in biomass reduction, decreased fertility, or even death, thus a prime reason of human concern, as they pose an increasing threat to food security by decreasing the quality or quantity of crop production. In order to combat such infections, plants have developed different response systems at molecular, cellular, organ levels through evolution for effective protection. Plants are predominantly resistant to the majority of pathogens. Primarily, they are protected against pathogens by various physical barriers, e.g., cuticle and cell wall. Cuticle is a combination of cutin and waxes and adcrusted on the epidermis of most aerial plant organs (Yeats and Rose 2013). Besides preventing water loss and protecting against UV radiation, it also forms a barrier against pathogens. Cell wall provides both skeletal support and protection. Fungal pathogens can penetrate these barriers by mechanical rupture or enzymatic degradation but bacteria cannot. The latter instead make entry into plant cells via natural openings, e.g., stomata, hydathodes, wound sites, etc. (Melotto et al. 2008).

In addition to physical barriers, plants also possess a variety of antimicrobial compounds that are produced during pathogen attack. Only a few successful microbes that can rupture the preformed barriers then have to confront the plant immune system which is comprised of pathogen recognition and defense. The first layer of the plant immune system involves pathogen perception via the identification of conserved pathogen-associated molecular patterns (PAMPs) by plant pathogen or pattern recognition receptors (PRRs). PRRs are usually plasma membrane-bound receptor-like kinases (RLKs) or receptor-like proteins with extracellular domains allowing MAMP perception (Bohm et al. 2014). Sometimes PAMPs are of nonpathogenic origin too, hence alternatively known as microbe-associated molecular patterns (MAMPs). The latter are microbe-derived particles that are important for microbes, but that can be identified by plants. MAMPs can be proteins (e.g., bacterial flagellin and elongation factor Tu), carbohydrates (e.g., fungal chitin), lipopolysaccharides, etc. (Felix et al. 1999; Kunze et al. 2004; Albert 2013). There is another set of molecules which are plant degradation products arising from the activity of invading pathogens following pathogen attacks. These are known as damage-associated molecular patterns (DAMPs), for example, cutin monomers (Yeats and Rose 2013; Serrano et al. 2014), cell wall damage products (Hamann 2012), or endogenous peptides, such as AtPep1, which is derived from its precursor PROPEP1 (Huffaker et al. 2006). The recognition of all these molecules by the plant incites the pathogen-triggered immunity or pattern-triggered immunity (PTI), a sophisticated set of reactions meant for resisting against a pathogen attack (Boller and Felix 2009; Yamaguchi and Huffaker 2011). Pathogen perception can also ensue via the detection of pathogen effectors, which are molecules manufactured by the pathogens and carried in the extracellular matrix or into the plant cell to augment pathogen competence by, for example, counteracting the induction of

PTI. Plants that are incapable of recognizing these effectors are susceptible to a pathogen, while plants that can recognize the effectors via disease resistance proteins (R proteins) can induce an immune response called effector-triggered immunity (ETI). The coevolution of plants and pathogens and particularly their range of effectors and R proteins created the so-called zigzag model (Jones and Dangl 2006).

Plants are proficient in inducing a number of defense mechanisms upon pathogen infection, including blockage of nutrient transfer from the cytosol to the apoplast to inhibit bacterial multiplication (Chen et al. 2010; Wang et al. 2012), closing of stomata to curb entry of bacteria (Melotto et al. 2008; Sawinski et al. 2013), production and secretion of antimicrobial compounds including phytoalexins (Cowan 1999; van Loon et al. 2006; Ahuja et al. 2012; Bednarek 2012), generation of reactive oxygen species (ROS), which have toxic effects on pathogens (O'Brien et al. 2012), and a programmed cell death (PCD), referred to as the hypersensitive response (HR), at the site of infection to terminate pathogen expansion (Mur et al. 2008).

The inception of these defense mechanisms upon pathogen recognition depends on an intricate network of signaling pathways. Nearly all of the signal transduction mechanisms are inflected by protein phosphorylation and dephosphorylation that is regulated by protein kinases and protein phosphatases (Zolnierowicz and Bollen 2000). This swift and effective signaling system performs all obligations desired to interconnect the signaling with gene expression processes in coordination with the physiological status of the cell. Several protein kinases (PKs) involved in cellular signal transduction services have been identified in plants. PKs transform proteins by catalyzing the addition of monophosphate groups to the side chains of amino acid residues (usually serine, threonine, and/or tyrosine) in the protein backbone. This process is reversible with the help of an enzyme, phosphatase which is capable of removing phosphate group (Zolnierowicz and Bollen 2000). The plant PK superfamily embodies different classes of kinases grouped on the basis of their amino acid sequence similarity. Mitogen-activated protein kinases (MAPKs) are one of the most conserved and best- characterized protein kinase signaling pathways. MAPK was first unveiled by Sturgill and Ray (1986) as a microtubule-associated protein kinase. It comprises a family of serine/threonine protein kinases. The extracellular signals are read by the uppermost MAPKKKs and are transferred to substrates through MAPKs (Hardie 1999). In plants, the first MAPK was observed in alfalfa, and pea succeeded by cloning of MAPKs from *Arabidopsis thaliana* (Mizoguchi et al. 1993) and *Nicotiana* (Wilson et al. 1996). Both environmental stresses and development signals are capable of triggering MAPK signaling cascade (Sinha et al. 2011; Xu and Zhang 2015), and the latter also have a significant role in plant disease resistance signaling (reviewed by Meng and Zhang (2013)). However, the molecular mechanisms of how signal perception is transduced to MAPK activation remain elusive. Recently, Cheng et al. (2015) reported a novel plant immune pathway where proteases secreted by pathogens stimulate a previously unknown signaling pathway in *A. thaliana* involving the Gα, Gβ, and Gγ subunits of heterotrimeric G-protein complexes, which function upstream of a MAPK cascade. This chapter mainly emphasizes the involvement of mitogen-activated protein kinases (MAPKs) in biotic stress.

6.2 General Framework of MAPK Cascade

MAPK cascade is one of the most widely studied signaling pathways, which eukaryotic cells use to adapt in response to extracellular stimuli. These are immensely conserved signaling modules. Typically, a MAPK cascade consists of three kinases, a MAPKKK (MAP3K or MEKK), a MAPKK (MAP2K or MEK or MKK), and a MAPK (MPK), which phosphorylate and therefore activate each other in a specific way. It is a three-tier system where each tier is phosphorylated by upper tier. The uppermost tier is composed of MAPKKKs that phosphorylate two amino acids in the S/T-X3−5-S/T motif of the MAPKK activation loop. The middle tier consists of MAPKKs that activate a MAPK through double phosphorylation of the T-X-Y motif in the activation loop. These converge all the signals from upstream MAP3Ks and target genome through downstream MAPKs. MAPKs are serine/threonine kinases capable of phosphorylating a wide range of substrates, including other kinases and/or transcription factors. The fourth level of kinases, named MAPKKKKs (or MEKKKs), may act as adaptors linking upstream signaling steps to the core MAPK cascades. Interactions between kinases within a MAPK cascade occur through docking sites present in the kinases and/or with the help of external scaffolding proteins (Hamel et al. 2006; Ichimura et al. 2006).

MAPK proteins possess a highly conserved kinase domain enclosing 11 subdomains. Depending on the amino acid motif present at the phosphorylation site between the sub-domains VII and VIII, they are grouped as TEY and TDY activation motifs. The TEY subtype can be further formulated into three groups, A, B, and C, whereas the TDY subtype forms a more distant group D (MAPK Group 2002). Group A MAPKs are mostly associated with developmental processes and are activated in response to biotic and abiotic stresses, whereas group B members are implicated in pathogen defense and abiotic stress responses (Rodriguez et al. 2010). Group D members possess a C-terminal common docking (CD) domain that may act as a docking site for MAPKKs. The former TEY subtype contains a phosphorylation motif in the activation loop, while TDY subtype has motif within its active site buried at its domain surface. The N-terminal domain of MAPK protein has 135 residues which are aligned largely in the form of β-sheets and a glycine-rich loop named as phosphate anchor ribbon, while the C-terminal domain is of 225 amino acids bearing a catalytic base, Mg^{2+} binding sites, and the phosphorylation lip (activation loop). The sequence of phosphorylation and activation loop affects the substrate specificity. The TXY(XD/E/P) motif is a dual phosphorylation site, and phosphorylation of both residues is a prerequisite for the activation of this cascade.

MAPKKs could be split into four groups (A, B, C, and D) based on sequence similarities. Members of groups A and B carry eight to nine exons, while groups C and D possess only one exon. Kinases in group B are depicted by a nuclear transfer factor (NTF) domain (MAPK group 2002) which increases the nuclear import of cargo proteins and hence actively involved in cytoplasmic-nuclear transport. Different members of the same group are activated in the presence of different stimuli (Xu et al. 2008a; Zhou et al. 2009).

MAPKKKs are highly divergent and constitute the largest and most complex group of MAPK cascade. Rao et al. (2010) categorized MAPKKKs from rice and *Arabidopsis* into three subgroups, i.e., Raf, ZIK, and MEKK. Raf family members are characterized by the presence of a long N-terminal regulatory domain and a C-terminal kinase domain. In contrast, ZIK family members largely have N-terminal kinase domain. MEKK family members possess relatively a less conserved protein structure with kinase domain located either at N- or C-terminal or central part of the protein. Ubiquitin-interaction motif and ACT domain, which has a regulatory role in a wide range of metabolic enzymes, are seen only in the members of Raf family from rice and *Arabidopsis* (Rao et al. 2010).

6.3 MAPK Substrates

Presently, it is a great challenge to link a given Protein Kinase (PK) to its substrate with high specificity. In higher eukaryotes, nearly one-third of all proteins are governed by phosphorylation by PKs. Although plants are known to contain more than 1000 PKs, the substrates of only a few of them have been correctly identified. MAPK substrates have been the most extensively studied. An estimated 50% of the MAPK substrates are transcription factors. It was reported that MAPK phosphorylation regulates transcription factors by altering their activity, localization, and/or stability. A regular MAPK has an active site and a common docking site. Both sites are placed nearby and are involved in recognition and binding of target proteins. MAPK phosphatases and scaffolding proteins play a regulatory role in MAPK signaling specificity, location, and duration.

To identify potential MAPK substrates, direct protein-protein interaction and activity screens were used in various studies (Bethke et al. 2009; Zhang et al. 2016a). Feilner et al. (2005) used protein microarray and identified 48 in vitro substrates for MPK3 and 39 substrates for MPK6. Twenty six substrates were found common to both MAPKs. Similarly, 570 in vitro substrates for 10 different MAPKs were identified by employing *Arabidopsis* protein microarrays (Popescu et al. 2009). There was an overlap of 30–40% between the MPK3 and MPK6 substrates suggesting some functional redundancies of the two MAPKs. In *Arabidopsis*, ACS6 (1-aminocyclopropane-1-carboxylic acid synthase), PHOS32, bHLH speechless, VIP1, EIN3, WRKY53, WRKY33, MKS1, and ERF104 have also been identified as MAPK substrates (Andreasson and Ellis 2010). Sheikh et al. (2016) characterized WRKY46 as a substrate of AtMPK3. The type 3 effector NopL of *Sinorhizobium* sp. strain NGR234 is a newly identified mitogen-activated protein kinase substrate (Ge et al. 2016). MKS1 was reported to interact with MPK4 in a yeast 2-hybrid analysis and has a putative role in plant defense (Andreasson et al. 2005). The MPK4-MKS1 interaction was validated in planta by co-immunoprecipitation, and it was found that MKS1 further interacted with two transcription factors, WRKY29 and WRKY33. Later on, it was reported that MPK4 is found in complexes in vivo with PAT1, a constituent of the mRNA decapping machinery (Roux et al. 2015). PAT1 is also phosphorylated by MPK4, and, upon flagellin PAMP treatment, PAT1 accumulates and localizes to cytoplasmic processing (P) bodies which are sites for mRNA decay. Pat1 mutants exhibit dwarfism and derepressed immunity dependent on the immune receptor SUMM2. Since mRNA decapping is a critical step in mRNA turnover, linking MPK4 to mRNA decay via PAT1 provides another mechanism by which MPK4 may rapidly instigate immune responses (Roux et al. 2015). There is another study by Li et al. (2015) which demonstrates the negative regulation of plant defense by MPK4 via ASR3 (*Arabidopsis* SH4-related3), a trihelix transcriptional repressor. ERF104, an ethylene response factor, is specifically phosphorylated by MPK6 in vitro and is involved in defense responses (Bethke et al. 2009).

6.4 MAPK Modules in Biotic Stress

As discussed previously, plants have multilayered defense mechanism to protect themselves from pathogens (Jones and Dangl 2006). The first mechanism involves pattern recognition receptors (PRRs), which recognize conserved PAMPs. Under stress, these receptors (PRRs) get activated and stimulate convergent intracellular signaling pathways, which initiates the establishment of PAMP-triggered immunity (PTI). Defense responses in PTI are, in general, transient and are not associated with HR cell death. The other mechanism is known as effector-triggered immunity (ETI) which is essentially quantitatively stronger and longer-lasting than PTI. It often results in HR death.

6.4.1 MAPKs in PAMP-Triggered Immunity (PTI)

MAPKs play key roles in PTI signal mechanism by transducing signals from PRRs to downstream targets. The bacterial flagellin receptor FLS2, which identifies a conserved 22-amino-acid peptide (flg22) from flagellin (Gomez-Gomez and Boller 2000); the bacterial elongation factor EF-Tu receptor EFR, which acknowledges a conserved 18-amino-acid epitope (elf18) of EF-Tu (Zipfel et al. 2006); and the chitin receptor CERK1 from *Arabidopsis* (Miya et al. 2007; Wan et al. 2008) are among few well-characterized plant PRRs. Both flg22 and elf18 can provoke a strong but transient stimulation of MAPKs in *Arabidopsis*, including MPK3, MPK6, MPK4, and MPK11 (Asai et al. 2002; Zipfel et al. 2006; Suarez-Rodriguez et al. 2007; Ranf et al. 2011; Roux et al. 2011; Bethke et al. 2012). Upon activation by PAMPs, both FLS2 and EFR form heterodimers with BRI1-associated kinase (BAK1), a positive regulator of PTI (Chinchilla et al. 2007). Mutation in BAK1 leads to significant decline of flg22- and elf18-triggered activation of MPK3, MPK6, and MPK4 (Heese et al. 2007; Roux et al. 2011). BKK1 was reported to be essential for flg22- and elf18-induced MAPK activation, and it is thought to act collaboratively with BAK1 in the PAMP signaling (Roux et al. 2011). At present, molecules connecting the receptor complexes and MAPK cascades are still enigmatic. One possible regulator downstream of the FLS2-BAK1 complex is the

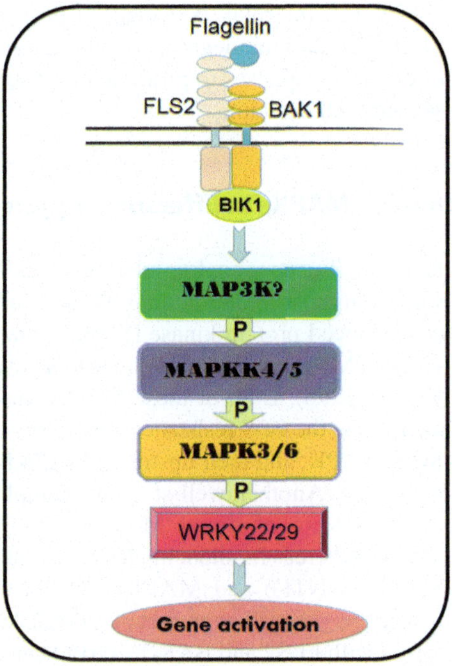

Fig. 6.1 Plant mitogen activated protein kinase (MAPK) cascade in response to flagellin in *Arabidopsis*

cytoplasmic protein kinase BIK1 (*Botrytis*-induced kinase 1), which interacts with and is phosphorylated by the FLS2-BAK1 complex and is needed for flg22-triggered PTI responses (Lu et al. 2010).

The first complete MAPK cascade in regulating plant defense against the bacterial pathogen, MAP3K1-MAPKK4/MAPKK5-MAPK3/MAPK6-WRKY22/WRKY29, was reported to be downstream of flagellin receptor kinase (FLS2 LRR) (Asai et al. 2002). Later studies revealed normal MPK3/6 activation in MAP3K1 mutant plants after flg22 treatment indicating that MAP3K1 is not upstream of MAPKK4/MAPKK5 (Ichimura et al. 2006; Nakagami et al. 2006; Suarez-Rodriguez et al. 2007). Instead, there are redundant orthologs of MAP3K1 in the flg22-elicited activation of MAPK3 and MAPK6. As mentioned above, an elicited FLS2 complex with BAK1 induces MAPK signaling cascade (Fig. 6.1). Another flg22-activated MAPK cascade, MAP3K1-MAPKK1/MAPKK2-MAPK4-MKS1, mediates jasmonate- and salicylate-dependent defense responses (Qiu et al. 2008). This cascade negatively regulates plant immunity as the mutants of either MAP3K1 or MAPK4 resulted in more resistance to pathogens. It was proposed that MAPK4 along with its substrate MAP kinase substrate 1 (MKS1) and WRKY33 controlled the expression of camalexin biosynthetic enzyme, phytoalexin deficient 3 (PAD3). Upon microbial attack, WRKY33 is released from this complex and attaches to a PAD3 promoter and activates its expression (Qiu et al. 2008). The production of camalexin was compromised in *mpk3* and *mpk6* mutants depicting a role of MPK3/6 in defense responses. MAPKK3 participates in jasmonate-mediated developmental signaling and pathogen defense responses through activation of MAPK6 (Takahashi et al.

2007). The interplay between the positive upregulation of defense responses via MAPK3/6 and the negative regulation via MAPK4 likely enables the tight control of defense responses to promote or restrict growth and make appropriate death versus survival decisions.

6.4.2 MAPKs in Effector-Triggered Immunity

ETI and *R* gene signaling also involves MAPK cascades as evidenced by various studies in tobacco and tomato. Wound-induced protein kinase (WIPK) and salicylic acid-induced protein kinase (SIPK) were found to get activated by tobacco mosaic virus (TMV) infection in plants that express the N resistance gene (Zhang and Klessig 1998). Both of these MAPKs also got induced in transgenic tobacco cells expressing the Cf-9 resistance gene. Loss-of-function experiments demonstrated that SIPK, WIPK, and their upstream MAPKK2 are required for N-gene-mediated TMV resistance. Another well-studied example is the tomato Pto-mediated ETI where overexpression of tomato MAP3K, MAPKKKα, increased the development of HR lesions, whereas its silencing led to the inhibition of HR (del Pozo et al. 2004). Later, MAPKK1-NTF6 and MAPKK2-SIPK were proposed to act downstream of Pto resistance gene (Ekengren et al. 2003). AvrPto and AvrPtoB effectors showed interaction with FLS2 and BAK 1. AvrPto can cease the binding of BAK1 to FLS2, while AvrPtoB mediates polyubiquitination and proteasome-dependent degradation of FLS2. In this way, both effectors negatively regulate MAPK signaling.

Many bacterial pathogens employ the HopAI1 effector, which has a unique phosphothreonine lyase activity that dephosphorylates MAPKs (Li et al. 2007). It was deciphered that inducible expression of HopAI1 inactivated the MAPK3 and MAPK6 resulting in the complete suppression of PAMP-induced genes and ROS burst. Another effector molecule HopPtoD$_2$ from *P. syringae* pv. tomato is known to possess phosphatase activity. Its expression in tobacco cells led to a decline in cell death induced by expression of the constitutively active MAPKK variant NtMEK2^{DD}. However, HopPtoD$_2$ does not inhibit flg22-mediated MAPK activation in *Arabidopsis*.

An interesting example of an opportunistic interaction between plants and bacteria is where pathogens manipulate host MAPK signaling and hijack AtMPK3 by *Agrobacterium tumefaciens*. The activation of MPK3 in response to flg22 or *Agrobacterium* results in the phosphorylation, subsequently followed by nuclear translocation of the host protein VIP1 (virE2-interacting protein 1). Agrobacteria hijack this nucleocytoplasmic shuttle system (VIP1) to transfer their T-DNA into the host nucleus, where it integrates into the host genome (Djamei et al. 2007). Because VIP1 does not only serve as a nuclear shuttle for the pathogenic T-DNA complex but can also induce the expression of defense genes, nuclear VIP1 would be counteracting *Agrobacterium* invasion. *Agrobacterium* targets nuclear VIP1 for proteasome degradation by secreting VirF effector, which helps in ceasing VIP1-induced plant defense genes activation.

Stomata play an essential role in plants as they are involved in various physiological functions like photosynthesis, respiration, and transpiration and also responsible for gas exchange. But sometimes, various pathogens use them as an entrance gate to enter into their hosts. Pathogen-induced stomatal closure restricts the invasion of various pathogens. Therefore, stomata are important in providing protection against both biotic and abiotic stresses. MAPKs have been implicated in both stomatal development and function. In drought stress, stomatal closure mediated by ABA involves MKK1, MPK3, and MPK6 (Hamel et al. 2006; Gudesblat et al. 2007). Previous studies have also indicated crosstalk between ABA and defense signaling. The MAPK cascade, comprising YODA-MAPKK4/MAPKK5-MAPK3/MAPK6, plays an important role in regulating stomatal development (Bergmann et al. 2004; Wang et al. 2007; Bayer et al. 2009; Popescu et al. 2009). MAPK9 and MAPK12 were reported to be downstream of H_2O_2 in regulating ABA-induced stomatal closure (Jammes et al. 2009). It was reported that mapk9-1/12-1 double mutants were highly susceptible to *Pseudomonas syringae* DC3000 compared to wild-type plants. These results suggested that the regulation of stomatal apertures by MAPK9 and MAPK12 contributed to the first line of defense against pathogens (Jammes et al. 2011). Whether pathogen-induced and ABA-induced stomatal closures are signaled via a common MAPK cascade remains to be elucidated.

Fungal pathogens are also known to induce MAPK cascades (Izumitsu et al. 2009). *Phytophthora infestans* attack led to the swift transcriptional induction of MAP3K19, MAPKK9, and MAPKK4, whereas *Botrytis cinerea* infection led to enhanced transcriptional activation of MAP3K18, MAP3K19, MAP3K20, Raf43, ZIK2, and ZIK8, suggesting the distinct signaling pattern in response to bacterial and fungal pathogen attack. Cardinale et al. (2002) described the activation of SIMK and SAMK by various fungal elicitors. Two alfalfa MAPKs, MMK2 and MMK3 (Medicago MAPK2 and MAPK3, respectively), were shown to be activated by fungal elicitors. Nevertheless, a fungal biocontrol agent *Trichoderma asperellum* was shown to induce systemic resistance in plants through a mechanism that employs jasmonic acid (JA) and ethylene (ET) signal transduction pathways resulting in activation of a Trichoderma-induced MAPK (TIPK) gene in cucumber (Shoresh et al. 2006). MAPKK1 is involved in defense responses including flg22-induced activation of MAPK4 (Asai et al. 2002; Meszaros et al. 2006). MAPKK7 is a positive regulator of systemic acquired response (Zhang et al. 2007). MAPKK7/9 is hypothesized to regulate cell death during pathogen defense. Several other groups reported proteins, e.g., NLPs (necrosis and ethylene-inducing peptide1-like proteins) triggering MAPK activation and inducing defense responses (Qutob et al. 2006).

6.5 Defense Studies from Different Plant Sources

Recently, a lot of information has been gathered where microbial attack led to the activation of MAPKs in different plant genera as described below. Ectopic expression of Cotton GhMPK11 decreased disease resistance through the gibberellin signaling pathway in transgenic *Nicotiana benthamiana* (Wang et al. 2016). In watermelon, transient expression of ClMPK1, ClMPK4-2, and ClMPK7 in *N. benthamiana* resulted in enhanced resistance to *B. cinerea* and upregulated expression of defense genes, while transient expression of ClMPK6 and ClMKK2-2 led to increased susceptibility to *B. cinerea* (Song et al. 2015). Approximately 74 MAP3Ks, 9 MAPKKs, and 19 MAPKs were identified in maize followed by functional studies (Kong et al. 2013; Liu et al. 2013). The role of ZmMKK1 in pathogen defense has been deciphered by Cai et al. (2013) where it played a differential function in necrotrophic versus biotrophic pathogen defense responses. Overexpression of ZmMPK5 in tobacco- induced increased resistance to viral pathogens and activated the expression of PR genes, e.g., PR1a, PR4, PR5, and EREBP (ethylene-responsive element binding protein) (Zhang et al. 2013). Pan et al. (2012) reported the involvement of ZmMPK17, a novel maize group D MAP kinase gene, in multiple stress responses. In *Gossypium*, GhMAPK6a played a negative role during pathogen infection as its overexpression increased sensitivity to the bacterial pathogen *Ralstonia solanacearum*. Another MAP2K from *G. hirsutum*, GhMKK1, showed increased susceptibility to the same pathogen by lowering the expression of PR genes, but no interaction between MAPK6a and MKK1 was observed (Lu et al. 2013).

Polyamines generated in plants play important roles in several developmental and physiological processes. AtMPK3 and AtMPK6 are known to play a positive role in the regulation of putrescine biosynthesis leading to bacterial pathogen defense in *Arabidopsis* (Kim et al. 2013). The *Hordeum vulgare* signaling protein MAP kinase 4 (HvMPK4) had a similar role to AtMPK4 as both showed negative regulation of SA in pathogen defense. HvMPK4 mutant lines detected elevated SA levels and were more resistant to hemibiotrophic fungal pathogen *Magnaporthe grisea* (Abass and Morris 2013). NtMKP1 (tobacco MAP kinase phosphatase) antisense lines exhibited increased resistance against a necrotrophic pathogen, *B. cinerea*, and lepidopteran herbivores, *Mamestra brassicae* and *Spodoptera litura*. It was suggested that NtMKP1 negatively regulates wound response and resistance against both necrotrophic pathogens and herbivorous insects through suppression of JA or ET pathways via inactivation of MAPK (Oka et al. 2013). Lately, NtMPK2 has been reported to positively regulate tobacco defense responses to *P. syringae* pv. tomato DC3000 (Zhang et al. 2016b).

6.6 Function of MAPK Activation in Plant Defense Responses

Plants respond differentially to invading pathogens and activate a variety of defense mechanisms that often includes stomatal closure, ROS generation, defense gene activation, phytoalexins accumulation, cell wall modification, and HR cell death. These defense responses are combined by a complex signaling network that involves MAPK cascades. In this section, the role of MAPKs in controlling the synthesis and/or signaling of defense hormones, reprogramming gene expression, and driving metabolic flow to antimicrobial metabolite synthesis, among other defense responses, will be discussed.

6.6.1 Defense Hormone Synthesis and/or Signaling

Phytohormones ethylene (ET), jasmonic acid (JA), and salicylic acid (SA) are an important class of signaling molecules involved in plant defense signaling (Wang et al. 2002; Broekaert et al. 2006; Browse 2009; Vlot et al. 2009; Spoel and Dong 2012). SA is usually involved against biotrophs or hemibiotrophs, while JA and ET are mostly important against necrotrophs. Upon microbial infection, plants activate MAPK cascade, which in turn modulates levels of these phytohormones. Plant MAPK cascades have been involved in both the regulation of defense hormone biosynthesis and the signaling events downstream of hormone sensing. Recent studies revealed that a subset of MAPKs in plants, represented by tobacco SIPK/Ntf4/WIPK and *Arabidopsis* MPK3/MPK6, plays key roles in regulating pathogen-induced ethylene biosynthesis (Liu and Zhang 2004; Han et al. 2010; Li et al. 2012). The ethylene signaling pathway is well characterized, and several works demonstrated that phosphorylation of ACS2 and ACS6 by MPK3 and MPK6 is a key step for ET production (Guo and Ecker 2004; Stepanova and Alonso 2009). Binding of ethylene to its receptors, ETR1, ETR2, ERS1, ERS2, and EIN4, results in the inactivation of the negative regulator CTR1, which leads to the derepression of the positive regulator EIN2, the stabilization of the EIN3/EIL transcription factors, ethylene-responsive gene expression, and eventually ethylene responses (Stepanova and Alonso 2009).

SA plays a pivotal role in plant defense responses, mainly through its downstream components, NPR1, and three redundant transcription factors, TGA2, TGA5, and TGA6 (Vlot et al. 2009; Spoel and Dong 2012). A current model advocates that NPR3 and NPR4 are SA receptors that govern NPR1 levels, ensuing cell death or survival depending on SA concentration (Yan and Dong 2014). AtMPK3 and, to a lesser extent, AtMPK6 were seen to play a principal role in SA-mediated stimulation of plants for disease resistance (Beckers et al. 2009). When plants were given benzothiadiazole (a functional analog of SA) treatment, it resulted in the accumulation of mRNA and inactive proteins of MPK3 and MPK6. Unlike MPK3/MPK6, the MPK4 is considered a negative regulator of SA signaling because *mpk4* mutant or mutation of its upstream MKK1/MKK2 and MEKK1showed elevated levels of SA, constitutive PR-1 expression, and SAR (Petersen et al. 2000; Kong et al. 2012). Lately, SUMM2, an R protein, was reported to be essential for activation of SA responses in the *mpk4* mutant, indicating that the constitutive SA responses in the *mpk4* mutant are induced by the SUMM2-mediated signaling pathway (Zhang et al. 2012). Additionally, MPK4 substrate 1 (MKS1), a substrate of MPK4, is also required for full activation of SA responses in the mpk4 mutant (Andreasson et al. 2005), but no interaction of MKS1 with SUMM2 has been detected (Zhang et al. 2012), proposing that SUMM2-mediated activation of SA responses may be independent of MKS1.

JA is another important defense hormone activated during pathogen infection, herbivore attack, and mechanical wounding (Browse 2009; Gfeller et al. 2010). In tobacco, WIPK and SIPK were shown to be vital for wounding- and herbivore-induced JA production, whereas overexpression of WIPK was seen to be sufficient to induce JA accumulation (Seo et al. 1999, 2007; Wu et al. 2007). However, in the conditional gain-of-function NtMEK2^{DD} plants, activation of WIPK and SIPK was insufficient to induce JA accumulation (Kim et al. 2003). It is possible that these two MAPKs are needed but not sufficient to induce JA synthesis. Later, AtMPK6 was reported to be an important regulator of the JA signaling pathway (Takahashi et al. 2007). AtMPK6 together with its upstream AtMKK3 was shown to be involved in JA-dependent negative regulation of JIN1/MYC2 expression and root growth. MPK4 was also implicated in the JA signaling pathway. The MPK4 deficient lines exhibited constitutive activation of SA-dependent defenses but compromised the expression of JA-responsive genes (Petersen et al. 2000). Interestingly, removing SA in the NahG/mpk4 double mutant failed to reverse the suppression of JA-inducible gene expression, suggesting that MPK4 positively regulates JA-inducible responses independent of its negative regulation of SA signaling (Petersen et al. 2000).

6.6.2 Activation of Defense Genes

MAPKs can activate defense genes via direct phosphorylation of downstream transcription factors. In *Arabidopsis*, MPK6 interacts with and phosphorylates ERF104 that activates defensin genes PDF1.2a and PDF1.2b (Bethke et al. 2009). Intriguingly, the interaction of MPK6 and ERF104 is rapidly lost in response to flg22, and this complex disruption requires MPK6 activity, suggesting that phosphorylation of ERF104 by MPK6 in response to flg22 leads to the release of ERF104 from MPK6 and defensin gene activation (Bethke et al. 2009). Meng et al. (2013) identified ERF6, another *Arabidopsis* ERF family member, as a new substrate of MPK3/MPK6. Phosphorylation of ERF6 by MPK3/MPK6 leads to stabilization of the ERF6 protein, which further activates the expression of multiple defense-related genes, including PDF1.1, PDF1.2a, PDF1.2b, ChiB, and HEL. The positive and negative roles for soybean MPK6 in regulating defense responses have also been reported (Liu et al. 2014).

6.6.3 Phytoalexin Biosynthesis

Phytoalexins are also known to play essential roles in plant defense (Hammerschmidt 1999; Dixon 2001; Ahuja et al. 2012). Camalexin (3-thiazol-2′-yl-indole) is the major phytoalexins produced in *Arabidopsis* and related Brassicaceae species (Tsuji et al. 1992; Ahuja et al. 2012). Conditional expression of constitutively active variants of MKK4 (MKK4^{DD}), MKK5 (MKK5^{DD}), or the functionally interchangeable NtMEK2 (NtMEK2^{DD}) from tobacco, all of which specifically activate MPK3/MPK6 in *Arabidopsis*, leads to camalexin induction and coordinated upregulation of multiple genes, including PAD2, CYP71A13, and PAD3, that encode enzymes in the camalexin biosynthetic pathway (Ren et al. 2008; Mao et al. 2011). Expression of constitutively active MKK9, which also activates MPK3/MPK6, induces camalexin accumulation (Xu et al. 2008b; Su et al. 2011). Consistent with the essential roles of MPK3/MPK6 in regulating camalexin biosynthesis, pathogen-induced camalexin accumulation is almost completely abolished in a rescued mpk3/mpk6 double mutant (Ren et al. 2008), whereas mutations in two MAPK phosphatases (MKP1 and PTP1), which target MPK3/MPK6, lead to constitutive camalexin accumulation (Bartels et al. 2009).

Biochemical data revealed that WRKY33 is a substrate of MPK3/MPK6, and the MPK3/MPK6 phosphorylation sites in WRKY33 are required for its function in vivo (Mao et al. 2011). Interestingly, MPK3/MPK6 also controls the pathogen-inducible expression of the WRKY33 gene, and the latter can bind to its own promoter, suggesting a potential MPK3/MPK6-mediated positive feedback regulatory loop that controls WRKY33 expression (Mao et al. 2011). WRKY33 was also reported to control camalexin production through its interaction with another MAPK, MPK4 (Qiu et al. 2008). It is likely that pathogen-responsive MAPK cascades in different families of plants are responsible for regulating the biosynthesis of different types of phytoalexins. The OsMKK4-OsMPK6 module in rice was shown to control the chitin elicitor-induced production of diterpenoid phytoalexins by regulating the expression of their biosynthetic genes (Kishi-Kaboshi et al. 2010). Tobacco SIPK/Ntf4 and WIPK also phosphorylate and thus activate the WRKY33-related NbWRKY8, which further induces the expression of 3-hydroxy-3-methylglutaryl CoA reductase2 (HMGR2), a key gene for the production of isoprenoid phytoalexins (Ishihama et al. 2011). Tobacco HMGR is one of the first few defense genes known to be highly induced by SIPK/WIPK activation (Yang et al. 2001; Kim and Zhang 2004).

6.6.4 Hypersensitive Response Cell Death

ETI frequently results in HR cell death, a process associated with MAPK activation, ROS generation, metabolic reprogramming, and SA accumulation (Greenberg and Yao 2004; Coll et al. 2011). Pharmacological studies using kinase inhibitors suggest that SIPK and WIPK activation in tobacco is involved in HR-like cell death (Zhang et al. 2000). Potential MAPKKKs upstream of the tobacco NtMEK2-SIPK/Ntf4/WIPK module include MAPKKKα and MAPKKKε (del Pozo et al. 2004; Melech-Bonfil and Sessa 2010). In the search for signaling events downstream of MAPKs in mediating HR cell death, ROS generation was found to be associated with MAPK-induced cell death (Ren et al. 2002).

6.6.5 PAMP/Pathogen-Induced Reactive Oxygen Species Burst

ROS is considered as an important and common messenger generated in various stresses and known to trigger the expression of many MAPKs. PTI is also associated with rapid ROS production, known as a ROS burst (Apel and Hirt 2004; Torres 2010). PAMP/pathogen-responsive MAPKs were believed to function downstream of early ROS burst in signaling plant immunity, because defense-related MAPKs, including *Arabidopsis* MPK3, MPK6, and MPK4 and tobacco SIPK and WIPK, can be activated by exogenously added H_2O_2 (Pitzschke et al. 2009). Recently, MAP3K7 was characterized to suppress the ROS burst downstream of FLS2. It was demonstrated that MAP3K7 negatively regulates flagellin-triggered signaling and basal immunity by attenuating MPK6 activity (Mithoe et al. 2016). The role of ROS-mediated MAPK signaling in plants has been reviewed by Siddhi et al. (2015) where they suggested that ROS can activate a similar MAPK cascade in different stresses and can exert differential responses accordingly.

6.6.6 Stomatal Immunity

Plants have an ability of closing their stomata as an immune response to pathogens. MAPKs have been cited in both ABA- and PAMP/pathogen-induced stomatal closure (Pitzschke and Hirt 2009). Gudesblat et al. (2007, 2009) suggested a role of MPK3 in stomatal immune response. Guard cell-specific silencing of MPK3 compromised PAMP/bacteria-induced stomatal closure but does not affect the stomatal closure induced by ABA. MPK6 was reported to modulate NO production by phosphorylation and activation of nitrate reductase NIA2 (Wang et al. 2010). Above results indicate that MPK3 regulates stomatal immunity by promoting NO synthesis, which has been implicated to be required for PAMP/pathogen-induced stomatal closure (Melotto et al. 2006).

6.7 Conclusion

In recent years, MAPK cascade has shown its universality among eukaryotes and has demonstrated its significance in multitude of stress elements. Multifunctionality and signaling specificity of MAPKs can be conferred by their ability to phosphorylate different substrates. For example, in *Arabidopsis*, MPK3/MPK6 can regulate camalexin production, ethylene biosynthesis, NO generation, and defensin gene expression by phosphorylating WRKY33, ACS2/ACS6, NIA2, and ERF104 proteins, respectively (Bethke et al. 2009; Han et al. 2010; Wang et al. 2010; Mao et al. 2011). The identification of additional MAPK substrates will extend our understanding of MAPK functions in plant disease resistance. One major void in our understanding of plant defense signaling is the connection(s) between the receptors/sensors and MAPK cascades. So, it's very crucial to identify these missing links in elucidating the functions and underlying molecular mechanisms of MAPKs in plant innate immunity. In light of these realities, new experimental strategies are needed to generate conditional mutant systems for functional analyses which might reveal the fine-tuning mechanisms in plant innate immunity system. The complete MAPK cascades have been identified in a few biotic and abiotic stresses, but many are still missing. So there is a great requisite to decode the complete signaling cascade with specific underlying mechanisms using novel approaches and strategies. This will help in engineering the MAPK cascades and their utilization in crop improvement.

Sawinski K, Mersmann S, Robatzek S, Bohmer M (2013) Guarding the green: pathways to stomatal immunity. Mol Plant-Microbe Interact 26:626–632
Seo S, Sano H, Ohashi Y (1999) Jasmonate -based wound signal transduction requires activation of WIPK, a tobacco mitogen -activated protein kinase. Plant Cell 11:289–298
Seo S, Katou S, Seto H, Gomi K, Ohashi Y (2007) The mitogen -activated protein kinases WIPK and SIPK regulate the levels of jasmonic and salicylic acids in wounded tobacco plants. Plant J 49:899–909
Serrano M, Coluccia F, Torres M, L'Haridon F, Metraux JP (2014) The cuticle and plant defense to pathogens. Front Plant Sci 5:274
Sheikh AH, Eschen-Lippold L, Pecher P, Hoehenwarter W, Sinha AK, Scheel D, Lee J (2016) Regulation of WRKY46 transcription factor function by mitogen -activated protein kinases in *Arabidopsis thaliana*. Front Plant Sci 7:61
Shoresh M, Gal-On A, Leibman D, Chet I (2006) Characterization of a MAPK gene from cucumber required for trichoderma-conferred plant resistance. Plant Physiol 142:1169–1179
Siddhi K. Jalmi, Alok K. Sinha (2015) ROS mediated MAPK signaling in abiotic and biotic stress-striking similarities and differences. Front Plant Sci 6
Sinha AK, Jaggi M, Raghuram B, Tuteja N (2011) Mitogen -activated protein kinase signaling in plants under abiotic stress. Plant Signal Behav 6:196–203
Song Q, Li D, Dai Y, Liu S, Huang L, Hong Y, Zhang H, Song F (2015) Characterization, expression patterns and functional analysis of the MAPK and MAPKK genes in watermelon (*Citrullus lanatus*). BMC Plant Biol 15:298
Spoel SH, Dong X (2012) How do plants achieve immunity? Defence without specialized immune cells. Nat Rev Immunol 12:89–100
Stepanova AN, Alonso JM (2009) Ethylene signaling and response: where different regulatory modules meet. Curr Opin Plant Biol 12:548–555
Sturgill TW, Ray LB (1986) Muscle proteins related to microtubule associated protein-2 are substrates for an insulin stimulatable kinase. Biochem Biophys Res Commun 134:565–571
Su T, Xu J, Li Y et al (2011) Glutathione -indole -3 -acetonitrile is required for camalexin biosynthesis in *Arabidopsis thaliana*. Plant Cell 23:364–380
Suarez-Rodriguez MC, Adams-Phillips L et al (2007) MEKK1 is required for flg22 -induced MPK4 activation in Arabidopsis plants. Plant Physiol 143:661–669
Takahashi F, Yoshida R, Ichimura K et al (2007) The mitogen -activated protein kinase cascade MKK3 -MPK6 is an important part of the jasmonate signal transduction pathway in Arabidopsis. Plant Cell 19:805–818
Torres MA (2010) ROS in biotic interactions. Physiol Plant 138:414–429
Tsuji J, Jackson EP, Gage DA, Hammerschmidt R, Somerville SC (1992) Phytoalexin accumulation in *Arabidopsis thaliana* during the hypersensitive reaction to *Pseudomonas syringae* pv syringae. Plant Physiol 98:1304–1309
van Loon LC, Rep M, Pieterse CM (2006) Significance of inducible defense -related proteins in infected plants. Annu Rev Phytopathol 44:135–162
Vlot AC, Dempsey DA, Klessig DF (2009) Salicylic acid, a multifaceted hormone to combat disease. Annu Rev Phytopathol 47:177–206
Wan J, Zhang XC, Neece D et al (2008) A LysM receptor -like kinase plays a critical role in chitin signaling and fungal resistance in Arabidopsis. Plant Cell 20:471–481
Wang KL, Li H, Ecker JR (2002) Ethylene biosynthesis and signaling networks. Plant Cell 14:S131–S151
Wang H, Ngwenyama N, Liu Y, Walker JC, Zhang S (2007) Stomatal development and patterning are regulated by environmentally responsive mitogen -activated protein kinases in Arabidopsis. Plant Cell 19:63–73
Wang P, Du Y, Li Y, Ren D, Song CP (2010) Hydrogen peroxide–mediated activation of MAP kinase 6 modulates nitric oxide biosynthesis and signal transduction in Arabidopsis. Plant Cell 22:2981–2998
Wang K, Senthil-Kumar M, Ryu CM, Kang L, Mysore KS (2012) Phytosterols play a key role in plant innate immunity against bacterial pathogens by regulating nutrient efflux into the apoplast. Plant Physiol 158:1789–1802

Wang F, Wang C, Yan Y, Jia H, Guo X (2016) Overexpression of cotton GhMPK11 decreases disease resistance through the gibberellin signaling pathway in transgenic *Nicotiana benthamiana*. Front Plant Sci 7:689

Wilson KP, Fitzgibbon MJ, Caron PR et al (1996) Crystal structure of p38 mitogen -activated protein kinase. J Biol Chem 271:27696–27700

Wu J, Hettenhausen C, Meldau S, Baldwin IT (2007) Herbivory rapidly activates MAPK signaling in attacked and unattacked leaf regions but not between leaves of *Nicotiana attenuata*. Plant Cell 19:1096–1122

Xu J, Zhang SQ (2015) Mitogen -activated protein kinase cascades in signaling plant growth and development. Trends Plant Sci 20:56–64

Xu H, Wang X, Sun X, Shi Q, Yang F, Du D (2008a) Molecular cloning and characterization of a cucumber MAP kinase gene in response to excess NO3− and other abiotic stresses. Sci Hortic 117:1–8

Xu J, Li Y, Wang Y et al (2008b) Activation of MAPK kinase 9 induces ethylene and camalexin biosynthesis and enhances sensitivity to salt stress in Arabidopsis. J Biol Chem 283:26996–27006

Yamaguchi Y, Huffaker A (2011) Endogenous peptide elicitors in higher plants. Curr Opin Plant Biol 14:351–357

Yan S, Dong X (2014) Perception of the plant immune signal salicylic acid. Curr Opin Plant Biol 20:64–68

Yang KY, Liu Y, Zhang S (2001) Activation of a mitogen -activated protein kinase pathway is involved in disease resistance in tobacco. Proc Natl Acad Sci U S A 98:741–746

Yeats TH, Rose JK (2013) The formation and function of plant cuticles. Plant Physiol 163:5–20

Zhang S, Klessig DF (1998) Resistance gene N -mediated de novo synthesis and activation of a tobacco mitogen -activated protein kinase by tobacco mosaic virus infection. Proc Natl Acad Sci U S A 95:7433–7438

Zhang S, Liu Y, Klessig DF (2000) Multiple levels of tobacco WIPK activation during the induction of cell death by fungal elicitins. Plant J 23:339–347

Zhang X, Dai Y, Xiong Y, Defraia C, Li J, Dong X, Mou Z (2007) Overexpression of Arabidopsis MAP kinase kinase 7 leads to activation of plant basal and systemic acquired resistance. Plant J 52:1066–1079

Zhang Z, Wu Y, Gao M et al (2012) Disruption of PAMP -induced MAP kinase cascade by a *Pseudomonas syringae* effector activates plant immunity mediated by the NB -LRR protein SUMM2. Cell Host Microbe 11:253–263

Zhang D, Jiang S, Pan J, Kong X, Zhou Y, Liu Y, Li D (2013) The overexpression of a maize mitogen -activated protein kinase gene (ZmMPK5) confers salt stress tolerance and induces defence responses in tobacco. Plant Biol (Stuttg) 16(3):558–570

Zhang T, Chen S, Harmon AC (2016a) Protein -protein interactions in plant mitogen -activated protein kinase cascades. J Exp Bot 67(3):607–618

Zhang X, Wang G, Gao J, Nie M, Liu W, Xia Q (2016b) Functional analysis of NtMPK2 uncovers its positive role in response to *Pseudomonas syringae* pv. Tomato DC3000 in tobacco. Plant Mol Biol 90(1–2):19–31

Zhou C, Cai Z, Guo Y, Gan S (2009) An Arabidopsis mitogen -activated protein kinase cascade, MKK9 -MPK6, plays a role in leaf senescence. Plant Physiol 150:167–177

Zipfel C, Kunze G, Chinchilla D et al (2006) Perception of the bacterial PAMP EF -Tu by the receptor EFR restricts agrobacterium -mediated transformation. Cell 125:749–760

Zolnierowicz S, Bollen M (2000) Protein phosphorylation and protein phosphatases. De Panne, Belgium, September 19–24, 1999. EMBO J 19:483–488

Role of Phytohormones in Plant Defense: Signaling and Cross Talk

7

Vibha Gulyani Checker, Hemant Ritturaj Kushwaha, Pragati Kumari, and Saurabh Yadav

Abstract

Plants, being sessile throughout their life cycle, are vulnerable to various kinds of abiotic and biotic stress conditions. They have evolved sophisticated mechanisms to detect precise environmental change and respond with an optimal response, thereby minimizing damage and conserving resources for growth and development. The response of plants towards these stresses are dynamic and complex. A defense response is initiated via modulation of molecular events, which involves interplay of signaling molecules including phytohormones. Phytohormones are small endogenous, low-molecular-weight molecules, which trigger an effective defense response against both biotic and abiotic stresses. Apart from defense signaling, these phytohormones are also regulators of growth, development, and physiological processes. The phytohormones such as auxins, cytokinins (CKs), gibberellins (GAs), salicylic acid (SA), jasmonic acid (JA), ethylene (ET), abscisic acid (ABA), and brassinosteroids (BRs) respond to stress via synergistic and antagonistic actions often referred to as signaling cross talk. These phytohormones coordinate with each other in a harmonious manner and respond to developmental and environmental cues. All defense response in plants are the result of interplay of many genes and gene families nicely orchestrated in

V. G. Checker
Department of Botany, Kirori Mal College, University of Delhi, New Delhi, India

H. R. Kushwaha
School of Biotechnology, Jawaharlal Nehru University, New Delhi 110067, New Delhi, India

P. Kumari
Department of Life Sciences, Singhania University, Rajasthan, India

Indian Agricultural Research Institute (IARI), New Delhi, India

S. Yadav (✉)
Department of Biotechnology, Hemvati Nandan Bahuguna Garhwal (Central) University, Srinagar Garhwal, Uttarakhand, India
e-mail: saurabhyadav40@rediffmail.com

A. Singh, I. K. Singh (eds.), *Molecular Aspects of Plant-Pathogen Interaction*,
https://doi.org/10.1007/978-981-10-7371-7_7

a network. Various phytohormones are known to play important role in almost all the process through the modulation of genes. Further, through the optimal mix of phytohormones, plants maintain homeostasis and adapt to the environmental changes. This is only possible by an efficient and systemic cross talk between various phytohormones which help plants to maintain a critical balance between growth and environmental response. This chapter would assist plant biologist in further understanding the ability of the plants to perceive, synthesize, and respond to phytohormones in response to environmental stresses.

Keywords
Signaling · Phytohormones · Jasmonates · Biotic stress · Plant defense · Cross talk

7.1 Introduction

Plants, throughout their life cycle, are entrenched to one place, which makes them vulnerable to various kinds of stress conditions. These stresses can be classified as abiotic and biotic stress. A high percentage of the crop production is said to be affected due to the persisting stresses (Boyer 1982). All stresses are often interrelated and cause physiological, biochemical, and molecular changes which are finally reflected in terms of its declined growth and productivity. Over the years, major progress has been made using tools and techniques of molecular biology, to understand the comprehensive view of abiotic and biotic stress perception, response, and tolerance in plants. Apart from various abiotic stresses, plants also have to face various biotic challenges such as pathogenic microbes and herbivorous insects. In order to defend against these challenges, plants rely on their immune system and preformed defense systems. Preformed defense system, which is the first hurdle for an invading organism, includes various chemical compounds (Osbourn 1996), cuticle layers, thick cell walls, needles, thorns, or trichomes. The principal components of cell walls are cross-linked high-molecular-weight polysaccharides which resist the physical penetration (Carpita and McCann 2000). Also, these cell walls are the dynamic reservoirs of antimicrobial proteins and secondary metabolites which prohibit the growth of many pathogens.

Plants also possess multiple layers of innate immune system which detects and limits pathogen expansion. One layer of this system uses pattern recognition receptors (PRRs) present on the surface of the plant cell to analyze for molecules containing signature patterns conserved in microbes known as pathogen-/microbe-associated molecular patterns (PAMPs/MAMPs). Detection of these PAMPs/MAMPs triggers pattern-triggered immunity (PTI). PTI prevents further pathogen establishment. However, some pathogens have developed effector proteins that overpower PTI. These pathogens are suppressed via other layers of innate immune system – effector-triggered immunity (ETI). Plants have evolved intracellular resistance (R) proteins which induce ETI to regain resistance. These R proteins recognize their pathogen-encoded effectors directly or indirectly. The responses initiated by ETI are more rapid and intense than PTI. ETI also turns local infection into global defense by forming necrotic lesions which restricts movement of pathogen from the infection site (Fu and Dong 2013). Both types of signaling mechanisms (PTI and ETI) activate a league of defense responses in the affected tissue, like generation of reactive oxygen species (ROS), increase in intracellular Ca^{2+} concentrations, activation of mitogen-activated protein kinases (MAPKs), increased expression of various defense-associated genes, and production of antimicrobial compounds (Chisholm et al. 2006). The rise in intracellular calcium is sensed by calcium sensors which in turn bind to cis-elements of promoters of genes conferring tolerance. These calcium sensors may also bind DNA-binding proteins controlling these stress-responsive genes. The increase in intracellular calcium further activates various kinases and phosphatases which can phosphorylate or dephosphorylate transcription factors controlling stress-responsive genes. Many defense-related proteins have been identified in various plant species like PR-1, β1,3-glucanase (PR-2), chitinases (PR-3, PR-4, PR-8, PR-11), thaumatin (PR-5), proteinase inhibitors (PR-9), ribonuclease like defensin (PR-10), lipid transfer protein (PR-14), NBS-LRR proteins, glycoproteins, WRKY proteins, and catalases.

Plants, in order to survive through biotic stresses, have evolved more sophisticated mechanisms in order to perceive external signals which allow them to prepare for an optimal response (Fig. 7.1). Small endogenous, low-molecular-weight molecules commonly known as phytohormones regulate the defensive response against biotic stresses. The phytohormones, such as salicylic acid (SA), jasmonic acid (JA), ethylene (ET), abscisic acid (ABA), brassinosteroids (BRs), auxins, cytokinins (CKs), and gibberellins (GAs), respond to stress via synergistic and antagonistic actions often referred to as signaling cross talk (Mauch-Mani and Mauch 2005). Over the years, analysis done using large-scale transcriptome analyses has supported the existence of cross talk between various signaling networks (Davletova et al. 2005). Apart from defense signaling, these phytohormones are also regulators of growth, development, and physiological processes. These phytohormones act while maintaining a balance in highly complex network in response to development and environmental cues. Like animal hormones, these phytohormones are transported from one location to the other for mediation in various processes (Davies 2010). Under stress conditions, the production, distribution, or signal transduction of these hormones is affected leading to morphological, molecular, and physiological changes which prepare plants to withstand the stress conditions (Eyidogan et al. 2012). The phytohormones, being lowest level transducers, lead to the stress signal activation of signaling cascade which further initiates the response mechanism (Harrison 2012).

We have summarized the roles of phytohormones in various signaling mechanisms, in response to various stress responses. Further, it also focuses on the crucial, delicate, and complex cross talk of these phytohormones in response to various biotic stresses, in various plant species. This chapter would assist plant biologist in further exploring the crucial role of phytohormones in response to biotic stresses.

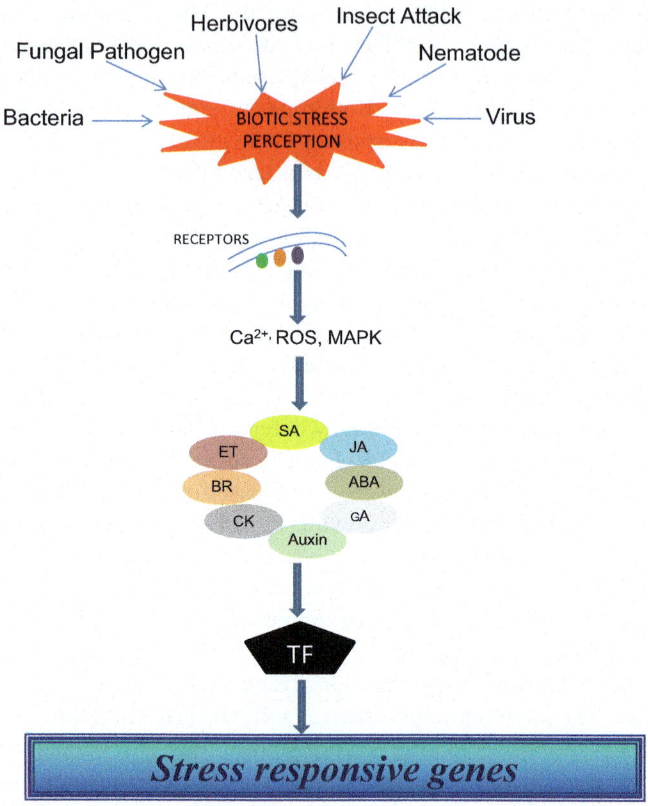

Fig. 7.1 Phytohormones are watchdogs of stress and cross talk to regulate the defense response against biotic stresses

7.2 Role of Phytohormones in Biotic Stress via Different Signaling Pathways

7.2.1 Brassinosteroids (BRs)

One of the most ubiquitously present steroids, brassinosteroid (BR), is an endogenous plant growth-promoting hormone. Initially reported as organic extract (Mitchell et al. 1970), brassinosteroids (BRs) are localized in almost all parts of the plants, namely, seeds, fruits, young vegetative tissues, and pollens, and are known to affect cellular proliferation and expansion (Clouse and Sasse 1998; Sakurai et al. 1999). The BRs on the basis of the number of carbons in their structure are classified as C27, C28, and C29 (Vardhini 2013a, b). Till date, 60 BR-related compounds have been identified (Haubrick and Assmann 2006), but brassinolide (BL), 28-homobrassinolide (28-HomoBL), and 24-epibrassinolide (24-EpiBL) are most widely known BRs, which are used in various physiological and experimental studies (Vardhini et al. 2006). These steroids are either present in free form or in conjugation with various sugars or fatty acids (Bajguz and Hayat 2009). BRs are involved in the vascular differentiation (Ashraf et al. 2010) and xylem formation in epicotyls (Zurek et al. 1994). Analyses suggest the role of BRs in providing the resistance against various biotic and abiotic stresses (Bajguz and Hayat 2009; Gomes 2011). In rice and tobacco, BR treatment has been observed to enhance the disease resistance (Nakashita et al. 2003). Thus, it has been hypothesized that BRs by triggering the accumulation of apoplastic H_2O_2 further upregulated the antioxidant system, thereby inducing the stress tolerance in plants (Jiang et al. 2012). A cell surface receptor kinase, brassinosteroid-insensitive 1 (BRI1), is known to perceive the BRs in plants. Various bioassays have revealed that BRs bind to the extracellular domain of the BRI1 receptor, which is plasma membrane-localized leucine-rich repeat (LRR) receptor of a serine/threonine (S/T) kinase (Friedrichsen et al. 2000). This binding initiates the dissociation of BRI1 from the negative regulator BIK1, activation of the co-receptor BRI1-associated receptor kinase 1 (BAK1) and its heterodimerization with *BRI1*, phosphorylation of the BRI1-interacting signaling kinase *(BSK1)*, and activation of the BSU1, a protein phosphatase (Lin et al. 2013). The signal is then transmitted to the cytoplasm where a protein kinase, brassinosteroid-insensitive 2 (BIN2), is inhibited, which is the negative regulator of BR biosynthetic pathway (Fariduddin et al. 2014), and transcription factors like BZR1 and BES1/BZR2 are activated. These transcription factors move to the nucleus and activate BR-responsive genes by binding to their promoter. Further BAK1 is involved in the regulation of microbe-induced cell death (Kemmerling et al. 2007), and even interacts with various pattern recognition receptors (PRRs), and is a part of PAMP-triggered immunity (PTI) (Fradin et al. 2009; Chaparro-Garcia et al. 2011). It is also suggested that BAK1 can function in control of cell death and innate immunity independent of BR (Chinchilla et al. 2009). The complexity of BR response pathway was studied in *Arabidopsis* and the authors demonstrate that this key player can act antagonistically and synergistically. Researchers fine-tuned the BR pathway and concluded that BAK1 acts as a mediator for synergistic activities of BR on PTI but also supports a scenario that BR can control plant defense independent of BAK1 (Belkhadir et al. 2012). Lozano-Duràn et al. (2013) and Shi et al. (2013) further provided insights into BR suppression of PTI responses independent of BAK1. Another feather to BR cap of signaling molecules was added as the study conducted by Lin et al. (2013) identified the receptor-like cytoplasmic kinase Botrytis-induced kinase 1 (BIK1) which shares the BR and PTI pathways.

BRs were regulated in plants leading to the various responses depending upon various environmental conditions. Studies in recent past have highlighted that exogenously given BRs exerted a positive effect on the resistance of various crops to a broad range of pathogens (Khripach 2000). This was further supported in small-scale disease trials by Nakashita et al. (2003). The group reported a varying (local, as well as systemic) positive effect of BL on disease tolerance to distinct leaf pathogens in tobacco and rice plants. Exogenous BR application was also shown to confer resistance to barley from several *Fusarium* diseases (Ali et al. 2013). Apart from these positive effects, BR can have negative impact on disease resistance. It was

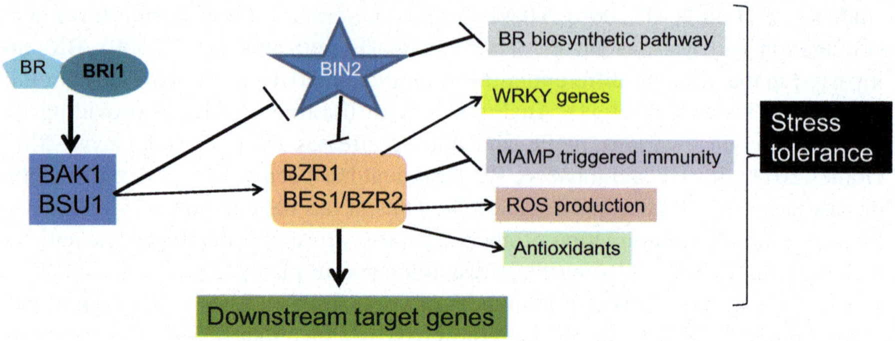

Fig. 7.2 Multifaceted roles of BR in plant pathogen interaction

seen that exogenous BL application either could not alter resistance of *Arabidopsis* to *Pseudomonas syringae* pv. tomato (Pto) and *Alternaria brassicicola* (Albrecht et al. 2012) or makes rice hypersusceptible to the *Pythium graminicola* and *Meloidogyne graminicola*, the root pathogens (De Vleesschauwer et al. 2012; Nahar et al. 2013). In the case of *P. graminicola*, it was even recommended that the pathogen seizes the host BR machinery, as virulence factors to promote infection (De Vleesschauwer et al. 2012). Thus, BRs are an important class of regulators that merge immune system function with normal growth and developmental programs.

The narrow range of BR concentrations can also influence the production of ROS (Baxter et al. 2014). BR-induced ROS production plays an important role in stress tolerance. BRs can also stimulate antioxidant production and scavenge ROS (Bajguz and Hayat 2009; Fariduddin et al. 2013). They can also influence disease resistance by fine-tuning secondary metabolite production. The repressive effect of BZR1 and BES1 on glucosinolate production in *Arabidopsis* was observed (Guo et al. 2013).

BR therefore plays multifaceted roles in plant-pathogen interactions (Fig. 7.2), ranging from modulation of PAMP perception, activation of stress-responsive gene, fine-tuning oxidative metabolism, and production of secondary metabolites. However its role in making the plant resistance or susceptible depends on plant-attacker combination. It further emphasizes the fact that plant hormone interaction in pathogen defense cannot be generalized (Bruyne et al. 2014). Recently, BRs are found to be the master regulators of gibberellic acids (GAs), which are related to the growth and development of vascular plants (Unterholzner et al. 2015). In rice, GA metabolism is found to be modulated by the BRs (Tong et al. 2014). Following stress perception, BRs can cross talk among defense signaling pathways with a range of hormones, such as SA, JA, ABA, auxins, and GA.

7.2.2 Abscisic Acid (ABA)

The phytohormone, abscisic acid (ABA), plays prominent role in various aspects of plant growth which include germination, dormancy, and seed development. Several analyses have shown its role on developing stress response in plants towards abiotic and biotic stresses. ABA accumulates after infection and can have positive and negative effect on defense responses depending on the environmental conditions. It mostly acts antagonistically with SA/JA/ethylene making the plant susceptible against disease and herbivore attack. ABA promotes stomatal closure as a physiological response to various environmental stresses. ABA thus induces preinvasive defense by inhibiting the entry of pathogens through these passive ports (Ton et al. 2009). This is proven by the analysis of susceptibility toward *Magnaporthe grisea* in rice plant due to ABA application (Koga et al. 2004). It has been found that viral infection in plants leads to the increased ABA concentration in tobacco (Whenham et al. 1986). Plants react to the infection once it is established. SA, JA, and ethylene might not be activated during initial exposure to stress, but at this time ABA acts as a key endogenous factor in playing important role by inducing stomatal closure. It can even antagonize their induction, thereby modifying future responses against potential pathogens. Postinfection, ABA signaling cascade initiates events to strengthen the resistant phenotype through callose accumulation to prevent pathogen invasion in attacked plants. The rise in ABA concentration has been shown to have negative effect on the ET production (LeNoble et al. 2004); they thus interact in antagonistic manner in plant-pathogen interaction. Similar response has been observed in case of jasmonic acid (JA) (Staswick et al. 1992). Overall, by interfering with the biotic stress signaling, ABA affects the resistance of plants toward various diseases in negative manner (Mauch-Mani et al. 2005). Exogenously applied ABA increased the virulence of *P. syringae* pv. tomato on *Arabidopsis* plants (de Torres et al. 2007). Furthermore, application of ABA suppressed the transcript accumulation of defense genes like *PDF1.2* (*plant defensin 1.2*), *CHI* (*basic chitinase*), *HEL* (*hevein-like protein*), and *LEC* (*lectin-like protein*), thereby increasing susceptibility of *Arabidopsis* plants to the fungus, *Fusarium oxysporum* (causal agent of wilt), and to the bacteria, *Erwinia chrysanthemi* (causal agent of bacterial wilt), infection (Asselbergh et al. 2008). Similarly the mutants defective in ABA biosynthesis and perception demonstrate the negative effect of ABA in disease resistance. For example, tomato mutants showed reduced pathogen growth after *B. cinerea* infection (Audenaert et al. 2002). Similarly, *ABA deficient Arabidopsis* mutant, *aba1-1*, induced HR-like defense response at the site of *Peronospora parasitica* inoculation. Such response was not observed in wild-type plants (Mohr and Cahill 2007). ABA-deficient mutant, *aba3-1*, failed to close stomata upon exogenous application of pathogen-derived elicitors suggesting the involvement of ABA-mediated signaling in closure of stomata. Evidences suggest that exogenous application of ABA bestowed resistance against fungal pathogens in *Arabidopsis* (Adie et al. 2007) and barley plants (Wiese et al. 2004). The increased resistance observed in these plants is because of reduced pathogen colonization as a result of ABA-mediated callose biosynthesis or inhibition of its degradation (Jacobs et al. 2003; Rezzonico et al. 1998). Also, there are evidences that show correlation between the nitrogen status and ABA levels in plants (Wilkinson and Davies 2002). Therefore,

ABA in plants, not only associated with growth and development but also with various levels of stress signaling.

7.2.3 Ethylene (ET)

In plants, the growth, development, and tolerance to many biotic stresses are dependent upon various hormones including ethylene. It regulates the flower fertilization, senescence, fruit ripening, and organ abscission in various plant species. The perception and signal transduction of ethylene are conserved among the plants, which show its relevance in the plant development and survival mechanisms. The ethylene induced during the stress acts as trigger due to its autocatalytic mechanism of ethylene synthesis in a stressed plant. The ethylene production is controlled by mitogen-activated protein kinase (MAPK) phosphorylation cascades (Takahashi et al. 2007). The ethylene pathway is a well-studied pathway in plants. It requires S-adenosylmethionine (SAM) which is also a precursor in other pathways and is widely found in plant tissues. SAM is converted to 1-aminocyclopropane-1-carboxylic acid (ACC) and 5′-methylthioadenosine (MTA) in the reaction catalyzed by ACC synthase in a rate-limiting step of the pathway. During the high rate of ethylene production, levels of l-methionine remain largely unchanged due to the presence of 5′-methylthioadenosine (MTA) (Abeles et al. 1992). The ACC produced is oxidized to ethylene and various other products in combination with oxygen, catalyzed by ACC oxidase. The presence of ethylene leads to the induced expression of gene encoding ethylene response sensors 1 and 2 (ERS1, ERS2) and ethylene resistant 2 (ETR2) proteins. Ethylene is sensed by five ER-localized receptors which are divided into two subfamilies. The subfamily I includes ethylene response 1 [ETR1] and ethylene response sensor 1 [ERS1], while subfamily II includes ETR2, ERS2, and ethylene insensitive 4 [EIN4]. Ethylene signaling also includes some downstream elements like constitutive triple response (CRT1), ethylene insensitive 2 (EIN2), ethylene insensitive 3 (EIN3)/ethylene insensitive-like protein 1(EIL1), and ethylene response factors (ERFs). The major factors governing ethylene signaling are ethylene response factors (ERFs). However, it is suggested that ethylene insensitive 3 (EIN3) induces ERF 1 gene expression to activate defense response. Thus ERF1 acts downstream of EIN3 (Solano et al. 1998). Several ERF subfamily members are associated with biotic stress tolerance, for example, it has been shown in rice and *Arabidopsis* that ERF1 binds to GCC box (AGCCGCC) in promoters of biotic stress-inducible genes (Xu et al. 2007; Cheng et al. 2013). In rice ERF 922 has been shown to be a negative regulator of defense against rice blast fungus *Magnaporthe oryzae* (Liu et al. 2012). It was observed that overexpression of ERF 922 enhanced ABA levels in rice which could be the cause of enhanced susceptibility to fungus (Koga et al. 2004). In several cases, ERFs are associated with long-distance leaf-to-leaf and leaf-to-root signaling to equip the plant against future pathogen attack (Dey et al. 2014). In a study conducted on rice roots, treatment with *Pseudomonas* isolate EA 105 induces the expression of ERF1 in distal uninfected leaves and enhances resistance to *M. oryzae* compared to noninfected plants (Spence et al. 2014). Further in the absence of ethylene, EIN2 is repressed by constitutive triple response (CTR1). The repression of EIN2 is relieved by ETR1 upon perception of ethylene. CTR1 interacts and phosphorylates the cytosolic C-terminal domain of EIN2 and prevents EIN2 from signaling in absence of ethylene (Ju et al. 2012). The rate of ethylene biosynthesis has been found to be enhanced under the biotic stress (Van Loon 1984; Abeles et al. 1992). Interestingly, ethylene signaling mechanism in plants shows novel negative feedback mechanism. This is because ethylene is essential for plant response to the stress and prepares the plants for tolerance mechanism which is essential for their survival. But excessive ethylene production due to persisting stress conditions leads to the inhibition of growth and development in plant, which if continued for long duration leads to the plant death. Hence, it is considered that a tight regulation of ethylene homeostasis is maintained for the plant survival during stress conditions and its recovery for growth later. Ethylene interacts with other phytohormones like jasmonic acid and salicylic acid and affects the signaling response. It may affect SA-mediated defense responses, positively and negatively depending on the different lifestyles of the pathogens (Derksen et al. 2013). The gain of function and loss of function of ethylene-related genes affect the susceptibility toward pathogens in plants. In rice, overexpression of ACS2 (1-aminocyclopropane-1-carboxylic acid synthase) resulted in high level of ethylene and increased resistance to *M. oryzae* and *R. solani* (Helliwell et al. 2013). RNAi silencing of OsEIN2b in rice plants increased susceptibility to *M. oryzae* infection (Seo et al. 2011).

7.2.4 Jasmonates

Jasmonates are one of the most widespread class of phytohormone in the plant kingdom. Jasmonic acid was first of all isolated as methyl jasmonate from the essential oil of *Jasminum grandifloram*. Vick and Zimmermann (1984) provided insights into biosynthesis of JA. The family, jasmonates, includes various other compounds, namely, 12-oxophytodienoic acid (12-OPDA), methyl jasmonate (MeJA), and amino acids conjugated to the jasmonic acid such as JA-leucine and JA-isoleucine. Jasmonic acid is the most studied and characterized member of the jasmonate family in plants (Avanci et al. 2010). The members of the jasmonate family have been analyzed for their role in development of fruits, senescence, reproductive process, various secondary metabolisms, and direct and indirect defense responses against pathogen (Wasternack 2007). Methyl jasmonate (MeJA) has been earlier shown to be involved as signaling molecule which mediates the intra- and interplant communications, thus altering the plant defense responses (Seo et al. 2011; Wasternack 2007). This is due to the ability of MeJA to diffuse through the membranes and its volatile nature. MeJA has also been observed to play major role in regulating the reproductive process in plants (von Malek et al. 2002). The biosynthesis and signaling pathways of the jasmonate have been studied in detail in *Arabidopsis* and tomato

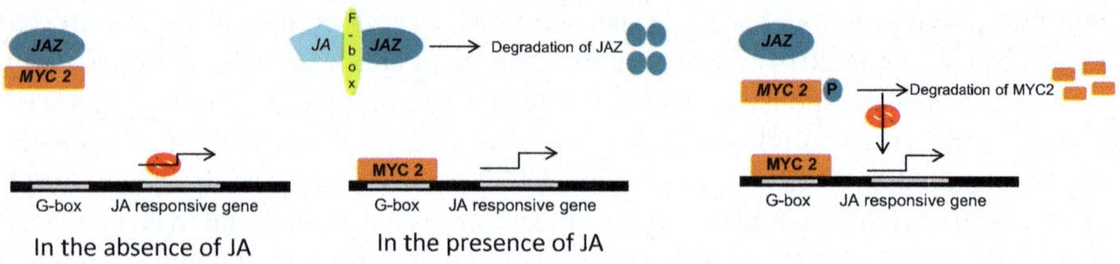

Fig. 7.3 (**a**, **b** and **c**) Role of JAZ and MYC2 in JA-responsive gene expression. In the absence of active JA, JAZ proteins interact with transcription factor JIN1/MYC2 and inhibit expression of JA-responsive genes. In the absence of active JA, JA binds to its receptor which is a F-box protein CORONATINE INSENSIVE 1 (CoI1) and mediated the 26 S-mediated degradation of JAZ, thereby allowing MYC2 to upregulate the expression of JA-responsive genes. (**c**) Fine-tuning of JA responses by MYC2

(Kazan and Manners 2008). Jasmonates play a role in the growth control, shoot and root tissue elongation, and floral bud formation (Maciejewska and Kopcewicz 2003). Apart from that, jasmonates also contribute to the regulation of the biosynthesis of many secondary metabolites such as phenylpropanoids, alkaloids, terpenoids, and antioxidants. During pathogen attack, especially necrotrophic fungal and bacterial infection, jasmonates lead to the expression of defense-related genes (Mei et al. 2006). JA biosynthesis has been found in the roots of the plants (Pedranzani et al. 2003), but evidence also suggests its presence in the leaves (Mueller et al. 1993). In sorbitol-treated tomato plant, endogenous levels of OPDA, JA, and MeJA were observed in leaves (Abdala et al. 2003). The activity of jasmonates is also associated with the slowing of activity of the photosynthetic apparatus. The hormone also plays an important role against attack by herbivores like caterpillars, mites, thrips, beetles, etc. The transcription factor JASMONATE INSENSITIVE/MYC2 (JIN1/MYC2) plays an important role in JA-responsive gene expression. JA-regulated stress responses are also mediated by ethylene-responsive factors. The JA-responsive marker gene PLANT DEFENSIN 1.2 (PDF 1.2) is regulated by ethylene-responsive factors ERF1, ERF2, and ERF5. A repressor protein JASMONATE-JIM DOMAIN (JAZ) plays a prominent role in JA-mediated stress response. In the absence of active JA, JAZ proteins interact with transcription factor JIN1/MYC2 and inhibit expression of JA-responsive genes. Once the JA pathway is activated, i.e., JA-Ile binds to its receptor which is a F-box protein CORONATINE INSENSIVE 1 (CoI1) and mediates the 26 S-mediated degradation of JAZ, thereby allowing MYC2 to upregulate the expression of JA-responsive genes. Recent studies have also highlighted the fine-tuning of JA responses by MYC2. A posttranslational modification of MYC2 at Thr 328 stimulates its transcription activity. The modified MYC2 is unstable and degraded by plant U-box protein (PUB 10), thereby facilitating turnover of MYC2 (Antico et al. 2012). The model depicting above interactions is shown in Fig. 7.3.

7.2.5 Salicylic Acid (SA)

SA is a plant phenolic compound synthesized by plants to regulate defense mechanism against biotrophic and hemibiotrophic pathogens and is derived from the shikimate-phenylpropanoid pathway (Sticher et al. 1997). Detection of phytopathogens stimulates the synthesis of SA. The pivotal role of salicylic acid in conferring disease resistance was first demonstrated in *Arabidopsis* and cucumber plants when levels of the hormone were detected prior to development of local and systemic resistance (Malamy et al. 1990). Since then its role in biotic stress has been established in all plant species. Activation of SA pathway at the site of infection triggers a defense response in distal parts of the plant.

The constitutive SA accumulation often leads to reduction of plant fitness (Ishihara et al. 2008), by mobilizing resources and energy away from growth and reproductive process. Hence, its biosynthesis and signaling are under tight control of the cell. SA is synthesized in the chloroplast and then transported to the cytosol where it signals immune responses. The export of the hormone is mediated by a chloroplast membrane-localized member of the multidrug and toxin (MATE) transporter family EDS5 (or SID1). Inside the cytoplasm, SA can be subjected to various modifications that usually render it inactive (Dempsey et al. 2011). These modifications are an important check to regulate the level of biologically active SA in the cytoplasm such as glucosylation of SA at its hydroxyl group generates a glucoside (SAG), whereas glucosylation at its carboxyl group produces an ester (SGE). SAG moves to the vacuole and behaves as a nontoxic storage from which it can be hydrolyzed following pathogen attack to release free active SA. Methylation of SA produces methyl SA (MeSA) which is a mobile phloem signal that travels from the infected leaf to the systemic tissues, where it activates defense mechanism after being converted back to SA (Fig. 7.4) (Park et al. 2007). Formation of SA-amino acid conjugates is another adjustment strategy (Dempsey et al. 2011) as salicyloyl-aspartate (SA-Asp) has been identified in plants. There are a number of physiological processes that are influenced by SA. In plants, SA is required for the initiation of various stress symptoms and hypersensitive-response (HR)-like cell death. The PR genes are a diverse class of genes encoding antimicrobial proteins and increasing resistance against broad spectrum of pathogens. The expression of various pathogenesis-related (PR) proteins encoding genes has also been observed with the exogenous application of the SA (Malamy et al. 1990). SA is also required for establishing the systemic acquired resistance (SAR) with its accumulation in the distant tissues in the condition of stress in plants (Vernooij et al. 1994). Earlier, significantly high number of leaves and increased biomass have been observed in wheat seedlings, when treated with SA (Hayat et al. 2005). Various analyses provided an indication that SA is related to several processes, and therefore it is assumed that it has a role in biotic stress conditions. Yalpani et al. (1994) have suggested that exposure to ozone and the ultraviolet light leads to the accumulation of SA. The ozone induces oxidative stress levels; thus, SA induces antioxidant defenses in plants showing it may have some role in other abiotic stress conditions. Also, analysis suggests that there is an overlap of ozone and pathogen-induced resistance

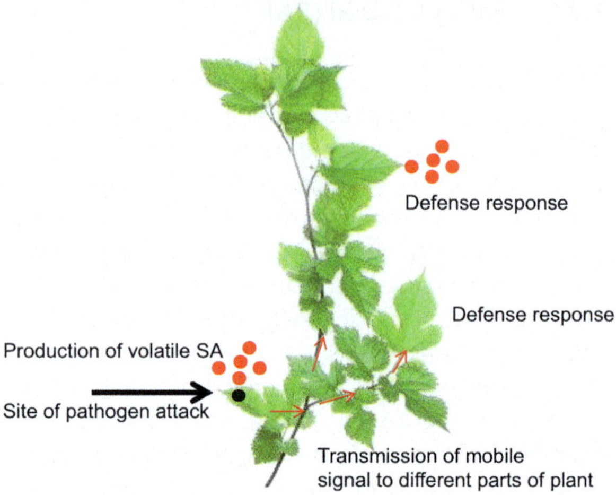

Fig. 7.4 Activation of SA pathway at the site of infection triggers a defense response in distal parts of the plant. Methylation of SA produces methyl SA (MeSA) which is a mobile phloem signal that travels from the infected leaf to the systemic tissues, where it activates defense mechanism after being converted back to SA

pathways which confirms the role of SA in plants (Sharma et al. 1996). SA leads to the induced expression of receptor protein kinase (RPK) which is known to initiate the response to the signal (physical or chemical) in plants (Bassett et al. 2005). Further, SA has been hypothesized to be the direct or indirect regulator of calcium-mediated signaling pathways. This is evident from the analysis in which SA was observed to induce expression of certain calcium-dependent protein kinases (CDPKs) in plants (Leclercq et al. 2005). Owing to the role of SA in inducing many crucial genes in signaling pathways, it is assumed that SA is a crucial part of the complex signaling transduction networks. The members of NON-EXPRESSOR OF PR GENE (NPR family) are proposed SA receptors. The implementation of SA action is monitored through action of SA marker gene PR-1 whose activation requires the positioning of the SA-dependent transcriptional enhanceosome to its promoter (Rochon et al. 2006). The enhanceosome contains the members of TGA2 clade of bZIP transcription factors (Zhang et al. 2003) and the transcriptional coactivator (NPR1) (Rochon et al. 2006). Recent studies provide insights into mode of action of this SA receptor-NPR1, which is central to activation of SA defense genes. TGA2 is a transcriptional repressor and its inactivation is brought about by NPR1. The N-terminal region of NPR1 contains a BTB/POZ domain that interacts with TGA2 repression domain and negates its function (Boyle et al. 2009). NPR1 harbors in its C-terminal region a transactivation domain which contains two cysteines (Cys^{521} and Cys^{529}) required for the activating function of the enhanceosome (Rochon et al. 2006). NPR1 binds specifically to SA, which in turn regulates conformation of NPR1 by depolymerizing it to a dimer. This causes release of the C-terminal transactivation domain from the N-terminal autoinhibitory BTB/POZ domain. SA binding to NPR1 relieves sequestration of NPR1 transactivation domain (Fig. 7.5). Further, nearly 30 SA-binding proteins (SABs) with variable affinities for SA have been identified. SA signals its effects by interacting with these SABs. SA research is even benefitting agriculture as pretreating plants with high concentration

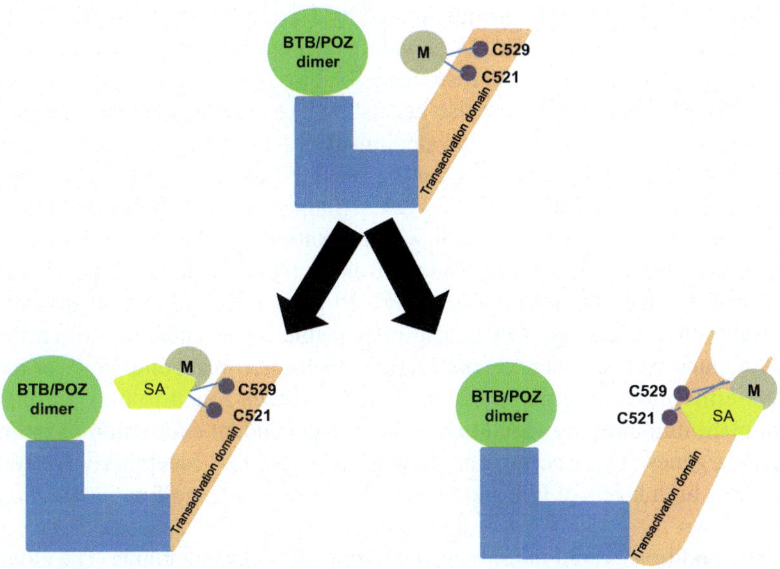

Fig. 7.5 SA binding to NPR1 relieves sequestration of NPR1 transactivation domain. The N-terminal region of NPR1 contains a BTB/POZ domain that interacts with TGA2 repression domain and negates its function. NPR1 harbors in its C-terminal region a transactivation domain which contains two cysteines (Cys^{521} and Cys^{529}). Binding of SA to the cysteines in the transactivation domain disrupts its interaction either due to steric hindrance or SA-induced conformational change

of SA can initiate defense response which can later be activated rapidly by subsequent infection (Dempsey and Klessig 2017).

7.2.6 Gibberellin (GA)

The plant phytohormone, gibberellins, is one of the large families of diterpenoid compounds. It is not only produced by higher plants but also by fungi and bacteria (MacMillan 2001). It is widely known to promote some of the essential processes related to floral development, flowering time, elongation of stem, trichome development, and seed germination (Davies 2010). Various environmental stimuli lead to the change in the GA concentrations in plants which further affects various processes (Davies 1995). The GAs are synthesized in the plastids by methylerythritol phosphate pathway from trans-geranylgeranyl diphosphate. Several analyses suggest that GA as a signaling component plays an important role in the disease susceptibility and resistance in plants. The first report for the role of GA came from the work of Zhu et al. (2005). They demonstrated that one of the capsid proteins of the rice dwarf virus interacts with plant ent-kaurene oxidases resulting in decreased GA levels in the infected plants. The DELLA proteins are a class of nuclear growth-repressing proteins. They are the negative regulators of GAs and have been found regulating the immune response modulated by JA and SA in plants. DELLAs proteins regulate the balance of SA/JA signaling during plant immunity, supporting JA perception and/or signaling, and restricting SA biosynthesis and signaling (Navarro et al. 2008). These proteins have also been shown to regulate the expression of various genes which are known for producing ROS detoxification enzymes (Achard et al. 2008). Thus, indirectly GAs can be assumed to play a role in the oxidative stress related to salinity and osmosis. GA signaling plays an important role in cell wall development and represses cell wall relaxation by altering the expression of xyloglucan endotransglucosylase/endohydrolases (XTHs) and expansins. Loosening of cell wall is undoubtedly important for cell growth, but it also allows pathogen entry and nutrient leakage. GA is also responsible for modification of carbon and energy metabolism and decrease production of antimicrobial compounds or increase nutrient efflux favoring microbes. DELLAs have been found to play an important role in plant immunity by regulating cell cycle-dependent expression of immunity-conferring genes. The rice mutants having defective GA receptors were found to accumulate high levels of GA and showed better resistance to fungus *Magnaporthe grisea* known for causing blast, than the wild types (Tanaka et al. 2006). Under flooding conditions, rapid internode elongation takes place in plants. The elongation is the result of upregulation of ethylene response factor (ERF) domain proteins SNORKEL1 and SNORKEL2 which leads to the direct or indirect increase in the levels of GAs in plants (Hattori et al. 2009). Interestingly, the GA signaling pathways have been found to be modulated by other phytohormones (Fu and Harberd 2003). Through many biological studies utilizing mutant lines and gain of function and loss of function and plants impaired in gibberellic acid signaling, it is observed that GAs are important regulators of plant growth including biotic stress (Colebrook et al. 2014). Exogenous application of GA in rice decreases resistance to the hemibiotrophic rice pathogens *Magnaporthe oryzae* and *Xanthomonas oryzae* pv. *oryzae* (Qin et al. 2013). Decreased level of GA by overexpressing a GA-deactivating enzyme in rice resulted in low levels of both salicylic acid and gibberellic acid, but the resistance to *Magnaporthe oryzae* and *Xanthomonas oryzae* pv. *oryzae* was enhanced and reverse result was obtained by loss of function and plants became more susceptible (Yang et al. 2008). Similar to the classic defense hormones jasmonic acid, salicylic acid, and ethylene, gibberellic acid acts as multifaceted regulators of plant immunity. This depends upon the plant species and pathogen type and also on plant pathogen interaction (Bruyne et al. 2014).

DELLA proteins interact with transcription factors and regulate plant growth and development. Many groups have shown that DELLA proteins compete with JAZ, a transcriptional repressor of JA to bind to MYC2. Binding of DELLA proteins relieves MYC2 from JAZ, which is now free to enhance JA-responsive gene action. This increases resistance against necrotrophic pathogens (Navarro et al. 2008). However, GA exerts its control as in the presence of GA, DELLA proteins are degraded, and JAZ proteins bind to MYC2, thereby inhibiting JA-responsive gene action. Figure 7.6 explains this interaction. Therefore GA may disable JA-mediated stress response against pathogen invasion. DELLAs seem to positively mediate JA-responsive gene expression by blocking JAZ proteins (Hou et al. 2010) and also

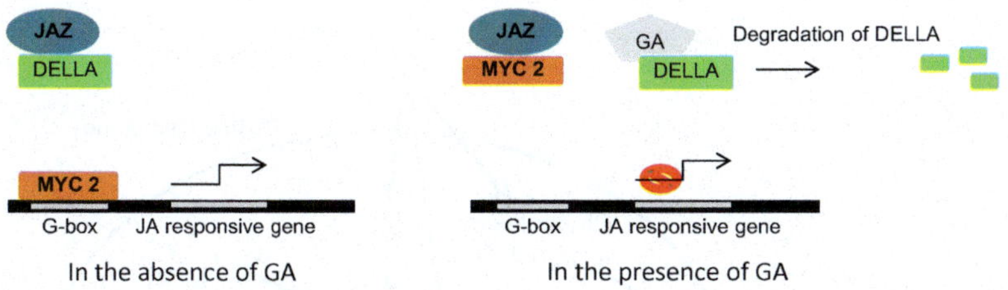

Fig. 7.6 Control of GA over JA-responsive gene action. In absence of GA, the DELLA proteins are stabilized and bind to JAZ. This frees MYC2 which can activate JA-responsive gene action. In presence of GA, DELLA proteins are degraded, JAZ proteins binds to MYC2, and JA-responsive gene action is blocked

by acting as JA-responsive regulators. To defend against pathogens, plants may co-opt for a JA signaling cascade to regulate DELLA levels in planta as a DELLA protein RGL3 is also known to be induced in a MYC2-dependent fashion by JA. DELLA proteins therefore form an important node of multiple hormone signaling pathways as they are main centers controlling multiple environmental signals and developmental cues. DELLA proteins also interact with BR-regulatory TF *brassinazole resistant 1* (BZR1). They inhibit BR signaling cascade by sequestering BZR1 into inactive protein complexes. This in turn regulates numerous defense-related genes controlled by BZR1 (Wang et al. 2014).

7.2.7 Cytokinins

The role of phytohormone, cytokinins, has been poorly understood in plants although there are certain reports which show its role in defense response against some pathogens. Apart from that, its role has been shown in the growth of root, shoot and inflorescence, seed development, senescence in leaves, and stress response (Muller and Sheen 2007). The cytokinins when applied with auxins have been found to trigger callus differentiation and induce cell proliferation in shoots. It also contributes to the sink strength, nutrient translocation, and grain yield of various plants. In developing tissues like cambium, shoot apex, and root tips, cytokinins have been found in abundance. Whole genome expression analysis *of Arabidopsis infected with Plasmodiophora brassicae* (known to cause clubroot disease in plants) showed downregulation of gene playing role in cytokinin homeostasis, namely, cytokinin synthases and cytokinin oxidases/dehydrogenases, which shows its role in the clubroot disease in *Arabidopsis* (Siemens et al. 2006). The phytohormone cytokinins have also been found to regulate the nitrogen metabolism in plants by enhancing the activity of nitrate reductase (Sykorová et al. 2008). The activities of the enzymes involved in the photosynthesis have been observed to be enhanced by the root-derived cytokinin signals (Sakakibara et al. 2006). In plants, cytokinins are perceived through a well-conserved multistep histidine-to-aspartate phosphorelay system which is known as two-component signaling systems (Schaller et al. 2011).

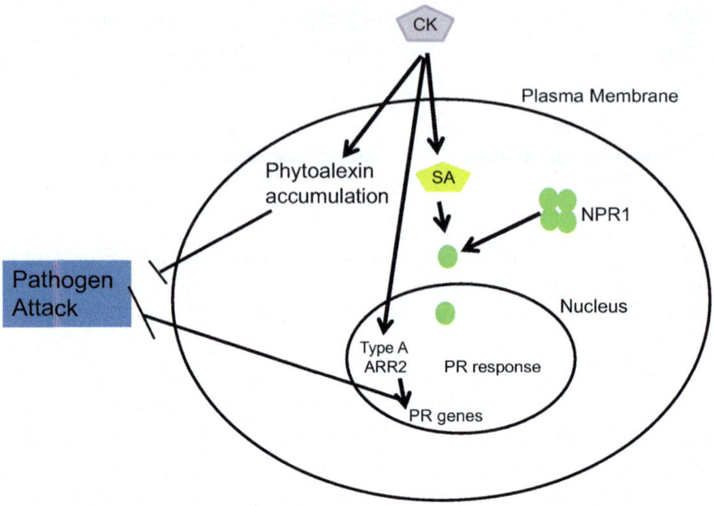

Fig. 7.7 Role of cytokinin in plant defense response. Arrows indicate positive regulation, and blocked arrows indicate negative regulation

The two-component system in plants is a well-studied and characterized signaling system specifically in *Arabidopsis* and rice. It is comprised of histidine-containing phosphotransfer proteins, receptors, and response regulators (Argueso et al. 2010). Specifically, type B response regulator, which functions as a transcription factor, regulates the expression of genes regulated by cytokinins. Analyses also suggest that cytokinins are the endogenous negative regulator of senescence (Singh et al. 1992). In *Glycine max*, the decrease in the cytokinin content has been observed with the onset of senescence (Noodén et al. 1990). Similar results have also been observed in *Arabidopsis* in high-throughput analysis (Breeze et al. 2011). These results suggest that cytokinins delay senescence in plants. Cytokinins have been proposed to induce synthesis and accumulation of phytoalexins in SA-independent manner. The hormone accumulated post pathogen attack and also helps the plant to cope against infection by affecting the priming response through SA-mediated signaling and inducing PR gene expression (PR-1, PR-3, PR-4, PR-5). Further a component of CK signaling pathway type B, ARR2 is required to regulate the activity of SA receptor NPR1 (O'Brien and Benkova 2013). The role of cytokinin in plant defense response is depicted in Fig. 7.7.

7.2.8 Auxins

Auxins are low-molecular-weight organic compounds, and it constitutes one of the most important and diverse groups of phytohormone generally found in all the plants. They are involved in large number of developmental processes such as shoot architecture control, vascular development and lateral root formation by controlling the cell division (Woodward and Bartel 2005). Auxins have also been shown to control the senescence, respond to various pathogens, and develop abiotic stress response in plants (Fu and Wang 2011; Wang et al. 2010). Apart from that, it also regulates the formation of fruits (De Jong et al. 2009). Indole acetic acid (IAA) is one of the most abundant endogenous auxins in plants and is synthesized via tryptophan (Trp)-independent and Trp-dependent pathways (Zhao 2010). In *Arabidopsis*, rice, and soybean, auxin stimulates the expression of auxin response genes (Javid et al. 2011). With the mediations and cross talk with other phytohormones, auxins play an important role in the biotic and abiotic stress response (Fahad et al. 2015). Apart from its positive role, auxins have been established to weakening of the defense response in plants. Analysis shows that its exogenous application promotes the diseases caused by Pst DC3000, *Pseudomonas savastanoi*, and *Agrobacterium tumefaciens* (Navarro et al. 2006; Chen et al. 2007). The signaling initiated by auxins regulated the response of cell to varied levels of auxins that are formed by auxin metabolism and transport. The TIR1/AFB family receptors are known for the auxin signaling (Dharmasiri et al. 2005; Kepinski and Leyser 2005). The phytohormone auxins can move long distances from highly active young tissues to the roots through the phloem (Marchant et al. 2002). Apart from that, it also moves to the short distances from cell to cell and is regulated by specific influx and efflux carrier proteins (Muday and Rahman 2008). Under various environmental stress conditions, inverse interactions between ROS and auxins have been observed, but the auxin-responsive gene is still unidentified. The impact of oxidative stress has been observed in auxin stability due to oxidization of IAA via peroxidase activity (Kawano 2003). Auxin acts in a mutually antagonistic fashion with SA and shares many commonalities with JA during plant defense. Recent evidences suggest that upon infection some pathogens either produce auxin themselves or increase auxin biosynthesis as a part of the plant's defense and developmental machinery (Valls et al. 2006; Kazan and Manners 2009; Bielach et al. 2017).

7.3 Cross Talk in Various Phytohormones

All the processes and responses in plants are the result of interplay of many genes and gene families suitably orchestrated in a network. In plants, growth, development, and response to various environmental cues go hand in hand. Various phytohormones are known to play important role in almost all the process through the modulation of genes. Further, through the balancing of phytohormones, plants maintain homeostasis and adapt to the environmental changes. This is only possible by an efficient and systemic cross talk between various phytohormones. SA, JA, and ET work together in a harmonious manner with BR, auxin, cytokinin, and GA in mediating plant response to biotic challenges. Plant defense response doesn't depend solely on any one hormone rather all the phytohormones work with each other regulating defense response positively or negatively (Fig. 7.8). There are evidences which suggest JA and SA signaling intersect in a negative manner against necrotrophic pathogens and herbivores (Glazebrook 2005). NPR1 is a key

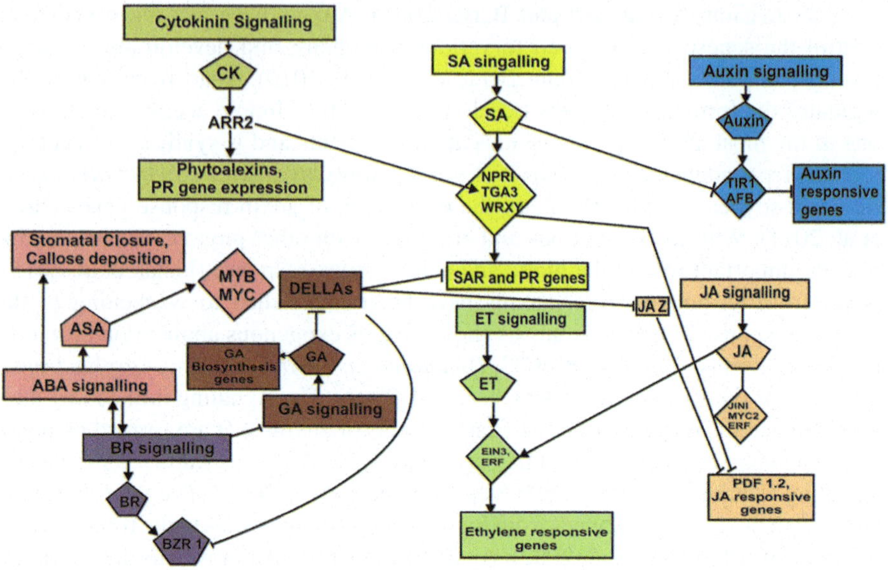

Fig. 7.8 Cross talk of phytohormones in biotic stress tolerance. ABA, SA, JA, and ET are key players in stress response. GA pathway produce DELLAs as an intersecting point. Auxins and CK pathway participate in biotic stress response in a SA-dependent manner. Arrows indicate positive regulation, and blocked arrows indicate negative regulation

intersecting player in the antagonistic talks of SA and JA. SA assisted in downregulation of JA-responsive genes like lipoxygenase 2 (LOX2), plant defensin 1.2, and vegetative storage protein (VSP). This suppression was abolished in *npr-1 mutants*. The WRKY transcription factor 70 family also plays prominent role in antagonistic interactions of SA and JA. The recombinant WRKY-70 overexpressing line displayed constitutive responsive expression of PR genes and repression of JA-responsive PDF1.2 (Li et al. 2004). Similar studies in *mpk4* (Map Kinase 4) mutants of *Arabidopsis* proved antagonistic interactions of SA and JA (Petersen et al. 2000). A few reports also prove their synergistic interactions.

JA and ET work together in a synergistic manner and regulate synthesis of defense genes postinfection. Both the phytohormones work cooperatively to induce or stabilize EIN3 and help the plant against necrotrophs by developing root hairs (Zhu et al. 2011). The synergy of JA interaction was observed in tomato plants where genes encoding proteinase inhibitors were induced in response to wounding (O'Donnell et al. 1996). Their cooperative action is even required for induction of ERF1-induced expression of PR genes (Lorenzo et al. 2003). Their antagonistic actions are also proved against insects and herbivores. The JA-activated *MYC2* represses the downstream functions of EIN3, and in a similar fashion, *MYC2* inhibits JA-regulated genes.

Auxins are known as key players in regulating plant growth and development, but several evidences suggest that auxins advance disease susceptibility and resistance against biotic challenges requires repression of auxin signaling. SAR reduces auxin-responsive genes. The treatment of *Arabidopsis* plants with a SA analog benzothiadiazole S-methyl ester (BTH) caused repression of auxin-responsive genes (Wang et al. 2007). Furthermore, SA signaling suppresses expression of transport inhibitor resistant 1 (TIR1) and auxin signaling F-box genes (AFB) and stabilizes auxin repressor protein (AUX/IAA) which represses further auxin action.

The pivotal role of cytokinins in biotic stress response has been demonstrated by several groups (Reusche et al. 2013; Wang et al. 2007). Recombinant *Arabidopsis* plants with stabilized CK levels displayed improved resistance against hemibiotrophic pathogen *Verticillium longisporum* (Reusche et al. 2013). CK and SA signaling cascades intersect to regulate plant defense as it has been seen that SA defense responses are promoted by cytokinin-activated transcription factor (Wang et al. 2007). The cooperativity between SA and CK was also observed in rice plants as increased resistance was observed against rice blast fungus in an *Os NPR1*- and *WRKY 45*-dependent manner (Jiang et al. 2013). GA and JA also interact as DELLA proteins are known to interact with key repressors of JA signaling JAZ1, thereby preventing JA-mediated suppression of transcription. JA is also known to regulate the expression of repressor of GA1–3 (RGL3), which positively regulates JA-mediated response against necrotrophs by competing with MYC2. BRs can cross talk among defense signaling pathways with a range of hormones such as SA, JA, ABA, auxins, and GA. BR is known to increase resistance against pathogens in a SA dependent as well as independent manner. It also negates JA-induced resistance in rice plants. BRs can also cross talk with auxins, and this BR-auxin bidirectional interplay can have effects on BR signaling cascade in disease and resistance. BR is also known to interact with GA. It suppresses GA biosynthesis genes and activates GA repressor genes. Known interactions also exist between BZR1 transcription factor and DELLA proteins further mediating cross talk between BR and GA.

There are several evidences which validate the cross-talk hypothesis of the plant phytohormones which seems to be true in the sense that plants maintain a critical balance between growth and environmental response in which all the phytohormones play important role. In conclusion, this is just the beginning of unraveling the complex phytohormones signaling cascades. The dissection of plant immunity web comprising of different hormones will help to understand the roles played by each one of them both as an antagonistic and synergistic way. So, elucidating the mechanisms of plant hormone interaction and solving the plant immune network will help fundamental understanding of how plants comprise of orchestrated immune system function. The phytohormone intricate web may be a target for crop improvement. The yield may be increased by the disease resistance plants by transgenic approaches. The methods employed will comprise of either strengthening resistance of signaling pathway by modulating the antagonistic pathway or synergistic pathways.

References

Abdala G, Miersch O, Kramell R, Vigliocco A, Agostini E, Forchetti G, Alemano S (2003) Jasmonate and octadecanoid occurrence in tomato hairy roots. Endogenous level changes in response to NaCl. J Plant Growth Regul 40:21–27

Abeles FB, Morgan PW, Saltveit ME Jr (1992) Ethylene in plant biology. Academic, New York

Achard P, Renou J-P, Berthomé R, Harberd NP, Genschik P (2008) Plant DELLAs restrain growth and promote survival of adversity by reducing the levels of reactive oxygen species. Curr Biol 18:656–660

Adie BAT, Perez-Perez J, Perez-Perez MM, Godoy M, Sanchez-Serrano JJ, Schmelz EA, Solano R (2007) ABA is an essential signal for plant resistance to pathogens affecting JA biosynthesis and the activation of defenses in *Arabidopsis*. Plant Cell 19:1665–1681

Albrecht C, Boutrot F, Segonzac C, Schwessinger B, Gimenez-Ibanez S, Chinchilla D, Rathjen JP, de Vries SC, Zipfel C (2003) Brassinosteroids inhibit pathogen-associated molecular pattern-triggered immune signaling independent of the receptor kinase BAK1. Proc Natl Acad Sci 109(1):303–308

Ali SS, Kumar GBS, Khan M, Doohan FM (2013) Brassinosteroid enhances resistance to fusarium diseases of barley. Ann Bot 86:441–447

Antico CJ, Colon C, Banks T, Ramonell KM (2012) Insights into the role of jasmonic acid-mediated defenses against necrotrophic and biotrophic fungal pathogens. Front Plant Biol 7:948–956

Argueso CT, Raines T, Kieber JJ (2010) Cytokinin signaling and transcriptional networks. Curr Opin Plant Biol 13:533–539

Ashraf M, Akram NA, Arteca RN, Foolad MR (2010) The physiological, biochemical and molecular roles of brassinosteroids and salicylic acid in plant processes and salt tolerance. Crit Rev Plant Sci 29:162–190

Asselbergh B, De Vleesschauwer D, Hofte M (2008) Global switches and fine-tuning-ABA modulates plant pathogen defense. Mol Plant-Microbe Interact 21:709–719

Audenaert K, De Meyer GB, Höfte MM (2002) Abscisic acid determines basal susceptibility of tomato to *Botrytis cinerea* and suppresses salicylic acid-dependent signaling mechanisms. Plant Physiol 128:491–501

Avanci NC, Luche DD, Goldman GH, Goldman MH (2010) Jasmonates are phytohormones with multiple functions, including plant defense and reproduction. Genet Mol Res 9(1):484–505

Bajguz, Hayat S (2009) Effects of brassinosteroids on the plant responses to environmental stresses. Plant Physiol Biochem 47(1):1–8

Bassett CL, Nickerson ML, Farrell RE, Artlip TS, El Ghaouth A, Wilson CL, Wisniewski ME (2005) Characterization of an S-locus receptor protein kinase-like gene from peach. Tree Physiol 25:403–411

Baxter A, Mittler R, Suzuki N (2014) ROS as key players in plant stress signalling. J Exp Bot 65:1229–1240

Beffa RS, Hofer R-M, Thomas M, Meins F Jr (1996) Decreased susceptibility to virus disease of β-1,3-glucanase-deficient plants generated by antisense transformation. Plant Cell 8:1001–1011

Belkhadir Y, Jaillais Y, Epple P, Balsemão-Pires E, Dangl JL (2012) ChoryJ Brassinosteroids modulate the efficiency of plant immune responses to microbe-associated molecular patterns. Proc Natl Acad Sci 109:297–302

Bielach A, Hrtyan M, Tognetti VB (2017) Plants under stress: involvement of auxin and cytokinin. Int J Mol Sci 18:1427

Boyer JS (1982) Plant productivity and environment. Science 218(4571):443–448

Boyle P, Le SE, Rochon A, Shearer HL, Murmu J, Chu JY et al (2009) The BTB/POZ domain of the *Arabidopsis* disease resistance protein NPR1 interacts with the repression domain of TGA2 to negate its function. Plant Cell 21:3700–3713

Breeze E, Harrison E, McHattie S, Hughes L, Hickman R, Hill C et al (2011) High-resolution temporal profiling of transcripts during *Arabidopsis* leaf senescence reveals a distinct chronology of processes and regulation. Plant Cell 23:873–894

Bruyne DL, Höfte M, De Vleesschauwer D (2014) Connecting growth and defense: the emerging roles of brassinosteroids and gibberellins in plant immunity. Mol Plant 7:943–959

Carpita NC, McCann M (2000) The cell wall. In: Buchanan B et al (eds) Biochemistry and molecular biology of plants. American Society of Plant Physiologists, Rockville, pp 52–108

Chaparro-Garcia A, Wilkinson RC, Gimenez-Ibanez S, Findlay K, Coffey MD, Zipfel C, Rathjen JP, Kamoun S, Schornack S (2011) The receptor-like kinase SERK3/BAK1 is required for basal resistance against the late blight pathogen *phytophthora infestan*s in *Nicotiana benthamiana*. PLoS One 6:e16608

Chen Z, Agnew JL, Cohen JD, He P, Shan L, Sheen J, Kunkel BN (2007) *Pseudomonas syringae* type III effector *AvrRpt2* alters *Arabidopsis thaliana* auxin physiology. Proc Natl Acad Sci 104:20131–20136

Cheng M-C, Liao P-M, Kuo W-W, Lin T-P (2013) The *Arabidopsis* ETHYLENE RESPONSE FACTOR1 regulates abiotic stress-responsive gene expression by binding to different cis-acting elements in response to different stress signals. Plant Physiol 162:1566–1582

Chinchilla D, Shan L, He P, de Vries S, Kemmerling B (2009) One for all: the receptor-associated kinase BAK1. Trends Plant Sci 14:535–541

Chisholm ST, Coaker G, Day B, Staskawicz BJ (2006) Host–microbe interactions: shaping the evolution of the plant immune response. Cell 12:803–814

Clouse S, Sasse J (1998) Brassinosteroids: essential regulators of plant growth and development. Annu Rev Plant Physiol Plant Mol Biol 49:427–451

Colebrook EH, Thomas SG, Phillips AL, Hedden P (2014) The role of gibberellin signalling in plant responses to abiotic stress. J Exp Bot 217:67–75

Davies PJ (1995) The plant hormone concept: concentration, sensitivity and transport. In: Davies PJ (ed) Plant hormones: physiology, biochemistry and molecular biology. Kluwer, Boston, pp 13–38

Davies PJ (2010) The plant hormones: their nature, occurrence, and functions. Plant hormones. Springer, Dordrecht, pp 1–15

Davletova S, Rizhsky L, Liang H, Shengqiang Z, Oliver DJ, Coutu J, Shulaev V, Schlauch K, Mittler R (2005) CYTOSOLIC ASCORBATE PEROXIDASE 1 is a central component of the reactive oxygen gene network of *Arabidopsis*. Plant Cell 17:268–281

De Jong M, Mariani C, Vriezen WH (2009) The role of auxin and gibberellin in tomato fruit set. J Exp Bot 60:1523–1532

de Torres Z, Mansfield JW, Grabov N, Brown IR, Ammouneh H, Tsiamis G et al (2007) *Pseudomonas syringae* effector *AvrPtoB* suppresses basal defence in *Arabidopsis*. Plant J 47:368–382

De Vleesschauwer D, Van Buyten E, Satoh K, Balidion J, Mauleon R, Choi IR, Vera-Cruz C, Kikuchi S, Höfte M (2012) Brassinosteroids antagonize gibberellin- and salicylate-mediated root immunity in rice. Plant Physiol 158:1833–1846

Dempsey DA, Klessig DF (2017) How does the multifaceted plant hormone salicylic acid combat disease in plants and are similar mechanisms utilized in humans? BMC Biol 15:23

Dempsey DM, Corina Vlot A, Wildermuth MC, Klessig DF (2011) Salicylic acid biosynthesis and metabolism. Arabidopsis Book 9:e0156

Derksen H, Rampitsch C, Daayf F (2013) Signaling cross-talk in plant disease resistance. Plant Sci 207:79–87

Dey S, Wenig M, Langen G, Sharma S, Kugler KG, Knappe C et al (2014) Bacteria-triggered systemic immunity in barley is associated with WRKY and ETHYLENE RESPONSIVE FACTORs but not with salicylic acid. Plant Physiol 166:2133–2151

Dharmasiri N, Dharmasiri S, Estelle M (2005) The F-box protein TIR1 is an auxin receptor. Nature 435:441–445

Eyidogan F, Oz MT, Yucel M, Oktem HA (2012) Signal transduction of phytohormones under abiotic stresses. In: Khan NA, Nazar R, Iqbal N, Anjum NA (eds) Phytohormones and abiotic stress tolerance in plants. Springer, Berlin, pp 1–48

Fahad SS, Hussain A, Matloob FA, Khan A, Khaliq S, Saud S, Hassan D, Shan F, Khan N, Ullah M, Faiq MR, Khan AK, Tareen A, Khan A, Ullah N, Ullah JL et al (2015) Phytohormones and plant responses to salinity stress: a review. Plant Growth Regul 75:391–404

Fariduddin Q, Khalil RRAE, Mir BA, Yusuf M, Ahmad A (2013) 24-Epibrassinolide regulates photosynthesis, antioxidant enzyme activities and proline content of *Cucumis sativus* under salt and/or copper stress. Environ Monit Assess 185(9):7845–7856

Fariduddin QM, Yusuf I, Ahmad A (2014) Brassinosteroids and their role in response of plants to abiotic stresses. Biol Plant 58:9–17

Fradin EF, Zhang Z, Juarez Ayala JC, Castroverde CDM, Nazar RN, Robb J, Liu C-M, Thomma BPH (2009) Genetic dissection of *Verticillium* wilt resistance mediated by tomato Ve1. Plant Physiol 150:320–332

Friedrichsen DM, Joazeiro CAP, Li J, Hunter T, Chory J (2000) Brassinosteroid-insensitive-I is a ubiquitously expressed leucine-rich repeat receptor serine/threonine kinase. Plant Physiol 123:1247–1255

Fu ZQ, Dong X (2013) Systemic acquired resistance: turning local infection into global defense. Annu Rev Plant Biol 64:839–863

Fu X, Harberd NP (2003) Auxin promotes *Arabidopsis* root growth by modulating gibberellin response. Nature 421(6924):740–743

Fu J, Wang S (2011) Insights into auxin signaling in plant–pathogen interactions. Front Plant Sci 2:74

Glazebrook J (2005) Contrasting mechanisms of defense against biotrophic and necrotphic pathogens. Annu Rev Phytopathol 43:205–227

Gomes MMA (2011) Physiological effects related to brassinosteroid application in plants. In: Hayat S, Ahmad A (eds) Brassinosteroids: a class of plant hormones. Springer, Dordrecht/Heidelberg/London/New York, p 193

Guo R, Qian H, Shen W, Liu L, Zhang M, Cai C, Zhao Y, Qiao J, Wang Q (2013) BZR1 and BES1 participate in regulation of glucosinolate biosynthesis by brassinosteroids in *Arabidopsis*. J Exp Bot 64:2401–2412

Harrison MA (2012) Cross-talk between phytohormone signaling pathways under both optimal and stressful environmental onditions. In: Khan NA, Nazar R, Iqbal N, Anjum NA (eds) Phytohormones and abiotic stress tolerance in plants. Springer, Berlin, pp 49–76

Hattori Y, Nagai K, Furukawa S, Song X-J, Kawano R, Sakakibara H, Wu J, Matsumoto T, Yoshimura A, Kitano H et al (2009) The ethylene response factors SNORKEL1 and SNORKEL2 allow rice to adapt to deep water. Nature 46:1026–1030

Haubrick LL, Assmann SM (2006) Brassinosteroids and plant function: some clues, more puzzles. Plant Cell Environ 29:446–457

Hayat S, Fariduddin Q, Ali B, Ahmad A (2005) Effect of salicylic acid on growth and enzyme activities of wheat seedlings. Acta Agron Hung 53:433–437

Helliwell EE, Wang Q, Yang Y (2013) Transgenic rice with inducible ethylene production exhibits broad-spectrum disease resistance to the fungal pathogens *Magnaporthe oryzae* and *Rhizoctonia solani*. Plant Biotechnol J 11:33–42

Hou X, Lee LYC, Xia K, Yan Y, Yu H (2010) DELLAs modulate jasmonate signaling via competitive binding to JAZs. Dev Cell 19:884–894

Ishihara T, Sekine KT, Hase S, Kanayama Y, Seo S, Ohashi Y et al (2008) Overexpression of the *Arabidopsis thaliana* EDS5 gene enhances resistance to viruses. Plant Biol 10:451–461

Jacobs AK, Lipka V, Burton RA, Panstruga R, Strizhov N, Schulze-Lefert P et al (2003) An *Arabidopsis* callose synthase, GSL5, is required for wound and papillary callose formation. Plant Cell 15(11):2503–2513

Javid MG, Sorooshzadeh A, Moradi F, Sanavy SAMM, Allahdadi I (2011) The role of phytohormones in alleviating salt stress in crop plants. Aust J Crop Sci 5:726–734

Jiang YP, Cheng F, Zhou YH, Xia XJ, Mao WH, Shi K, Chen Z, Yu JQ (2012) Cellular glutathione redox homeostasis plays an important role in the brassinosteroid-induced increase in CO2 assimilation in *Cucumis sativus*. New Phytol 194:932–943

Jiang CJ, Shimono M, Sugano S, Kojima M, Liu X, Inoue H, Sakakibara H, Takatsuji H (2013) Cytokinins act synergistically with salicylic acid to activate defense gene expression in rice. Mol Plant-Microbe Interact 26:287–296

Ju C, Yoon GM, Shemansky JM, Lin DY, Ying ZI, Chang J et al (2012) CTR1 phosphorylates the central regulator EIN2 to control ethylene hormone signalling from the ER membrane to the nucleus in Arabidopsis. Proc Natl Acad Sci 109:19486–19491

Kawano T (2003) Roles of the reactive oxygen species-generating peroxidase reactions in plant defense and growth induction. Plant Cell Rep 21:829–837

Kazan K, Manners JM (2008) Jasmonate signaling: toward an integrated view. Plant Physiol 146:1459–1468

Kazan K, Manners JM (2009) Linking development to defense: auxin in plant–pathogen interactions. Trends Plant Sci 14:373–382

Kemmerling B, Schwedt A, Rodriguez P, Mazzotta S, Frank M, Qamar SA, Mengiste T, Betsuyaku S, Parker JE, Müssig C et al (2007) The BRI1-associated kinase 1, BAK1, has a brassinolide-independent role in plant cell death control. Curr Biol 17:1116–1122

Kepinski S, Leyser O (2005) The *Arabidopsis* F-box protein *TIR1 is an auxin receptor*. Nature 435:446–451

Khripach V (2000) Twenty years of brassinosteroids: steroidal plant hormones warrant better crops for the XXI century. Ann Bot 86:441–447

Koga H, Dohi K, Mori M (2004) Abscisic acid and low temperatures suppress the whole plant-specific resistance reaction of rice plants to the infection with *Magnaporthe grisea*. Physiol Mol Plant Pathol 65:3–9

Leclercq J, Ranty B, Sanchez-Ballesta MT, Li ZG, Jones B, Jauneau A, Pech JC, Latche A, Ranjeva R, Bouzayen M (2005) Molecular and biochemical characterization of LeCRK1, a ripening-associated tomato CDPK-related kinase. J Exp Bot 56:25–35

LeNoble ME, Spollen WG, Sharp RE (2004) Sharp maintenance of shoot growth by endogenous ABA: genetic assessment of the involvement of ethylene suppression. J Exp Bot 55:237–245

Li J, Brader G, Palva ET (2004) The WRKY70 transcription factor: a node of convergence for jasmonate-mediated and salicylate-mediated signals in plant defense. Plant Cell 16:319–331

Lin W, Lu D, Gao X, Jiang S, Ma X, Wang Z, Mengiste T, He P (2013) Inverse modulation of plant immune and brassinosteroid signaling pathways by the receptor-like cytoplasmic kinase BIK1. Proc Natl Acad Sci U S A 110:12114–12119

Liu D, Chen X, Liu J, Ye J, Guo Z (2012) The rice ERF transcription factor OsERF922 negatively regulates resistance to *Magnaporthe oryzae* and salt tolerance. J Exp Bot 63:3899–3912

Lorenzo O, Piqueras R, Sánchez-Serrano JJ, Solano R (2003) ETHYLENE RESPONSE FACTOR1 integrates signals from ethylene and jasmonate pathways in plant defense. Plant Cell 15:165–178

Lozano-Durán R, Macho AP, Boutrot F, Segonzac C, Somssich IE, Zipfel C (2013) The transcriptional regulator BZR1mediates trade-off between plant innate immunity and growth. elife 2:e00983

Maciejewska B, Kopcewicz J (2003) Inhibitory effect of methyl jasmonate on flowering and elongation growth in pharbitis nil. J Plant Growth Regul 21:216–223

MacMillan J (2001) Occurrence of gibberellins in vascular plants, fungi, and bacteria. J Plant Growth Regul 20:387–442

Malamy J, Carr JP, Klessig DF, Raskin I (1990) Salicylic acid: a likely endogenous signal in the resistance response of tobacco to viral infection. Science 250:1002–1004

Marchant A, Bhalerao R, Casimiro I, Eklof J, Casero PJ, Bennett M, Sandberg G (2002) AUX1 promotes lateral root formation by facilitating indole-3-acetic acid distribution between sink and source tissues in the *Arabidopsis* seedling. Plant Cell 14:589–597

Mauch-Mani B, Mauch F (2005) The role of abscisic acid in plant–pathogen interactions. Curr Opin Plant Biol 8:409–414

Mei C, Qi M, Sheng G, Yang Y (2006) Inducible overexpression of a rice allene oxide synthase gene increases the endogenous jasmonic acid level, PR gene expression, and host resistance to fungal infection. Mol Plant-Microbe Interact 19:1127–1137

Mitchell JW, Mandava NB, Worley JF, Plimmer JR, Smith MV (1970) Brassins: a new family of plant hormones from rape pollen. Nature 225:1065–1066

Mohr P, Cahill D (2007) Suppression by ABA of salicylic acid and lignin accumulation and the expression of multiple genes, in *Arabidopsis* infected with *Pseudomonas syringae* pv. tomato. Funct Integr Genomics 7:181–191

Muday GK, Rahman A (2008) Auxin transport and the integration of gravitropic growth. In: Gilroy S, Masson P (eds) Plant tropisms. Blackwell Publishing, Oxford, pp 47–68

Mueller MJ, Brodschelm W, Spannagl E, Zenk MH (1993) Signalling in the elicitation process is mediated through the octadecanoid pathway leading to jasmonic acid. Proc Natl Acad Sci U S A 90:7490–7494

Muller B, Sheen J (2007) Advances in cytokinin signaling. Science 318:68–69

Nahar K, Kyndt T, Hause B, Höfte M, Gheysen G (2013) Brassinosteroids suppress rice defense against root-knot nematodes through antagonism with the jasmonate pathway. Mol Plant Microbe Interact 26:106–115

Nakashita H, Yasuda M, Nitta T, Asami T, Fujioka S, Arai Y, Sekimata K, Takatsuto S, Yamaguchi I, Yoshida S (2003) Brassinosteroid functions in a broad range of disease resistance in tobacco and rice. Plant J 33(5):887–898

Navarro L, Dunoyer P, Jay F, Arnold B, Dharmasiri N, Estelle M, Voinnet O, Jones JD (2006) A plant miRNA contributes to antibacterial resistance by repressing auxin signaling. Science 312:436–439

Navarro L, Bari R, Achard P, Lison P, Nemri A, Harberd NP, Jones JD (2008) DELLAs control plant immune responses by modulating the balance of jasmonic acid and salicylic acid signaling. Curr Biol 18:650–655

Noodén LD, Singh S, Letham DS (1990) Correlation of xylem sap cytokinin levels with monocarpic senescence in soybean. Plant Physiol 93:33–39

O'Brien JA, Benkova E (2013) Cytokinin cross-talking during biotic and abiotic stress responses. Front Plant Sci 4:451

O'Donnell PJ, Calvert C, Atzorn R, Wasternack C, Leyser HMO, Bowles DJ (1996) Ethylene as a signal mediating the wound response of tomato plants. Science 274:1914–1917

Osbourn AE (1996) Preformed antimicrobial compounds and plant defense against fungal attack. Plant Cell 8(10):1821–1831

Park SW, Kaimoyo E, Kumar D, Mosher S, Klessig DF (2007) Methyl salicylate is a critical is a critical mobile signal for plant systemic acquired resistance. Science 318:113

Pedranzani H, Racagni G, Alemano S, Miersch O, Ramírez I, Peña-Cortés H, Taleisnik E, Machado-Domenech E, Abdala G (2003) Salt tolerant tomato plants show increased levels of jasmonic acid. Plant Growth Regul 41:149–158

Petersen M, Brodersen P, Naested H, Andreasson E, Lindhart U, Johansen B, Nielsen HB, Lacy M, Austin MJ, Parker JE, Sharma SB, Klessig DF, Martienssen R, Mattsson O, Jensen AB, Mundy J (2000) Arabidopsis MAP kinase 4 negatively regulates systemic acquired resistance. Cell 103:1111–1120

Qin X, Liu JH, Zhao WS, Chen XJ, Guo ZJ, Peng YL (2013) Gibberellin 20-oxidase gene OsGA20ox3 regulates plant stature and disease development in rice. Mol Plant-Microbe Interact 26:227–239

Reusche M, Kláskova J, Thole K, Truskina J, Novák O, Janz D, Strnad M, Spichal L, Lipka V, Teichmann T (2013) Stabilization of cytokinin levels enhances *Arabidopsis* resistance against *Verticillium longisporum*. Mol Plant-Microbe Interact 26:850–860

Rezzonico E, Flury N, Meins F, Beffa RS (1998) Transcriptional down-regulation by abscisic acid of pathogenesis-related beta-1,3-glucanase genes in tobacco cell cultures. elife 2:e00983

Rochon A, Boyle P, Wignes T, Fobert PR, Despres C (2006) The coactivator function of *Arabidopsis* NPR1 requires the core of its BTB/POZ domain and the oxidation of C-terminal cysteines. Plant Cell 18:3670–3685

Sakakibara H, Takei K, Hirose N (2006) Interactions between nitrogen and cytokinin in the regulation of metabolism and development. Trends Plant Sci 11:440–448

Sakurai A, Yokota T, Clouse SD (eds) (1999) Brassinosteroids: steroidal plant hormones. Springer, Tokyo, pp 1–253

Schaller GE, Shiu SH, Armitage JP (2011) Two-component systems and their co-option for eukaryotic signal transduction. Curr Biol 21:R320–R330

Seo S, Mitsuhara I, Feng J, Iwai T, Hasegawa M, Ohashi Y (2011) Cyanide, a coproduct of plant hormone ethylene biosynthesis, contributes to the resistance of rice to blast fungus. Plant Physiol 155:502–514

Sharma YK, León J, Raskin I, Davis KR (1996) Ozone-induced responses in *Arabidopsis thaliana*: the role of salicylic acid in the accumulation of defense-related transcripts and induced resistance. Proc Natl Acad Sci U S A 93:5099–5104

Shi H, Shen Q, Qi Y, Yan H, Nie H, Chen Y, Zhao T, Katagiri F, Tang D (2013) BR-signaling kinase1physically associates with flagellin sensing 2 and regulates plant innate immunity in *Arabidopsis*. Plant Cell 25:1143–1157

Siemens J, Keller I, Sarx J et al (2006) Transcriptome analysis of Arabidopsis club roots indicate a key role for cytokinins in disease development. Mol Plant-Microbe Interact 19:480–494

Singh S, Letham DS, LMS P (1992) Cytokinin biochemistry in relation to leaf senescence. Endogenous cytokinin levels and exogenous applications of cytokinins in relation to sequential leaf senescence of tobacco. Physiol Plant 86:388–397

Solano R, Stepanova A, Chao Q, Ecker JR (1998) Nuclear events in ethylene signaling: a transcriptional cascade mediated by ETHYLENE-INSENSITIVE3 and ETHYLENE-RESPONSE-FACTOR1. Genes Dev 12(23):3703–3714

Spence C, Alff E, Johnson C, Ramos C, Donofrio N, Sundaresan V et al (2014) Natural rice rhizospheric microbes suppress rice blast infections. BMC Plant Biol 14:130–110

Staswick PE, Su W, Howell SH (1992) Methyl jasmonate inhibition of root growth and induction of leaf protein are decreased in an *Arabidopsis thaliana* mutant. Proc Natl Acad Sci U S A 89:6837–6840

Sticher L, MauchMani B, Metraux JP (1997) Systemic acquired resistance. Annu Rev Plant Pathol 35:235–270

Sykorová B, Kuresová G, Daskalova S, Trcková M, Hoyerová K, Raimanová I, Motyka V, Trávnícková A, Elliott M, Kamínek M (2008) Senescence-induced ectopic expression of the *A. tumefaciens*iptgene in wheat delays leaf senescence, increases cytokinin content, nitrate influx, and nitrate reductase activity, but does not affect grain yield. J Exp Bot 59(2):377–387

Takahashi F, Yoshida R, Ichimura K, Mizoguchi T, Seo S, Yonezawa M, Maruyama K, Yamaguchi-Shinozaki K, Shinozaki K (2007) The mitogen-activated protein kinase cascade MKK3-MPK6 is an important part of the jasmonate signal transduction pathway in Arabidopsis. Plant Cell 19:805–818

Tanaka N, Matsuoka M et al (2006) gid1, a gibberellin-insensitive dwarf mutant, shows altered regulation of probenazole-inducible protein (PBZ1) in response to cold stress and pathogen attack. Plant Cell Environ 29:619–631

Ton J, Flors V, Mauch-Mani B (2009) The multifaceted role of ABA in disease resistance. Trends Plant Sci 14:310–317

Tong H, Xiao Y, Liu D, Gao S, Liu L, Yin Y, Jin Y, Qian Q, Chu C (2014) Brassinosteroid regulates cell elongation by modulating gibberellin metabolism in rice. Plant Cell 26(11):4376–4393

Unterholzner SJ, Rozhon W, Papacek M, Ciomas J, Lange T, Kugler KG, Mayer KF, Sieberer T, Poppenberger B (2015) Brassinosteroids are master regulators of gibberellin biosynthesis in Arabidopsis. Plant Cell 27(8):2261–2272

Valls M, Genin S, Boucher C (2006) Integrated regulation of the type III secretion system and other virulence determinants in *Ralstonia solanacearum*. PLoS Pathog 2:e82

Van Loon LC (1984) Regulation of pathogenesis and symptom expression in diseased plants by ethylene. In: Fuchs Y, Chalutz E (eds) Ethylene: biochemical, physiological and applied aspects. MartinusNijhoff/Dr. W. Junk, The Hague, pp 171–180

Vardhini BV (2013a) Brassinosteroids role for amino acids, peptides and amines modulation in stressed plants – a review. In: Anjum NA, Gill SS, Gill R (eds) Plant adaptation to environmental change: significance of amino acids and their derivatives. CAB International, Wallingford, pp 300–316

Vardhini BV (2013b) Comparative study of *Sorghum vulgare* Pers. Grown in two experimental sites by brassinolide application at vegetative, flowering and grain filling stage. Proc Andhra Pradesh Akad Sci 15:75–79

Vardhini BV, Anuradha S, Rao SSR (2006) Brassinosteroids – a great potential to improve crop productivity. Indian J Plant Physiol 11:1–12

Vernooij B, Friedrich L, Morse A, Reist R, Kolditz-Jawhar R, Ward E, Uknes S, Kessmann H, Ryals J (1994) Salicylic acid is not the translocated signal responsible for inducing systemic acquired resistance but is required in signal transduction. Plant Cell 6:959–965

Vick BA, Zimmerman DC (1984) Biosynthesis of jasmonic acid by several plant species. Plant Physiol 75(2):458–461

von Malek B, van der Graaff E, Schneitz K, Keller B (2002) The Arabidopsis male-sterile mutant dde2-2 is defective in the ALLENE OXIDE SYNTHASE gene encoding one of the key enzymes of the jasmonic acid biosynthesis pathway. Planta 216:187–192

Wang D, Pajerowska-Mukhtar K, Culler AH, Dong X (2007) Salicylic acid inhibits pathogen growth in plants through repression of the auxin signalling pathway. Curr Biol 17:1784–1790

Wang S, Bai Y, Shen C, Wu Y, Zhang S, Jiang D, Guilfoyle TJ, Chen M, Qi Y (2010) Auxin-related gene families in abiotic stress response in Sorghum bicolor. Funct Integ Genomics 10:533–546

Wang W, Bai MY, Wang ZY (2014) The brassinosteroid signalling network-a paradigm of signal integration. Curr Opin Plant Biol 2:147–153

Wasternack C (2007) Jasmonates: an update on biosynthesis, signal transduction and action in plant stress response, growth and development. Ann Bot 100:681–697

Whenham RJ, Fraser RSS, Brown LP, Payne JA (1986) Tobacco mosaic virus-induced increase in abscisic acid concentration in tobacco leaves: intracellular location in light and dark green areas, and relationship to symptom development. Planta 168(1986):592–598

Wiese J, Kranz T, Schubert S (2004) Induction of pathogen resistance in barley by abiotic stress. Plant Biol 6(5):529–536

Wilkinson S, Davies WJ (2002) ABA-based chemical signalling: the co-ordination of responses to stress in plants. Plant Cell Environ 25:195–210

Woodward AW, Bartel B (2005) Auxin: regulation, action, and interaction. Ann Bot 95:707–735

Xu ZS, Xia LQ, Chen M, Cheng XG, Zhang RY, Li LC et al (2007) Isolation and molecular characterization of the *Triticumaestivum* L. ethylene-responsive factor 1 (TaERF1) that increases multiple stress tolerance. Plant Mol Biol 65:719–732

Yalpani N, Enyedi AJ, León J, Raskin I (1994) Ultraviolet light and ozone stimulate accumulation of salicylic acid, pathogenesis-related proteins and virus resistance in tobacco. Planta 193:372–376

Yang DL, Li Q, Deng YW, Lou YG, Wang MY, Zhou GX et al (2008) Altered disease development in the *eui*mutants and *Eui*over expressors indicates that gibberellins negatively regulate rice basal disease resistance. Mol Plant 1:528–537

Zhang Y, Tessaro MJ, Lassner M, Li X (2003) Knockout analysis of *Arabidopsis* transcription factors TGA2 TGA5 and TGA6 reveals their redundant and essential roles in systemic acquired resistance. Plant Cell 15:2647–2653

Zhao Y (2010) Auxin biosynthesis and its role in plant development. Annu Rev Plant Biol 61:49–64

Zhu J, Verslues PE, Zheng X, Lee BH, Zhan X, Manabe Y et al (2005) *HOS10* encodes an R2R3-type MYB transcription factor essential for cold acclimation in plants. Proc Natl Acad Sci U S A 102:9966–9971

Zhu Z, An F, Feng Y, Li P, Xue L, Jiang Z, Kim JM, To TK, Li W et al (2011) Derepression of ethylene-stabilized transcription factors (eiN3/eil1) mediates jasmonate and ethyl-enesignaling synergy in Arabidopsis. Proc Natl Acad Sci U S A 108:12539–12544

Zurek DM, Rayle DL, McMorris TC, Clouse SD (1994) Investigation of gene expression, growth kinetics, and wall extensibility during brassinosteroid regulated stem elongation. Plant Physiol 104:503–513

8 Insights into the Role of WRKY Superfamily of Protein Transcription Factor in Defense Response

Preeti Singh, Shashi Shekhar, Anjana Rustagi, Vinay Sharma, and Deepak Kumar

Abstract

Plants are constantly challenged by a variety of biotic and abiotic stresses. To combat these challenges, plants have developed intricate mechanisms to perceive external signals and respond with the proper physiological and morphological changes. Generally, plants regulate the expression of many stress-related genes by activating or repressing their transcription upon signal perception and transduction of the external stimuli. The WRKY transcription factors comprise a large family of plant-specific zinc-finger-type regulatory proteins and regulate many plant defense responses to diverse biotic and abiotic stresses. WRKY proteins possess either one or two WRKY domains, a 60-amino-acid region that contains the amino acid sequence WRKYGQK, and a zinc-finger-like motif. In spite of the strong conservation of their DNA-binding domain, the overall structures of WRKY proteins are highly divergent and can be categorized into distinct groups, which might reflect their different functions. Based on the number of conserved WRKY domains and the features of the zinc-finger motif, the WRKY superfamily can be divided into three distinct groups: I, II, and III. Previous studies have demonstrated that WRKY transcription factors participate in regulating defense gene expression at various levels, partly by directly modulating immediate downstream target genes, by activating or repressing other TF genes, and by

P. Singh · S. Shekhar
School of Life Sciences, Jawaharlal Nehru University, New Delhi, India

A. Rustagi
Department of Botany, Gargi College, New Delhi, India

V. Sharma
Department of Bioscience and Biotechnology, Banasthali University, Banasthali, Rajasthan, India

D. Kumar (✉)
Department of Botany, Central University of Jammu, Jammu, J&K, India
e-mail: deepakkumar@cujammu.ac.in

A. Singh, I. K. Singh (eds.), *Molecular Aspects of Plant-Pathogen Interaction*,
https://doi.org/10.1007/978-981-10-7371-7_8

regulating WRKY genes. WRKY proteins also seem to be involved in other plant-specific processes, such as trichome development and the biosynthesis of secondary metabolites. In this chapter, we will focus our attention to the role of WRKY TFs in plant defense response.

Keywords
Biotic stress · Plant defense · Transcription factor · WRKY · Zinc-finger protein

8.1 Introduction

The global climate change leads to the dynamic interaction between climatic and biological factors. It is not only confined up to modification of physiology and resistance of plants, but rather it also modifies the rates and stages of pathogen development, which will ultimately lead to the shifting of host-pathogen physiological interactions. Pathogen would be following the hosts and may infect vegetation of natural plant communities which were previously not exposed to the more aggressive strains. Therefore, new combinations of species are evolving. So the emphasis is shifted to develop new strategies to cope up with these biotic constraints in present scenario. However, the nature orchestrated by plants with an inherent defense system can be generally divided into two levels: the first is PAMP-triggered immunity (PTI), which confers resistance to most pathogens, and the second begins in the cytoplasm and mainly relies on the recognition of microbial effectors which is called effector-triggered immunity (ETI). Bostock (2005) reported that both PTI and ETI activate local as well as systemic defense responses, modulated by jasmonic acid (JA) and salicylic acid (SA). Activation of these two pathways extensively shares downstream signaling pathway, which in turn induces expression of defense gene and their corresponding defense responses (Tsuda et al. 2013). The response leads to adaptive plasticity of plants, which is mainly achieved by enforcement of a network of various transcription factors (TFs). TFs are multimers of polypeptide mediating different cellular responses through recognizing the specific cis-regulatory DNA sequences at the promoters of their targets genes (Franco-Zorrilla et al. 2014) and the rearrangement of the multimeric subunits leading to different functions through their differential expression patterns (Berk and Schmidt 1990). The binding of TFs with *cis*-elements of stress-related genes results in either overexpression or suppression of these genes, which may improve the plant's tolerance potential against different biotic stresses. Approximately 6% of the plant genome encodes for TFs, and among all, WRKY TFs are one of the largest families of transcriptional regulators in plants (Eulgem and Somssich 2007; Bakshi and Oelmüller 2014), involved in regulation of various physiological processes. WRKY TFs are emerging players in plant signaling, which regulate diverse cellular programs by relaying extracellular signals to intracellular responses and involved in multiple defense responses, development, metabolism, etc. The reprogramming of

WRKY network under biotic stress efficiently deteriorates the pathogens, and at the same time, it restricts defense responses, which can be detrimental for plant growth, development, and reproductive fitness. However, in the contemporary time of scientific advancement, enormous role of WRKY TFs in abiotic stresses has also been revealed.

In the present chapter, we emphasize on WRKY TFs and their action on downstream regulation of different molecular switches under biotic stress. This will provide important insights in understanding of regulatory networks and its associated functions to develop strategies for crop improvement and value addition in plants, which could be useful to the humankind.

8.2 WRKY Domains and Classification

WRKY proteins constitute a novel family of plant-specific TFs and are characterized by the presence of WRKY domain which consists of ~60 amino acid residues at the N-terminus and a zinc-finger-like motif C-C-H-H/C at the C-terminus (Rushton et al. 1996). The WRKY domains contain the conserved heptapeptide "WRKYGQK" also referred to as the "signature sequence" at the N-terminus of DNA-binding domain. WRKY proteins bind to W-box elements containing the consensus motif TGACC/T, which occur either as single hexamers, TTGACC/T; as palindromic sequence, TGACC/T-A/GTCA; or as tandem repeats, TGACC/C--TGACC/T in the promoter of target genes (Eulgem et al. 1999; Yang et al. 1999). WRKY proteins have been categorized into three groups based on the number of WRKY domains and the type of their zinc-finger-like motif (Kumar et al. 2016). Generally, group I member contains two WRKY domains both at N- and C-terminal and C2H2-type zinc-finger motif (C–X4–5–C–X22–23–H–X1–H), and group II has one WRKY domain with C2H2-type zinc-finger motif. Group II members have been further divided into subgroups a–e based upon additional amino acid motifs present outside the WRKY domain. Group III also has one WRKY domain but with C2HC-type zinc-finger motif (C–X7–C–X23–H–X–C) at C-terminal (Eulgem et al. 2000). It has been reported for group I WRKY proteins from *Arabidopsis thaliana*, parsley (*Petroselinum crispum*), and sweet potato (*Ipomoea batatas*) that sequence-specific DNA binding occurs at the C-terminal of WRKY domain, but not the N-terminal domain (Ishiguro and Nakamura 1994; Agarwal et al. 2011) (Table 8.1).

The first WRKY TF has been identified as DNA-binding protein (SPF1) from *Ipomoea batatas* and shown to regulate gene expression in sucrose inducibility (Ishiguro and Nakamura 1994). WRKY proteins have been identified in a wide range of plants due to successive duplication events, resulting in large gene families including up to 74 members in *Arabidopsis* (Ülker and Somssich 2004), >109 in rice (Shimono et al. 2012), 197 in *Glycine max* (Schmutz et al. 2010), 66 in papaya, 104 in *Populus*, 68 in sorghum (Pandey and Somssich 2009), and 45 in barley (Mangelsen et al. 2008). WRKY TFs play a broad-spectrum regulatory role as a positive and negative regulator to control gene expression (Eulgem and Somssich 2007).

Table 8.1 WRKY TFs involved in plant defense and its associated regulatory pathways

S. No.	Gene/pathway	Action/role	Plant species	Tolerance/resistance	References
1.	*AtWRKY18*	Modulation of defense-related genes	*Arabidopsis thaliana*	*Pseudomonas syringae*	Chen and Chen. (2002)
2.	*AtWRKY3* and *AtWRKY4*	Regulation of SA and JA-/ET-mediated signaling pathway	*Arabidopsis thaliana*	*Pseudomonas syringae* and *Botrytis cinerea* *A*	Lai et al. (2008)
3.	*AtWRKY33*	Reduced expression of the salicylate and jasmonate defense genes	*Arabidopsis thaliana*	*Botrytis cinerea*, *Pseudomonas syringae*	Zheng et al. (2006)
4.	*AtWRKY6*	Regulation of genes associated with leaf senescence	*Arabidopsis thaliana*	*Pseudomonas syringae*	Robatzek and Somssich. (2001)
5.	*AtWRKY7*	Reduced expression of defense-related genes	*Arabidopsis thaliana*	*Pseudomonas syringae*	Kim et al. (2006)
6.	*AtWRKY70*	Involvement of R-gene	*Arabidopsis thaliana*	*Hyaloperonospora parasitica*	Knoth et al. (2007)
7.	*BnWRKY33*	Induction of salicylic acid, jasmonic acid, ethylene, and glucosinolate synthesis pathway	Oilseed rape	*Sclerotinia sclerotiorum*	Wang et al. (2014)
8.	*BrWRKY12*	By upregulation of defense-related proteins and jasmonic acid signaling pathway	*Brassica rapa*	*Pectobacterium carotovorum*	Kim et al. (2013)
9.	*FaWRKY1* and *WRKY75*	Uncoupled to PATHOGENESIS-RELATED (PR) gene expression but strongly associated with oxidative burst and glutathione S-transferase (GST) induction	Strawberry	*Colletotrichum acutatum*	Encinas-Villarejo et al. (2009)
10.	*GhWRKY15*	Through increased RNA expression of pathogen-related genes, NONEXPRESSOR OF PR1, ET biosynthesis, and antioxidant enzymes POD and APX	*Gossypium hirsutum*	*C. gossypii*, *Fusarium oxysporum*, and *Rhizoctonia solani*	Yu et al. (2012)
11.	*GhWRKY25*	Through co-expression of large number of resistance- and defense-related genes	*Gossypium hirsutum*	*Magnaporthe grisea, Blumeria graminis, and M. oryzae*	Liu et al. (2015)

12.	*GhWRKY27a*	Through increased RNA expression of pathogen-related genes, NONEXPRESSOR OF PR1, ET biosynthesis, and antioxidant enzymes POD and APX	*Gossypium hirsutum*	*Rhizoctonia solani*	Yan et al. (2015)
13.	*MaWRKY1* and *MaWRKY2*	Upregulation of defense-associated genes	*Poplar tomentosa* Carr	*Dothiorella gregaria* Sacc	Ye et al. (2014)
14.	*MrWRKY30*	Uncoupled to PATHOGENESIS-RELATED (PR) gene expression but strongly associated with oxidative burst and glutathione S-transferase (GST) induction	*Muscadinia rotundifolia*	*Peronospora parasitica*	Jiang et al. (2015)
15.	*OsWRKY22*	Through co-expression of large number of resistance- and defense-related genes	Rice	*Magnaporthe grisea, Blumeria graminis, and M. oryzae*	Abbruscato et al. (2012)
16.	*OsWRKY45*	By upregulation of defense-related proteins and jasmonic acid signaling pathway	*Oryza sativa*	Brown planthopper (BPH, *Nilaparvata lugens*)	Huangfu et al. (2016)
17.	*OsWRKY53*	Upregulation of defense-related genes	*Oryza sativa*	*Magnaporthe grisea*	Chujo et al. (2007)
18.	*OsWRKY77*	Through enhanced expression of defense-related PR1, PR2, and PR5 genes	*Oryza sativa* L.	*Pseudomonas syringae*	Lan et al. (2013)
19.	*PtoWRKY60*	Upregulation of defense-associated genes	*Poplar tomentosa* Carr	*Dothiorella gregaria* Sacc	Ye et al. (2014)
20.	*PtrWRKY89*	Through enhanced expression of defense-related PR1, PR2, and PR5 genes and downregulating the marker gene of SA and JA pathways	*Populus trichocarpa*	*Pseudomonas syringae*	Jiang et al. (2016)
21.	*SpWRKY1*	Enhanced expression of SA- and JA-associated genes (*NtPR1*, *NtPR2*, *NtPR4*, *NtPR5*, and *NtPDF1.2*), as well as various defense-related genes (*NtPOD*, *NtSOD*, and *NtPAL*)	*Solanum pimpinellifolium* L3708	*Phytophthora nicotianae*	Li et al. (2015)
22.	*VvWRKY1*	Involvement of SA- and JA-/ET-mediated signaling pathway	*Vitis vinifera*	*Downy mildew, Plasmopara viticola*	Marchive et al. (2013)

8.3 Structure of DNA-Binding Domain of WRKY Proteins

The first structure of the C-terminal WRKY domains of AtWRKY1 and AtWRKY4 protein of *Arabidopsis thaliana* revealed that the WRKY domain consists of a four-stranded antiparallel β-sheet and zinc-binding pocket formed by the coordination of zinc atom with the conserved two cysteine and histidine residues (Yamasaki et al. 2005). The crystal structure of the C-terminal of WRKY domain of AtWRKY1 consists of five β-strands, with DNA-binding residues located at *β2* and *β3* strands (Duan et al. 2007). The N-terminal region of the β-strand consisting of WRKY signature sequence partly protrudes from one surface of the protein, thereby enabling access to the major DNA groove, and binds to its cognate W-box. Recently, solution structure of WRKY domain with W-box-binding site has been determined, and it revealed that four-stranded β-sheet enters the major groove of the DNA in an atypical mode where the plane of the sheet is nearly perpendicular to the helical axis of DNA (Yamasaki et al. 2012). In the WRKYGQK signature sequence, the tryptophan residue forms the core of the structure, while all the other amino acids (RKYGQK) are directly involved in DNA binding. The glycine residue helps in bending of the strand and thus enables deep penetration into the DNA groove. Recognition of the W-box sequence occurs mainly through the hydrophobic interaction with the methyl groups of thymine (T) bases of the DNA strand. Mutations in the residues involved either in DNA binding or in Zn binding significantly impaired the DNA-binding activity due to the disruption of the tertiary structure, which is important in DNA binding (Duan et al. 2007; Yamasaki et al. 2013).

8.4 Regulation of WRKY TFs

8.4.1 Autoregulation and Cross-Regulation

WRKY proteins are involved in diverse pathways and regulate the expression of downstream target genes either as a positive or negative regulator. The regulation of WRKY-dependent signaling pathways is very extensive and complex. In response to the internal or external stimuli, WRKY TFs trigger the expression of the target genes by binding to their W-box elements in the promoter regions. W-box elements are also present in the promoters of the majority of the WRKY genes, and this suggests that they are regulated via specific feedback mechanisms (autoregulation by themselves or cross-regulation by other WRKY TFs) (Eulgem and Somssich 2007; Rushton et al. 2010). For example, chromatin immunoprecipitation (ChIP) analysis of PcWRKY1 of parsley (*Petroselinum crispum*) revealed that it binds not only to the W-box of its own promoter but also has affinity toward binding the promoters of PcWRKY3 and marker gene PcPR1 (Eulgem et al. 1999; Turck et al. 2004). Likewise, WRKY33 expression is activated by the MAPK3/6, and it autoregulates its expression via a positive feedback loop by binding to its own promoter (Mao et al. 2011). WRKY18, WRKY40, and WRKY60 act as a negative regulator of ABA signaling and could directly bind to the W-box in the promoter region of their respective genes and thus repress the expression of all three WRKY genes (Chen et al. 2010; Yan et al. 2013). The above finding suggests the importance of autoregulation and cross-regulation of WRKY TFs in maintaining the homeostasis of WRKY protein expression in the cell during abiotic and biotic stress conditions.

8.4.2 Regulation of WRKY TFs by MAP Kinases

Some WRKY TFs are also regulated via MAPK (mitogen-activated protein kinase) pathway (Adachi et al. 2015). WRKY TFs act downstream of various MAPKs to regulate defense-related plant genes (Phukan et al. 2016). The AtMPK3, AtMPK6, and AtMPK4 get activated during both biotic and abiotic stresses (Banerjee and Roychoudhury 2015). Group I WRKY TFs are the first protein, which gets phosphorylated by MAP kinases in response to PAMP-triggered MAPK signaling. Two WRKY proteins AtWRKY22 and AtWRKY29 act downstream of the bacterium flagellin receptor FLS2, are upregulated by a PAMP-induced MAPK cascade, and contain multiple W-boxes within their respective promoters. AtWRKY33, involved in the production of phytoalexin during pathogen attack, forms a complex with MPK4-MKS1 (MPK4 substrate) in the nucleus. Upon infection MPK, MKK (MAP kinase kinase), and MEKK (MAP kinase kinase kinase) are activated. The activated MPK4 phosphorylate MKS1, which lead to the dissociation of the MPK4-MKS1-WRKY33 complex, and AtWRKY33 was released. Then AtWRKY33 binds to the promoter of the target gene PAD3 (phytoalexin deficient 3) that is required for the synthesis of antimicrobial compound camalexin (Qiu et al. 2008). WRKY33 could be phosphorylated by two other MPKs, MPK3 and MPK6, which led to binding to its own and the PAD3 promoters in response to *B. cinerea*. Mao et al. (2011) had shown that in *wrky33* mutant, camalexin production was abolished and mutation in the phosphorylation sites of WRKY33 also had the same effect. Taken together, these results suggested that AtWRKY33 works downstream of the MPK3/MPK6 and phosphorylation of WRKY is important for the production of camalexin upon bacterial infection. In rice, OsWRKY33 is phosphorylated by BWMK1 (blast- and wounding-activated MAP kinase 1) and binds to the promoter of PR genes during salicylic acid-dependent defense responses (Koo et al. 2009). OsWRKY53 suppresses herbivore-induced defense in rice by negative feedback modulation of MPK3/MPK6 activity (Hu et al. 2015). Therefore, phosphorylation and activation of WRKY proteins by MPKs is an important regulatory mechanism which increases the capacity of WRKYs to bind to the promoters of target gene which are involved in the plant defense responses.

8.4.3 Regulation of WRKY TFs via Histone Modification

A few WRKY TFs have been shown to be regulated by histone-modifying complex. *AtWRKY70* gets activated by the *Arabidopsis* homolog of trithorax (ATX1) leading to nucleosomal histone H3K4 trimethylation which results in the activation of

SA-responsive gene PR1 and JA-responsive gene *THI2.1* (*THIONIN2.1*). This finding suggests that *PR1* and *THI2.1* genes are the downstream targets of WRKY70 and regulated epigenetically (Alvarez-Venegas et al. 2007). In response to senescence, H3K4 dimethylation and H3K4 trimethylation by histone methyltransferase occur at 5′ end and coding regions of *AtWRKY53*gene (Ay et al. 2009). Another example of histone modification has been shown in two type III WRKY TFS, *AtWRKY38* and *AtWRKY62*. During bacterial infection, HDA19 (histone deacetylase 19) removes acetyl group from histone tails and represses transcription of *AtWRKY38* and *AtWRKY62* and thus negatively regulates basal defense (Kim et al. 2008). Similarly, methylation at the promoter of *AtWRKY40* inhibits expression of *ABI5* and negatively regulates ABA signaling in seed germination and post-germination growth (Shang et al. 2010). Wang et al. (2012) showed that the protein encoded by chromatin remodeling linker histone H1gene (*MaHIS1*) and MaWRKY1 could interact and regulate physiological processes like fruit ripening and stress responses in banana. MaHIS1 has also been shown to be induced by other factors like JA, ABA, and hydrogen peroxide and under cold stress.

8.4.4 Interaction of WRKY TFs with Other Factors

8.4.4.1 VQ Proteins

It has been reported in the literature that many interacting partners like coactivators regulate the expression of many WRKY TFs. One of the interacting partners is VQ protein, which is a group of cofactors containing a short conserved VQ-related motif (FxxxVQxLTG). The conserved valine and glutamine residues in the conserved motif are important and required for the interaction with the C-terminal domain of WRKY TFs. In *Arabidopsis* and rice, 34 and 40 VQ members were identified, respectively, and shown to be involved in disease resistance and in the plant response to environmental stresses (Cheng et al. 2012; Kim et al. 2013). The first VQ proteins were identified as a MPK4 substrate (MKS) in *Arabidopsis* by using a yeast two-hybrid assay. The VQ protein MKS has been shown to form complex with AtWRKY25 and AtWRKY33, which are involved in the regulation of plant defense responses (Andreasson et al. 2005). Binding of VQ proteins with WRKY TFs changes the binding affinity of the latter for the nucleotides flanking the conserved W-box. It has been shown that C-terminal domain of AtWRKY33 interacts with two VQ proteins, SIGMA FACTORBINDINGPROTEIN 1 (SIB1) and SIB2, to regulate plant defense response against necrotrophic pathogens Lai et al. (2011). These results demonstrate that VQ proteins are crucial cofactors in regulating WRKY-mediated gene expression (Cheng et al. 2012; Chi et al. 2013).

8.4.4.2 Calmodulin (CaM) Proteins

CaM binds to the conserved Ca^{+2}-dependent calmodulin-binding domain (CaBD) (DxxVxKFKxVISLLxxxR) present in WRKY group II members like *AtWRKY7* (Park et al. 2005). Increasing concentration of calcium triggers the interaction of CaM and WRKY members over WRKY-WRKY interaction (Chi et al. 2013).

8.4.4.3 14-3-3 Proteins

14-3-3 proteins specifically bind to phosphoserine and phosphothreonine and regulate many processes like plant development, plant defense, and stress responses (Roberts 2003; Denison et al. 2011). They function as homo- or heterodimers and each dimer binds two substrates. In *Arabidopsis*, seven WRKY members including WRKY6, WRKY16, WRKY18, WRKY19, WRKY27, WRKY32, and WRKY40 have been identified as putative interacting partners for 14-3-3 proteins by tandem affinity purification tag assay (Chang et al. 2009). 14-3-3 proteins interact and phosphorylate AtWRKY18 and AtWRKY40 to regulate ABA signaling (Shang et al. 2010; Shen et al. 2003). These results suggest that 14-3-3 proteins might have potential roles in regulating biotic and abiotic stress responses via WRKY TFs (Chang et al. 2009; Rushton et al. 2010; Chi et al. 2013).

8.5 WRKY TFs in Defense Response

The plant innate immunity is mainly responsive to two interconnected pathways termed PTI or ETI (Jones and Dangl 2006). PTI is initiated by the recognition of molecular patterns of pathogens and activates MAP kinase cascade pathway and defense-related genes, while ETI is associated with plant disease resistance (R) proteins that activate defense reactions upon specific recognition of pathogen effectors (Chisholm et al. 2006). PTI and ETI activate local as well as long-distance defense reactions like systemic acquired resistance (SAR) (Durrant and Dong 2004; Bostock 2005).

8.5.1 Interaction of WRKY TFs with SA and JA Signaling Pathway

SA and JA are two important signaling molecules in defense response. JA-dependent plant defense pathways are activated by necrotrophic pathogens, whereas SA-dependent defenses are triggered by biotrophic pathogens. JA and SA signaling pathway act antagonistically in regulating defense response (Koornneef and Pieterse 2008). During the past few years, much attention has been focused on TFs involved in the regulation of gene expression upon pathogen challenge. Expression profiling studies have revealed that a large set of the *WRKY TF* gene family members are responsive to pathogen challenge and regulate plant defense responses either as a positive or negative regulator (Eulgem and Somssich 2007). Expression of WRKY genes has been shown to get induced by pathogen infection and pathogen elicitors or by SA treatment in a number of plants (Agarwal et al. 2011). In *Arabidopsis* and rice, more than 75 and 109 WRKY genes have been reported (Shimono et al. 2012). Upon infection, pathogens induce SAR leading to accumulation of SA. Many WRKYs are positively regulated by SA through the receptors NPR1 and its paralogues NPR3 and NPR4 (Wang et al. 2006; Fu et al. 2012; Wu et al. 2012). A few WRKY genes including WRKY18, WRKY38, WRKY53, WRKY54, WRKY58, WRKY59, WRKY66, and WRKY70 bind to the W-box sequences in the promoter region of NPR1 genes in *Arabidopsis*; this suggests that WRKY genes act upstream of NPR1 genes and involved in the positive regulation of WRKY TFs during pathogen-induced signaling (Wang et al. 2006; Ishihama and Yoshioka 2012). Many WRKY TFs are common component in the SA-/JA-mediated plant defense pathway (Koornneef and Pieterse 2008; Thaler et al. 2012). For example, WRKY70 works at a convergence point for maintaining balance between SA- and JA-mediated signaling pathways as well as also plays a crucial role for R-gene-mediated resistance. Overexpression of AtWRKY70 induces the expression of SA-induced PR genes and acts as a positive transcriptional regulator of SA signaling while for JA-responsive pathways acts as a negative regulator. Overexpression of AtWRKY70 improved resistance to biotrophic pathogen *Erysiphe cichoracearum* and necrotrophic bacteria *Erwinia carotovora* (Ecc) but reduced resistance to fungal necrotroph *Alternaria brassicicola*. Similar dual roles have also been observed for WRKY53. It positively regulates plant defense response during *P. syringae* infection, while its mutant displayed delayed symptom development toward *Ralstonia solanacearum* (Murray et al. 2007; Hu et al. 2008). Moreover, during *P. syringae* infection, WRKY11 and WRKY17 have shown to positively regulate the JA biosynthesis pathway genes, *LOX2* and *AOS*, while negatively regulate the expression of WRKY70 (Li et al. 2004, 2006; Journot-Catalino et al. 2006). AtWRKY53 was reported to positively regulate the basal defense response during *P. syringae* infection while negatively regulate during JA and ethylene signaling pathway (Murray et al. 2007).

Three WRKY TFs of subgroup IIa, WRKY18, WRKY40, and WRKY60, function in a partly redundant way in regulating plant disease resistance. Xu et al. (2006) showed that double mutants *wrky18wrky40* and *wrky18wrky60* and the triple mutant *wrky18wrky40wrky60* were found to be more resistant to *P. syringae* infection but susceptible to *B. cinerea*. In other studies, *Atwrky18/Atwrky40* double mutants showed resistance toward avirulent powdery mildew fungus *Golovinomyces orontii*, and complementation of WRKY40 in this mutant partially restored susceptibility (Pandey et al. 2010). The HvWRKY1 and HvWRKY2 homologs of AtWRKY18 and AtWRKY40 in barley act as a suppressor of PAMP-induced basal defense, leading to resistance against virulent pathogen *B. graminis*. During infection, fungal effector AVR10 is recognized by the resistance protein MLA (mildew resistance locus A) in the cytoplasm followed by interaction of HvWRKY 1 and 2 with activated MLA10 in the nucleus (Shen et al. 2007). In addition, AtWRKY33 is another example and was known to act as a positive regulator of resistance to the necrotrophic pathogens *Botrytis cinerea* and *Alternaria brassicicola*, while overexpression leads to susceptibility to *Pseudomonas syringae* DC3000. However, loss of function mutant of AtWRKY33 showed increased resistance toward *R. solanacearum* (Zheng et al. 2006; Birkenbihl et al. 2012). Similarly, WRKY3 and WRKY4, which are structurally similar proteins, confer resistance to necrotrophic pathogens (Lai et al. 2008).

Few of the WRKY TF members act as negative regulator of defense signaling including AtWRKY7, AtWRKY38, AtWRKY62, and AtWRKY52. AtWRKY11, AtWRKY17, AtWRKY38, and AtWRKY62 negatively regulate basal defense response toward bacterial pathogen *P. syringae*. Interaction of AtWRKY38 and AtWRKY62 with HDA19, a positive regulator of plant basal disease resistance, leads to inactivation of defense repressing WRKY38 and WRKY62 TFs (Journot-Catalino et al. 2006; Kim et al. 2008). Expression of AtWRKY62 is induced by SA and JA in a NPR1-dependent manner. Loss of function mutant of AtWRKY62 resulted in enhanced expression of JA-response genes, while overexpression of AtWRKY62 inhibited JA-response gene expression (Mao et al. 2007). In other study, overexpression of WRKY62 leads to elevated transcript levels of PR1 gene, whereas in Atwrky62 mutant, PR1 gene is downregulated (Kim et al. 2008). In addition, AtWRKY48 and AtWRKY8 also negatively regulate basal resistance to *P. syringae* (Xing et al. 2008; Chen et al. 2010). Additionally, WRKY8 was also involved in negative regulation of crucifer-infecting tobacco mosaic virus (TMV-cg) (Chen et al. 2013). AtWRKY48 mutants showed increased expression of PR1 genes found to be associated with reduced bacterial growth, whereas *AtWRKY48* overexpressor lines showed the opposite phenotypes. *AtWRKY58* acts downstream ofNPR1, negatively regulating SAR (Wang et al. 2006). Some WRKY proteins exist as chimeric proteins like AtWRKY52 which possesses TIR-NBS-LRR (Toll/interleukin-1 receptor-nucleotide-binding site-leucine-rich repeat) domain in combination with group III-type WRKY domain and mediates R-gene-based resistance toward bacterial wilt *Ralstonia solanacearum*. The physical interaction of *AtWRKY52*/RRS1 with its cognate bacterial effector PopP2 within the plant cell nucleus has been suggested to inactivate the WRKY domain of RRS1 to activate defense mechanisms by derepression (Deslandes et al. 2003). *AtWRKY16* and *AtWRKY19* also contain NBS-LRR domain reported in *Arabidopsis*.

8.5.2 Overexpression/Downregulation of WRKY TFs for Biotic Stress Tolerance

Till date WRKY TFs have been reported from many plant species suggesting its importance in regulating plant defense response. Overexpression of many WRKY TFs from rice like *OsWRKY13*, *OsWRKY31*, *OsWRKY45*, *OsWRKY53*, and *OsWRKY47* showed enhanced resistance to fungal pathogen *Magnaporthe grisea*, the causal agent of the devastating rice blast disease (Ryu et al. 2006; Wei et al. 2013). In rice, OsWRKY13, an ortholog of AtWRKY70, is reported to have similar functions. Overexpression of *OsWRKY13* activates the genes related to SA pathways but reduces the expression of genes in JA pathway (Qiu et al. 2007). Overexpression of *OsWRKY3* led to elevated expression of NPR1, PR1b, phenylalanine ammonia-lyase (ZB8), and peroxidase (POX22.3), suggesting that it works as a transcriptional regulator in SA- or JA-dependent defense signaling pathway (Liu et al. 2005). Overexpressor transgenic lines of OsWRKY53 showed resistance against blast disease and induced the expression of PR proteins and peroxidase enzymes (Chujo et al. 2007). Overexpression of *OsWRKY89* showed more tolerance to the rice blast fungus white-backed planthopper (*Sogatella furcifera*), a rice herbivore (Wang et al. 2007). Lan et al. (2013) showed that overexpression of *OsWRKY77* in *Arabidopsis* led to enhanced resistance toward *P. syringae* suggesting its function as a positive regulator of plant defense. The OsWRKY45 showed improved resistance to rice blast fungus and might work independent of NPR1-mediated SA signaling (Shimono et al. 2007). In addition OsWRKY45 is found to negatively modulate the resistance of rice to the brown planthopper *Nilaparvata lugens* (Huangfu et al. 2016). Silencing of *NaWRKY3* and *NaWRKY6* in *Nicotiana attenuata* made plants highly susceptible to lepidopteran herbivore (Skibbe et al. 2008). *CaWRKY1* from pepper (*Capsicum annuum*) negatively regulates plant defense, as silencing of this gene led to decreased growth *of Xanthomonas* (Oh et al. 2008), whereas constitutive overexpression of *CaWRKY40* resulted in enhanced resistance toward *Ralstonia solanacearum* (Dang et al. 2013). Shi et al. (2014) suggested that the overexpression of *GhWRKY39* may positively regulate the plant response against bacterial *R. solanacearum* and bacterial pathogen *R. solani.* Transgenic tobacco plants overexpressing *GhWRKY15* displayed more resistance toward viral and fungal infections and showed induced expression of NPR1 gene (Yu et al. 2012). *GhWRKY25* overexpression resulted in enhanced sensitivity to the fungal pathogen *Botrytis cinerea* by reducing the expression of SA or ET signaling-related genes and inducing the expression of genes involved in the JA signaling pathway (Liu et al. 2015). GhWRKY27a-overexpressing plants conferred reduced resistance to *R. solani* infection as demonstrated by severe disease symptoms in transgenic lines (Yan et al. 2015). Additionally, a number of WRKY TFs which are important players of plant immunity have been found in different plant species, for example, VvWRKY1 and VvWRKY2 from grapevine (*Vitis vinifera*), *PtrWRKY89* from *Populus trichocarpa*, and MaWRKY1 and MaWRKY2 from *Musa* spp. (Marchive et al. 2013; Jiang et al. 2014; Shan et al. 2016).

8.6 Conclusion and Future Perspectives

In this chapter, we have focused on the most recent advances on WRKY TFs. Over the last two decades, significant progress has been made in order to understand the role of WRKY TFs. Current information suggests that the WRKY superfamily of TFs is composed of different types of proteins that have been implicated in plant developmental processes and pathogen-induced defense response. New finding illustrates that they participate in regulating a plethora of genes at various levels, by working as positive or negative regulator, by direct activation of downstream target genes, and by activating or repressing other TF genes. WRKY TFs are itself regulated by a highly intricate mechanism in plants, and they are required to maintain normal cellular homeostasis under normal condition. One WRKY protein is found to regulate several plant processes at a time, and the mechanisms of regulation are not yet clear. Extensive study of these TF families is needed for better understanding of the signaling pathways involved in WRKY-mediated regulation of defense and developmental processes. In the future it would be exciting to explore "how WRKY TF networks exert their functions on DNA/chromatin level" which will certainly allow us to open new vistas of diverse metabolic pathways, their cross-linking, and overall cellular physiology of plants under biotic stress conditions.

References

Abbruscato P, Nepusz T, Mizzi L, Del Corvo M, Morandini P, Fumasoni I, Michel C, Paccanaro A, Guiderdoni E, Schaffrath U, Morel JB, Piffanelli P, Faivre-Rampant O (2012) OsWRKY22, a monocot WRKY gene, plays a role in the resistance response to blast. Mol Plant Pathol 3:828–841

Adachi H, Nakano T, Miyagawa N, Ishihama N, Yoshioka M, Katou Y et al (2015) WRKY transcription factors phosphorylated by MAPK regulate a plant immune NADPH oxidase in *Nicotiana benthamiana*. Plant Cell 27:2645–2663

Agarwal P, Reddy MP, Chikara J (2011) WRKY: its structure, evolutionary relationship, DNA-binding selectivity, role in stress tolerance and development of plants. Mol Biol Rep 38:3883–3896

Alvarez-Venegas R, Abdallat AA, Guo M, Alfano JR, Avramova Z (2007) Epigenetic control of a transcription factor at the cross section of two antagonistic pathways. Epigenetics 2:106–113

Andreasson E, Jenkins T, Brodersen P, Thorgrimsen S, Petersen NHT, Zhu S, Qiu J-L, Micheelsen P, Rocher A, Petersen M, Newman M-A et al (2005) The MAP kinase substrate MKS1 is a regulator of plant defense responses. EMBO J 24:2579–2589

Ay N, Irmler K, Fischer A, Uhlemann R, Reuter G, Humbeck K (2009) Epigenetic programming via histone methylation at WRKY53 controls leaf senescence in *Arabidopsis thaliana*. Plant J 58:333–346

Bakshi M, Oelmüller R (2014) WRKY transcription factors: Jack of many trades in plants. Plant Signal Behav 9:e27700

Banerjee A, Roychoudhury A (2015) WRKY proteins: signaling and regulation of expression during abiotic stress responses. Sci World J 2015:807560

Berk AJ, Schmidt MC (1990) How do transcription factors work? Genes Dev 4:151–155

Birkenbihl RP, Diezel C, Somssich IE (2012) *Arabidopsis* WRKY33 is a key transcriptional regulator of hormonal and metabolic responses toward *Botrytiscinerea* infection. Plant Physiol 159:266–285

Bostock RM (2005) Signal crosstalk and induced resistance: straddling the line between cost and benefit. Annu Rev Phytopathol 43:545–580

Chang IF, Curran A, Woolsey R, Quilici D, Cushman JC, Mittler R et al (2009) Proteomic profiling of tandem affinity purified 14-3-3 protein complexes in *Arabidopsis thaliana*. Proteomics 9:2967–2985

Chen C, Chen Z (2002) Potentiation of developmentally regulated plant defense response by AtWRKY18, a pathogen-induced *Arabidopsis* transcription factor. Plant Physiol 129:706–716

Chen H, Lai Z, Shi J, Xiao Y, Chen Z, Xu X (2010) Roles of *Arabidopsis* WRKY18, WRKY40 and WRKY60 transcription factors in plant responses to abscisic acid and abiotic stress. BMC Plant Biol 10:281

Chen L, Zhang L, Li D, Wang F, Yu D (2013) WRKY8 transcription factor functions in the TMV-cg defense response by mediating both abscisic acid and ethylene signaling in Arabidopsis. Proc Natl Acad Sci 110:E1963–E1971

Cheng Y, Zhou Y, Yang Y, Chi YJ, Zhou J, Chen JY et al (2012) Structural and functional analysis of VQ motif-containing proteins in *Arabidopsis* as interacting proteins of WRKY transcription factors. Plant Physiol 159:810–825

Chi Y, Yang Y, Zhou Y, Zhou J, Fan B, Yu JQ et al (2013) Protein-protein interactions in the regulation of WRKY transcription factors. Mol Plant 6:287–300

Chisholm ST, Coaker G, Day B, Staskawicz BJ (2006) Host-microbe interactions: shaping the evolution of the plant immune response. Cell 124:803–814

Chujo T, Takai R, Akimoto-Tomiyama C, Ando S, Minami E, Nagamura Y, Kaku H, Hibuya N, Yasuda M, Nakashita H, Umemura K, Okada A, Okada K, Nojiri H, Yamane H (2007) Involvement of the elicitor-induced gene OsWRKY53 in the expression of defense-related genes in rice. BBA Gene Struct Expr 1769:497–505

Dang FF, Wang YN, Yu L, Eulgem T, Lai Y, Liu ZQ, Wang X, Qiu AL, Zhang TX, Lin J, Chen YS, Guan DY, Cai HY, Mou SL, He SL (2013) CaWRKY40, a WRKY protein of pepper, plays an important role in the regulation of tolerance to heat stress and resistance to *Ralstonia solanacearum* infection. Plant Cell Environ 36:757–774

Denison FC, Paul A-L, Zupanska AK, Ferl RJ (2011) 14-3-3 proteins in plant physiology. Semin Cell Dev Biol 22:720–727

Deslandes L, Olivier J, Peeters N, Feng DX, Khounlotham M, Boucher C, Somssich I, Genin S, Marco Y (2003) Physical interaction between RRS1-R, a protein conferring resistance to bacterial wilt, and PopP2, a type III effector targeted to the plant nucleus. Proc Natl Acad Sci U S A 100:8024–8029

Duan M-R, Nan J, Liang YH, Mao P, Lu L, Li L, Wei C, Lai L, Li Y, Su XD (2007) DNA binding mechanism revealed by high resolution crystal structure of *Arabidopsis thaliana* WRKY1 protein. Nucleic Acids Res 35:1145–1154

Durrant WE, Dong X (2004) Systemic acquired resistance. Annu Rev Phytopathol 42:185–209

Encinas-Villarejo S, Maldonado AM, Amil-Ruiz F, de los Santos B, Romero F, Pliego-Alfaro F, Muñoz-Blanco J, Caballero JL (2009) Evidence for a positive regulatory role of strawberry (Fragaria x ananassa) FaWRKY1 and *Arabidopsis* AtWRKY75 proteins in resistance. J Exp Bot 60:3043–3065

Eulgem T, Somssich IE (2007) Networks of WRKY transcription factors in defense signaling. Curr Opin Plant Biol 10:366–371

Eulgem T, Rushton PJ, Schmelzer E, Hahlbrock K, Somssich IE (1999) Early nuclear events in plant defense signalling: rapid gene activation by WRKY transcription factors. EMBO J 18:4689–4699

Eulgem T, Rushton PJ, Robatzek S, Somssich IE (2000) The WRKY superfamily of plant transcription factors. Trends Plant Sci 5:199–206

Franco-Zorrillaa JM, López-Vidrieroa I, Carrascob JL, Godoya M, Verab P, Solanoc R (2014) DNA-binding specificities of plant transcription factors and their potential to define target genes. PNAS 111(6):2367–2372

Fu ZQ, Yan S, Saleh A, Wang W, Ruble J, Oka N, Mohan R, Spoel SH, Tada Y, Zheng N, Dong X (2012) NPR3 and NPR4 are receptors for theimmune signal salicylic acid in plants. Nature 486:228–232

Hu J, Barlet X, Deslandes L, Hirsch J, Feng DX, Somssich I, Marco Y (2008) Transcriptional responses of *Arabidopsis thaliana* during wilt diseasecaused by the soil-borne phytopathogenic bacterium, *Ralstonia solanacearum*. PLoS One 3:e2589

Hu L, Ye M, Li R, Zhang T, Zhou G, Wang Q et al (2015) The rice transcription factor WRKY53 suppresses herbivore-induced defenses by acting as a negative feedback modulator of mitogen-activated protein kinase activity. Plant Physiol 169:2907–2921

Huangfu J, Li J, Li R, Ye M, Kuai P, Zhang T, Lou Y (2016) The transcription factor OsWRKY45 negatively modulates the resistance of rice to the brown planthopper *Nilaparvata lugens*. Int J Mol Sci 17:697. https://doi.org/10.3390/ijms17060697

Ishiguro S, Nakamura K (1994) Characterization of a cDNA encoding a novel DNA-binding protein, SPF1, that recognizes SP8 sequences in the 50 upstream regions of genes coding for sporamin and β-amylase from sweet potato. Mol Gen Genet 244:563–571

Ishihama N, Yoshioka H (2012) Post-translational regulation of WRKY transcription factors in plant immunity. Curr Opin Plant Biol 15:431–437

Jiang Y, Duan Y, Yin J, Ye S, Zhu J, Zhang F, Lu W, Fan D, Luo K (2014) Genome-wide identification and characterization of the Populus WRKY transcription factor family and analysis of their expression in response to biotic and abiotic stresses. J Exp Bot 65:6629–6644

Jiang W, Wu J, Zhang Y, Yin L, Lu J (2015) Isolation of a WRKY30 gene from *Muscadinia rotundifolia* (Michx) and validation of its function under biotic and abiotic stresses. Protoplasma 252:1361–1374

Jiang Y, Guo L, Liu R, Jiao B, Zhao X, Ling Z, Luo K (2016) Overexpression of poplar PtrWRKY89 in transgenic Arabidopsis leads to a reduction of disease resistance by regulating defense-related genes in salicylate- and jasmonate-dependent signaling. PLoS One 28(11):e0149137

Jones JDG, Dangl JL (2006) The plant immune system. Nature 444:323–329

Journot-Catalino N, Somssich IE, Roby D, Kroj T (2006) The transcription factors WRKY11 and WRKY17 act as negative regulators of basal resistance in *Arabidopsis thaliana*. Plant Cell 18:3289–3302

Kim KC, Fan B, Chen Z (2006) Pathogen-induced Arabidopsis WRKY7 is a transcriptional repressor and enhances plant susceptibility to *Pseudomonas syringae*. Plant Physiol 142:1180–1192

Kim KC, Lai Z, Fan B, Chen Z (2008) *Arabidopsis* WRKY38 and WRKY62 transcription factors interact with histone deacetylase 19 in basal defense. Plant Cell 20:2357–2371

Kim DY, Kwon SI, Choi C, Lee H, Ahn I, Park SR et al (2013) Expression analysis of rice VQ genes in response to biotic and abiotic stresses. Gene 529:208–214

Kim HS, Park YH, Nam H, Lee YM, Song K, Choi C, Ahn I, Park SR, Lee YH, Hwang DJ (2014) Overexpression of the *Brassica rapa* transcription factor WRKY12 results in reduced soft rot symptoms caused by *Pectobacterium carotovorum* in *Arabidopsis* and Chinese cabbage. Plant Biol (Stuttg) 16:973–981

Knoth C, Ringler J, Dangl JL, Eulgem T (2007) *Arabidopsis* WRKY70 is required for full RPP4-mediated disease resistance and basal defense against *Hyaloperonospora parasitica*. Mol Plant-Microbe Interact 20(2):120–128

Koo SC, Moon BC, Kim JK, Kim CY, Sung SJ, Kim MC et al (2009) OsBWMK1 mediates SA-dependent defense responses by activating the transcription factor OsWRKY33. Biochem Biophys Res Commun 387:365–370

Koornneef A, Pieterse CMJ (2008) Cross talk in defense signaling. Plant Physiol 146:839–844

Kumar K, Srivastava V, Purayannur S, Kaladhar VC, Cheruvu PJ, Verma PK (2016) WRKY domain-encoding genes of a crop legume chickpea (*Cicer arietinum*): comparative analysis with *Medicago truncatula* WRKY family and characterization of group-III gene(s). DNA Res 23:225–239

Lai Z, Vinod K, Zheng Z, Fan B, Chen Z (2008) Roles of *Arabidopsis* WRKY3 and WRKY4 transcription factors in plant responses to pathogens. BMC Plant Biol 8:68

Lai Z, Li Y, Wang F, Cheng Y, Fan B, Yu JQ et al (2011) *Arabidopsis* sigma factor binding proteins are activators of the WRKY33 transcription factor in plant defense. Plant Cell 23:3824–3841

Lan A, Huang J, Zhao W, Peng Y, Chen Z, Kang D (2013) A salicylic acid-induced rice (*Oryza sativa* L.) transcription factor OsWRKY77 is involved in disease resistance of *Arabidopsis thaliana*. Plant Biol (Stuttg) 15:452–461

Li J, Brader G, Palva ET (2004) The WRKY70 transcription factor: a node of convergence for jasmonate-mediated and salicylate-mediated signals in plant defense. Plant Cell 16:319–331

Li J, Brader G, Kariola T, Palva ET (2006) WRKY70 modulates the selection of signaling pathways in plant defense. Plant J 46:477–491

Li JB, Luan YS, Liu Z (2015) Overexpression of SpWRKY1 promotes resistance to *Phytophthora nicotianae* and tolerance to salt and drought stress in transgenic tobacco. Physiol Plant 155:248–266

Liu XQ, Bai XQ, Qian Q, Wang XJ, Chen MS, Chu CC (2005) OsWRKY03, a rice transcriptional activator that functions in defense signaling pathway upstream of OsNPR1. Cell Res 15:593–603

Liu X, Song Y, Xing F, Wang N, Wen F, Zhu C (2015) GhWRKY25, a group I WRKY gene from cotton, confers differential tolerance to abiotic and biotic stresses in transgenic *Nicotiana benthamiana*. Protoplasma. https://doi.org/10.1007/s00709-015-0885-3

Mangelsen E, Kilian J, Berendzen KW, Kolukisaoglu UH, Harter K, Jansson C, Wanke D (2008) Phylogenetic and comparative gene expression analysis of barley (*Hordeum vulgare*) WRKY transcription factor family reveals putatively retained functions between monocots and dicots. BMC Genomics 9:194

Mao P, Duan M, Wei C, Li Y (2007) WRKY62 transcription factor acts downstream of cytosolic NPR1 and negatively regulates jasmonate-responsive gene expression. Plant Cell Physiol 48:833–842

Mao G, Meng X, Liu Y, Zheng Z, Chen Z, Zhang S (2011) Phosphorylation of a WRKY transcription factor by two pathogen-responsive MAPKs drives phytoalexin biosynthesis in *Arabidopsis*. Plant Cell 23:1639–1653

Marchive C, Léon C, Kappel C, Coutos-Thévenot P, Corio-Costet MF, Delrot S, Lauvergeat V (2013) Over-expression of VvWRKY1 in grapevines induces expression of jasmonic acid pathway-related genes and confers higher tolerance to the downy mildew. PLoS One 8:e54185

Murray SL, Ingle RA, Petersen LN, Denby KJ (2007) Basal resistance against *Pseudomonas syringae* in *Arabidopsis* involves WRKY53 and a protein with homology to a nematode resistance protein. Mol Plant-Microbe Interact 20:1431–1438

Oh SK, Baek KH, Park JM, Yi SY, Yu SH, Kamoun S, Choi D (2008) Capsicum annuum WRKY protein CaWRKY1 is a negative regulator of pathogen defense. New Phytol 177:977–989

Pandey SP, Somssich IE (2009) The role of WRKY transcription factors in plant immunity. Plant Physiol 150:1648–1655

Pandey SP, Roccaroz M, Schön M, Logemann E, Somssich IE (2010) Transcriptional reprogramming regulated by WRKY18 and WRKY40 facilitates powdery mildew infection of *Arabidopsis*. Plant J 64:912–923

Park CY, Lee JH, Yoo JH, Moon BC, Choi MS, Kang YH (2005) WRKY Group IId transcription factors interact with calmodulin. FEBS Lett 579:1545–1550

Phukan UJ, Jeena GS, Shukla RK (2016) WRKY transcription factors: molecular regulation and stress responses in plants. Front Plant Sci 7:760. https://doi.org/10.3389/fpls.2016.00760

Qiu D, Xiao J, Ding X, Xiong M, Cai M, Cao Y, Li X, Xu C, Wang S (2007) OsWRKY13 mediates rice disease resistance by regulating defense-related genes in salicylate- and jasmonate-dependent signaling. Mol Plant-Microbe Interact 20:492–499

Qiu JL, Fiil BK, Petersen K, Nielsen HB, Botanga CJ, Thorgrimsen S et al (2008) *Arabidopsis* MAP kinase 4 regulates gene expression through transcription factor release in the nucleus. EMBO J 27:2214–2221

Ramamoorthy R, Jiang SY, Kumar N, Venkatesh PN, Ramachandran S (2008) A comprehensive transcriptional profiling of the WRKY gene family in rice under various abiotic and phytohormone treatments. Plant Cell Physiol 49:865–879

Robatzek S, Somssich IE (2001) A new member of the Arabidopsis WRKY transcription factor family, AtWRKY6, is associated with both senescence- and defense-related processes. Plant J 28:123–133

Roberts MR (2003) 14-3-3 proteins find new partners in plant cell signalling. Trends Plant Sci 8:218–223

Rushton PJ, Torres JT, Parniske M, Wernert P, Hahlbrock K, Somssich IE (1996) Interaction of elicitor-induced DNA-binding proteins with elicitor response elements in the promoters of parsley PR1 genes. EMBO J 15:5690e700

Rushton PJ, Somssich IE, Ringler P, Shen QJ (2010) WRKY transcription factors. Trends Plant Sci 15:247–258

Ryu H-S, Han M, Lee S-K, Cho JI, Ryoo N, Heu S, Lee YH, Bhoo SH, Wang GL, Hahn TR, Jeon JS (2006) A comprehensive expressionanalysis of the *WRKY* gene superfamily in rice plants during defense response. Plant Cell Rep 25:836–847

Schmutz J, Cannon SB et al (2010) Genome sequence of the palaeopolyploid soybean. Nature 463:178–183

Shan W, Chen JY, Kuang JF, Lu WJ (2016) Banana fruit NAC transcription factor MaNAC5 cooperates with MaWRKYs to enhance the expression of pathogenesis-related genes against *Colletotrichum musae*. Mol Plant Pathol 17:330–338

Shang Y, Yan L, Liu ZQ, Cao Z, Mei C, Xin Q et al (2010) The Mg-chelatase H subunit of *Arabidopsis* antagonizes a group of WRKY transcription repressors to relieve ABA-responsive genes of inhibition. Plant Cell 22:1909–1935

Shen YH, Godlewski J, Bronisz A, Zhu J, Comb MJ, Avruch J et al (2003) Significance of 14–3–3 self-dimerization for phosphorylation-dependent target binding. Mol Biol Cell 14:4721–4733

Shen QH, Saijo Y, Mauch S, Biskup C, Bieri S, Keller B, Seki H, Ulker B, Somssich IE, Schulze-Lefert P (2007) Nuclear activity of MLA immune receptors links isolate-specific and basal disease-resistance responses. Science 315:1098–1103

Shi W, Liu D, Hao L, Wu C, Guo X, Li H (2014) GhWRKY39, a member of the WRKY transcription factor family in cotton, has a positive role in disease resistance and salt stress tolerance. Plant Cell Tissue Organ Cult 118:17–32

Shimono M, Sugano S, Nakayama A, Jiang CJ, Ono K, Toki S, Takatsuji H (2007) Rice WRKY45 plays a crucial role in benzothiadiazole-inducible blast resistance. Plant Cell 19:2064–2076

Shimono M, Koga H, Akagi A, Hayashi N, Goto S, Sawada M, Kurihara T, Matsushita A, Sugano S, Jiang CJ, Kaku H, Inoue H, Takatsuji H (2012) Rice WRKY45 plays important roles in fungal and bacterial disease resistance. Mol Plant Pathol 13:83–94

Skibbe M, Qu N, Galis I, Baldwin IT (2008) Induced plant defenses in the natural environment: *Nicotiana attenuate* WRKY3 and WRKY6 coordinate responses to herbivory. Plant Cell 20:1984–2000

Song Y, Jing S, Yu D (2010) Overexpression of the stress-induced OsWRKY08 improves osmotic stress tolerance in *Arabidopsis*. Chin Sci Bull 54:4671–4678

Thaler JS, Humphrey PT, Whiteman NK (2012) Evolution of jasmonate and salicylate signal crosstalk. Trends Plant Sci 17:260–270

Tsuda K, Mine A, Bethke G, Igarashi D, Botanga CJ, Tsuda Y, Glazebrook J, Sato M, Katagiri F (2013) Dual regulation of gene expression mediated by extended MAPK activation and salicylic acid contributes to robust innate immunity in *Arabidopsis thaliana*. PLoS Genet 9:e1004015

Turck F, Zhou A, Somssich IE (2004) Stimulus-dependent, promoter-specific binding of transcription factor WRKY1 to its native promoter and the defense-related gene PcPR1-1 in parsley. Plant Cell 16:2573–2585

Ulker B, Somssich IE (2004) WRKY transcription factors: from DNA binding towards biological function. Curr Opin Plant Biol 7:491–498

Wang D, Amornsiripanitch N, Dong X (2006) A genomic approach to identify regulatory nodes in the transcriptional network of systemic acquired resistance in plants. PLoS Pathog 2:e123

Wang H, Hao J, Chen X, Hao Z, Wang X, Lou Y, Peng Y, Guo Z (2007) Overexpression of rice *WRKY89* enhances ultraviolet B tolerance and disease resistance in rice plants. Plant Mol Biol 65:799–815

Wang JN, Kuang JF, Shan W, Chen J, Xie H, Lu WJ et al (2012) Expression profiles of a banana fruit linker histone H1gene MaHIS1 and its interaction with a WRKY transcription factor. Plant Cell Rep 31:1485–1494

Wang Z, Fang H, Chen Y, Chen K, Li G, Gu S, Tan X (2014) Over-expression of BnWRKY33 in oilseed rape enhances resistance to *Sclerotinia sclerotiorum*. Mol Plant Pathol 15:677–689

Wei T, Ou B, Li J, Zhao Y, Guo D, Zhu Y, Chen Z, Gu H, Li C, Qin G, Qu LJ (2013) Transcriptional profiling of rice early response to *Magnaporthe oryzae* identified OsWRKYs as important regulators in rice blast resistance. PLoS One 8:e59720

Wei W, Hu Y, Han YT, Zhang K, Zhao FL, Feng JY (2016) The WRKY transcription factors in the diploid woodland strawberry *Fragaria vesca*: identification and expression analysis under biotic and abiotic stresses. Plant Physiol Biochem 105:129–144

Wu Y, Zhang D, Chu JY, Boyle P, Wang Y, Brindle ID, De Luca V, Després C (2012) The *Arabidopsis* NPR1 protein is a receptor for the plant defense hormone salicylic acid. Cell Rep 1:639–647

Xing DH, Lai ZB, Zheng ZY, Vinod KM, Fan BF, Chen ZX (2008) Stress and pathogen-induced *Arabidopsis* WRKY48 is a transcriptional activator that represses plant basal defense. Mol Plant 1:459–470

Xu X, Chen C, Fan B, Chen Z (2006) Physical and functional interactions between pathogen-induced *Arabidopsis WRKY18*, *WRKY40*, and *WRKY60* transcription factors. Plant Cell 18:1310–1326

Yamasaki K, Kigawa T, Inoue M, Tateno M, Yamasaki T, Yabuki T, Aoki M, Seki E, Matsuda T, Tomo Y et al (2005) Solution structure of an *Arabidopsis* WRKY DNA binding domain. Plant Cell 17:944–956

Yamasaki K, Kigawa T, Watanabe S, Inoue M, Yamasaki T, Seki M, Shinozaki K, Yokoyama S (2012) Structural basis for sequence-specificDNA recognition by an *Arabidopsis* WRKY transcription factor. J Biol Chem 287:7683–7691

Yamasaki K, Kigawa T, Seki M, Shinozaki K, Yokoyama S (2013) DNA-binding domains of plant-specific transcription factors: structure, function, and evolution. Trends Plant Sci 18:267–276

Yan L, Liu Z-Q, Xu Y-H, Lu K, Wang X-F, Zhang D-P (2013) Auto and cross-repression of three *Arabidopsis* WRKY transcription factors WRKY18, WRKY40, and WRKY60 negatively involved in ABA signaling. J Plant Growth Regul 32:399–416

Yan Y, Jia H, Wang F, Wang C, Liu S, Guo X (2015) Overexpression of GhWRKY27a reduces tolerance to drought stress and resistance to *Rhizoctonia solani* infection in transgenic *Nicotiana benthamiana*. Front Physiol 24(6):265

Yang P, Chen C, Wang Z, Fan B, Chen Z (1999) A pathogen and salicylic acid-induced WRKY DNA-binding activity recognizes the elicitor response element of the tobacco class I chitinase gene promoter. Plant J 18:141–149

Ye S, Jiang Y, Duan Y, Karim A, Fan D, Yang L, Zhao X, Yin J, Luo K (2014) Constitutive expression of the poplar WRKY transcription factor PtoWRKY60 enhances resistance to *Dothiorella gregaria* Sacc. in transgenic plants. Tree Physiol 34(10):1118–1129

Yu FF, Huaxia YF, Lu WJ, Wu C, Guo XQ (2012) GhWRKY15, a member of the WRKY transcription factor family identified from cotton (*Gossypium hirsutum* L.), is involved in disease resistance and plant development. BMC Plant Biol 12:144

Zheng Z, Qamar SA, Chen Z, Mengiste T (2006) *Arabidopsis* WRKY33 transcription factor is required for resistance to necrotrophic fungal pathogens. Plant J 48:592–605

Small Noncoding RNA-Based Regulation of Plant Immunity

9

Om Prakash Gupta, Vanita Pandey, Nand Lal Meena, Suhas G. Karkute, Sagar Banerjee, and Anil Dahuja

Abstract

Plant pathogens trigger massive changes in plant gene expression in the host as a result of transcriptional reprogramming. This activates several defense-related pathways such as hormonal imbalances, signal transduction, induction of defense-related proteins, ROS generation, small RNA expression, etc.; small RNA regulates myriad biological processes in several eukaryotes constituting a vital group of gene expression regulators. Among all, plants utilize small non-coding RNA machinery as a crucial means to respond and defend against pathogens by regulating immune-responsive genes. In turn, phytopathogens have evolved various effector molecules such as proteins and recently discovered sRNAs of fungal origin delivered into host cells to suppress plant immunity, to counter-defend the effect of host small RNA machinery. The significance of the small RNA-mediated plant defense response during plant-pathogen interaction

O. P. Gupta (✉)
Division of Biochemistry, ICAR-Indian Agricultural Research Institute, New Delhi, India

Division of Quality and Basic Sciences, ICAR-Indian Institute of Wheat and Barley Research, Karnal, Haryana, India
e-mail: OP.Gupta@icar.gov.in; opguptaiari@gmail.com

V. Pandey
Division of Quality and Basic Sciences, ICAR-Indian Institute of Wheat and Barley Research, Karnal, Haryana, India

N. L. Meena
Plant Biochemistry, ICAR-Indian Institute of Millets Research, Hyderabad, Telangana, India

S. G. Karkute
Division of Vegetable Improvement, ICAR-Indian Institute of Vegetable Research, Varanasi, Uttar Pradesh, India

S. Banerjee · A. Dahuja
Division of Biochemistry, ICAR-Indian Agricultural Research Institute, New Delhi, India

A. Singh, I. K. Singh (eds.), *Molecular Aspects of Plant-Pathogen Interaction*,
https://doi.org/10.1007/978-981-10-7371-7_9

have been well-established. Here, we discuss findings on noncoding small RNAs (sRNAs) from plants and pathogens, which regulate host immunity and pathogen virulence.

Keywords
Small RNA · Plant immunity · Effector molecules · Phytopathogen · miRNAs

9.1 Introduction

Multicellular organisms (plants and animals) are evolved with various immune systems against invading pathogens (Ausubel 2005). PAMP-triggered immunity (PTI) is the first line of defense mediated by recognition of microbial- or pathogen-associated molecular patterns (MAMPs or PAMPs) via transmembrane pattern recognition receptors (PRRs). Effector-triggered susceptibility (ETS) is achieved by microbes which provide effector molecules into the host cells to successfully suppress PTI. Consecutively, several plant species have developed a category of immunity activated by resistance (R) proteins that acts in response to pathogen effector proteins and attenuates PTI seizure. The disease resistance developed as a consequence of hypersensitive response (HR) at the site of infection is a type of immunity called effector-triggered immunity (ETI). In this coevolutionary perspective, natural selection compels pathogens and plants to diversify their effector and resistance genes, respectively (Ronald and Beutler 2010). Recently, host-encoded small RNAs are shown to be involved in PTI and ETI (Jin 2008). These tiny regulatory small RNAs act by causing either transcriptional gene silencing (TGS) or posttranscriptional gene silencing (PTGS) to a set of pathogen or host genes (Baulcombe 2004). The sRNA-based TGS is triggered by DNA methylation and histone modifications, while PTGS is mediated by sRNA involved in mRNA cleavage or translational repression (Schramke and Allshire 2004). MicroRNAs (miRNAs), natural antisense transcript-derived small interfering RNAs (nat-siRNAs), trans-acting small interfering RNAs (ta-siRNAs), heterochromatic small interfering RNAs (hc-siRNAs) or repeat-associated small interfering RNAs (ra-siRNAs), and long small interfering RNAs (lsiRNAs) (Vazquez et al. 2010) are the major classes of sRNA reported so far in eukaryotes. The current chapter focuses on the role of sRNAs during plant immunity upon exposure to various phytopathogens.

9.2 sRNA Biogenesis Pathway

sRNAs are 20–40 nucleotide long, noncoding RNAs existing in most of the eukaryotic organisms and control expression of the genes either at transcription or posttranscription level. The biosynthetic mechanism of sRNA generation has been broadly investigated in the model plant *Arabidopsis*. Rising reports have recommended the significance of both forward and reverse genetics for outlining the cellular proteins that are implicated in the biogenesis and function of small RNAs (miRNAs and siRNAs). The noncoding genomic regions generally harbor the plant miRNAs. This is contradictory to animal miRNAs, which are occasionally processed from introns of protein-coding genes (Baskerville and Bartel 2005). RNA polymerase II transcribes miRNA genes to generate imperfect fold-back structure called primary miRNA (pri-miRNA). The pri-miRNA is processed into a stem-loop precursor miRNA (pre-miRNA) and then diced as a duplex encompassing the mature miRNA and a passenger strand called miRNA. DDL (DAWDLE) protein together with HYL1 (HYPONASTIC LEAVES 1) and zinc finger protein SE (SERRATE) acts upon the pri-miRNA to form pre-miRNA. The DCL1/DCL4 protein further mediates the processing of pre-miRNAs to form miRNA duplex (Fahlgren et al. 2007). miRNA duplex is then 2′O methylated at 3′ end by HEN1 (HUA ENHANCER 1) and is exported to the cytoplasm by an exportin homolog, HST (HASTY) (Fig. 9.1). Processed mature miRNA strand is selectively integrated into AGO-containing RISC complex to cause either cleavage or translational repression of target mRNA (Brodersen et al. 2008).

The major information of the process of posttranscriptional gene silencing (PTGS) was furnished by Boulcombe and Hamilton who identified the degraded RNA products as small RNA species (siRNA) of ~25 nucleotides. siRNAs form and accumulate as double-stranded RNA molecules and were first detected in plants undergoing either co-suppression or virus-induced gene silencing and were also undetectable in control plants. In contrast to miRNAs, siRNAs are derived from perfectly paired double-stranded RNA (dsRNA) precursors. These dsRNA precursors are produced either from antisense transcription or by the action of a cellular RNA-dependent RNA polymerase (RDR). Four different types of siRNAs have been reported in plants till now, viz., natural antisense transcript (NAT)-derived siRNAs (nat-siRNAs), heterochromatic siRNAs (hc-siRNAs) or repeat-associated siRNAs (ra-siRNAs), trans-acting siRNAs (ta-siRNAs), and long siRNAs (lsiRNAs).

9.3 Role of Noncoding sRNA in Bacterial, Viral, and Fungal Pathogenicity

Absence of DCL proteins makes bacterial sRNA different from other eukaryotic sRNAs. Bacterial sRNAs are heterogeneous in length, varying from 50 to 300 nucleotides, and regulate the stability and translation efficiency of target mRNAs through short and imperfect base pairing (10–25). Although in the recent past, several high-throughput RNA-seq studies have identified potential sRNAs in phytopathogenic bacteria such as *Agrobacterium tumefaciens* (Wilms et al. 2012), *Pst* (Filiatrault et al. 2010), *Xanthomonas campestris* (Schmidtke et al. 2012), and *Xanthomonas oryzae* pv. *oryzae* (Liang et al. 2011), however their defined function in pathogenesis is still unclear. Recently, genome-wide transcriptome analysis identified sRNA in *X. campestris* pv. *vesicatoria*, as the causal agent of bacterial spot disease in tomato and pepper. Gene deletion analysis demonstrated that the noncoding sRNAs

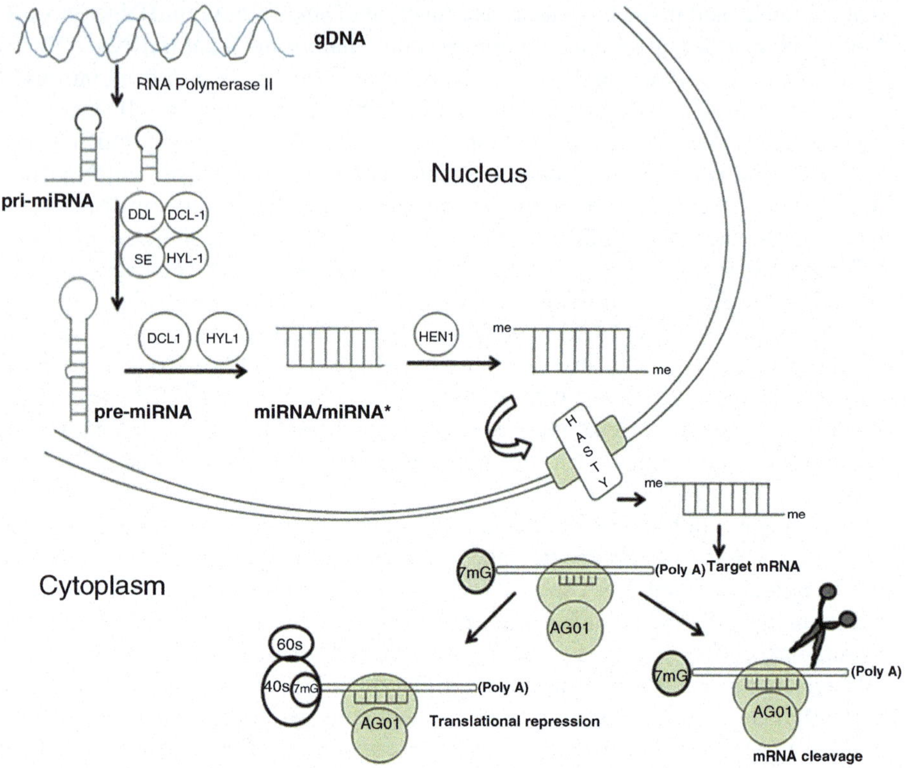

Fig. 9.1 miRNA biogenesis pathway in plants

sX12 and sX13 contribute to virulence (Schmidtke et al. 2013). sX13 promotes synthesis of HrpX and regulates the expression of other proteins putatively involved in motility, signal transduction, posttranscriptional and transcriptional regulation, and virulence. Bacterial noncoding sRNAs operate with RNA-binding protein complexes such as clustered regularly interspaced short palindromic repeats (CRISPR)-CRISPR-associated (Cas) system, regulatory protein Hfq, and CsrA/RsmA RNA-binding protein (Wiedenheft et al. 2012; Karkute et al. 2017). CRISPR-Cas is an exclusive prokaryotic adaptive immune system, similar to eukaryotic RNAi defense (Horvath and Barrangou 2010). CRISPR-Cas system produces small crRNAs to identify target DNA/RNA by short base pairing in the presence of PAM sequence. Hfq is a RNA-binding protein that acts as a global posttranscriptional regulator by binding to bacterial sRNAs to inhibit translation or promote degradation of target mRNAs (Vogel and Luisi 2011). Hfq protein is reported in many bacteria such as *Pectobacterium carotovorum*, *A. tumefaciens*, *P. syringae*, *Ralstonia solanacearum*, and *Xanthomonas spp*. In *A. tumefaciens*, Hfq binds to sRNA AbcR1, which regulates expression of atu2422, an mRNA component of ABC transporter. The *hfq* mutant exhibits overproduction of several other ABC transporter components and shows ectopic phenotypes of delayed growth, reduced motility, altered cell morphology, and, most significantly, attenuated virulence (Wilms et al. 2012).

In addition, numerous reports have revealed that plant sRNAs are directly involved in bacterial disease responses. Among all, miRNAs were the first to be recognized to contribute in plant immunity during bacterial infections. Plants tested with pathogenic bacteria showed differential changes in miRNA accumulation (Jagadeeswaran et al. 2009). Virus-responsive sRNAs have also been investigated during several plant-virus interactions (Pumplin and Voinnet 2013). Nonetheless, in contradiction to bacterial infections, direct proof for the specific role of endogenous sRNAs in plant antiviral immunity has been incomplete due to interruption of the function and the biogenesis of plant sRNAs by viruses. Similar to a bacterial suppressor that persuaded transcriptional subjugation of miR393, viral infections may also change miRNAs at the transcription level (Table 9.1) (Bazzini et al. 2009). Additionally, other viral proteins may interrupt miRNA (miR156, miR167, miR171, and miR390) accumulation (Feng et al. 2012; Jay et al. 2011). Nevertheless, variations in accumulation of miRNA-derived tasiRNAs have also been linked with phenotypic variations during plant viral infections (Yifhar et al. 2012). Viral infections can alter hc-siRNAs production and modify the RNA-directed DNA methylation (RdDM) pathway which could reactivate transposons and transcription of silenced genes leading to negative plant defense responses (Dowen et al. 2012). In spite of the fact that viruses cause extensive effects on host sRNAs, some specific miRNA families seem to be directly implicated in antiviral immunity. Markedly, numerous host sRNAs are reported to be involved in regulation of host resistance genes.

Flagellin is perceived by *Arabidopsis* which limits *Pseudomonas* invasion, although detailed mechanism is unclear. Gene expression profiling of *flg22*-exposed *Arabidopsis* seedlings leads to build up of three auxin receptors such as TIR1, AFB2, and AFB3, which are targets of miR393 (Navarro et al. 2006). This suggests that miR393 play a role in regulating defense responses during *Pseudomonas invasion*. miR393a-overexpressing lines showed enhanced *P. syringae* pv. *tomato* (Pst) DC3000 resistance (Table 9.1) (Navarro et al. 2006). During infiltration with *Agrobacterium tumefaciens*, increase of miR393 was also observed. Together, these results suggest that miR393a is clearly involved in ETI via auxin signaling. In addition to miR393, miR160 and miR167 were also induced upon nonpathogenic *Pst* DC3000 hrcC inoculation and *flg22* treatment (Li et al. 2010). miR398 which targets a cytochrome c oxidase subunit V (COX5) and two superoxide dismutases (CSD1,CSD2) was shown to be reduced in the plants exposed to avirulent strains such as PstDC3000 avrRpm1 and PstDC3000 avrRpt2 (Jagadeeswaran et al. 2009). Expression of miR398 is downregulated in oxidative stress, promoting accumulation of CSD1 and CSD2 (Sunkar et al. 2006). miR773 is also reported to be involved in PTI which targets the mRNA coding for DNA methyltransferase 2 (DMT2) (Li et al. 2010). During *Agrobacterium* infection, reduced tumor formation results from RNAi-mediated gene silencing of DMT2 and a different DNA methyltransferase (DMT1) (Crane and Gelvin 2007). In tomato and legumes, a group of miRNA families have been reported to be directly involved in ETI by controlling many R genes of the NBS-LRR class (Zhai et al. 2011). In an attempt to discover other sRNAs

Table 9.1 Plant small RNAs involved in immunity

Small RNA	Target(s)	Host(s)	Pathogen(s)	Reference(s)
miR156	Ta3711, Ta7012, TaGAMYB1, and TaGAMYB2	*Triticum aestivum*	Fungus	Xin et al. (2010)
miR158	PPR gene	*Brassica napus* and *B. rapa*	Viruses and fungus	He et al. (2008)
miR159	MYB33, MYB65, and MYC101	*Arabidopsis* and *T. aestivum*	Bacteria and fungus	Zhang et al. (2011a) and Xin et al. (2010)
miR160	ARF10, ARF16, ARF17, ARF16, and a B3 DNA binding, Domain-containing protein	*Arabidopsis*, *M. esculenta*, and *O. sativa*	Bacteria and fungus	Li et al. (2010, 2014) and Pinweha et al. (2015)
miR162	DCL1	*Arabidopsis*	Viruses	Azevedo et al. (2010)
miR164	NAC1	*O. sativa* and *T. aestivum*	Fungus	Li et al. (2013) and Xin et al. (2010)
miR166	Ta3711, Ta7012	*T. aestivum*	Fungus	Xin et al. (2010)
miR167	ARF8, ARF6	*Arabidopsis* and *T. aestivum*	Bacteria and fungus	Fahlgren et al. (2007) and Gupta et al. (2012)
miR168	AGO1	*Arabidopsis*, *N. benthamiana*, and *O. sativa*	Bacteria and viruses	Varallyay et al. (2010) and Wu et al. (2015)
miR169	Nuclear transcription factor Y subunit A-3 (putative)	*O. sativa*	Fungus	Li et al. (2013)
miR171	SCL	*T. aestivum*	Fungus	Gupta et al. (2012)
miR172	MADS box (putative)	*O. sativa*	Fungus	Li et al. (2013)
miR390	TAS3	*Arabidopsis*	Bacteria	Zhang et al. (2011a)
miR393	TIR1, AFB2, AFB3, AFB1	*Arabidopsis* and *T. aestivum*	Bacteria	Navarro et al. (2006)
miR393b	MEMB12	*Arabidopsis* and *N. benthamiana*	Bacteria	Zhang et al. (2011b)
miR396a-5p	GRF	Solanaceae	Fungus	Chen et al. (2015)
miR398	CSD1, CSD2, COX5, SOD1, and SOD2	*Arabidopsis*, *Hordeum vulgare*, and *O. sativa*	Bacteria and fungus	Jagadeeswaran et al. (2009), Li et al. (2010, 2014)

(continued)

Table 9.1 (continued)

Small RNA	Target(s)	Host(s)	Pathogen(s)	Reference(s)
miR399	PHO2	*Citrus, Solanum melongena*	Bacteria and fungus	Yang et al. (2013)
miR403	AGO protein genes	*Glycine max*	Fungus	Guo et al. (2011)
miR408	Copper protein	*Arabidopsis* and *T. aestivum*	Bacteria and fungus	Zhang et al. (2011), Feng et al. 2012) and Gupta et al. (2012)
	Plantacyanin			
	Laccase copper			
	Protein and copper			
	Ion-binding protein			
	Genes			
miR444	MADS box	*T. aestivum*	Fungus	Gupta et al. (2012)
miR472	CC-NBS-LRR	*Arabidopsis*	Bacteria	Boccara et al. (2014)
miR482	NBS-LRR	*S. lycopersicum* and *Gossypium hirsutum*	Viruses and fungus	Shivaprasad et al. (2012) and Zhu et al. (2013)
miR773	DMT2, MET2	*Arabidopsis*	Bacteria	Li et al. (2010)
miR825	Remorin, zinc finger, Homeobox family, Frataxin-related	*Arabidopsis*	Bacteria	Fahlgren et al. (2007)
miR827/miR1138	eIF-4b	*T. aestivum*	Fungus	Gupta et al. (2012)
miR1507	NBS-LRR	*M. truncatula*	–	Zhai et al. (2011)
miR1510	AGO protein genes	*Glycine max*	Fungus	Guo et al. (2011)
miR1535	AGO protein genes	*Glycine max*	Fungus	Guo et al. (2011)
miR1885	TIR-NBS-LRR	*Brassica napus*	Virus	Wroblewski et al. (2007)
miR2001/miR2005/miR2006/miR2008/miR2011/miR2012	Unknown	*T. aestivum*	Fungus	Xin et al. (2010)
miR2013	Receptor N	*N. tabacum* and *T. aestivum*	Viruses	Li et al. (2012)
miR2109	NBS-LRR	*Medicago*		Zhai et al. (2011)
miR2118	NBS-LRR	*Medicago, S. lycopersicum, G. hirsutum*	Viruses	Shivaprasad et al. (2012)
miR5300	Solyc05g008650, tm-2	*S. lycopersicum*	Fungus	Ouyang et al. (2014)
miR6019/miR6020	TIR-NBS-LRR	*N. tabacum*	Viruses	Li et al. (2012)
miR7695	OsNramp6	*O. sativa*	Fungus	Campo et al. (2013)

(continued)

Table 9.1 (continued)

Small RNA	Target(s)	Host(s)	Pathogen(s)	Reference(s)
miR9863	Mla1	*H.vulgare*	Fungus	Liu et al. (2014)
nat-siRNAATGB2	PPRL	*Arabidopsis*	Bacteria	Katiyar-Agarwal et al. (2006)
AtlsiRNA-1	AtRAP	*Arabidopsis*	Bacteria	Katiyar-Agarwal et al. (2007)
Bc-siR3.1	PRXIIF	*Arabidopsis* and *S. lycopersicum*	Fungus	Weiberg et al. (2013)
Bc-siR3.2	MPK2 and MPK1	*Arabidopsis* and *S. lycopersicum*	Fungus	Weiberg et al. (2013)
Bc-siR5	WAK	*Arabidopsis* and *S. lycopersicum*	Fungus	Weiberg et al. (2013)
TMV vsiRNA	CPSF30, TRAPa	*Arabidopsis*	Viruses	Qi et al. (2009)
Y-Sat siRNA	CHLI	*N. tabacum*	Viruses	Shimura et al. (2011) and Smith et al. (2011)
PC-sRNA8a/PC-sRNA8b	HSP90	*P. persica*	Viruses	Navarro et al. (2012)
vd39/vd40	CalS11-like and CalS12-like	*S. lycopersicum*	Viruses	Adkar-Purushothama et al. (2015)
vdsiRNA	SolWD40	*S. lycopersicum*	Viruses	Avina-Padilla et al. (2015)

specifically induced in a pathogen response (*P. syringae*), Katiyar-Agarwal and co-workers (2007) identified a class of small RNAs (lsiRNAs). Of the six lsiRNAs identified, five were induced in response to *Pst* (avrRpt2) infection where AtlsiRNA-1 is the most functionally characterized lsiRNA.

In the recent past, more focus has been given to identify and characterize sRNAs in eukaryotic plant pathogens such as fungus involved in pathogenic development and virulence. An inclusive sRNA expression study in *M. oryzae* has identified two class of sRNAs linked with pathogenesis (Nunes et al. 2011). The sRNAs of one class were mapped to transfer RNA (tRNA) loci which are enriched in the appressoria region, while the other classes of sRNAs were mapped to several kinds of genomic. A second study reported that 24 nts is the primary size of sRNAs spotted from *M. oryzae* under physiological stress conditions and *in planta* during infection of rice (Raman et al. 2013). *M. oryzae* sRNAs regulate a subset of mRNAs post-transcriptionally, including an effector gene, ACE1. A hybrid between a polyketide synthase and a nonribosomal peptide synthetase is encoded putatively by ACE1, which probably functions in secondary metabolite production, and it is tightly controlled during the start of appressorial penetration (Fudal et al. 2007). Similarly, sRNAs from the entomopathogenic fungus *Metarhizium anisopliae* (Zhou et al. 2012b) and the white mold fungus *Sclerotinia sclerotiorum* (Zhou et al. 2012a) showed differential regulation during sclerotia conidiogenesis and assembly, respectively. These results lend support to the belief that fungal endogenous sRNAs play a vital role in regulating virulence and developmental processes of fungal pathogens.

Additionally, plant miRNAs regulate phytohormones homeostasis by regulating the expression of their target transcripts (Sunkar and Zhu 2004). miR408 has been identified as negative controller of plantacyanins and laccase (Abdel-Ghany and Pilon 2008). Although the accurate function of plantacyanins in plants is unidentified, they have been, however, proposed to function in lignin formation, cell-to-cell signaling, and stress responses (Kim et al. 2003). Therefore, differential regulation of miR408 in susceptible and resistant wheat cultivars infected with *Puccinia graminis* f. sp. *tritici* pathotype 62G29-1 might lead to plantacyanin-mediated perturberance of lignin biosynthesis as a result of HR (Table 9.1) (Gupta et al. 2012). Similarly, miR2118 targets TIR-NBS-LRR in cotton infected with *Verticillium dahliae* (Yin et al. 2012), while pbe-SR23 and pbe-SR3 have been predicted to target TIR-LRR in *Populus* upon infection with *Dothiorella gregaria* (Chen et al. 2012). Very recently, Campo and co-workers (2013) have shown Osa-miR7695-mediated negative regulation of natural resistance-associated macrophage protein 6 (OsNramp6) in disease resistance, while illustrating the presence of a novel regulatory network that integrates miRNA function and mRNA processing in plant immunity. Overexpression of Osa-miR7695 in rice has resulted in resistance to blast fungus.

Yin and co-workers (2012) carried out studies on inclusive identification of miRNAs in two cotton cultivars, viz., Yi-11 (*Gossypium hirsutum*, *Verticillium*-sensitive cultivar) and Hai-7124 (*Gossypium barbadense*, *Verticillium*-tolerant cultivar), after infection with *Verticillium* fungus. Among identified miRNAs, three miRNAs, viz., Ptc-miR1444, Ptc-miR1448, and Ptc-miR482, target the cleavage of PPO gene (polyphenol oxidase) and disease resistance protein genes which regulate biotic and abiotic stress resistance in plants (Lu et al. 2008). Chen and co-workers (2012) suggested that there were as many as 74 conserved miRNA belonging to 37 miRNA families and 27 novel miRNAs in *Populus* infected with *D. gregaria*. In contrast to the results obtained in galled loblolly pine stems infected with the fungus *Cronartium* where miR156 were significantly repressed (Lu et al. 2007), pbe-miR156aee was found to be induced upon infection with *D. gregaria* in *Populus*. The distinctive expression pattern of miRNAs in the same family can be linked to different function in different species under different sets of pathogens. Consequently, to develop an improved understanding of the regulatory role of miRNAs on their target genes during pathogen stress, further experimental confirmation of miRNAs is essential. Taking together the previous studies, all the fungi found responsive to miRNAs target several genes simultaneously, and each target gene is involved in controlling numerous physiological and biochemical processes. Therefore, regulation and cross talk of gene expression during pathological development is an actively growing area to develop better understanding of disease pathogenesis. Nevertheless, taking into account the cases uncovered here of siRNAs related to plant immunity during bacterial and viral infections, together with the emergent information generated around sRNAs, more siRNAs could also be involved in regulating viral and bacterial stress responses.

9.4 Pathogen sRNAs Act as Effectors to Suppress Host Immunity

Pathogen effectors are molecules of pathogen origin, which are conveyed into host cells to overpower host immunity. Till now, almost all the studied effectors are protein in nature. Interestingly, a recent study in *Botrytis cinerea* (Bc) pathogen revealed that sRNA could also act as effector. This pathogen is a devastating pathogen having wide host range and can infect 200 or more different plant species. It was observed in *Arabidopsis thaliana* and tomato that during infection the Bc-sRNAs are delivered into host cells where they silence plant immunity genes. In both *Arabidopsis* and tomato hosts, more than 70 Bc-sRNAs have been identified to be probable effectors based on *in planta* expression and target gene predictions, for which three sRNA effectors have been demonstrated experimentally to silence host plant immunity genes by overpowering host RNAi machinery. The successful infection of *B. cinerea* in host plants can be ensured by silencing of host immune genes (Weiberg et al. 2013). The Bc-sRNA effectors share similar features with host sRNAs and are favorably sorted into *Arabidopsis* AGO1 (AtAGO1) protein and thus exploit the host RNAi machinery by loading into host AGO1 to silence host immunity genes. In support of this, the *Arabidopsis* mutant ago1-27 was less susceptible to *B. cinerea*, because the Bc-sRNA effectors were no longer functional in guiding the host gene silencing without the suitable AGO protein (Weiberg et al. 2013). It was the first report of inhibition of host immunity where pathogenic sRNAs act as effectors. Further research will reveal whether this novel sRNA-based virulence pathway exists in other plant eukaryotic pathogens or not. Undeniably, an aggressive fungal pathogen, *Verticillium dahliae*, might have developed a comparable tactic of hijacking the host plant RNAi machinery to repress host immunity. Similar to that observed during *B. cinerea* infection, the *Arabidopsis* ago1-27 mutant was more resistant against *Verticillium* spp., whereas several other *Arabidopsis* RNAi mutants exhibited improved vulnerability (Ellendorff et al. 2009).

9.5 sRNA Biogenesis Proteins in Eukaryotic Plant Pathogens

The key components of the RNAi pathway are the proteins DCL, AGO, and RDR which are encoded by most of the eukaryotic pathogens. Several RNAi-dependent phenomena in *N. crassa* and other fungal species have been studied. Quelling, meiotic silencing of unpaired DNA (MSUD), the silencing of repeat-induced point (RIP) mutation, sex-induced silencing (SIS), DNA repair and homologous recombination, and DNA methylation are examples of RNAi-dependent phenomena. However, it is unclear whether RNAi pathway machinery are directly essential for pathogenicity in eukaryotic plant pathogens. The *M. oryzae* genome encodes three putative AGOs (MoAGO1–MoAGO3), two DCLs (MoDCL1 and MoDCL2), and three RDRs (MoRdRP1–MoRdRP3). MoDCL2 is required for transgene-induced gene silencing and sRNA-directed transposon silencing (Kadotani et al. 2004).

Although there are no reports of any RNAi mutant strains, including *Mordrp*, *Modcl1*, *Modcl2*, and *Modcl1dcl2*, *Modcl2* shows somewhat reduced growth—that illustrates obvious phenotypes associated with pathogenicity (Kadotani et al. 2004). In contrast, obvious defects in sporulation and growth have been observed in the zygomycete fruit rot pathogen *Mucor circinelloides dcl2* and *dcl1* mutants, respectively (Nicolas et al. 2007). An *ago1* mutant of *M. circinelloides* depicts defects in asexual spore production.

Very divergent RNAi pathways are active in fungus; for example, in plants and animals, the miRNA pathways are generally conserved but not in fungi. Instead, at least four different milRNA biogenesis pathways, including both DCL-dependent and DCL-independent pathways, are proposed in *N. crassa*, and miRNA-like RNAs (milRNA) have been identified in fungi (Zhou et al. 2012). It is imperative to deduce the existence of similar sRNA biogenesis pathways in oomycete and fungal pathogens. Moreover, DCLs appear to be functionally superfluous in sRNA processing in the majority of fungus (Weiberg et al. 2013). The probable existence of varied sRNA biogenesis pathways in these organisms represents a challenge for researchers in the dissection of RNAi pathways and characterization of regulatory sRNAs in pathogenicity.

9.6 Conclusion and Future Perspectives

RNAi pathways and sRNAs play a vital role in regulating plant immunity. sRNAs from eukaryotic and bacterial pathogens are also significant regulators of pathogenicity. Since the RNA molecules are known to act as effectors in suppressing host immunity, so it would not be surprising if in plants some of them also serve as PAMPs to trigger PTI. It would be fascinating to examine whether and how extensive noncoding RNAs are in plant pathogens that can activate host innate immune responses or control pathogen virulence in plants. Several reports support the belief that sRNAs can translocate in between organisms, including between the pathogens and host cells, and can even stimulate cross-kingdom RNAi. Hence, this will continue to be a dynamic and pertinent research topic in near future.

References

Abdel-Ghany SE, Pilon M (2008) MicroRNA-mediated systemic down-regulation of copper protein expression in response to low copper availability in Arabidopsis. J Biol Chem 283:15932e45

Adkar-Purushothama CR, Brosseau C, Giguere T, Sano T, Moffett P, Perreault JP (2015) Small RNA derived from the virulence modulating region of the potato spindle tuber viroid silences callose synthase genes of tomato plants. Plant Cell 27:2178–2194

Ausubel FM (2005) Are innate immune signaling pathways in plants and animals conserved? Nat Immunol 6:973–979

Avina-Padilla K, de la Vega OM, Rivera-Bustamante R, Martinez-Soriano JP, Owens RA, Hammond RW, Vielle-Calzada JP (2015) In silico prediction and validation of potential gene targets for pospiviroid-derived small RNAs during tomato infection. Gene 564:197–205

Azevedo J, Garcia D, Pontier D, Ohnesorge S, Yu A, Garcia S, Braun L, Bergdoll M, Hakimi MA, Lagrange T, Voinnet O (2010) Argonaute quenching and global changes in Dicer homeostasis caused by a pathogen-encoded GW repeat protein. Genes Dev 24:904–915

Baskerville S, Bartel DP (2005) Microarray profiling of microRNAs reveals frequent coexpression with neighboring miRNAs and host genes. RNA 11:241–247

Baulcombe D (2004) RNA silencing in plants. Nature 431:356–363

Bazzini AA, Almasia NI, Manacorda CA, Mongelli VC, Conti G, Maroniche GA et al (2009) Virus infection elevates transcriptional activity of miR164a promoter in plants. BMC Plant Biol 9:152

Boccara M, Sarazin A, Thiebeauld O, Jay F, Voinnet O, Navarro L, Colot V (2014) The Arabidopsis miR472-RDR6 silencing pathway modulates PAMP- and effector-triggered immunity through the post-transcriptional control of disease resistance genes. PLoS Pathog 10:e1003883

Brodersen P, Sakvarelidze-Achard L, Bruun-Rasmussen M, Dunoyer P, Yamamoto YY et al (2008) Widespread translational inhibition by plant miRNAs and siRNAs. Science 320:1185–1190

Campo S, Peris-Peris C, Sire C, Moreno AB, Donaire L, Zytnicki M et al (2013) Identification of a novel microRNA (miRNA) from rice that targets an alternatively spliced transcript of the Nramp6 (Natural resistance-associated macrophage protein 6) gene involved in pathogen resistance. New Phytol 199:212e27

Chen L, Ren Y, Zhang Y, Xu J, Zhang Z, Wang Y (2012) Genome-wide profiling of novel and conserved Populus microRNAs involved in pathogen stress response by deep sequencing. Planta 235:873e83

Chen L, Luan Y, Zhai J (2015) Sp-miR396a-5p acts as a stress-responsive genes regulator by conferring tolerance to abiotic stresses and susceptibility to *Phytophthora nicotianae* infection in transgenic tobacco. Plant Cell Rep 34(12):2013–2025

Crane YM, Gelvin SB (2007) RNAi-mediated gene silencing reveals involvement of Arabidopsis chromatinrelated genes in agrobacterium-mediated root transformation. Proc Natl Acad Sci U S A 104:15156–15161. https://doi.org/10.1073/pnas.0706986104

Dowen RH, Pelizzola M, Schmitz RJ, Lister R, Dowen JM, Nery JR et al (2012) Widespread dynamic DNA methylation in response to biotic stress. Proc Natl Acad Sci 109:2183–2191

Ellendorff U, Fradin EF, de Jonge R, Thomma BP (2009) RNA silencing is required for Arabidopsis defence against Verticillium wilt disease. J Exp Bot 60:591–602

Fahlgren N, Howell MD, Kasschau KD, Chapman EJ, Sullivan CM et al (2007) High-throughput sequencing of *Arabidopsis* microRNAs: evidence for frequent birth and death of *MIRNA* genes. PLoS ONE 2:e219

Feng J, Lai L, Lin R, Jin C, Chen J (2012) Differential effects of cucumber mosaic virus satellite RNAs in the perturbation of microRNA-regulated gene expression in tomato. Mol Biol Rep 39:775–784

Filiatrault MJ, Stodghill PV, Bronstein PA, Moll S, Lindeberg M et al (2010) Transcriptome analysis of Pseudomonas syringae identifies new genes, noncoding RNAs, and antisense activity. J Bacteriol 192:2359–2372

Fudal I, Collemare J, Bohnert HU, Melayah D, Lebrun MH (2007) Expression of Magnaporthe grisea avirulence gene ACE1 is connected to the initiation of appressorium-mediated penetration. Eukaryot Cell 6:546–554

Guo N, Ye WW, Wu XL, Shen DY, Wang YC, Xing H, Dou DL (2011) Microarray profiling reveals microRNAs involving soybean resistance to Phytophthora sojae. Genome 54:954–958

Gupta OP, Permar V, Koundal V, Singh UD, Praveen S (2012) MicroRNA regulated defense response in *Triticum aestivum* L. during *Puccinia graminis* f.sp. *tritici* infection. Mol Biol Rep 39(2):817–824

He XF, Fang YY, Feng L, Guo HS (2008) Characterization of conserved and novel microRNAs and their targets, including aTuMV-induced TIR-NBS-LRR class R gene-derived novel miRNA in *Brassica*. FEBS Lett 582:2445–2452

Horvath P, Barrangou R (2010) CRISPR/Cas, the immune system of bacteria and archaea. Science 327:167–170

Jagadeeswaran G, Saini A, Sunkar R (2009) Biotic and abiotic stress down-regulate miR398 expression in Arabidopsis. Planta 229:1009–1014

Jay F, Wang Y, Yu A, Taconnat L, Pelletier S, Colot V et al (2011) Misregulation of AUXIN RESPONSE FACTOR 8 underlies the developmental abnormalities caused by three distinct viral silencing suppressors in Arabidopsis. PLoS Pathog 7:e1002035

Jin H (2008) Endogenous small RNAs and antibacterial immunity in plants. FEBS Lett 582:2679–2684. https://doi.org/10.1016/j.febslet.2008.06.053

Kadotani N, Nakayashiki H, Tosa Y, Mayama S (2004) One of the two Dicer-like proteins in the filamentous fungi Magnaporthe oryzae genome is responsible for hairpin RNA-triggered RNA silencing and related small interfering RNA accumulation. J Biol Chem 279:44467–44474

Karkute SG, Singh AK, Gupta OP, Singh PM, Singh B (2017) CRISPR/Cas9 mediated genome engineering for improvement of horticultural crops. Front Plant Sci 8:1635

Katiyar-Agarwal S, Gao S, Vivian Smith A, Jin H (2007) A novel class of bacteria-induced small RNAs in Arabidopsis. Genes Dev 21:3123–3134

Kim S, Mollet JC, Dong J, Zhang K, Park SY, Lord EM (2003) Chemocyanin, a small basic protein from the lily stigma, induces pollen tube chemotropism. Proc Natl Acad Sci 100:16125e30

Li Y, Zhang Q, Zhang J, Wu L, Qi Y, Zhou JM (2010) Identification of microRNAs involved in pathogen-associated molecular pattern-triggered plant innate immunity. Plant Physiol 152:2222–2231

Li F, Pignatta D, Bendix C, Brunkard JO, Cohn MM, Tung J, Sun H, Kumar P, Baker B (2012) MicroRNA regulation of plant innate immune receptors. Proc Natl Acad Sci U S A 109:1790–1795

Li Z, Peng J, Wen X, Guo H (2013) Ethylene-insensitive3 is a senescence-associated gene that accelerates age-dependent leaf senescence by directly repressing miR164 transcription in Arabidopsis. Plant Cell 25(9):3311–3328

Li Y, Lu YG, Shi Y, Wu L, Xu YJ, Huang F, Guo XY, Zhang Y, Fan J, Zhao JQ, Zhang HY, Xu PZ, Zhou JM, Wu XJ, Wang PR, Wang WM (2014) Multiple rice microRNAs are involved in immunity against the blast fungus *Magnaporthe oryzae*. Plant Physiol 164:1077–1092

Liang H, Zhao YT, Zhang JQ, Wang XJ, Fang RX, Jia YT (2011) Identification and functional characterization of small non-coding RNAs in *Xanthomonas oryzae* pathovar *oryzae*. BMC Genomics 12:87

Liu J, Cheng X, Liu D, Xu W, Wise R, Shen QH (2014) The miR9863 family regulates distinct Mla alleles in barley to attenuate NLR receptor-triggered disease resistance and cell-death signaling. PLoS Genet 10:e1004755

Lu S, Sun YH, Amerson H, Chiang VL (2007) MicroRNAs in loblolly pine (*Pinus taeda* L.) and their association with fusiform rust gall development. Plant J 51:1077e98

Lu S, Sun YH, Chiang VL (2008) Stress-responsive microRNAs in Populus. Plant J 55:131e51

Navarro L, Dunoyer P, Jay F, Arnold B, Dharmasiri N, Estelle M et al (2006) A plant miRNA contributes to antibacterial resistance by repressing auxin signalling. Science 312:436–439

Navarro B, Gisel A, Rodio ME, Delgado S, Flores R, Seri FD (2012) Small RNAs containing the pathogenic determinant of a chloroplast-replicating viroid guide the degradation of a host mRNA as predicted by RNA silencing. Plant J 70:991–1003

Nicolas FE, de Haro JP, Torres-Martinez S, Ruiz-Vazquez RM (2007) Mutants defective in a *Mucor circinelloides* dicer-like gene are not compromised in siRNA silencing but display developmental defects. Fungal Genet Biol 44:504–516

Nunes CC, Gowda M, Sailsbery J, Xue M, Chen F et al (2011) Diverse and tissue-enriched small RNAs in the plant pathogenic fungus, *Magnaporthe oryzae*. BMC Genomics 12:288

Ouyang S, Park G, Atamian HS, Han CS, Stajich JE, Kaloshian I, Borkovich KA (2014) MicroRNAs suppress NB domain genes in tomato that confer resistance to *Fusarium oxysporum*. PLoS Pathog 10:e1004464

Pinweha N, Asvarak T, Viboonjun U, Narangajavana J (2015) Involvement of miR160/miR393 and their targets in cassava responses to anthracnose disease. J Plant Physiol 174:26–35

Pumplin N, Voinnet O (2013) RNA silencing suppression by plant pathogens: defence, counter-defence and counter-counter-defence. Nat Rev Microbiol 11:745–760

Qi X, Bao FS, Xie Z (2009) Small RNA deep sequencing reveals role for Arabidopsis thaliana RNAdependent RNA polymerases in viral siRNA biogenesis. PLoS One 4:e4971. https://doi.org/10.1371/journal.pone.0004971

Raman V, Simon SA, Romag A, Demirci F, Mathioni SM et al (2013) Physiological stressors and invasive plant infections alter the small RNA transcriptome of the rice blast fungus, Magnaporthe oryzae. BMC Genomics 14:326

Ronald PC, Beutler B (2010) Plant and animal sensors of conserved microbial signatures. Science 330:1061–1064

Schmidtke C, Findeiss S, Sharma CM, Kuhfuss J, Hoffmann S et al (2012) Genome-wide transcriptome analysis of the plant pathogen *Xanthomonas* identifies sRNAs with putative virulence functions. Nucleic Acids Res 40:2020–2031

Schmidtke C, Abendroth U, Brock J, Serrania J, Becker A, Bonas U (2013) Small RNA sX13: a multifaceted regulator of virulence in the plant pathogen *Xanthomonas*. PLoS Pathog 9:e1003626

Schramke V, Allshire R (2004) Those interfering little RNAs! Silencing and eliminating chromatin. Curr Opin Genet Dev 14:174–180

Shimura H, Pantaleo V, Ishihara T, Myojo N, Inaba J, Sueda K, Burgyan J, Masuta C (2011) A viral satellite RNA induces yellow symptoms on tobacco by targeting a gene involved in chlorophyll biosynthesis using the RNA silencing machinery. PLoS Pathog 7:e1002021

Shivaprasad PV, Chen HM, Patel K, Bond DM, Santos BA, Baulcombe DC (2012) A microRNA superfamily regulates nucleotide binding site-leucine-rich repeats and other mRNAs. Plant Cell 24:859–874

Smith NA, Eamens AL, Wang MB (2011) Viral small interfering RNAs target host genes to mediate disease symptoms in plants. PLoS Pathog 7:e1002022

Sunkar R, Zhu JK (2004) Novel and stress-regulated microRNAs and other small RNAs from Arabidopsis. Plant Cell 16:2001e19

Sunkar R, Kapoor A, Zhu JK (2006) Posttranscriptional induction of two Cu/Zn superoxide dismutase genes in Arabidopsis is mediated by down regulation of miR398 and important for oxidative stress tolerance. Plant Cell 18:2051–2065

Varallyay E, Valoczi A, Agyi A, Burgyan J, Havelda Z (2010) Plant virus-mediated induction of miR168 is associated with repression of ARGONAUTE1 accumulation. EMBO J 29:3507–3519

Vazquez F, Legrand S, Windels D (2010) The biosynthetic pathways and biological scopes of plant small RNAs. Trends Plant Sci 15:337–345

Vogel J, Luisi BF (2011) Hfq and its constellation of RNA. Nat Rev Microbiol 9:578–589

Weiberg A, Wang M, Lin FM, Zhao H, Zhang Z, Kaloshian I, Huang HD, Jin H (2013) Fungal small RNAs suppress plant immunity by hijacking host RNA interference pathways. Science 342:118–123

Wiedenheft B, Sternberg SH, Doudna JA (2012) RNA-guided genetic silencing systems in bacteria and archaea. Nature 482:331–338

Wilms I, Overloper A, Nowrousian M, Sharma CM, Narberhaus F (2012b) Deep sequencing uncovers numerous small RNAs on all four replicons of the plant pathogen Agrobacterium tumefaciens. RNA Biol 9:446–457

Wroblewski T, Piskurewicz U, Tomczak A, Ochoa O, Michelmore RW (2007) Silencing of the major family of NBS-LRR-encoding genes in lettuce results in the loss of multiple resistance specificities. Plant J 51:803–818

Wu J, Yang Z, Wang Y et al (2015) Viral-inducible Argonaute18 confers broad-spectrum virus resistance in rice by sequestering a host microRNA. elife 4:e05733

Xin M, Wang Y, Yao Y, Xie C, Peng H, Ni Z, Sun Q (2010) Diverse set of microRNAs are responsive to powdery mildew infection and heat stress in wheat (*Triticum aestivum* L.) BMC Plant Biol 10:123

Yang L, Jue D, Li W, Zhang R, Chen M, Yang Q (2013) Identification of MiRNA from eggplant (*Solanum melongena* L.) by small RNA deep sequencing and their response to *Verticillium dahliae* infection. PLoS One 8(8):e72840

Yifhar T, Pekker I, Peled D, Friedlander G, Pistunov A, Sabban M et al (2012) Failure of the tomato trans-acting short interfering RNA program to regulate AUXIN RESPONSE FACTOR3 and ARF4 underlies the wiry leaf syndrome. Plant Cell 24:3575–3589

Yin Z, Li Y, Han X, Shen F (2012) Genome-wide profiling of miRNAs and other small non-coding RNAs in the *Verticillium dahliae* inoculated cotton roots. PLoS One 7(4):e35765

Zhai J, Jeong DH, De Paoli E, Park S, Rosen BD, Li Y et al (2011) MicroRNAs as master regulators of the plant NB-LRR defense gene family via the production of phased, trans-acting siRNAs. Genes Dev 25:2540–2553

Zhang W, Gao S, Zhou X, Chellappan P, Chen Z, Zhang X, Fromuth N, Coutino G, Coffey M, Jin H (2011a) Bacteria-responsive microRNAs regulate plant innate immunity by modulating plant hormone networks. Plant Mol Biol 75:93–105

Zhang XM, Zhao HW, Gao S, Wang WC, Katiyar-Agarwal S, Huang HD, Raikhel N, Jin HL (2011b) Arabidopsis argonaute 2 regulates innate immunity via miRNA393_-mediated silencing of a golgi-localized SNARE gene, MEMB12. Mol Cell 42:356–366

Zhou J, Fu Y, Xie J, Li B, Jiang D et al (2012a) Identification of micro RNA-like RNAs in a plant pathogenic fungus *Sclerotinia sclerotiorum* by high-throughput sequencing. Mol Gen Genomics 287:275–282

Zhou Q, Wang Z, Zhang J, Meng H, Huang B (2012b) Genome-wide identification and profiling of microRNA-like RNAs from *Metarhizium anisopliae* during development. Fungal Biol 116:1156–1162

Zhu QH, Fan L, Liu Y, Xu H, Llewellyn D, Wilson I (2013) miR482 regulation of NBS-LRR defense genes during fungal pathogen infection in cotton. PLoS ONE 8:e84390

10 Transcriptomic Studies Revealing Enigma of Plant-Pathogen Interaction

Zahoor Ahmed Wani and Nasheeman Ashraf

Abstract

Plants being sessile organisms encounter numerous attacks by pathogens and pests with different lifestyles and modes of attack. In response, plants undergo cellular reprogramming in order to perceive these attacks and activate specific defense pathways. Plants possess extensive regulatory mechanisms which come into play during defense responses so as to coordinate the perception and activation of pathways specific to the type of pathogen in question. Further, many small molecule hormones play pivotal role in defense pathways and cross communicate with each other, thereby helping plant to finely regulate its response. This suggests that plant defense is controlled by intricate transcriptional regulatory network, therefore urging the need to develop genome- and transcriptome-based strategies to unravel these mechanisms. Transcriptomics has fuelled a better understanding of many biological processes and can therefore be used for understanding the host-pathogen interactions as well. Transcriptome analysis can provide more comprehensive picture of the pathways that come into play in response to different pathogens and also decipher the cascade of transcriptional events involved. This may also help in identifying the regulatory nodes in the transcriptional networks and understanding the hierarchical relationship between them. These resources in turn will help in understanding of the complex architecture of plant/host defense system which will have a long-term impact and value for crop improvement.

Z. A. Wani · N. Ashraf (✉)
Plant Biotechnology Division, CSIR—Indian Institute of Integrative Medicine, Srinagar, Jammu and Kashmir, India

Academy of Scientific and Innovative Research, CSIR—Indian Institute of Integrative Medicine, Jammu Tawi, India
e-mail: nashraf@iiim.ac.in

A. Singh, I. K. Singh (eds.), *Molecular Aspects of Plant-Pathogen Interaction*,
https://doi.org/10.1007/978-981-10-7371-7_10

Keywords
Pathogen · Transcriptomics · Metabolomics · Defense · Signaling

10.1 Introduction

Plant-pathogen interactions are battles of attack and counterattack which are fought with highly sophisticated means for the survival of an individual. During this discourse plants respond by dynamic structural rearrangements within and around the attacked cells followed by reprogramming of cellular metabolism. In animals there are two forms of immune responses called as innate and adaptive immunity. Innate immunity forms a first line of defense against invading pathogens and is also a key element for the deployment of adaptive immunity. Plants lack adaptive immune system and depend entirely on innate immunity for defending themselves against pathogen attack.

Plants have developed a two-tier innate immune system to combat the invading pathogenic microbes. The first-tier immune system is designated as PTI (PAMP/pattern-triggered immunity), which is activated upon perception of conserved molecular structures called microbe-/pathogen-associated molecular patterns (MAMPs/PAMPs) by plasma membrane-bound receptors known as plasma membrane-localized pattern recognition receptors (PRRs). For example, PRRs in *Arabidopsis thaliana* are AtFLS2 (FLAGELLIN SENSING 2) and AtEFR (EF-TU receptor) that recognize bacterial flagellin and elongation factor-Tu (EF-Tu), respectively (Zipfel et al. 2006). The pathogen/microbe in turn acquires a number of mechanisms imparting them virulence to suppress the host immune system by activating various effector proteins (Dou and Zhou 2012). To counter this acquired virulence mechanism of pathogen/microbe, the plants have evolved the second tier of the innate immune system known as effector-triggered immunity (ETI). In this system the host/plant recognizes – directly or indirectly – such effector proteins, resulting in initiation of effector-triggered immunity (ETI). During PTI and ETI, plants come up with an array of immune responses such as reactive oxygen species (ROS) generation, cellular Ca^{2+} release, MAP kinase (MAPK) activation, phytohormone production, and reprogramming of transcriptional mechanism, which collectively contribute to immunity. The signaling components are similar in both the immune systems, PTI and ETI, but with distinct activation dynamics and amplitudes (Tsuda and Katagiri 2010; Tsuda and Somssich 2015).

Earlier, knowledge about various cellular processes including pathogen stress was gained by working on individual genes in context of a particular process. But recent developments have shown that cellular processes are controlled by highly connected gene networks. Therefore, the function of an individual gene should be understood in the context of its complex interplay with other gene products (Dittrich et al. 2008). Also existence of cross talk between various processes and pathways has been revealed. For example, many biotic and abiotic stress pathways have been shown to have some overlap. Further, many development-related genes are shown to play a role in other cellular processes like stress responses (Chung et al. 2008). Therefore, in order to gain a broader understanding of biological processes, we cannot study genes in isolation but in the context of other genes. We have to move beyond the single gene approach and pave way to the study of genome or at least transcriptome of organisms as a whole. This will allow developing a wide picture of gene characteristics. Further, more needs to be achieved in less period of time. Food- and health-related problems are appealing the scientific community to do better and come up with more comprehensive understanding of biology and provide solutions to the problems. Here genomics and transcriptomics come to our rescue as these have the potential to address many such problems.

In recent years transcriptomic studies have changed the whole scenario of our understanding of the molecular approaches/mechanisms of cells and tissues in health and diseases. This provides essential tools to fully understand the molecular basis of various agronomic traits and to manipulate them for human benefits (Harlizius et al. 2004). The field of transcriptomics allows the simultaneous analysis of thousands of genes and their interactive networks to understand the architecture of genomes. Recent technological advances in the field of transcriptomics have seen a paradigm shift enabling the analysis of organisms in terms of genome organization, expression networks, and interaction (Hocquette 2005). This has led to and will further extend substantial and rapid advances in our understanding of the molecular basis of various processes including stresses (Mathers 2004). Such approaches may assist in illuminating the mechanism as it enables the simultaneous discovery and study of many biological processes and genes involved in such processes. Another importance is that it helps to capture the structure-function relationships of genes. Also, the genomic, proteomic, and metabolomic studies in combination will provide the link between the relatively static genome and the highly dynamic physiological processes. In this chapter we will focus on how transcriptomics leads to resource generation and subsequently development of understanding about how plants perceive and respond to pathogen stress. It will also throw light on how transcriptomic studies lead to unraveling the plant signaling and defense pathways.

10.2 Tools and Techniques Used in Transcriptomic Studies

There are various techniques used for the transcriptome-based studies, viz., real-time quantitative PCR (qPCR), northern blot, serial analysis of gene expression (SAGE), massively parallel signature sequencing (MPSS), microarrays, bead arrays, and RNA sequencing. The qPCR and northern blot techniques are generally best for analyzing relatively small number of transcripts in a large set of samples; however, the RNA sequencing and microarrays offer genome-wide surveys of the transcriptome. The RNA sequencing technique generates expressed sequence tags (ESTs) from a given RNA sample without prior knowledge of the genes. In RNA sequencing, ESTs are created by sequencing the extreme ends of randomly isolated transcript cDNA, while as in Serial analysis of gene expression (SAGE) and Massive parallel signature sequencing (MPSS), 15-nucleotide tags and 17- to 20-nucleotide tags are used, respectively. In case of microarrays and bead arrays, mRNAs of interest are hybridized with a large number of probes spotted on a suitable substrate. These can be in the form of rectangular chips for microarrays or beads in fiber-optic bundles in the case of bead arrays (Schena et al. 1995; Kuhn et al. 2004). These strategies are complementary to each other.

The development of EST libraries associated with "differential gene expression" (DGE) technologies is increasing and provides an idea about many biological processes (Marshall 2004). However, to generate ESTs related to a particular biological function like defense-related, development of suppression subtractive hybridization (SSH) technique (Diatchenko et al. 1996) might be helpful. In addition to these, there are various bioinformatics tools for *in silico* analysis of the transcriptomic database available to elucidate the identity and function of genes, promoter analysis of genes, interactome analysis to study the co-expressed networks of genes and transcription factors, etc. to analyze plant development as well as stress response.

10.3 Resource Generation

10.3.1 Gene Inventory and Function

Genome sequencing is a key step which may lead to our understanding of genetic organization and functions of genes. However, the genomes of most plants are quite large making genome sequencing very expensive. In contrast the sequencing of purified mRNA (mRNA-seq) can be used for the identification of genes and for providing information such as gene expression pattern and indications of epigenetic regulation. Thus mRNA-seq datasets can provide information about gene sequences and gene expression as well (Bancroft 2013). Transcriptome, as we define it, represents the complete set of RNAs encoded by a genome of a specific cell or an organism at a specific time or under specific conditions. Thus generation of sequence tags by transcriptomic techniques including next-generation sequencing (NGS) by 454 and Illumina platforms can be used as an effective method of gene discovery. cDNA libraries constructed from various tissues and the sequencing followed by annotation of the genes obtained thereof can lead to the discovery of genes associated with the specific tissues (Wellmer et al. 2006). Also construction of subtracted libraries from plants subjected to various stresses has led to the identification of many stress-related genes. Many studies have been reported where genes associated with perception of pathogen signals and other signaling pathway components were identified using transcriptomic approach.

Over the last few years, the plant EST database generation is observing a surge at an exponential rate, and these EST databases have become a major source of plant sequence data. At present there are more than two million plant-derived ESTs from various species available at public databases. These data have provided a rich resource for gene discovery and annotation (Rudd 2003). Additional information can be obtained from these collections by comparing ESTs from multiple species (Fulton et al. 2002; Vincentz et al. 2004). A large number of databases from

Table 10.1 Major resources/database for plant EST data

Name of database/ organization	URL	References
NCBI dbEST	http://www.ncbi.nlm.nih.gov/dbEST/	Boguski et al. (1993)
NCBI UniGene	http://www.ncbi.nlm.nih.gov/entrez/query.fcgi?db=unigene	Pontius et al. (2003)
PlantGDB	http://www.plantgdb.org	Dong et al. (2004)
Sputnik	http://sputnik.btk.f	Rudd et al. (2003)
TIGR	http://www.tigr.org/tdb/tgi/plant.shtm	Ouyang and Buell (2004)
Genoplante	http://genoplante-info.infobiogen.fr/	Samson et al. (2003)
Gene Expression Omnibus	http://www.ncbi.nlm.nih.gov/geo/	Edgar et al. (2002)
ArrayExpress	http://www.ebi.ac.uk/microarray-as/ae/	Brazma et al. (2003)
Genevestigator	http://www.genevestigator.com/	Zimmermann et al. (2004) and Hruz et al. (2008)
The Botany Array Resource (BAR)	http://bbc.botany.utoronto.ca/	Toufighi et al. (2005)
ATTED-II	http://atted.jp/	Obayashi et al. (2007)
AtGenExpress/PRIME	http://prime.psc.riken.jp/	Akiyama et al. (2008)
PlantTFDB	http://planttfdb.cbi.pku.edu.cn/	Jin et al. (2014)

different sources with different experimental conditions per dataset have been developed in the last couple of years (see Table 10.1).

Recently van Verk et al. (2011) developed gene co-expression networks by using publicly available *Arabidopsis* microarray datasets. They reported that many genes previously reported in literature to be relevant for stress responses fit current models of stress gene regulation. Despite this potential, the resources available for comparative EST analyses in plants remain limited and dispersed. Further studies on different plant-pathogen model systems based on bioinformatics approach to increase the resource database will benefit further characterization of the genes, regulatory factors, and signal transduction pathways involved in plant defense.

10.3.2 Generation of Functional Molecular Markers

The RNA-seq or transcriptome data has helped in the development of molecular markers from the transcribed regions of the genome. Among the important molecular markers that can be developed from transcriptome are single-nucleotide polymorphisms (SNPs) (Rafalski 2002) and simple sequence repeats (SSRs) also called as microsatellites (Varshney et al. 2005). Putative functions of molecular markers can be deduced using homology searches (BLASTX) with protein databases and are therefore known as "functional markers" (FMs) (Gupta and Rustgi 2004). FMs have some advantages over random markers (RMs) because FMs are completely linked to the desired trait allele. Functional markers can be derived from the gene responsible for the trait of interest and target the functional polymorphism in the gene, thus allowing selection in different genetic backgrounds. These markers will be helpful to identify loci controlling important traits like disease resistance and can be used for the development of improved or disease-resistant cultivars.

Simple sequence repeat (SSR) markers are widely used in plants and have various uses including development of linkage map, quantitative trait loci (QTL) mapping, and marker-assisted selection. They are also used in evolutionary studies (Cavagnaro et al. 2010; Zhu et al. 2012). SSRs are tandem arrays of one to six nucleotides and occur in genomes of all prokaryotes and eukaryotes (Buschiazzo and Gemmell 2006; Kelkar et al. 2008). These tandem arrays have a high rate of mutation, and therefore, the number of repeat units varies, thereby resulting in highly polymorphic SSRs. Isolation of SSRs by traditional methods is a costly and labor-intensive process. With the development of NGS technologies using 454 or Illumina platforms, the sequencing of large portions of plant genomes can be done easily at low costs, thus allowing the rapid development of molecular markers, including microsatellites or SSRs (Ekblom and Galindo 2011). Recent studies have shown the efficient use of NGS technology for discovery of SSR loci in plants. Further, it has an advantage of developing SSR markers associated with functional genes and therefore represent specific phenotypes (Li et al. 2002).

SSRs are widely used in crop breeding programs, as large numbers of molecular markers linked to disease resistance traits are available in most crop species (Miah et al. 2013). For example, SSRs and SNPs also form an important class of functional molecular markers. NGS technology has potential to identify biologically significant SNPs. These techniques identify SNPs from plant populations (different varieties/genotypes) associated with a particular trait like disease resistance. Hence these can be used as toolbox aiding in selection and management of important traits like disease resistance in plant populations.

10.4 Unraveling the Defense Signaling Pathways Using Transcriptomics

The advent of whole transcriptome sequencing techniques presented a new dimension to biological studies and brought a paradigm shift from single gene analysis to whole transcriptome analysis across a wide spectrum of biological systems. Instead of focusing on single gene studies, transcriptomics allows understanding of whole transcriptome changes across a spectrum of biological conditions, resulting in a massive accumulation of gene database. Transcriptomics helps in gaining information about the default and responsive expression states of the individual cells and tissues so as to understand how these states collectively form a functioning organism. Transcriptional profiling has become a component of a biologist's toolbox (Brady et al. 2006). Plants establish multiple interactions with different microorganisms in natural environments; therefore multi-species transcriptomics may lead to the discovery of key plant and microbial genes regulating these multi-species cross talks. Such transcriptomic studies will be pivotal in plant defense signaling research in elucidating the key mechanisms by which plants respond to microbial infections and the counterresponse of microbes to plant defense signals. Defense responses are regulated by various signaling molecules including phytohormones [salicylic acid (SA), ethylene, jasmonic acid (JA)], nitric oxide (NO), reactive oxygen species (ROS), phytoalexins, etc. which vary as per the type of organism that interacts with the plant. The activation of a specific set of genes resulting in defense response is very much determined by the type of attacker encountered (De Vos et al. 2005; Mur et al. 2006; Ranjan et al. 2014). ROS and NO are important signaling molecules during the hypersensitive response (HR) and work together to bring about localized plant cell death as a defense response. SA has been observed to be involved in defense response against many plant pathogens including viruses, bacteria, biotrophic fungi, and phloem-feeding insects (De Vos et al. 2005; Glazebrook 2005). Ethylene and JA signaling pathways work synergistically against insects, necrotrophic fungi, and bacteria as a defense response (Reymond et al. 2004; De Vos et al. 2005; Bodenhausen and Reymond 2007; Ranjan et al. 2014). Reymond et al. (2004) estimated that 67–84% of *Arabidopsis thaliana*'s transcriptional responses to *Pieris rapae* were JA-mediated. Several of these genes code for proteins involved in anti-insect defenses. However, when *Arabidopsis* was given combined stress of drought and herbivory, the plant defense genes against biotrophic pathogens (e.g., *PR2*, *PR5*, *RLP39*, *RLP41*, *WAK3*) were downregulated (Olivas et al. 2016). This was in conformity with various reports that abiotic stresses have a negative impact on plant defense against pathogens (Suzuki et al. 2014; Ramegowda and Senthil-Kumar 2015). For plants exposed to combined stress imposed by *B. cinerea* and *P. rapae*, there was initial upregulation of *ERF104* and *BAP1* gene. ERF104 encodes a transcription factor that is involved in ET-mediated responses through interaction with MPK6 (Bethke et al. 2009), and BAP1 encodes a negative regulator of plant defenses and is required for growth homeostasis under normal conditions (Yang et al. 2007). Several key regulatory proteins functioning as a molecular switch between SA and JA cross talk have been identified in *Arabidopsis* (Li et al. 2004). Furthermore, PYL4 and PYL5 have been identified as components of the cross talk between the JA and ABA signaling pathways (Lackman et al. 2011). Thus in depth understanding of the type of response and the genes expressed has been achieved with the help of transcriptomic approaches. Various defense genes involved in different signaling pathways as elucidated by various transcriptomic-based studies are given in Table 10.2.

Further, transcriptomic studies have shown that plants have evolved a powerful regulatory mechanism by cross talk among hormonal signaling pathways and pathways involving metabolism, development, and reactive oxygen species synthesis, to effectively adapt to the complex stress situation (Narusaka et al. 2003; Brady et al. 2006; van Verk et al. 2011; Schenk et al. 2012; Maleck et al. 2000). However, unraveling the complexity of the mechanisms underlying these molecular cross talks and their role in the plant's response to the hostile environment is a challenge in the field of molecular plant-microbe interactions, and transcriptomics hold an important portfolio in this aspect.

Table 10.2 Defense signaling genes involved in various signaling pathways

Signaling pathways	Genes involved	References
MAPK pathway	MAPKKK, MEKK1, MKK4/MKK5, MPK3/MPK6, WRKY22, WRKY29, etc.	Asai et al. (2002)
JA pathway	MYC2, JAZ2, JAZ3, JMT, AOS, WRKY18/53, WRKY54/70, OPR3, etc.	McGrath et al. (2005) and Thines et al. (2007)
ET pathway	ACS2, ACS4, ACS5, ACS6, ACO, ETR1, EIN2, EIN3, EOL1, ETO1, etc.	Chang (2003), Leon-Reyes et al. (2009), and An et al. (2010)
SA pathway	EDS1, PADA, ICS1, PBS3, WRKY28, WRKY46, etc.	Broderson et al. (2006) and van Verk et al. (2011)

10.5 Identification of Promoter Elements and Their Targets

Promoters are the regions on the DNA that regulate the gene expression at the transcriptional level and identification of the promoter elements, and their targets are crucial for improving the understanding of gene regulation. Promoters of protein-encoding genes often contain a "core promoter," which is a region located ~40 bp upstream of the transcriptional initiation site and comprises the TATA box, which is the binding site for the transcription initiation factor TFIID TBP (TATA-box-binding protein) subunit (Molina and Grotewold 2005). Upstream of the core promoter lie the proximal and distal regions of the promoter containing different regulatory sequences such as enhancers, silencers, insulators, and *cis*-acting elements that act as binding sites for the basic transcriptional machinery involved in the initiation and regulation of transcription (Lee and Young 2000; Hernandez-Garcia and Finer 2014). The modulation of gene expression during transcription by proximal promoter elements is straightforward due to their close proximity to the core promoter. The distal promoter elements involve DNA folding mediated by conformational changes in the three-dimensional structure of DNA and chromatin (Hernandez-Garcia and Finer 2014). Numerous genes induced in response to pathogen attack have been identified, and their promoters have also been characterized. These promoters contain specific cis-regulatory elements {W boxes, Box S (GCC-like elements), and D box} which are involved in inducing an anti-pathogen molecular cascade (Rushton et al. 2002). In recent times, many regulatory interactions between transcription factors (TFs) and the promoters of their target genes have been determined by various methods, for example, in vitro (by electrophoretic mobility shift assay and yeast one-hybrid) (Cai et al. 2008) or in vivo (by chromatin immune precipitation) (Aerts et al. 2008; Calo and Wysocka 2013). Recently two pathogen-responsive *cis*-elements, PRE2 and PRE4, were identified from the promoter region of OsWRKY13 in rice. The two *cis*-elements negatively regulate gene expression without pathogen challenge and positively regulate gene expression after pathogen-induced protein binding (Cai et al. 2008). However, it has been observed that a large number of genes in eukaryotes are not regulated by single promoters but multiple alternative promoters. For instance the MAP kinase gene *OsBWMK1* and the *LAGGING GROWTH DEVELOPMENT 1* (LGD1) genes in rice are differentially

Table 10.3 Pathogen-inducible promoters isolated from plants

Promoters	Plant species	References
Defensin promoters	All plants	Kovalchuk et al. (2010)
CaMV 35S	Cauliflower mosaic virus	Odell et al. (1985)
OsPR10a	Rice (*Oryza sativa*)	Hwang et al. (2008)
Germin-like GER4	Barley (*Hordeum vulgare* L.)	Himmelbach et al. (2010)
PPP1	Tobacco (*Nicotiana tabacum*)	Peng et al. (2004)
hsr203J	Tobacco (*Nicotiana tabacum*)	Pontier et al. (1994)
str246C	Tobacco (*Nicotiana tabacum*)	Gough et al. (1995)
gst1	Potato (*Solanum tuberosum*)	Martini et al. (1993)
sgd24	Tobacco (*Nicotiana tabacum*)	Malnoy et al. (2003)
Ypr10	*Malus domestica*	Pühringer et al. (2000)
VpSTS promoter	Chinese wild (*Vitis pseudoreticulata*)	Xu et al. (2010)
PR2	Parsley (*Petroselinum crispum*)	van de Löcht et al. (1990)
PRE2 and PRE4	Rice (*Oryza sativa*)	Cai et al. (2008)
EL117	Parsley (*Petroselinum crispum*)	Kirsch et al. (2001)

expressed due to usage of alternative promoters (Koo et al. 2009; Thangasamy et al. 2012). Synthetic promoters developed by combinatorial engineering of *cis*-elements of the native promoters have been pivotal in signaling and transcriptional activation. Synthetic promoters responsive to pathogen attack demonstrate that defense signaling is largely conserved across species at the promoter level (Rushton et al. 2002). An extensive review on the importance of synthetic promoters reported *in planta* has been recently published by Dey et al. (2015). Similarly minimal promoters containing pathogenesis-related elements (PR1), salicylic acid-responsive elements (SARE), jasmonic acid-responsive elements (JARE), and ethylene-responsive elements (ERE) involved in pathogen stress response were characterized using *Agrobacterium*-mediated transient expression assay (Hernandez-Garcia and Finer 2014).

Defense response promoters have great biotechnological applications in developing transgenic crop plants which show disease resistance. Though constitutive promoters, like CaMV 35S promoter from the cauliflower mosaic virus (Odell et al. 1985), have been commonly used in developing transgenic plants, they can have metabolic cost associated which may eventually impact the traits of interest like yield and biomass. The ideal pathogen-inducible promoters would be rapidly activated by a wide array of pathogens and deactivated under disease-free conditions. So far various pathogen-inducible promoters have been isolated from numerous pathogen-responsive genes in plants as given in Table 10.3.

A large database of plant promoters is available from different sources like PlantCARE (http://bioinformatics.psb.ugent.be/webtools/plantcare/html/) (Lescot et al. 2002), PlantProm (http://mendel.cs.rhul.ac.uk/) (Shahmuradov et al. 2003), PLACE (http://www.dna.affrc.go.jp/PLACE/) (Higo et al. 1998), PlantPAN (http://PlantPAN. mbc. nctu.edu.tw.) (Chang et al. 2008), TRANSFAC (http://transfac.gbf-braunschweig.de) (Wingender et al. 1996), and Eukaryotic Promoter Database (EPD) (http://www.epd.isb-sib.ch) (Perier et al. 1998). However, there is a need to increase the current toolbox of promoters and cis-elements to gain an understanding of the gene regulation and action in basic and applied studies. Genome-wide transcriptome analyses using high-throughput sequencing technologies will not only lead to better understanding of regulation of gene expression but also to the identification of novel promoters and *cis*-elements in plants.

10.6 Transcription Factors as Drivers of Transcriptional Regulatory Networks

Transcription factors (TFs) and transcriptional regulatory networks play key roles in both development and stress responses by acting as on/off switches for gene transcription. Multiple signaling pathways regulate stress response, and there is significant overlap between the gene expression patterns induced in response to different stresses (van Verk et al. 2011; Jin et al. 2015). TFs act as molecular switches in these multiple signaling pathways and by integrating and rewiring these pathways (Jin et al. 2015). There are almost 1500 genes encoding transcription factors (Czechowski et al. 2004) in *Arabidopsis*, and among these about 350 genes are involved in regulation of these defense-signaling pathways as per the *Arabidopsis* transcriptional regulatory map (ATRM) (Jin et al. 2015). The four prominent families of TFs involved in stress response are ethylene-responsive-element-binding factors (ERF), basic leucine zipper (bZIP) domain, WRKY proteins, and MYB proteins.

ERF proteins belong to subfamily of the APETALA2 (AP2)/ethylene-responsive-element-binding protein (EREBP). ERF proteins share a conserved 58–59-amino acid domain (the ERF domain) that binds to two similar cis-elements: the GCC box. This box is found in several PR (pathogenesis-related) gene promoters and confers ethylene responsiveness. The ERFs are also involved in dehydration and cold responsive gene expression moderated by the C-repeat (CRT)/dehydration-responsive element (DRE) motif. There are about 124 ERF proteins in *Arabidopsis* (Riechmann et al. 2000). It is reported that octadecanoid-responsive-Catharanthus-APETALA2-domain proteins (ORCAs) form a link between JA and the production of secondary metabolites for defense (van der Fits and Memelink 2000).

bZIPs form a large family of transcription factors in plants, and there are 75 members in *Arabidopsis* (Jakoby et al. 2002). TGA/*octopine synthase (ocs)*-element-binding factor (OBF) proteins (a member of bZIP) bind to the *activation sequence-1 (as-1)/ocs* element and regulate the expression of some stress-responsive genes such as the *PR-1* and *GLUTATHIONE S-TRANSFERASE 6* (*GST6*) genes (Lebel et al. 1998; Chen and Singh 1999). *Arabidopsis* has seven members of the TGA/OBF family, which play roles in different stress responses.

WRKY proteins form a large family with 74 WRKY proteins identified in *Arabidopsis* and 109 in rice, 66 in papaya, 104 in poplar, 68 in sorghum, and 38 in the moss *Physcomitrella patens* (Eulgem et al. 2000; Robatzek and Somssich 2002; Qu and Zhu 2006; Pandey and Somssich 2009). WRKY proteins share a conserved 60-amino acid domain (WRKY domains) containing the amino acid sequence WRKYGQK and a zinc fingerlike motif. WRKY family members show enhanced expression in response to a range of pathogens, defense signals, and wounding (reviewed in Eulgem et al. 2000). WRKY transcription factors bind to W box [(T)TGAC(C/T)] (Rushton et al. 1996) which forms a major class of cis-acting elements responsible for the pathogen inducibility of many plant genes (Rushton et al. 1996; Wang et al. 1998). It was reported that the expression profiling of *Arabidopsis* WRKY genes revealed that majority of them were differentially regulated in response to SA treatment or infection by a bacterial pathogen (Dong et al. 2003). Recently co-expression analysis showed that the WRKY transcription factors involved in SA-mediated defense signaling and the co-expression network comprises W22/29, W11/25, W18/53, W30/33/40, W18/53, W48, W28/46, and W22, while in JA-mediated defense signaling, the co-expression network comprises W25/33, W54, W46, W11/48, W28, W18/40, and W53/70 (van Verk et al. 2011). The role of WRKY transcription factors in plant defense has been extensively reviewed by Pandey and Somssich (2009). Some WRKY genes involved in biotic stress responses in plants are given in Table 10.2.

MYB TF family is involved in regulating many processes like responses to biotic and abiotic stresses, development, differentiation, metabolism, defense, etc. MYB proteins have a highly conserved MYB DNA-binding domain at N-terminus and 50–53-amino acid repeats encoding three a-helix structures (Lipsick 1996). However, C-terminus is the activation domain and varies significantly between MYB proteins. This variation results in a wide range of regulatory roles of MYB gene family (Jin and Martin 1999; Dubos et al. 2010; Muthamilarasan et al. 2014). Co-expression analysis of *Arabidopsis* transcription factors showed involvement of MYB15 and MYB32 in SA signaling pathway; MYB2, MYB29, and MYB15/95 in JA signaling pathway; and MYB36/38, MYB39/65, MYB43/55, and MYB61 in ET signaling pathway (van Verk et al. 2011). Some MYB genes involved in biotic stress responses in plants reported so far are given in Table 10.4.

10.7 Role of Small RNAs in Plant Immunity

Small RNAs are 20–40-nucleotide-long noncoding RNA molecules, generated by endoribonucleases DICER or DICER-like (DCL), and are present in most eukaryotic organisms. They regulate gene expression at either transcriptional or posttranscriptional level (Katiyar-Agarwal and Jin 2010) in many biological processes including development, metabolism, and biotic and abiotic stress responses (Katiyar-Agarwal and Jin 2010; Westermann et al. 2016). Recently, mobile small RNAs (sRNAs) have been indicated to have a critical role in the regulation of transportation of small regulatory molecules across the cellular boundaries, between the host and its interacting microbial partner in plant-microbe interactions (Westermann et al. 2016). Some of the sRNA and their targets reported so far, derived from the plant and microbial partner (pathogen) under different pathosystems, are given in Table 9.1 (Chap. 9).

Study of regulatory mechanisms involving small RNAs in plant defense is an emerging field, and with the advent of new technologies like high-throughput

Table 10.4 Some WRKY and MYB proteins and their plant defense response in plants

Name of the transcription factors	Plant species	Functions	References
AtWRKY16	*Arabidopsis thaliana*	Increase the expression of PR genes	Chen and Chen (2002)
AtWRKY18	*Arabidopsis thaliana*	Increase the expression of PR genes	Robatzek and Somssich (2002)
AtWRKY70	*Arabidopsis thaliana*	SA- and JA-induced resistance pathways	Li et al. (2004)
AtWRKY38 and AtWRKY62	*Arabidopsis thaliana*	Contribute negatively to basal resistance toward this bacterial pathogen	Kim et al. (2008)
AtWRKY3 and AtWRKY4	*Arabidopsis thaliana*	Induce plant resistance toward necrotrophic pathogens	Lai et al. (2008)
OsWRKY31	*Oryza sativa*	Enhanced resistance to fungal blast	Zhang et al. (2008)
OsWRKY13	*Oryza sativa*	Activates SA-biosynthesis and SA-response genes while suppressing JA signaling	Qiu et al. (2008)
OsWRKY03	*Oryza sativa*	Defense regulation	Liu et al. (2005)
PcWRKY1	*Petroselinum crispum*	Induces the expression of PR10 class gene, PcPR1-1, and represses PcWRKY3	Turck et al. (2004)
OsWRKY13	*Oryza sativa*	Regulates rice resistance to bacterial blight and fungal blast	Wen et al. (2003)
MYB1	*Nicotiana tabaccum*	Plant defense response against TMV	Liu et al. (2004)
TiMYB2R-1	*Thinopyrum intermedium*	Plant defense response against *Gaeumannomyces graminis*	Liu et al. (2013)
AtMYB44	*Arabidopsis thaliana*	Plant defense response against aphid	Liu et al. (2010)
AtMYB060/and AtMYB094	*Arabidopsis thaliana*	Biotic stress response	Cominelli et al. (2005)
MTF1	*Arabidopsis thaliana*	Regulates susceptibility against *Agrobacterium*	Sardesai et al. (2014)
AtMYB30, AtMYB44, AtMYB108/BOSI1	*Arabidopsis thaliana*	Plant defense response	Buscaill and Rivas (2014)
Yellow seed1 (y1)	*Zea mays*	Plant defense response against *Colletotrichum sublineolum*	Ibraheem et al. (2015)
AtMYB108	*Arabidopsis thaliana*	Biotic stress response	Mengiste et al. (2003)
TaPIMP1	*Triticum aestivum*	Plant defense response against *Bipolaris sorokiniana*	Zhang et al. (2012)
AtMYB15, AtMYB34, AtMYB51, and AtMYB75	*Arabidopsis thaliana*	Plant defense response against insect herbivore	Cheong et al. (2002) and Johnson and Dowd (2004)

(continued)

Table 10.4 (continued)

Name of the transcription factors	Plant species	Functions	References
AtMYB102	*Arabidopsis thaliana*	Plant defense response against insect herbivore *Pieris rapae*	De Vos et al. (2006)
OsJAMyb	*Oryza sativa*	Plant defense response against *Magnaporthe oryza*	Cao et al. (2015)
AtMYB96	*Arabidopsis thaliana*	Biotic stress response	Seo and Park (2010)
AtMYB72	*Arabidopsis thaliana*	Induced systemic resistance mediated by beneficial fungi and bacteria	Segarra et al. (2009)

next-generation sequencing, many more endogenous plant sRNA and pathogen-derived sRNAs will be identified in the future. Characterization of these small RNAs and their target genes will help in revealing new dimensions in plant defense signaling pathways and will ultimately lead to the development of effective tools for controlling diseases in plants.

10.8 Conclusion

Transcriptional reprogramming is important in the context of plant defense system, major challenge being to discriminate between gene expression associated with MTI and that orchestrated by effectors (Lewis et al. 2015). It is quite important to understand the responses of plant hosts to microbial infections so as to develop strategies for disease control. Since plant responses are complex, system-level transcriptomic studies will help in understanding these responses (Van Verk et al. 2013; Jin et al. 2015; Lewis et al. 2015). This will lead to understanding of whole transcriptome changes across a spectrum of biological conditions, resulting in development of comprehensive gene database. The latest transcriptomic tools able to determine gene expression pattern would result in development of knowledge-base about the up- or downregulation of genes in response to various stresses (Brady et al. 2006). Co-expression gene networks and transcriptional regulatory networks could be built using bioinformatics approach, therefore adding to this knowledge-base (Van Verk et al. 2011; Jin et al. 2015). Fifty years ago no one could have guessed that we would have the tools to answer these questions. We need to follow global approach well suited for the analysis of plant-pathogen interactions, and in this context integration of genomics, proteomics, and metabolomics with statistics will refine our understanding of how the transcriptome gives rise to biological form and function.

References

Aerts S, Haeussler M, van Vooren S, Griffith OL, Hulpiau P, Jones SJ, Montgomery SB, Bergman CM (2008) Text-mining assisted regulatory annotation. Genome Biol 9:R31

Akiyama K, Chikayama E, Yuasa H, Shimada Y, Tohge T, Shinozaki K, Hirai MY, Sakurai T, Kikuchi J, Saito K (2008) PRIMe: a web site that assembles tools for metabolomics and transcriptomics. In Silico Biol 8:0027

An F, Zhao Q, Ji Y, Li W, Jiang Z, Yu X, Zhang X, Zhang C, Han Y, He W, Liu Y, Zhang S, Ecker JR, Guo H (2010) Ethylene-induced stabilization of ETHYLENE INSENSITIVE3 and EIN3-LIKE1 is mediated by proteasomal degradation of EIN3 binding F-box 1 and 2 that requires EIN2 in Arabidopsis. Plant Cell 22:2384–2401

Asai T, Tena G, Plotnikova J, Willmann MR, Chiu WL, Gomez-Gomez L, Boller T, Ausubel FM, Sheen J (2002) MAP kinase signalling cascade in Arabidopsis innate immunity. Nature 415:977–983

Bancroft I (2013) Association genetics and more from crop transcriptome sequences ISB news report

Bethke G, Unthan T, Uhrig JF, Poschl Y, Gust AA, Scheel D, Lee J (2009) Flg22 regulates the release of an ethylene response factor substrate from MAP kinase 6 in Arabidopsis thaliana via ethylene signaling. PNAS, USA 106:8067–8072

Bodenhausen N, Reymond P (2007) Signaling pathways controlling induced resistance to insect herbivores in Arabidopsis. Mol Plant Microb Interact 20:1406–1420

Boguski MS, Lowe TM, Tolstoshev CM (1993) dbEST – database for "expressed sequence tags". Nat Genet 10:369–371

Brady SM, Long TA, Benfey PN (2006) Unraveling the dynamic transcriptome. Plant Cell 18(9):2101–2111

Brazma A, Parkinson H, Sarkans U, Shojatalab M, Vilo J, Abeygunawardena N, Holloway E, Kapushesky M, Kemmeren P, Lara GG, Oezcimen A, Rocca-Serra P, Sansone SA (2003) ArrayExpress – a public repository for microarray gene expression data at the EBI. Nucleic Acids Res 31:68–71

Brodersen P, Petersen M, Nielsen HB, Zhu S, Newman MA, Shokat KM, Rietz S, Parker J, Mundy J (2006) Arabidopsis MAP kinase 4 regulates salicylic acid- and jasmonic acid/ethylene-dependent responses via EDS1 and PAD4. Plant J 47:532–546

Buscaill P, Rivas S (2014) Transcriptional control of plant defence responses. Curr Opin Plant Biol 20C:35–46

Buschiazzo E, Gemmell NJ (2006) The rise, fall and renaissance of microsatellites in eukaryotic genomes. BioEssays 28:1040–1050

Cai M, Qiu D, Yuan T, Ding X, Li H, Duan L, Xu C, Li X, Wang S (2008) Identification of novel pathogen-responsive cis-elements and their binding proteins in the promoter of OsWRKY13, a gene regulating rice disease resistance. Plant Cell Environ 31(1):86–96

Calo E, Wysocka J (2013) Modification of enhancer chromatin: what, how, and why? Mol Cell 49:825–837

Cao WL, Chu RZ, Zhang Y, Luo J, Su YY et al (2015) OsJAMyb, a R2R3-type MYB transcription factor, enhanced blast resistance in transgenic rice. Physiol Mol Plant Pathol 83:79–99

Cavagnaro PF, Senalik DA, Yang L, Simon PW, Harkins TT, Kodira CD, Huang S, Weng Y (2010) Genome-wide characterization of simple sequence repeats in cucumber (*Cucumis sativus* L.) BMC Genomics 11:569–586

Chang C (2003) Ethylene signaling: the MAPK module has finally landed. Trends Plant Sci 8:365–368

Chang WC, Lee TY, Huang HD, Huang HY, Pan RL (2008) PlantPAN: plant promoter analysis navigator, for identifying combinatorial cis-regulatory elements with distance constraint in plant gene groups. BMC Genomics 9:561

Chen C, Chen Z (2002) Potentiation of developmentally regulated plant defense response by AtWRKY18, a pathogen-induced *Arabidopsis* transcription factor. Plant Physiol 129:706–716

Chen W, Singh KB (1999) The auxin, hydrogen peroxide and salicylic acid induced expression of the *Arabidopsis* GST6 promoter is mediated in part by an *ocs* element. Plant J 19:667–677

Cheong YH, Chang HS, Gupta R, Wang X, Zhu T, Luan S (2002) Transcriptional profiling reveals novel interactions between wounding, pathogen, abiotic stress, and hormonal responses in Arabidopsis. Plant Physiol 129:661–677

Chung HS, Koo AJK, Gao X, Jayanty S, Thines B, Jones AD et al (2008) Regulation and function of *Arabidopsis*JASMONATE ZIM-domain genes in response to wounding and herbivory. Plant Physiol 146:952–964. https://doi.org/10.1104/pp.107.115691

Cominelli E et al (2005) A guard-cell-specific MYB transcription factor regulates stomatal movements and plant drought tolerance. Curr Biol 15:1196–1200

Czechowski T, Bari RP, Stitt M, Scheible W, Udvardi MK (2004) Real-time RT-PCR profiling of over 1400 Arabidopsis transcription factors: unprecedented sensitivity reveals novel root- and shoot-specific gene. Plant J 38:366–379

De Vos M, Van Oosten VR, Van Poecke RMP et al (2005) Signal signature and transcriptome changes of Arabidopsis during pathogen and insect attack. Mol Plant Microb Interact 18:923–937

De Vos M, Van Zaanen W, Koornneef A, Korzelius JP, Dicke M, Van Loon LC, Pieterse CMJ (2006) Herbivore-induced resistance against microbial pathogens in Arabidopsis. Plant Physiol 142:352–363

Dey N, Sarkar S, Acharya S, Maiti IB (2015) Synthetic promoters *in planta*. Planta 242:1077–1094

Diatchenko L, Lau YF, Campbell AP, Chenchik A, Moqadam F, Huang B, Lukyanov S, Lukyanov K, Gurskaya N, Sverdlov ED, Siebert PD (1996) Suppression subtractive hybridization: a method for generating differentially regulated or tissue-specific cDNA probes and libraries. Proc Natl Acad Sci U S A 93:6025–6030

Dittrich MT, Klau GW, Rosenwald A, Dandekar T, Müller T (2008) Identifying functional modules in protein–protein interaction networks: an integrated exact approach. Bioinformatics 24(13):i223–i231

Dong J, Chen C, Chen Z (2003) Expression profiles of the *Arabidopsis* WRKY gene superfamily during plant defense response. Plant Mol Biol 51(1):21–37

Dong Q, Schlueter SD, Brendel V (2004) PlantGDB, plant genome database and analysis tools. Nucleic Acids Res 32:354–359

Dou D, Zhou JM (2012) Phytopathogen effectors subverting host immunity: different foes, similar battleground. Cell Host Microbe 12(4):484–495

Dubos C, Stracke R, Grotewold E, Weisshaar B, Martin C et al (2010) MYB transcription factors in Arabidopsis. Trends Plant Sci 15:573–581

Edgar R, Domrachev M, Lash AE (2002) Gene expression omnibus: NCBI gene expression and hybridization array data repository. Nucleic Acids Res 30:207–210

Ekblom R, Galindo J (2011) Applications of next generation sequencing in molecular ecology of non-model organisms. Heredity 107(1):1–15

Eulgem T, Rushton PJ, Robatzek S, Somssich IE (2000) The WRKY superfamily of plant transcription factors. Trends Plant Sci 5:199–206

Fulton TM, Van der Hoeven R, Eannetta NT, Tanksley SD (2002) Identification, analysis, and utilization of conserved ortholog set markers for comparative genomics in higher plants. Plant Cell 14(7):1457–1467

Gough C, Hemon P, Tronchet M, Lacomme C, Marco Y, Roby D (1995) Developmental and pathogen-induced activation of an msr gene, str246C, from tobacco involves multiple regulatory elements. Mol Gen Genet 247(3):323–337

Glazebrook J (2005) Contrasting mechanisms of defense against biotrophic and necrotrophic pathogens. Annu Rev Phytopathol 43:205–227

Gupta PK, Rustgi S (2004) Molecular markers from the transcribed/expressed region of the genome in higher plants. Funct Integr Genom 4(3):139–162

Harlizius B, van Wijk R, Merks JW (2004) Genomics for food safety and sustainable animal production. J Biotechnol 113(1–3):33–42

Hernandez-Garcia CM, Finer JJ (2014) Identification and validation of promoters and cis-acting regulatory elements. Plant Sci 217–218:109–119

Hocquette JF (2005) Where are we in genomics. J Physiol Pharmacol 56:37–70

Higo K, Ugawa Y, Iwamoto M, Higo H (1998) PLACE: a database of plant *cis*-acting regulatory DNA elements. Nucleic Acids Res 26(1):358–359

Himmelbach A et al (2010) Promoters of the barley germin-like GER4 gene cluster enable strong transgene expression in response to pathogen attack. Plant Cell 22:937–952

Hruz T, Laule O, Szabo G, Wessendrop F, Bleuler S, Oertle L, Widmayer P, Gruissem W, Zimmermann P (2008) Genevestigator V3: a reference expression database for the meta-analysis of transcriptomes. Adv Bioinform e420747. doi:https://doi.org/10.1155/2008/420747

Hugot K, Riviere MP, Moreilhon C, Dayem MA, Cozzitorto J, Arbiol G, Barbry P, Weiss C, Galiana E (2004) Coordinated regulation of genes for secretion in tobacco at late developmental stages: association with resistance against oomycetes. Plant Physiol 134:858–870

Hwang SH, Lee IA, Yie SW, Hwang DJ (2008) Identification of an OsPR10a promoter region responsive to salicylic acid. Planta 227:1141–1150

Ibraheem F, Gaffoor I, Tan Q, Shyu CR, Chopra SA (2015) Sorghum MYB transcription factor induces 3-deoxyanthocyanidins and enhances resistance against leaf blights in maize. Molecules 20(2):2388–2404

Jakoby M, Weisshaar B, Droge-Laser W, Vicente-Carbajosa J, Tiedemann J, Kroj T, Parcy F (2002) bZIP transcription factors in *Arabidopsis*. Trends Plant Sci 7:106–111

Jin H, Martin C (1999) Multifunctionality and diversity within the plant MYB gene family. Plant Mol Biol 41:577–585

Jin J, Zhang H, Kong L, Gao G, Luo J (2014) PlantTFDB 3.0: a portal for the functional and evolutionary study of plant transcription factors. Nucleic Acids Res 42:D1182–D1187

Jin J, He K, Tang X, Li Z, Lv L, Zhao Y, Luo J, Gao G (2015) An Arabidopsis transcriptional regulatory map reveals distinct functional and evolutionary features of novel transcription factors. Mol Biol Evol 32(7):1767–1773

Johnson ET, Dowd PF (2004) Differentially enhanced insect resistance, at a cost, in Arabidopsis thaliana constitutively expressing a transcription factor of defensive metabolites. J Agric Food Chem 52:5135–5138

Katiyar-Agarwal S, Jin H (2010) Role of small RNAs in host-microbe interactions. Annu Rev Phytopathol 48:225–246

Kelkar YD, Tyekucheva S, Chiaromonte F, Makova KD (2008) The genome-wide determinants of human and chimpanzee microsatellite evolution. Genome Res 18:30–38

Kim K-C, Lai Z, Fan B, Chen Z (2008) *Arabidopsis* WRKY38 and WRKY62 transcription factors interact with histone deacetylase 19 in basal defense. Plant Cell 20:2357–2371

Kirsch C, Takamiya-Wik M, Schmelzer E, Hahlbrock K, Somssich IEA (2001) novel regulatory element involved in rapid activation of parsley ELI7 gene family members by fungal elicitor or pathogen infection. Mol Plant Pathol 1(4):243–251

Koo S et al (2009) Identification and characterization of alternative promoters of therice MAP kinase gene OsBWMK1. Mol Cells 27:467–473

Kovalchuk N, Li M, Wittek F, Reid N, Singh R, Shirley N, Ismagul A, Eliby S, Johnson A, Milligan AS, Hrmova M, Langridge P, Lopato S (2010) Defensin promoters as potential tools for engineering dis-ease resistance in cereal grains. Plant Biotechnol J 8:47–64

Kuhn K, Baker SC, Chudin E, Lieu MH, Oeser S, Bennett H, Rigault P, Barker D, McDaniel TK, Chee MS (2004) A novel, high-performance random array platform for quantitative gene expression profiling. Genome Res 14:2347–2356

Lackman P, Gonzalez-Guzman M, Tilleman S et al (2011) Jasmonate signaling involves the abscisic acid receptor PYL4 to regulate metabolic reprogramming in Arabidopsis and tobacco. PNAS, USA 108:5891–5896

Lai Z, Vinod KM, Zheng Z, Fan B, Chen Z (2008) Roles of Arabidopsis WRKY3 and WRKY4 transcription factors in plant responses to pathogens. BMC Plant Biol 8:68

Lebel E, Heifetz P, Thorne L, Uknes S, Ryals J, Ward E (1998) Functional analysis of regulatory sequences controlling *PR-1* gene expression in *Arabidopsis*. Plant J 16:223–233

Lee TI, Young RA (2000) Transcription of eukaryotic protein-coding genes. Ann Rev Genet 34:77–137

Leon-Reyes A, Spoel SH, De Lange ES, Abe H, Kobayashi M, Tsuda S, Millenaar FF, Welschen RAM, Ritsema T, Pieterse CMJ (2009) Ethylene modulates the role of NONEXPRESSOR OF PATHOGENESIS-RELATED GENES1 in cross talk between salicylate and jasmonate signaling. Plant Physiol 149:1797–1809

Lescot M, Déhais P, Gert Thijs G, Marchal K, Moreau Y, Van de Peer Y, Rouzé P, Rombauts S (2002) PlantCARE, a database of plant *cis*-acting regulatory elements and a portal to tools for in silico analysis of promoter sequences. Nucleic Acids Res 30(1):325–327

Lewis LA, Polanski K, Torres-Zabala M et al (2015) Transcriptional dynamics driving MAMP-triggered immunity and pathogen effector-mediated immunosuppression in Arabidopsis leaves following infection with *Pseudomonas syringae* pv tomatoDC3000. Plant Cell 27:3038–3064

Li YC, Korol AB, Fahima T, Beiles A, Nevo E (2002) Microsatellites: genomic distribution, putative functions and mutational mechanisms: a review. Mol Ecol 11:2453–2465

Li J, Brader G, Palva ET (2004) The WRKY70 transcription factor: a node of convergence for jasmonate-mediated and salicylate-mediated signals in plant defense. Plant Cell 16:319–331

Lipsick JS (1996) One billion years of Myb. Oncogene 13:223–235

Liu Y, Schiff M, Dinesh-Kumar SP (2004) Involvement of MEK1 MAPKK, NTF6 MAPK, WRKY/MYB transcription factors, COI1 and CTR1 in N-mediated resistance to tobacco mosaic virus. Plant J 38(5):800–809

Liu XQ, Bai XQ, Qian Q, Wang XJ, Chen MS, Chu CC (2005) OsWRKY03, a rice transcriptional activator that functions in defense signaling pathway upstream of OsNPR1. Cell Res 15:593–603

Liu R, Lü B, Wang X, Zhang C, Zhang S, Qian J, Chen L, Shi H, Dong H (2010) Thirty-seven transcription factor genes differentially respond to a hairpin protein and affect resistance to the green peach aphid in Arabidopsis. J Biosci 35:435–450

Liu X, Yang L, Zhou X, Zhou M, Lu Y, Ma L, Ma H, Zhang Z (2013) Transgenic wheat expressing *Thinopyrum intermedium* MYB transcription factor TiMYB2R-1 shows enhanced resistance to the take-all disease. J Exp Bot 64(8):2243–2253

Liu WD, Liu JL, Triplett L, Leach JE, Wang GL (2014) Novel insights into rice innate immunity against bacterial and fungal pathogens. Annu Rev Phytopathol 52:213–241

Maleck K, Levine A, Eulgem T, Morgan A, Schmid J, Lawton KA, Dangl JL, Dietrich RA (2000) The transcriptome of Arabidopsis thaliana during systemic acquired resistance. Nat Genet 26:403–410

Malnoy M, Venisse JS, Reynoird JP, Chevreau E (2003) Activation of three pathogen-inducible promoters of tobacco in transgenic pear (*Pyrus communis* L.) after abiotic and biotic elicitation. Planta 216(5):802–814

Marshall E (2004) Getting the noise out of gene arrays. Science 306:630–631

Martini N, Egen M, Rüntz I, Strittmatter G (1993) Promoter sequences of a potato pathogenesis-related gene mediate transcriptional activation selectively upon fungal infection. Mol GenGenet 236(2):179–186

Mathers JC (2004) What can we expect to learn from genomics? Proc Nutr Soc 63:1–4

McGrath KC, Dombrecht B, Manners JM, Schenk PM, Edgar CI, Maclean DJ, Scheible WR, Udvardi MK, Kazan K (2005) Repressor- and activator-type ethylene response factors functioning in jasmonate signaling and disease resistance identified via a genome-wide screen of Arabidopsis transcription factor gene expression. Plant Physiol 139:949–959

Mengiste T et al (2003) The BOTRYTIS SUSCEPTIBLE1 gene encodes an R2R3MYB transcription factor protein that is required for biotic and abiotic stress responses in Arabidopsis. Plant Cell 15:2551–2565

Miah G, Rafii MY, Ismail MR, Puteh AB, Rahim HA, KhN I, Latif MA (2013) A review of microsatellite markers and their applications in rice breeding programs to improve blast disease resistance. Int J Mol Sci 14:22499–22528

Molina C, Grotewold E (2005) Genome wide analysis of Arabidopsis core promoters. BMC Genomics 6:25. https://doi.org/10.1186/1471-2164-1186-1125

Mur LAJ, Kenton P, Atzorn R, Miersch O, Wasternack C (2006) The outcomes of concentration-specific interactions between salicylate and jasmonate signaling include synergy, antagonism, and oxidative stress leading to cell death. Plant Physiol 140:249–262

Muthamilarasan M, Khandelwal R, Yadav CB, Bonthala VS, Khan Y, Prasad M (2014) Identification and molecular characterization of MYB transcription factor superfamily in C4 model plant foxtail millet (*Setaria italica* L.) PLoS One 9(10):e109920

Narusaka Y et al (2003) The cDNA microarray analysis using an Arabidopsis pad3 mutant reveals the expression profiles and classification of genes induced by *Alternaria brassicicola* attack. Plant Cell Physiol 44:377–387

Obayashi T, Kinoshita K, Nakai K, Shibaoka M, Hayashi S, Saeki M, Shibata D, Saito K, Ohta H (2007) ATTED-II: a database of co-expressed genes and cis elements for identifying co-regulated gene groups in Arabidopsis. Nucleic Acids Res 35:D863–D869

Odell JT, Nagy F, Chua NH (1985) Identification of DNA sequences required for activ-ity of the cauliflower mosaic virus 35S promoter. Nature 313:810–812

Ouyang S, Buell CR (2004) The TIGR plant repeat databases: a collective resource for the identification of repetitive sequences in plants. Nucleic Acids Res 32:D360–D363

Olivas NHD, Coolen S, Huang P et al (2016) Effect of prior drought and pathogen stress on Arabidopsis transcriptome changes to caterpillar herbivory. New Phytol 210(4):1344–1356

Pandey SP, Somssich IE (2009) The role of WRKY transcription factors in plant immunity. Plant Physiol 50(4):1648–1655

Peng J, Bao Z, Li P, Chen G, Wang J, Dong H (2004) HarpinXoo and its functional domains activate pathogen-inducible plant promoters in Arabidopsis. Acta Bot Sin 46:1083–1090

Périer RC, Junier T, Bucher P (1998) The eukaryotic promoter database EPD. Nucleic Acids Res 26:353–357

Pontier D, Godiard L, Marco Y, Roby D (1994) hsr203J, a tobacco gene whose activation is rapid, highly localized and specific for incompatible plant/pathogen interactions. Plant J 5(4):507–521

Pontius JU, Wagner L, Schuler GD (2003) UniGene: a unified view of the transcriptome. In: The NCBI handbook. Bethesda. http://www.ncbi.nlm.nih.gov/

Puhringer H, Dieter Moll D, Hoffmann-Sommergruber K, Watillon B, Katinger H, Machado MLC (2000) The promoter of an apple Ypr10 gene, encoding the major allergen Mal d 1, is stress-and pathogen-inducible. Plant Sci 152:35–50

Qiu D, Xiao J, Xie W, Liu H, Li X, Xiong L, Wang S (2008) Rice gene network inferred from expression profiling of plants overexpressing OsWRKY13, a positive regulator of disease resistance. Mol Plant 1:538–551

Qu LJ, Zhu YX (2006) Transcription factor families in Arabidopsis: major progress and outstanding issues for future research. Curr Opin Plant Biol 9(5):544–549

Rafalski A (2002) Applications of single nucleotide polymorphisms in crop genetics. Curr Opin Plant Biol 5(2):94–100

Ramegowda V, Senthil-Kumar M (2015) The interactive effects of simultaneous biotic and abiotic stresses on plants: mechanistic understanding from drought and pathogen combination. J Plant Physiol 176:47–54

Ranjan A, Ichihashi Y, Farhi M, Zumstein K, Townsley B, David-Schwartz R, Sinha NR (2014) De novo assembly and characterization of the transcriptome of the parasitic weed dodder identifies genes associated with plant parasitism. Plant Physiol 166:1186–1199

Reymond P, Bodenhausen N, Van Poecke RMP, Krishnamurthy V, Dicke M, Farmer EE (2004) A conserved transcript pattern in response to a specialist and a generalist herbivore. Plant Cell 16:3132–3147

Riechmann JL, Heard J, Martin G, Reuber L, Jiang C, Keddie J, Adam L, Pineda O, Ratcliffe OJ, Samaha RR et al (2000) *Arabidopsis* transcription factors: genome-wide comparative analysis among eukaryotes. Science 290:2105–2110

Robatzek S, Somssich IE (2002) Targets of AtWRKY6 regulation during plant senescence and pathogen defense. Genes Dev 16:1139–1149

Rudd S (2003) Expressed sequence tags: alternative or complement to whole genome sequences? Trends Plant Sci 8(7):321–329

Rudd S, Mewes HW, Mayer KF (2003) Sputnik: a database platform for comparative plant genomics. Nucleic Acids Res 31:128–132

Rushton PJ, Torres JT, Parniske M, Wernert P, Hahlbrock K, Somssich IE (1996) Interaction of elicitor-induced DNA-binding proteins with elicitor response elements in the promoters of parsley PR1 genes. EMBO J 15(20):5690–5700

Rushton PJ, Reinstadler A, Lipka V, Lippok B, Somssich IE (2002) Synthetic plant promoters containing defined regulatory elements provide novel insights into pathogen- and wound-induced signaling. Plant Cell 14:749–762

Samson D, Legeai F, Karsenty E, Reboux S, Veyrieras JB, Just J, Barillot E (2003) GénoPlante-info (GPI): a collection of databases and bioinformatics resources for plant genomics. Nucleic Acids Res 31(1):179–182

Sardesai N, Laluk K, Mengiste T, Gelvin S (2014) The Arabidopsis Myb transcription factor MTF1 is a unidirectional regulator of susceptibility to Agrobacterium. Plant Signal Behav 30:9

Schena M, Shalon D, Davis RW, Brown PO (1995) Quantitative monitoring of gene-expression patterns with a complementary- DNA microarray. Science 270:467–470

Schenk PM, Carvalhais LC, Kazan K (2012) Unraveling plant-microbe interactions: can multi-species transcriptomics help? Trends Biotechnol 30(3):177–184

Segarra G, Van der Ent S, Trillas I, Pieterse CMJ (2009) MYB72, a node of convergence in induced systemic resistance triggered by a fungal and a bacterial beneficial microbe. Plant Biol 11:90–96

Seo PJ, Park CM (2010) MYB96-mediated abscisic acid signals induce pathogen resistance response by promoting salicylic acid biosynthesis in Arabidopsis. New Phytol 186:471–483

Shahmuradov IA, Gammerman AJ, Hancock JM, Bramley PM, Solovyev VV (2003) PlantProm: a database of plant promoter sequences. Nucleic Acids Res 31(1):114–117

Suzuki N, Rivero RM, Shulaev V, Blumwald E, Mittler R (2014) Abiotic and biotic stress combinations. New Phytol 203:32–43

Thangasamy S, Chen PW, Lai MH, Chen J, Jauh GY (2012) Rice LGD1 contain-ing RNA binding activity affects growth and development through alternative promoters. Plant J 71:288–302

Thines B, Katsir L, Melotto M, Niu Y, Mandaokar A, Liu G, Nomura K, He SY, Howe GA, Browse J (2007) JAZ repressor proteins are targets of the SCFCOI1 complex during jasmonate signalling. Nature 448:661–665

Toufighi K, Brady SM, Austin R, Ly E, Provart NJ (2005) The botany array resource: e-Northerns, expression angling, and promoter analyses. Plant J 43:153–163

Tsuda K, Katagiri F (2010) Comparing signaling mechanisms engaged in pattern-triggered and effector-triggered immunity. Curr Opin Plant Biol 13:459–465

Tsuda K, Somssich IE (2015) Transcriptional networks in plant immunity. New Phytol 206(3):932–947

Turck F, Zhou A, Somssich IE (2004) Stimulus-dependent, promoter-specific binding of transcription factor WRKY1 to its native promoter and the defense related gene *PcPR1-1* in parsley. Plant Cell 16:2573–2585

van de Löcht U, Meier I, Hahlbrock K, Somssich IE (1990) A 125 bp promoter fragment is sufficient for strong elicitor-mediated gene activation in parsley. EMBO J 9(9):2945–2950

van der Fits L, Memelink J (2000) ORCA3, a jasmonate-responsive transcriptional regulator of plant primary and secondary metabolism. Science 289:295–297

van Verk MC, Bol JF, Linthorst HJM (2011) Prospecting for genes involved in transcriptional regulation of plant defenses, a bioinformatics approach. BMC Plant Biol 11:88

Van Verk MC, Hickman R, Pieterse CMJ, Van Wees SCM (2013) RNA-seq: revelation of the messengers. Trends Plant Sci 18:175–179

Varshney RK, Graner A, Sorrells ME (2005) Genic microsatellite markers in plants: features and applications. Trends Biotechnol 23(1):48–55

Vincentz M, Cara FAA, Okura VK et al (2004) Evaluation of monocot and eudicot divergence using the sugarcane transcriptome. Plant Physiol 134:951–959. https://doi.org/10.1104/pp.103.033878

Wang Z, Yang P, Fan B, Chen Z (1998) An oligo selection procedure for identification of sequence-specific DNA-binding activities associated with plant defense. Plant J 16:515–552

Wellmer F, Alves-Ferreira M, Dubois A, Riechmann JL, Meyerowitz EM (2006) Genome-wide analysis of gene expression during early *Arabidopsis* flower development. PLoS Genet 2(7):e117. https://doi.org/10.1371/journal.pgen.0020117

Wen N, Chu Z, Wang S (2003) Three types of defense-responsive genes are involved in resistance to bacterial blight and fungal blast diseases in rice. Mol Genet Genomics 269:331–339

Westermann AJ, Förstner KU, Amman F, Barquist L, Chao Y, Schulte LN, Müller L, Reinhardt R, Stadler PF, Vogel J (2016) Dual RNA-seq unveils noncoding RNA functions in host-pathogen interactions. Nature 529(7587):496–501

Wingender E, Dietze P, Karas H, Knüppel R (1996) TRANSFAC: a database on transcription factors and their DNA binding sites. Nucleic Acids Res 24(1):238–241

Xu W, Yu Y, Ding J, Hua Z, Wang Y (2010) Characterization of a novel stilbene synthase promoter involved in pathogen- and stress-inducible expression from Chinese wild *Vitis pseudoreticulata*. Planta 231:475–487

Yang HJ, Yang SH, Li YQ, Hua J (2007) The Arabidopsis BAP1 and BAP2 genes are general inhibitors of programmed cell death. Plant Physiol 145:135–146

Zhang J, Peng Y, Guo Z (2008) Constitutive expression of pathogen-inducible *OsWRKY31* enhances disease resistance and affects root growth and auxin response in transgenic rice plants. Cell Res 18:508

Zhang Z, Liu X, Wang X, Zhou M, Zhou X, Ye X, Wei X (2012) An R2R3 MYB transcription factor in wheat, TaPIMP1, mediates host resistance to *Bipolaris sorokiniana* and drought stresses through regulation of defense- and stress-related genes. New Phytol 196(4):1155–1170

Zhu H, Senalik D, McCown BH, Zeldin EL, Speers J, Hyman J, Bassil N, Hummer K, Simon PW, Zalapa JE (2012) Mining and validation of pyrosequenced simple sequence repeats (SSRs) from American cranberry (*Vaccinium macrocarpon* Ait.) Theor Appl Genet 124(1):87–96

Zimmermann P, Hirsch-Hoffmann M, Hennig L, Gruissem W (2004) GENEVESTIGATOR: Arabidopsis microarray database and analysis toolbox. Plant Physiol 136:2621–2632

Zipfel C, Kunze G, Chinchilla D, Caniard A, Jones JD, Boller T, Felix G (2006) Perception of the bacterial PAMP EF-Tu by the receptor EFR restricts agrobacterium-mediated transformation. Cell 125:749–760

Proteomic Studies Revealing Enigma of Plant–Pathogen Interaction

11

Anjana Rustagi, Garvita Singh, Shachi Agrawal, and Prem Kumar Gupta

Abstract

Pathogen attack is an intricate stimulus that induces stepwise defence response, namely, pathogen recognition, signal transduction and accomplishment of resistance/defense. These steps employ an array of proteins, interacting among themselves to sense the pathogen and produce antimicrobials antagonistic to pathogen growth. In order to gain insights in molecular mechanism of plant–pathogen interaction at the biochemical and cellular level, deciphering the proteins that are involved in this cellular medley is a prerequisite. Proteomics, one of the important subjects of "OMICS" generation, has played a principal role in the identification of these proteins. Proteomics aims at identification and quantification of the proteins mediating a specific cellular process. While the current proteomic studies give valid information about these processes, they also emphasize upon the significance of post-translational modifications. The information on sequence and post-translational modifications of proteins is then used to further decipher the biological processes using bioinformatics, genomics, cell biology, biochemistry and other areas of life sciences. We present a brief overview of the proteomic studies related to host–virus, host–bacteria and host–fungus interaction. We also provide the current stage of information on the techniques applied in proteomics and also the future challenges in this area of biological science.

Anjana Rustagi and Garvita Singh contributed equally

A. Rustagi (✉) · G. Singh · S. Agrawal
Department of Botany, Gargi College, New Delhi, India
e-mail: anjana.chaurasia@gmail.com

P. K. Gupta
Ionics Clinical Laboratory, Gurgaon, India

A. Singh, I. K. Singh (eds.), *Molecular Aspects of Plant-Pathogen Interaction*,
https://doi.org/10.1007/978-981-10-7371-7_11

Keywords
Proteomics · Plant–pathogen interaction · Two-dimensional polyacrylamide gel electrophoresis (2D-PAGE) · Fluorescent two-dimensional differencial gel electrophoresis (2D-DIGE) · ICAT (Isotope-coded affinity tags) · ITRAQ (Isobaric tagged for relative and absolute quantitation) · MudPIT (Multidimensional protein identification technology) · Mass spectrometry (MS) · Protein microarray

11.1 Introduction

All plant species, whether wild or cultivated, are persistently challenged by a variety of phytopathogens. Understanding the plant immune system has been a challenging task for plant biologists. Plants have multiple pathways that function in defence system against pathogens. The recognition of the pathogen and early signalling events are extremely rapid, consistent and specific. Subsequently, signalling cascades comprising of a large number of proteins and signalling pathways are activated which lead to production of antimicrobial compounds (Lodha et al. 2013). Recent results reveal that cellular responses to pathogen attack are dependent on phytohormone signals that are highly regulated and complex. Both positive and negative crosstalk occurs among themselves, in order to regulate appropriate defence pathway. Plants deploy two modes of defence: constitutive and inducible. Inducible defence covers a wide range of molecules, including various chemicals like secondary metabolites, inhibitors of necessary cellular pathways and digestive enzyme inhibitors like protease inhibitors. Although the basic phases in plants during pathogen attack remains to be the same, the response can alter greatly when host and pathogen combinations are different. The exact mechanism and strategies adopted by various plant species to ward off pathogens is not very well understood. However, the complicated mechanisms of plant–pathogen interactions and plant defence are being revealed anew due to considerable progress made in recent years on identifying differentially regulated genes and proteins in pathogen attack. Proteomic studies of plant–pathogen interaction to measure differential expression of proteins and their function in defence have opened up Pandora's box of novel information for plant pathologists.

In the past two decades, proteomics has emerged as a powerful tool for investigation of biological systems. Descriptive proteomics aims at identification of proteins present in an organism under a given condition. Thus structural and functional changes in an organism under different sets of conditions could be explained in terms of differential abundance of proteins under those conditions. Proteomics also studies the post-translational modifications of proteins which fine-tune the functions and structures of many proteins. With advancements in the techniques as well as expansion of our knowledge about protein functions and protein–protein interactions, other subfields of proteomic such as interactomics and secretomics have emerged out to be very crucial and are providing interesting information regarding protein structure and function (Gonzalez-Fernandez and Jorrin 2010). While interactomics aims to determine protein–protein interactions, secretomics focuses upon only those proteins which are secreted from an organism. This chapter discusses the modern techniques of proteomics and its contributions unravelling the complex mechanism of different plant–pathogen interactions.

11.2 Tools of Proteomics

Proteins are structurally and functionally diverse biomolecules and are of immense importance in biocatalysis. The term proteomics in simplest terms is the study of the "proteome" (Wilkins et al. 1996). The proteome is the total protein of a defined biological space at a given time under specific conditions. Thus, proteomic studies explain the quantity, time, location and purpose of the proteins that are synthesized in an organism. Proteomics is performed by separating the proteins present in a biological specimen on two-dimensional polyacrylamide gels and then identification of differentially expressed spots using Sanger sequencing or mass spectrometry-based sequencing. In recent times it has also been possible to directly separate proteins on high-pressure liquid chromatography. Then protein identification and sequencing may be done by mass spectrometry (Fig. 11.1).

11.2.1 Two-Dimensional Polyacrylamide Gel Electrophoresis (2D-PAGE)

2D-PAGE is a high-impact and popular method for separation of individual protein from a complex mixture. It is based on the two independent properties of proteins, isoelectric point (pI) and molecular size. Initially, proteins are separated on gels on the basis of their isoelectric points. Thus the technique is called as isoelectric focusing (IEF). All proteins have positively and negatively charged amino acids, and at one particular pH value, the net charge on the protein becomes zero. This is called isoelectric point of that protein. In isoelectric focussing, a pH gradient is made across the gel using electric current, and each protein comes to lie at a particular position in the gel where pH is equal to the isoelectric point at which the net charge on protein molecule is zero and it cannot move in the electric field.

The proteins are solubilized generally in highly concentrated urea solution, reducing agents and chaotrophs. Because of variable ion contents among different types of samples, the optimization of IEF buffer and the electrical profile is required for each type of sample. The pH gradient in gel is created using either carrier ampholytes, or alternatively immobilized pH gradient (IPG) gels can be used. Most modern researches use commercial IPGs for highly reproducible results. In the second dimension, protein molecules are separated on the basis of molecular weights (MW). This technique is called SDS-polyacrylamide gel electrophoresis (SDS-PAGE). In order to visualize the separated protein spots, the gel is stained with different stains viz. Coomassie Brilliant Blue (CBB), silver stain or SYPRO

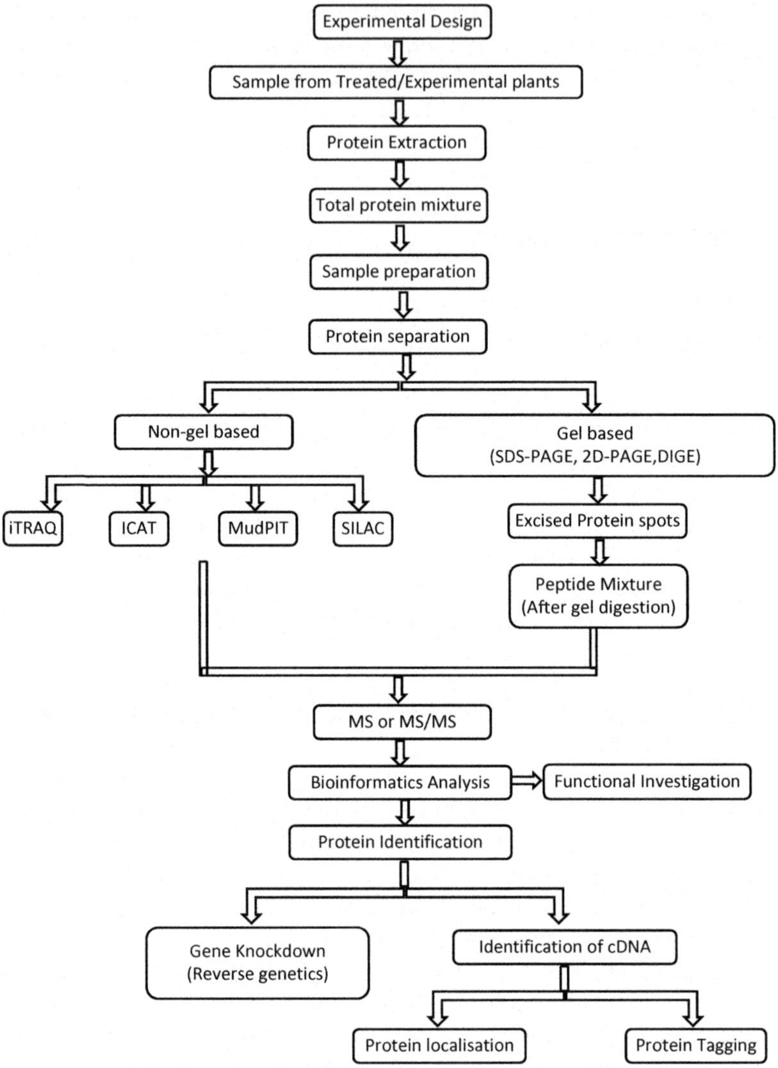

Fig. 11.1 Schematic flow chart to show steps involved in proteomic studies of plant–phytopathogen interaction

stain (Duley and Grover 2001; Nat et al. 2007). Although silver stain is highly sensitive and can detect proteins in the nanogram range, it has few drawbacks also. Silver staining is incompatible with the protein's microchemical preparation and identification by mass spectrometric techniques, staining of proteins is nonstoichiometric, and the assessment of silver staining is highly subjective. Another stain, SYPRO Ruby stain, which is ruthenium complex based is used for luminescent detection of protein molecules. It is reproducible, linear and highly compatible with the protein identification by mass spectrometry (White et al. 2004).

Digital images obtained from 2D-PAGE could be analysed using softwares such as Melanie, PDQuest, Phoretix, Progenesis, Z3 or Z4000 (Righetti et al. 2004). The spots of interest are excised and digested with sequence-specific proteases before subjecting to identification by mass spectrometry (Zhu et al. 2003; Rose et al. 2004). Thus with the help of 2D-PAGE, proteins could be quantified, and their molecular weight, isoelectric focusing point and post-translational modifications (PTMs) could be characterized (Gorg et al. 2004; Wittmann-Liebold et al. 2006). The 2D-PAGE holds the potential in thrust areas of research such as de novo sequencing and protein identification from unsequenced organisms, identification of modified proteins and protein isoforms (Rogowska-Wrzesinska et al. 2013). 2D-PAGE is especially useful for studying protein modifications and to find prognostic or diagnostic biomarkers in various disease states. The protein modifications can be identified on 2D gels by looking for protein spots close to other protein spots with isoelectric point (pI) and molecular weight shift characteristic of a post-translational modification.

11.2.2 Fluorescent Two-Dimensional Difference Gel Electrophoresis (2D-DIGE)

2D DIGE is a versatile and advantageous technique of protein separation. It is an advanced form of 2D PAGE in which two or three protein samples can be compared simultaneously on the same gel. It can be used to study protein regulation between control and experimental samples. In this technique, the proteins in each sample are first covalently tagged using spectrally distinct, fluorescent cyanine dyes (e.g. Cy2, Cy3 and Cy5, etc.). These dyes are designed by matching charge and size to nullify its effect on the relative migration of proteins during electrophoresis. These dyes are pH and photostable and are highly sensitive for detection (Westermeier 2006). Each dye reacts with the N-terminal amino group present in proteins or with the amino group of lysine. After labelling with two different dyes, the two protein mixtures used for comparison are mixed together and run on a single two-dimensional (2D) gel. The dye-labelled samples are then analysed individually by gel scanning at different wavelengths. Thus every protein in one sample superimposes with its differentially labelled but identical counterpart in other sample. Proteins that are common in both samples appear as "spots" with a fixed ratio of fluorescent signals, whereas proteins that differ between the samples differ in fluorescence intensity. Scanning of the gel at two different wavelengths indicates that whether any individual spot is associated with molecule of only one dye rather than two (Unlu et al. 1997). Then various softwares, specifically designed for 2D-DIGE analysis, are used to analyse the resulting images (Marouga et al. 2005). Using these image analysis programmes, volume ratios are generated for each spot. Volume ratios explain the intensity of a particular spot in each test sample, and thus changes in the protein abundance level can be identified and quantified.

The 2D-DIGE technique possesses all advantages of 2D and additionally eliminates gel-to-gel variation, provides high resolution and is more sensitive and reproducible (Gao 2014). Also the samples can be multiplexed on the same gel thereby reducing the number of gels required and limiting the experimental variations. The 2D-DIGE uses an internal standard; therefore, differences as low as 10% in protein expression can be quantified reliably (McGregor and Dunn 2006). The analysis of digital images of 2D-DIGE significantly improves the statistical assessment of proteome variation.

11.2.3 ICAT (Isotope-Coded Affinity Tags)

Isotope-coded affinity tag (ICAT) is a gel-free technology used for quantitative proteomics. In this technique the proteins present in samples are collected from two different experimental conditions (e.g. pathogen-challenged plant tissue extracts and control) and can be identified and quantified on the basis of chemical labelling agents (Gygi et al. 1999; Nat et al. 2007). The two different samples are first tagged with two different ICAT reagents. ICAT reagents consist of three functional elements: (i) a thiol-reactive group used for the selective labelling of reduced Cys residues, (ii) a biotin affinity tag that allows selective isolation of labelled peptides and (iii) a linker synthesized in either an isotopically normal ("light") or "heavy" form (utilizing ^{2}H or ^{13}C). Linker incorporates the stable isotope tags. Under denaturing conditions, protein disulphide bridges are reduced, and the free sulphhydryl groups of the proteins from the samples are labelled/tagged, respectively, with the isotopically "light" or "heavy" forms of the reagent. The samples are then mixed and cleaved enzymatically to generate peptide fragments. Then isolation of tagged peptides is done using avidin-affinity chromatography, and their analysis is carried out by microcapillary liquid chromatography-electrospray tandem mass spectrometry (Gygi et al. 1999; Nat et al. 2007). Peptides/proteins are identified using MS/MS analyses of all the individual fractions, and then observed MS/MS spectra are searched in protein sequence database. Thereafter the relative abundances of the peptide is obtained by observing the ratio between the signal intensities for the unfragmented isotopically "light" and "heavy" forms of the same peptide. Hence the protein can be identified from which it was derived, in the original samples. In this technique, relative quantities of the protein components present in mixtures can be determined in a single automated operation. This technique gives a measure of changes in protein levels induced by different stress conditions both quantitatively as well as qualitatively. ICAT is widely used to identify proteins associated with centrosomes that accumulate in abnormal ways in cancer cells and tumours, ICAT. This technique eliminates the need for 2-DE; however, major limitations of this technique are (i) selective detection of proteins which have relatively high cysteine content and (ii) difficulties in the detection of acidic proteins (Gygi et al. 2000; Zhou et al. 2002).

11.2.4 ITRAQ (Isobaric Tagged for Relative and Absolute Quantitation)

ITRAQ technology is a variation of ICAT. In ITRAQ proteins can be quantified from different sources in a single experiment (Ross et al. 2004; Agarwal et al. 2006; Zieske 2006; Lund et al. 2007). In this technique, the proteins are differentially labelled with different isotope tags that can be up to four, at the peptide level so that every peptide generated from the digestion of a complex sample is labelled. This method is based on covalent labelling of N-terminus and side chain amine of peptides from protein digestion with isotopic tag of varying mass. The labelled peptides are then separated by nanochromatography and analysed by tandem mass spectrometry (MS/MS). The data obtained after fragmentation is used in database search to identify the labelled peptides and hence the corresponding proteins. The low molecular mass reporter ions also generate from the fragmentations of attached tag. This data is used in quantification of peptides and the proteins from which they originate by using softwares such as i-tracker and jTraqX.

The iTRAQ technology is more advantageous as it includes the ability to multiplex several samples, quantification and simplified analysis with more analytical precision and accuracy (Agarwal et al. 2006; Lund et al. 2007; Zieske 2006). Two to four different samples can be used for comparisons in one MS-based experiment. The iTRAQ labelling strategy is not dependent on cysteine, so it eliminates the limitation of ICAT technology. The iTRAQ potentially covers a large range of the proteome by tagging tryptic peptides, and generally all of them possess primary amine groups (Agarwal et al. 2006; Ross et al. 2004; Zieske 2006).

11.2.5 MudPIT (Multidimensional Protein Identification Technology)

Another alternative to gel electrophoresis is MudPIT, an online 2D ion-exchange/reversed-phase HPLC method. In this technology peptides are separated systematically, depending on charge present on the molecule in the first dimension and on hydrophobicity of the molecule in the second (Veenstra and Smith 2003). The protein samples are first subjected to digestion using sequence-specific enzymes such as trypsin and endoproteinase lysC. Secondly, the mixtures of peptides obtained are separated by two orthogonal separation systems – (i) strong cation exchange (SCX) and (ii) reversed-phase high-performance liquid chromatography (RP-HPLC) (Issaq et al. 2005; Washburn et al. 2001). Peptides from the RP column are then subjected to MS analysis and searched for similarities in the protein databases (Washburn et al. 2001). Using this method, high-complex peptide mixtures can be analysed in a single experiment. Thus MudPIT technique provides a complete list of proteins present in a specific protein sample. It is fast, sensitive and highly reproducible. This method has added advantage of analysis of proteins of all functional and physical classes. Therefore, it is used for identification of protein complexes, cataloguing of proteins in cells and organisms at a large scale, profiling of proteins in membranes and organelles, determination of post-translational modifications (PTMs), protein ubiquitination in diverse plant species and quantitative analysis of protein expression (Yates et al. 2005, Cantin et al. 2006, Speers and Wu 2007; Maor et al. 2007). MudPIT can catalogue proteins in pathogens such as protozoa, bacteria and viruses if their sequences appear in the databases. Using the MudPIT approach, more than 2000 proteins can be identified in any particular sample (Hernandez et al. 2012). Although MudPIT experiments have several advantages over gel-based methods, it is simultaneously a relatively lengthy process as the number of fractions produced take too much time to analyse by MS using the reverse-phase gradient (Anguraj-Vadivel 2015).

11.2.6 Mass Spectrometry (MS)

MS is a well-accepted analytical tool used to distinguish molecules on the basis of their mass-to-charge ratios (m/z) and thus plays a central role in the field of proteomics (Zhu et al. 2009, 2010). Mass spectrometers consist of an ion source, a mass analyser and a detector. Ion source converts analyte molecules into gas-phase ions, and mass analyser separates ionized analytes on the basis of m/z ratio. Then the number of ions at each m/z value becomes recorded in the detector. In advanced MS analysis, two soft ionization methods, i.e. electrospray ionization (ESI) (Fenn et al. 1989) and matrix-assisted laser desorption/ionization (MALDI) (Karas and Hillenkamp 1988; Tanaka et al. 1988) are used. In both ESI and MALDI, ionization of large and nonvolatile analytes, such as proteins and peptides, is done with minimal fragmentation. Thus, biomacromolecules may be analysed with high throughput and much better sensitivity (Feng et al. 2015).

In protein samples, proteins or peptides are first fragmented using enzyme trypsin, and then liquid chromatography is used for separation of fragments obtained. After separation, the samples undergo ionization by ESI or MALDI method on a high-resolution mass spectrometer where peptide masses can be measured to three or four decimal places (exact mass) with high degree of accuracy. Gaseous ions are separated using mass analyser, and mass peaks are registered by the detector. The pattern of mass peaks is then searched against the database using search algorithms to identify similar patterns already reported in other proteins, and depending on the degree of similarity of peptide mass peak pattern, protein sequence and function can be predicted. This technique is called "peptide mass fingerprinting".

If required, the peptides can also be sequenced de novo, i.e. in any peptide, the sequence of amino acids can be determined without referring to any database. For this, first, a peptide having a specific mass is fragmented by using collision-induced dissociation, and then it is sent through another mass analyser which generates a set of fragment peaks. The amino acid sequences of the peptides are inferred using this exact mass of these peaks, using de novo sequence search algorithms or software. There are two categories of algorithms: database search algorithms and de novo search algorithms. The database algorithms include ProLUCID, SEQUEST, Mascot, PEAKS Protein ID, Phenyx OMSSA, MyriMatch, ByOnic, SIMS, MassWiz, etc.

(Lodha et al. 2013; Xu et al. 2015), while DeNoS, PEAKS, CycloBranch and Novor are de novo search programmes (Lodha et al. 2013; Ma 2015; Novak et al. 2015). Some other softwares used in protein identification are ESI prot 1.0, Medicwave Bioinformatics Suite, ProteoIQ, VIPER and Decon2LS, PatternLab for proteomics, MSight and Spectromania (Lodha et al. 2013).

Mass spectrometry has been much improved by the invention of the time-of-flight mass spectrometry (TOF-MS), which is most commonly used with MALDI and relatively non-destructive method for the conversion of proteins into volatile ions. MALDI is an ionization technique which is used for analyses of biomolecules and large organic molecules based on embedding samples in a matrix from which they are desorbed by laser light. Nowadays this technique has been an indispensable tool to analyse biological samples and has various applications in the field of bio-analysis, diagnostics, drug discovery, environmental analysis etc. In proteomics, the main applications of MS are (i) identification of a protein from its peptide fragments, (ii) determination of protein folding and interactions, (iii) cataloguing protein expression, (iv) identification of sites of protein modification, (v) detection of PTMs in complex biological proteins and (vi) quantification of a given sample protein (Han et al. 2008; Lodha et al. 2013).

11.2.7 Protein Microarray

Protein microarray, also known as a protein chip, is a powerful technology for the study of hundreds or thousands of proteins simultaneously as it possesses the potential for giving fundamental information on proteins, analytes, ligands, receptors, various interactions based on antibody affinity and partners involved in binding. Protein microarray thus permits high-throughput analysis (Romanov et al. 2014). In protein microarrays, antibodies or other affinity reagents, e.g. short peptides, polysaccharides, aptamers, allergens or synthetic small molecules, are first arrayed on a chip surface that may be of glass or plastic or silicon. Then the cell lysate is passed over this surface resulting in the binding of antigens to their cognate antibodies. The bound antigens are screened either using fluorescently tagged or radioactively labelled proteins. Secondary antibodies against each antigen of interest can also be used for screening. There are three types of protein microarrays, (i) analytical, (ii) functional and (iii) reverse-phase microarrays, which are used in biochemical activity studies of proteins. Analytical microarrays are used in protein profiling of a complex mixture. In analytical microarray, binding affinities, specificities and the expression levels of the various proteins present in the mixture can be measured. The most common analytical microarrays are antibody microarrays (Bertone and Snyder 2005).Various interactions such as interaction of protein molecules with protein or DNA or RNA or phospholipid or with some other small molecules can be studied using functional microarray analysis (Hall et al. 2004; Zhu et al. 2001). Reverse-phase protein microarray (RPA) is used to determine the presence of altered proteins that may occur in diseased tissue. RPAs can also be used, specifically, for characterization of post-translational modifications occurred in the tissues as a result of disease (Speer et al. 2005). The major applications of microarrays are in protein–protein interaction analysis, host–microbe interaction analysis, biomarker identification, biochemical pathway mapping, detection of infectious diseases, drug screening, development of vaccines, enzyme–substrate profiling and immunoprofiling (Zhu et al. 2012; Romanov et al. 2014; Moore et al. 2016). Thus it has become the significant tool for both classical as well as functional proteome analysis (Lodha et al. 2013).

11.2.8 Gel-Based Versus Non-gel-Based Proteomic Tools: Advantages and Limitations

Since two decades of the proteomics era, gel-based proteomics has gained immense importance for various studies pertaining to (i) the proteomic changes during plant growth and development and (ii) analysis of responses to various biotic and abiotic stimuli. The most widely used gel-based technique is 2D gel electrophoresis, in which ~2000 protein spots can be distinguished and processed before the identification by mass spectrometry (Vadivel 2015). Using combination of narrow range pH gels, more than 5000 distinct protein spots can be resolved in 2-D gels (Hoving et al. 2000; Fey and Larsen 2001). These gels provide an exhaustive information about the protein sample analysed by providing quantitative maps of intact proteins. This technique can be used for visualization of small sized sample of proteins (even less than 0.1 ng of protein per spot) (Smith 2009), detection of differential protein expression between two or more biologically relevant conditions and separation of protein isoforms (Rogowska-Wrzesinska et al. 2013). Although gel-based techniques are widely used, still there are some limitations, e.g. insensitivity to low-abundant proteins, inability to characterize the entire proteome in one gel and poor reproducibility. Because any given proteome may be much complex and 2D PAGE techniques have separation limitations, only fraction of the proteome can be analysed (Zhu et al. 2003). Instead, 2D-PAGE provides a map of intact proteins, and changes in protein expression level, isoforms or post-translational modifications can be detected easily. Due to post-translational modifications, isoelectric point (e.g. in phosphorylations) or relative mass (e.g. glycosylation or truncation) may change, and thus mobility of protein molecules may vary on a 2DE gel. Therefore, different isoforms of the same protein may be visualized as different spots on the 2DE gel. Gel-based methods are time-consuming and expensive too. Most of these limitations are addressed by gel-free proteomics techniques.

In proteomics, gel-free technology is more suitable for the analysis of proteins found in less abundance in complex samples. With liquid chromatography (LC) system, proteins and peptides can be separated in complex samples, very efficiently. Multidimensional chromatographic separation has further advantages in separation and identification of peptides. The advanced MS systems considerably improve the protein identification qualitatively as they are more sensitive and give more accurate protein quantitation results. However, gel-free MS-based proteomics requires considerable investment in expensive MS instruments and associated infrastructure and requires expert personnel to run the facility. Thus both gel and non-gel based of these approaches are of great importance, but no single approach can provide the whole information (qualitative as well as quantitative) of all protein components in any given complex mixture (Abdallah et al. 2012). So both approaches have their own advantages and disadvantages and are complementary to each other. They may be used in parallel so that one can get a more exhaustive information of protein expression and interactions in a specific physiological condition.

11.3 Understanding Plant–Pathogen Interaction in Light of Proteomics

In order to understand the disease resistance mechanism, the specific proteins expressed during plant–pathogen interaction must be identified. Use of tools of genomics, proteomics and high-throughput sensitive instruments has remarkably increased the knowledge of such interactions. Proteomics for studying plant–pathogen interactions started with the pioneering work of Ekramoddoullah and Hunt in 1993. They studied protein profiles of two varieties of *Pinus lambertiana* (sugar-pine) (susceptible or resistant variety to pine blister rust fungus *Cronartium ribicola*). Since then various plant–pathogen interactions have been studied using the proteomic tools. Here we summarize the important plant–virus, plant–bacteria and plant–fungus interaction studies carried out using proteomics.

In a successful plant defence response, plants induce different complex pathways against pathogen to inhibit the growth of pathogen. In general, two types of defence responses are triggered in plants during pathogenic attack, local and systemic (Lodha and Basak 2012; Hammond-Kosack and Jones 2000; Schenk et al. 2009). Local response recognizes the pathogens on the cell surface (Zipfel 2008; Hammond-Kosack and Jones 2000; Schenk et al. 2009; Mur et al. 2008). The second type of recognition pattern is a systematic or long-distance response. It induces defence signals that are not just located locally but spread to distant systemic tissues and is thus called as systemic acquired resistance (SAR). Activation of SAR takes place by salicylic acid (SA)-mediated pathway which subsequently elevates the level of other stress-responding molecules such as ethylene, jasmonic acid (JA), nitric oxide (NO) and pathogenesis-related (PR) proteins (Lodha et al. 2013). The signalling pathways which are activated by different stress factors have been widely documented (Pieterse and Van-Loon 2004). Efforts are on to elucidate this complex defence system by comparative analysis of the proteins present in specific tissues, cells or cellular compartments during control conditions or under phytopathogenic stress.

11.3.1 Plant–Virus Interactions

Plant–virus interaction is one of the most studied biological relationships and is still an open area of research due to lacunae in our understanding of such interactions. Viruses are true parasites as they lack the cellular machinery necessary for independent survival and replication outside a living cell. Proteomics has emerged as a promising tool to identify different class of proteins involved in plant–virus interactions. Cooper et al. (2003) and Brizard et al. (2006) isolated several virus–host protein complexes using size exclusion chromatography from the rice yellow mottle virus (RYMV)-infected rice plants (*Oryza sativa).* Further identification of various types of proteins participating in metabolism functioning (e.g. glycolysis, malate and Krebs cycles), plant defence strategy (e.g. peroxidase, superoxide-dismutase) and in protein synthesis (e.g. molecular chaperones, elongation factors) were studied with the help of mass spectrometry.

The plant–virus interactions start with transmission of the viruses into the plant cells through mechanical means which brings a healthy plant in close proximity to a diseased (infected) plant or transmission through an effective vector (highly mobile elements) playing a major role in virus infection, followed by replication. They subsequently travel through plasmodesmata to nearby cells by local movement (cell to cell) and ultimately they circulate, systemically after reaching vascular tissues. This initiates circulation through vascular movement starting from phloem tissues to the sink tissues of the host leading viruses to establish systemic infection with the help of several cycles of replication. Viruses are dependent on plant proteins to carry out mode of infection and also are influenced by the counteraction against the infection emerging from plant host proteins. By means of several transcriptional tools, genes encoding these plant host proteins have been charted out from several plant–virus interactions (Table 11.1) and are extensively studied by researchers in this area. Diaz-Vivancos et al. (2006) reported about the variation in antioxidative system mapping levels of enzymatic activity and protein expression within leaf apoplast of *Prunus persica* cv. GS305 (peach) infected with on plum pox potyvirus (PPV). De-Blasio et al. (2015) studied interaction of host plant potato and its pathogen potato leafroll virus (PLRV) which produces a read-through protein (RTP). They carried out translational read-through of the amber stop codon to better understand the interaction via affinity purification along with quantitative MS (mass spectrometry) to form protein networks for a PLRV mutant (incapable of producing read-through domain (RTD)) in order to compare it with known wild-type PLRV protein interaction network.

As a result, PLRV–plant interactions were classified into four different categories: category I includes plant proteins along with nonstructural viral proteins interacting with assembled coat proteins; category II elaborates about plant proteins in association with both RTD and coat protein; category III includes plant proteins in close association with RTD; and category IV explains those plant protein which show close affinity for virion but lacking RTD.

Several other studies were carried out on plant–virus interaction which demonstrate that photosynthetic electron transport in photosystem II (PSII) was reduced which ultimately affects oxygen-evolving complex (Rahoutei et al. 1999, 2000; Perez-Bueno et al. 2004). This reduction in level of functioning of oxygen-evolving complex leads to large multiplication and accumulation of virus in the plant host. Another proteomic study on interaction of rice yellow mottle virus with its natural host showed expression of proteins involved in maintaining oxidoreduction

Table 11.1 Proteomic studies on plant–virus interaction

Plant species	Virus	Proteomic approach	References
Capsicum annuum	Tobacco mosaic virus	2D-PAGE, MALDI-TOF MS	Lee et al. (2006)
N. tabacum	Pepper mild mottle virus	2D-PAGE, N-terminal sequencing	Perez-Bueno et al. (2004)
Oryza sativa	Rice yellow mottle virus (RYMV)	SDS-PAGE, Nano-LC-MS/MS, 2D-PAGE, MALDI, LC-MS/MS	Ventelon-Debout et al. (2004)
Prunus persica, and *P. serotina*	Plum pox virus	SDS-PAGE, IEF, MALDI-TOF	Diaz et al. (2006)
Vigna mungo	Mungbean yellow mosaic virus	MALDI-TOF TOF-MS	Kundu et al. (2011)
Lycopersicon esculentum	Tomato yellow leaf curl virus	LC-MS/MS, 2D-PAGE, MALDI, LC-MS/MS	Pakkianathan and Murad (2014)
Beta vulgaris	*Beet necrotic yellow vein virus* (BNYVV)	MS/MS spectra	Kimberley et al. (2014)
Arabidopsis thaliana	*Tobacco etch virus (TEV, genus Potyvirus)*	Twin-strep-tag and identification by affinity purification followed by mass spectrometry analysis (AP-MS)	Martinez et al. (2016)
Lycopersicon esculentum	PMMoV-S	2D-PAGE, MALDI	Casado-Vela et al. (2006)
Nicotiana benthamiana	PMMoV-S	2D-PAGE, MALDI	Perez-Bueno et al. (2004)

environment; the generation and detoxification of reactive oxygen species were identified and were presumed to maintain an oxidoreduction environment which favours viral replication (Brizard et al. 2006). Several other categories of differentially expressed proteins which are actively involved in translation, elongation factors, chaperones and chaperonins and proteins involved in proteomic turnover were reported in other plant–virus interaction study (Brizard et al. 2006). Proteomic analysis performed in order to identify proteins involved to study the interaction between *Oryza sativa* (rice) and rice yellow mottle sobemovirus (RYMV) (Delalande et al. 2005) showed expression of a phenylalanine ammonia-lyase, chaperonin-60 (mitochondrial) along with aldolase C; however, their exact role and mechanism of these proteins in defense is still not clear. Recently a proteomic study has identified approximately 16 proteins from the interaction of host tomato fruits (*Lycopersicon esculentum*) with TMV. Most of the proteins involved in interaction were pathogenesis related (PR), and several others were antioxidative enzymes, and their expression probably helped in resistance of plant against viral infection (Casado-Vela et al. 2006).

Although proteomic approaches have identified several different categories of proteins, their probable role and mechanism in plant virus interaction is still unachieved. Proteomic assay of various cell type involved in viral movement either it be leaf parenchyma or phloem tissue will only resolve this issue.

11.3.2 Plant–Bacterium Interactions

Plants are attacked by several groups of bacteria. Phylogenetic diversity of bacteria, on the basis of their interaction with plants, can be categorized into commensals (acquire nutrients in a manner the host remains unaffected), mutualists (helpful in plant growth) and pathogenic (acquire nutrients and plant health is negatively affected) (Newton et al. 2010). All three classes of bacteria have developed exceptionally modified physiology to interact with plants that reflects their requirement (Martin et al. 2003; Boller and Felix 2009). Environmental stresses result in changes in patterns of structural as well as transporter proteins, toxins and enzymes of bacteria, thus enabling them to adapt to the changed environment (Boller and Felix 2009). The cell wall degradation in plants aggravates plant defense strategy as signals received through cell surface proteins, polysaccharides, lipopolysaccharides and degradative enzymes after infection (Newton et al. 2010). Proteins of plant-associated bacteria (PAB) are studied in plants, in the form of bacterial responses to active biomolecules or natural plant extracts, or secretome analysis in order to study the virulence (vir) factor of the pathogenic bacteria (Guerreiro et al. 1997; Corbett et al. 2005; Gourion et al. 2006; Chung et al. 2007). As per Mehta and Rosato (2001), differentially expressed proteins, including a sulphate-binding protein, were studied by NH_2 terminal sequencing when *Xanthomonas axonopodis* pv. *citri* was cultivated in the host *Citrus sinensis.* Similar studies helped in identification of differentially expressed proteins, including a sulphate-binding protein, by NH_2 terminal sequencing (Table 11.2). Jones et al. 2004, analysed the proteomic and transcriptomic profiles of*Arabidopsis thaliana* during early response to *Pseudomonas syringae pv. Tomato* strain DC3000. Their experiments suggested that bacterial challenge generally induce the enzymes peroxiredoxins (PrxA, B and IIE) and the antioxidants glutathione S-transferases (GSTs F2, F6, F7 and F8) which potentially lead to specific post-translational modifications. They also concluded that individual members of these families may be specifically modified depending upon the degree of virulence of the DC3000 strain and outcome of the interaction. Similar studies were carried out in *Medicago truncatula* (model legume) in order to detect its response to pathogenic bacterium *Pseudomonas aeruginosa.* This proteomic analysis revealed significant changes in expression of 154 proteins, out of which 21 are related to defense and stress responses. Genome-based identification of types of proteins and effective toxins which are directly related with plants are referred as effectors and are quite significant to researchers in this area (Hogenhout et al. 2009).

In innate immune response, membrane proteins known as pattern recognition receptors (PRRs) are capable to identify pathogen-associated molecular patterns (PAMPs) of the invading pathogens (Gomez-Gomez and Boller 2000). As a result of PAMP recognition, there is activation of systemic acquired resistance (SAR), and this leads to production of resistance (R) proteins which results in effector-triggered

Table 11.2 Proteomic studies on plant–bacteria interaction

Plant species	Bacterial pathogen	Proteomic approach	References
Arabidopsis thaliana	*Pseudomonas syringae*	2D-PAGE, MALDI-TOF MS/MS	Jones et al. (2004), (2006)
Oryza sativa	*Xanthomonas oryzae* pv. *Oryzae*	2D-DIGE, MALDI-TOF MS	Mahmood et al. (2006) and Chen et al. (2007)
Oryzae sativa	*Xanthomonas oryzae*	2D-PAGE/MudPIT and MALDI-TOF/MS or nESI-LC-MS/MS	Wang et al. (2013)
Lycopersicon hirsutum	*Clavibacter michiganensis* ssp. *michiganensis*	2D-PAGE, MALDI-TOF MS/MS	Coaker et al. (2004)
Medicago truncatula	*Streptomyces meliloti*	2D-PAGE, MALDI-TOF MS/MS	Mathesius et al. (2003)
Brassica oleracea	*Xanthomonas campestris* pv. *Campestris*/parasite)	2D-PAGE, MALDI-TOF MS, MS/MS	Andrade et al. (2008)
Glycine max	*Bradyrhizobium japonicum*/symbiont	2D-PAGE,LCMS/MS, LTQ-Orbitrap MS	Delmotte et al. (2010)
A. thaliana	*Methylobacterium extorquens*/epiphyte	2D-PAGE, LC-MS/MS, MALDI-QqTOF MS/MS	Gourion et al. (2006)
Pisum sativum and *Vicia cracca*	*Rhizobium leguminosarum Biovar viciae* (symbiont)	2D-PAGE, microarray	Karunakaran et al. (2009)
Chrysanthemum	*Erwinia chrysanthemi* (soft rot)	2D-PAGE, MALDI-TOF-MS, LCQ ion trap MS	Kazemi-Pour et al. (2004)
Potato	*Pectobacterium atrosepticum*	2D-PAGE, MALDI-TOF-MS	Mattinen et al. (2007)
Medicago truncatula	*Streptomyces meliloti* strain	LC-MS/MS	Larrainzar et al. (2007)
Rice (var. Co43)	*P. fluorescens* KH- 1	2-DE, MS, LC-MS/MS	Kandasamy et al. (2009)

immunity (ETI) followed by hypersensitive response (HR) and leading to programmed cell death (Jones and Dangl 2006). Many reference data sets of proteomics have been established which identifies proteins for various PAB, via tools as two-dimensional gel electrophoresis (2-DGE) or liquid chromatography-tandem mass spectrometry (LC-MS/MS) techniques which identify proteins for various PAB (both pathogenic and mutualist) (Rosen et al. 2004; Chung et al. 2007; Anderson et al. 2006; Bosch et al. 2008). Basal defense system in the form of PAMPs and plant responses are induced, and such responses are in the form of flagellin protein, cold shock proteins and elongation factor (EFTu), chaperones and chaperonins (Newton et al. 2010; Rosen et al. 2004; Jacobs et al. 2012). A broad study of transcriptomics elaborates about pathogenic bacteria involved in pathogenicity and hypersensitivity and shows different secretion systems for colonization leading to host cell death (Buttner and Bonas 2010). Proteomics is important to understand the reaction mechanism behind pathogenesis as well as symbiosis (mutualism). This will further open up new areas of research in protein-based plant–microbe crosstalk.

11.3.3 Plant–Fungus Interactions

Proteomics has been extensively used to understand and characterize the key elements involved in plant–fungus interactions. In recent past, quite a few studies have been performed, which describe proteomic studies of fungi-infected plants. The main focus of such studies had been to analyse, identify and quantify the proteins present in the host plant during successful or unsuccessful resistance to fungal pathogen. In the recent years, a wide variety of fungal proteins involved in successful pathogenesis have also been explored (Gonzalez-Fernandez and Jorrin-Novo 2010, 2012).

The pathogenic fungal/host plant protein after separation in gels can be identified using bioinformatics tools. Since availability of high-quality genome data is limited to only those organisms which have been sequenced so far, the proteomic analysis and identification of proteins are also limited to these plants. A list of important plant–fungus studies using proteomics is enlisted in Table 11.3.

The pathogenic fungi adopt different strategies for infecting the host plant. The necrotrophic fungi kill the host tissue and thereafter obtain nutrients from necrotic host cells. The biotrophic fungi colonize the living host cells either intracellularly or intercellularly and obtain nutrients from living host tissues. Hemibiotrophic fungi exhibit both biotrophy and necrotrophy in a two-phasic manner (Lo Presti et al. 2015). Due to difference in mode of infection, different types of proteins are involved in different plant–fungus interactions. Therefore, it is imperative to study individual interactions separately. The host plant in response to the invading pathogenic fungi elicits a host defense response. Both local as well as systemic immunity could be triggered by the host plant (Schwessinger and Ronald 2012). The plants show a two-tier innate immune response involving the localized plasma membrane proteins as well as the intracellular receptors (Lo Presti 2006; Dodds and Rathjen 2010; Asai and Shirasu 2015). During pathogenesis, intracellular proteins and the secreted proteins of fungi are either upregulated or downregulated, thereby enabling the fungi for pathogenesis. Considering this, numerous proteomic efforts have been done to identify and characterize these proteins. Bohmer et al. (2007) carried out pioneering research to create a proteome map from mycelia of phytopathogenic fungus *Ustilago maydis*, during its morphological transition from bud stage to filamentous stage. Greenville-Briggs et al. (2005) performed a parallel study of transcriptome and proteome of *Phytophthora infestans* so as to identify proteins/genes upregulated during appressorium formation in the germinating cysts of the fungus. Five different protein-expressing genes were identified which are methionine synthase (*Pi-met1*), a ketol-acid reductoisomerase (*Pi-kari1*), a tryptophan synthase (*Pi-trp1*), an acetolactate synthase (*Pi-als1*) and a threonine synthase (*Pi-ts1*). The

Table 11.3 Proteomic studies on plant–fungal pathogen interaction

Plant species	Fungal pathogen	Proteomic approach	References
Zea mays	*F. verticillioides*	2D-PAGE, MALDI-TOF-MS, LCQ ion trap MS	Campo et al. (2004)
O. sativa	*Rhizoctonia solani*	2D-PAGE, ESI Q-TOF MS, MS/MS	Lee et al. (2006)
Triticum aestivum	*Puccinia triticina*	2D-PAGE, MALDI-QqTOF MS/MS	Rampitsch et al. (2006)
A. thaliana	*Plasmodiophora brassicae*	2D-PAGE, MALDI-TOF-MS	Devos et al. (2006)
A. thaliana	*Fusarium elicitor*	2D-DIGE, MALDI-TOF MS	Chivasa et al. 2006
Triticum aestivum	*Fusarium graminearum*	2D-PAGE, LC-MS/MS	Zhou et al. (2006)
Brassica napus	*Sclerotinia sclerotiorum*	2D-PAGE, ESI-q-TOF MS/MS	Liang et al. (2008)
Fagus sylvatica	*Phytophthora citricola*	SDS-PAGE, 2D-PAGE, ESI, LC-ESI	Valcu et al. (2009)
Arabidopsis thaliana	*Alternaria brassicicola*	2D-PAGE, LC-MS/MS	Mukherjee et al. (2010)
Humulus lupulus	*Verticillium albo-atrum*	2D-PAGE, de novo sequencing	Mandelc et al. (2013)
Triticum aestivum	*Septoria tritici*	HPLC, MS/MS	Yang et al. (2013)
Triticum aestivum	*Rhizoctonia solani*	Nano-LC-MS/HPLC	Anderson et al. (2016)
Humulus lupulus	*Verticillium* nonalfalfae	LC-MS/MS	Flajsman et al. (2016)
Triticum aestivum	*Puccinia striiformis* f. sp. *tritici*	Nano-LC-ESI-MS/MS	Demirci et al. (2016)
Solanum tuberosum	*Phytophthora infestans*	LC-MS/MS	Larsen et al. (2016)
A. thaliana	*PGPR Paenibacillus Polymyxa*	2D-PAGE, LC-MS/MS	Kwon et al. (2016)

proteomes of both the pathogen and its host plant potato were evaluated during disease development.

Rampitsch et al. (2006) compared the proteome of a susceptible line of wheat which was infected with leaf rust to mock-inoculated wheat using 2DE followed by MS analysis. They observed 22 differently expressed proteins which included proteins with known and hypothetical functions.

Another approach is the study of fungal proteins for the analysis of exoproteome, also called secretome. In this context, Phalip et al. (2005) identified 84 fungal secreted proteins. They grew *Fusarium graminearum*, on hop (*Humulus lupulus*), and analysed the secretome using 2D-PAGE/MS. Some of the fungal proteins involved in plant–pathogen interaction are methionine synthase and threonine synthase (*Phytophthora infestans*/*Solanum tuberosum*); chitinase, serine proteinase, leucine aminopeptidase, lipase, pectate lyase, α-arabinofuranosidase, ceramidase, chitin deacetylase, β-glucosidase, polygalacturonidase, trypsin, aspartyl proteinase, xyloglucanase, carboxypeptidase and α-amylase (*Fusarium graminearum*/*Humulus lupulus*); and mucin, transglutaminase and glucanase (*Phytophthora ramorum*/Oak) (Grenville-Briggs et al. 2005; Phalip et al. 2005; Meijer et al. 2006). Yajima and Kav (2006) cultured *Sclerotinia sclerotiorum* and identified four different fungal proteins viz. exopolygalacturonase, cellobiohydrolase-1-catalytic domain, acid protease and aspartyl proteinase using proteomic tools. In these studies, various proteins involved in defense, and stress responses, signal transduction, photosynthesis, electron transport and metabolism, have been identified.

Among the many proteins identified in plants during plant–fungus interactions are different classes of pathogenesis-related (PR) proteins such as chitinase, β-1,3 glucanase, peroxidases and thaumatin-like protein (*Oryza sativa*, *Triticum aestivum* and *Lycopersicon esculentum*). The PR proteins β-1,3 glucanase and peroxidases have also been identified in *Zea mays* and *Arabidopsis thaliana*, respectively. Several other proteins reported from plants are glutathione S-transferase, glyceraldehyde 3-phosphate dehydrogenase, fructose-bisphosphate aldolase, probenazole-induced protein, adenosine kinase, superoxide dismutase (SOD), glutamate dehydrogenase, thioredoxin, 20S proteasome β unit, chaperonin 60 β precursor, disease-resistance-response protein pi 49, receptor-like protein kinase, 14-3-3-like protein, etc. (Mehta et al. 2008).

Babich and Katam (2016) investigated molecular changes in the grape leaf during the process of berry development and tolerance to anthracnose. Proteins extracted from leaf of different cultivars were separated using 2D-PAGE and characterized using MS and later compared with *Vitis* database. They observed a total of 56 differentially expressed proteins, which are known to be involved in the processes of pathogen response, photosynthesis and metabolism. The tolerant grape cultivars showed more abundant proteins in comparison to susceptible cultivars.

Flajsman et al. (2016) did pioneering research using proteomic tools to identify proteins secreted in xylem sap following infection with *Verticillium nonalfalfae* spp. VnaPRX1.1277 of hop plants (*Humulus lupulus*). Three fungal proteins of *V. nonalfalfae* were found to be present in abundance, α-N-arabinofuranosidase (VnaAbf4.216), peroxidase (VnaPRX1.1277) and a hypothetical protein (VnaSSP4.2). These three proteins are the first ever secreted proteins which have been identified in xylem sap upon infection with *Verticillium* spp. They also reported that in planta expression of the two protein genes VnaPRX1.1277 and VnaSSP4.2 increase with the progress in colonization, thereby establishing their significance in fungal virulence. Also the deletion mutants of *V. nonalfalfae* for these two genes showed compromised pathogenicity, and hence it is imperative to consider VnaPRX1.1277 and VnaSSP4.2 as virulence factors necessary for colonization of hop plants by *V. nonalfalfae*.

During the last 10 years, comparative molecular profiles of compatible and incompatible plant–pathogen interactions have been studied. Demirci et al. (2016) analysed proteome profiles of interacting wheat and *Puccinia striiformis* f. sp. *tritici* (Pst). Proteins were isolated from infected and control samples and separated using 2D-LC system. The proteome of the two samples were compared. The differentially expressed proteins were excised, eluted and identified with the help of nano-LC-ESI-MS/MS. Out of the total differentially expressed proteins identified, 62% belonged to the wheat database and 38% were Pst proteins. The subcellular localization and signal peptide motifs of these identified proteins were ascertained with the help of bioinformatics tools. All the identified wheat proteins were categorized in seven functional groups, viz. defense, stress responsive, gene expression, signal transduction, metabolism, electron transport and photosynthesis. Most of the identified proteins belonged to defense and stress-responsive groups. The defense group proteins included the antioxidant and detoxification proteins as dehydroascorbate reductase (DHAR), glutathione S-transferase (GST), superoxide dismutases (SODs), ascorbate peroxidase (APX), catalase (CAT), peroxidases (PX), peroxiredoxins (PRX), etc. About 64 proteins were identified from the pathogen *Puccinia striiformis* f. sp. *tritici.* Out of these, 30 proteins were categorized in five different functional groups, namely, structure, attack, metabolism, gene expression and signal transduction. No function could be assigned to the remaining 34 proteins, and thus they were considered hypothetical.

Larsen et al. (2016) established label-free proteomics to study the pathogenic interaction of *P. infestans* with three cultivars (Kuras, Bintje and Sarpo Mira) of potato plant. About 3248–3529 unique proteins were detected and identified from each cultivar of the potato plant. From the pathogen (*P. infestans)*, nearly 758 proteins were identified. The data set detailing the information about these proteins is available via ProteomeXchange, with the identifier PXD002767.

11.4 Apoplast Proteomics: An Emerging Field to Uncover Plant Pathogen Crosstalk

The extracellular matrix of the plant cell is called apoplast. The apoplastic fluid circulates in the intercellular spaces. The cell wall, the apoplast, the apoplastic fluid and the extracellular space outside the plasma membrane are the first compartmental venue in the plant body, which faces the pathogen challenge (Agrawal et al. 2010). It potentially perceives and transduces signals from the external environment to the symplast. The apoplastic space is the first physiological compartment for the interaction of plant and the pathogen, and the key processes that occur there determine whether a successful parasitism will be established or not (Doehlemann and Hemetsberger 2013). The apoplastic fluid constituents are nutrients, polysaccharides, secondary metabolites and secreted proteins. It facilitates intercellular communications and has significant role in regulation of growth, biotic and abiotic stress responses and cell wall maintenance (Ellis et al. 2007; Tseng et al. 2009). The mechanism of reactive oxygen species (ROS) accumulation and altered synthesis of extracellular protein takes place in apoplast as the first line of defense. A fine regulation of expression of apoplastic proteins is necessary for pathogen perception and for maintaining integrity of cell wall (Pechanova et al. 2010).

In the recent few years, proteomic tools have been harnessed to study the apoplast during pathogenesis process. Yang et al. (2015) reported the results of a survey of leaf apoplastic proteome in the resistant and susceptible varieties of wheat in response to the pathogen *Zymoseptoria tritici*. This pathogen inhibits the apoplast of the wheat plant and is responsible for causing *Septoria tritici* blotch (STB) on the foliage. They demonstrated that resistance of plants to *Z. tritici* has correlation with responses at proteome level.

They concluded that factors such as carbohydrate metabolism, reinforcement of cell wall, and synthesis of PR proteins in apoplast are linked with disease resistance. The pathogen has to overcome all these armours of host defense responses in order to achieve successful colonization.

Delaunois et al. (2014) provided an insight to highlight the key molecules involved in plant–pathogen interaction using proteomic tools to study the apoplast. The apoplastic proteome under biotic stress still remains poorly characterized. Thus the future studies should aim at identification of apoplastic proteins during pathogen infection so that the mechanism of perception of pathogen stressors and regulation of stress could be understood.

Currently, apoplast proteomics is an emerging topic of research among scientists studying the complex phenomenon of plant–pathogen interaction. More light will be thrown upon this intricate relationship by studying the apoplast proteome in the coming years.

11.5 Conclusion

In recent years, the various gel- and non-gel-based protein separation methods coupled with advanced spectrometry techniques have evolved as the prominent tools for protein identification and characterization.

As the proteomic tools have evolved and become more sensitive, the number of proteins that can be observed, identified and characterized has increased. The high-volume protein information, when synergized with high-throughput genomic information, provides a window for protein complement to be characterized in silico. Thus the modern bioinformatics increase the efficacy of the proteomic studies immensely.

Numerous proteins have been identified so far both from host and the pathogen, and their specific roles are being explored. Since proteomics provides a rapid insight into the plant–pathogen intricacies, it is expected to be one of the imminent and integrative tools in biological research.

Acknowledgement The authors would like to acknowledge the kind help of Dr. Shalu Jain, Plant Pathology Division, North Dakota State University, Fargo, USA, for critical comments and suggestions.

References

Abdallah C, Dumas-Gaudot E, Renaut J et al (2012) Gel-based and gel-free quantitative proteomics approaches at a glance. Int J Plant Genom 2012:Article ID 494572., 17 pages. https://doi.org/10.1155/2012/ 494572

Agarwal K, Choe LH, Lee KH (2006) Shotgun proteomics using the iTRAQ isobaric tags. Brief Funct Genom Proteomics 5(2):112–120

Agrawal GK, Jwa NS, Lebrun MH et al (2010) Plant secretome: unlocking secrets of the secreted proteins. Proteomics 10:799–827

Anderson DC, Campbell EL, Meeks JC (2006) A soluble 3D LC/MS/MS proteome of the filamentous cyanobacterium *Nostoc punctiforme*. J Proteome Res 5:3096–3104

Andrade AE, Silva LP, Pereira JL et al (2008) In vivo proteome analysis of *Xanthomonas campestris* pv. *campestris* in the interaction with the host plant *Brassica oleracea*. FEMS Microbiol Lett 281:167–174

Anguraj-Vadivel AK (2015) Gel-based proteomics in plants: time to move on from the tradition. Front Plant Sci 6:369

Asai S, Shirasu K (2015) Plant cells under siege: plant immune system versus pathogen effectors. Curr Opin Plant Biol 28:1–8

Babich R, Katam R (2016) Leaf proteome profiling and their interactions to determine disease resistance in Grape. Book of Abstract 'Plant and Animal genome conference XXIV' January 08–13, 2016 San Diego, CA

Bertone P, Snyder M (2005) Advances in functional protein microarray technology. FEBS J 272:5400–5411

Bohmer M, Colby T, Bohmer C et al (2007) Proteomic analysis of dimorphic transition in the phytopathogenic fungus *Ustilago maydis*. Proteomics 7:675–685

Boller T, Felix G (2009) A renaissance of elicitors: perception of microbe associated molecular patterns and danger signals by pattern-recognition receptors. Annu Rev Plant Biol 60:379–406

Bosch G, Skovran E, Xia Q et al (2008) Comprehensive proteomics of methylobacterium extorquens AM1 metabolism under single carbon and non methylotrophic conditions. Proteomics 8:3494–3505

Brizard JP, Carapito C, Delalande F et al (2006) Proteome analysis of plant–virus interactome: comprehensive data for virus multiplication inside their hosts. Mol Cell Proteomics 5:2279–2297

Buttner D, Bonas U (2010) Regulation and secretion of *Xanthomonas* virulence factors. FEMS Microbiol Rev 34:107–133

Campo S, Carrascal M, Coca M et al (2004) The defence response of germinating maize embryos against fungal infection: a proteomics approach. Proteomics 4(2):383–396

Cantin GT, Venable JD, Cociorva D et al (2006) Quantitative phosphoproteomic analysis of the tumor necrosis factor pathway. J Proteome Res 5(1):127–134

Casado-Vela J, Selles S, Martinez RB (2006) Proteomic analysis of tobacco mosaic virus-infected tomato (*Lycopersicon esculentum* M.) fruits and detection of viral coat protein. Proteomics 6(Suppl. 1):S196–S206

Chen F, Yuan Y, Li Q et al (2007) Proteomic analysis of rice plasma membrane reveals proteins involved in early defense response to bacterial blight. Proteomics 7:1529–1539

Chivasa S, Hamilton JM, Pringle RS et al (2006) Proteomic analysis of differentially expressed proteins in fungal elicitor-treated *Arabidopsis* cell cultures. J Expt Bot 57(7):1553–1562

Chung WJ, Shu HY, Lu C et al (2007) Qualitative and comparative proteomic analysis of Xanthomonas campestris pv. Campestris17. Proteomics 7:2047–2058

Coaker GL, Willard B, Kinter M et al (2004) Proteomic analysis of resistance mediated by Rcm 2.0 and Rcm 5.1, two loci controlling resistance to bacterial canker of tomato. Mol Plant-Microbe Interact 17:1019–1028

Cooper B, Clarke JD, Budworth P et al (2003) A network of rice genes associated with stress response and seed development. Proc Natl Acad Sci U S A 100:4945–4950

Corbett M, Virtue S, Bell K et al (2005) Identification of a new quorum-sensing controlled virulence factor in *Erwinia carotovora* subsp. Atroseptica secreted via the type II targeting pathway. Mol Plant-Microbe Interact 18:334–342

De-Blasio SL, Johnson R, Sweeney MM et al (2015) The potato leafroll virus structural proteins manipulate overlapping, yet distinct protein interaction networks during infection. Proteomics 15(12):2098–2112

Delalande F, Carapito C, Brizard JP et al (2005) Multigenic families and proteomics: extended protein characterization as a tool for paralog gene identification. Proteomics 5:450–460

Delaunois B, Jeandet P, Clément C et al (2014) Uncovering plant-pathogen crosstalk through apoplastic proteomic studies. Front Plant Sci 5(249):1–18

Demirci YE, Inan C, Gunel A et al (2016) Proteome profiling of the compatible interaction between wheat and stripe rust. Eur J Plant Pathol 145(4):941–962

Devos S, Laukens K, Deckers P et al (2006) A hormone and proteome approach to picturing the initial metabolic events during Plasmodiophora brassicae infection on *Arabidopsis*. Mol Plant-Microbe Interact 19:1431–1443

Diaz-Vivancos P, Rubio M, Mesonero V et al (2006) The apoplastic antioxidant system in Prunus: response to long-term plum pox virus infection. J Exp Bot 57:3813–3824

Dodds PN, Rathjen JP (2010) Plant immunity: towards an integrated view of plant– pathogen interactions. Nat Rev Genet 11(8):539–548

Doehlemann G, Hemetsberger C (2013) Apoplastic immunity and its suppression by filamentous plant pathogens. New Phytol 198:1001–1016

Duley H, Grover A (2001) Current initiatives in proteomics research: the plant perspective. Curr Sci 80(2):262–269

Ekramoddoullah AKM, Hunt RS (1993) Changes in protein profile of susceptible and resistant sugar-pine foliage infected with the Whitepine blister rust fungus *Cronartium ribicola*. Can J Plant Pathol 15(4):259–264

Ellis JG, Dodds PN, Lawrence GJ (2007) The role of secreted proteins in diseases of plants caused by rust, powdery mildew and smut fungi. Curr Opin Microbiol 10:326–331

Feng X, Liu BF, Li J et al (2015) Advances in coupling microfluidic chips to mass spectrometry. Mass Spectrom Rev 34(5):535–557

Fenn JB, Mann M, Meng CK et al (1989) Electrospray ionization for mass spectrometry of large biomolecules. Science 246:64–71

Fey SJ, Larsen PM (2001) 2D or not 2D two-dimensional gel electrophoresis. Curr Opin Chem Biol 5:26–33

Flajsman M, Mandelc S, Radisek S et al (2016) Identification of novel virulence-associated proteins secreted to xylem by *Verticillium nonalfalfae* during colonization of hop plants. Mol Plant Microbe Interact 29(5):362–373

Gao W (2014) Analysis of protein changes using two-dimensional difference gel electrophoresis. Mol Toxicol Protocol 1105:17–30

Gomez-Gomez L, Boller T (2000) FLS2: an LRR receptor-like kinase involved in the perception of the bacterial elicitor flagellin in *Arabidopsis*. Mol Cell 5:1003–1011

Gonzalez-Fernandez R, Jorrin-Novo JV (2010) Proteomics of fungal plant pathogens: the case of *Botrytis cinerea*. In: Méndez-Vilas A (ed) Current research, technology and education topics in applied microbiology and microbial biotechnology. FORMATEX, Badajoz, pp 205–217

Gorg A, Weiss W, Dunn M (2004) Current two-dimensional electrophoresis technology for proteomics. Proteomics 4(12):3665–3685

Gourion B, Rossignol M, Vorholt JA (2006) A proteomic study of Meth*ylobacterium extorquens* reveals a response regulator essential for epiphytic growth. Proc Natl Acad Sci U S A 103:13186–13191

Grenville-Briggs LJ, Avrova AO, Bruce CR et al (2005) Elevated amino acid biosynthesis in *Phytophthora infestans* during appressorium formation and potato infection. Fungal Genet Biol 42:244–256

Guerreiro N, Redmond JW, Rolfe BG et al (1997) New *Rhizobium leguminorum* flavonoid induced proteins revealed by proteome analysis of differentially displayed proteins. Mol Plant-Microbe Interact 10:506–516

Gygi SP, Rist B, Gerber SA et al (1999) Quantitative analysis of complex protein mixtures using isotope-coded affinity tags. Nat Biotechnol 17(10):994–999

Hall DA, Zhu H, Zhu X et al (2004) Regulation of gene expression by a metabolic enzyme. Science 306:482–484

Hammond-Kosack KE, Jones JDG (2000) Responses to plant pathogens. In: Buchanan BB, Gruissem W, Jones RL (eds) Biochemistry and molecular biology of plants. American Society of Plant Physiology, Rockville, pp 1102–1156

Han X, Aslanian A, Yates JR (2008) Mass spectrometry for proteomics. Curr Opin C Biol 12(5):483–490

Hernandez LG, Lu B, Da Cruz GC et al (2012) Worker honeybee brain proteome. J Proteome Res 11:1485–1493

Hogenhout SA, Van der Hoorn RA, Terauchi R et al (2009) Emerging concepts in effector biology of plant-associated organisms. Mol Plant-Microbe Interact 22:115–122

Hoving S, Voshol H, Oostrum J (2000) Towards high performance two-dimensional gel electrophoresis using ultrazoom gels. Electrophoresis 21:2617–2621

Issaq HJ, Chan KC, Janini GM et al (2005) Multidimensional separation of peptides for effective proteomic analysis. J Chromatogr B 817(1):35–47

Jacobs JM, Babujee L, Meng F et al (2012) The in planta transcriptome of *Ralstonia solanacearum*: conserved physiological and virulence strategies during bacterial wilt of tomato. MBio 3:112–114

Jones JDG, Dangl JL (2006) The plant immune system. Nature 444:323–329

Jones AME, Thomas V, Truman B et al (2004) Specific changes in the Arabidopsis proteome in response to bacterial challenge: differentiating basal and R-gene mediated resistance. Phytochemistry 65:1805–1816

Jones AME, Thomas V, Bennett MH et al (2006) Modifications to the *Arabidopsis* defence proteome occur prior to significant transcriptional change in response to inoculation with *Pseudomonas syringae*. Plant Physiol 142:1603–1620

Kandasamy S, Loganathan K, Muthuraj R et al (2009) Understanding the molecular basis of plant growth promotional effect of *Pseudomonas fluorescens* on rice through protein profiling. Proteome Sci 7:47

Karas M, Hillenkamp F (1988) Laser desorption ionization of proteins with molecular masses exceeding 10,000 daltons. Anal Chem 60:2299–2301

Karunakaran R, Ramachandran VK, Seaman JC et al (2009) Transcriptomic analysis of *Rhizobium leguminosarum biovar viciae* in symbiosis with host plants *Pisum sativum* and *Vicia cracca*. J Bacteriol 191:4002–4014

Kazemi-Pour N, Condemine G, Hugouvieux-Cotte-Pattat N (2004) The secretome of the plant pathogenic bacterium *Erwinia chrysanthemi*. Proteomics 4:3177–3186

Kundu S, Chakraborty D, Pal A (2011) Proteomic analysis of salicylic acid induced resistance to Mungbean yellow mosaic India virus in *Vigna mungo*. J Proteome 74:337–349

Kwon YS, Lee DY, Rakwal R et al (2016) Proteomic analyses of the interaction between the plant-growth promoting *Rhizobacterium paenibacillus polymyxa* E681 and *Arabidopsis thaliana*. Proteomics 1:122–135

Larrainzar E, Wienkoop S, Weckwerth W et al (2007) Medicago truncatula root nodule proteome analysis reveals differential plant and bacteroid responses to drought stress. Plant Physiol 144:1495–1507

Larsen MKG, Jorgensen MM, Bennike TB et al (2016) Time-course investigation of *Phytophthora infestans* infection of potato leaf from three cultivars by quantitative proteomics. Elsevier Data Brief 6:238–248

Lee BJ, Kwon SJ, Kim SK et al (2006) Functional study of hot pepper 26S proteasome subunit RPN7 induced by *tobacco mosaic virus* from nuclear proteome analysis. Biochem Biophys Res Commun 351:405–411

Liang Y, Srivastava S, Rahman MH et al (2008) Proteome changes in leaves of *Brassica napus* L. as a result of *Sclerotinia sclerotiorum* challenge. J Agric Food Chem 56(60):1963–1976
Lo Presti L, Lanver D, Schweizer G et al (2015) Fungal effectors and plant susceptibility. Annu Rev Plant Biol 66:513–545
Lodha TD, Basak J (2012) Plant-pathogen interaction: what microarray tells about it? Mol Biotechnol 50(1):87–97
Lodha TD, Hembram P, Tep N et al (2013) Proteomics: a successful approach to understand the molecular mechanism of plant-pathogen interaction. Am J Plant Sci 4:1212–1226
Lund TC, Anderson LB, McCullar V (2007) iTRAQ is a useful method to screen for membrane-bound proteins differentially expressed in human natural killer cell types. J Proteome Res 6:644–653
Ma B (2015) Novor: real-time peptide de novo sequencing software. J Am Soc Mass Spectrom 26:1885–1894
Mahmood T, Jan A, Kakishima M et al (2006) Proteomic analysis of bacterial-blight defence responsive proteins in rice leaf blades. Proteomics 6:6053–6065
Mandelc S, Timperman I, Radisek S et al (2013) Comparative proteomic profiling in compatible and incompatible interactions between hop roots and *Verticillium alboatrum*. Plant Physiol Biochem 68:23–31
Maor R, Jones A, Nühse TS et al (2007) Multidimensional protein identification technology (MudPIT) analysis of ubiquitinated proteins in plants. Mol Cell Proteomics 6(4):601–610
Marouga R, David S, Hawkins E (2005) The development of the DIGE system: 2D fluorescence difference gel analysis technology. Anal Bio Anal Chem 382(3):669–678
Martin GB, Bogdanove AJ, Sessa G (2003) Understanding the functions of plant disease resistance proteins. Annu Rev Plant Biol 54:23–61
Mathesius U, Mulders S, Gao M et al (2003) Extensive and specific responses of a eukaryote to bacterial quorum-sensing signals. Proc Natl Acad Sci U S A 100:1444–1449
Mattinen L, Nissinen R, Riipi T et al (2007) Host-extract induced changes in the secretome of the plant pathogenic bacterium *Pectobacterium atrosepticum*. Proteomics 7:3527–3537
McGregor E, Dunn MJ (2006) Proteomics of the heart unraveling disease. Circ Res 98(3):309–321
Mehta A, Rosato YB (2001) Differentially expressed proteins in the interaction of *Xanthomonas axonopodis pv. citri* with leaf extract of the host plant. Proteomics 1:1111–1118
Mehta A, Brasileiro ACM, Souza DSL et al (2008) Plant–pathogen interactions: what is proteomics telling us? FEBS J 275:3731–3746
Meijer HJ, Van-de-Vondervoort PJ, Yin QY et al (2006) Identification of cell wall-associated proteins from *Phytophthora ramorum*. Mol Plant-Microbe Interact 19:1348–1358
Moore CD, Ajala OZ, Zhu H (2016) Applications in high-content functional protein microarrays. Curr Opin Chem Biol 30:21–27
Mukherjee AK, Carp MJ, Zuchman R et al (2010) Proteomics of the response of Arabidopsis thaliana to infection with *Alternaria brassicicola*. J Proteome 73:709–720
Mur LA, Kenton P, Lloyd AJ et al (2008) The hypersensitive response; the centenary is upon us but how much do we know? J Expt Bot 59:501–520
Nat NVK, Srivastava S, Yajima W et al (2007) Application of proteomics to investigate plant-pathogen interactions. Curr Proteomics 4(1):28–43
Newton A, Fitt BDL, Atkins SD et al (2010) Pathogenesis, parasitism and mutualism in the trophic space of microbe–plant interactions. Trends Microbiol 18:365–373
Novak J, Lemr K, Schug KA et al (2015) CycloBranch: de novo sequencing of nonribosomal peptides from accurate product ion mass spectra. J Am Soc Mass Spectrom 10:1780–1786
Pakkianathan BC, Murad G (2014) Recent advances on interactions between the whitefly *Bemisia tabaci* and *begomoviruses*, with emphasis on *Tomato yellow leaf curl virus*. In: Gaur RK, Hohn T, Sharma P (eds) Plant virus-host interaction. Elsevier, Amsterdam, pp 79–103
Perez-Bueno ML, Rahoutei J, Sajnani C et al (2004) Proteomic analysis of the oxygen evolving complex of photosystem II under biotic stress: studies on *Nicotiana benthamiana* infected with *tobamoviruses*. Proteomics 4:418–425

Phalip V, Delalande F, Carapito C et al (2005) Diversity of the exoproteome of *Fusarium graminearum* grown on plant cell wall. Curr Genet 48:366–379
Pieterse CMJ, Van Loon LC (2004) NPR1: the spider in the web of induced resistance signaling pathways. Curr Opin Plant Biol 4:456–464
Rahoutei J, Baron M, Garcia-Luque I et al (1999) Effect of tobamovirus infection on the thermoluminescence characteristics of chloroplast from infected plants. Z Naturforsch Teil C 54:634–639
Rahoutei J, Garcia-Luque I, Baron M (2000) Inhibition of photosynthesis by viral infection: effect on PSII structure and function. Physiol Plant 110:286–292
Rampitsch C, Bykova NV, Mccallum B et al (2006) Analysis of the wheat and *Puccinia triticina* (leaf rust) proteomes during a susceptible host pathogen interaction. Proteomics 6:1897–1907
Righetti PG, Castagna A, Antonucci F et al (2004) Critical survey of quantitative proteomics in two-dimensional electrophoretic approaches. J Chromatogr A 1051:3–17
Rogowska-Wrzesinska A, Bihan MCL, Thaysen-Andersen M et al (2013) 2D gels still have a niche in proteomics. J Proteome 88:4–13
Romanov V, Davidoff SN, Miles AR et al (2014) A critical comparison of protein microarray fabrication technologies. Analyst 139(6):1303–1326
Rose JKC, Bashir S, Giovannoni JJ et al (2004) Tackling the plant proteome: practical approaches, hurdles and experimental tools. Plant J 5:715–733
Rosen R, Sacher A, Shechter N et al (2004) Two dimensional reference map of *Agrobacterium tumefaciens* proteins. Proteomics 4:1061–1073
Ross PL, Huang YN, Marchese JN et al (2004) Multiplexed protein quantitation in *Saccharomyces cerevisiae* using amine-reactive isobaric tagging reagents. Mol Cell Proteomics 3:1154–1169
Schenk PM, Choo JH, Wong CL (2009) Microarray analyses to study plant defense and rhizosphere microbe interaction. CAB Rev: Perspect Agric Vet Sci Nutr Nat Resour 45:1–14
Schwessinger B, Ronald PC (2012) Plant innate immunity: perception of conserved microbial signatures. Annu Rev Plant Biol 63:451–482
Smith R (2009) Two-dimensional electrophoresis: an overview. In: Tyther R, Sheehan D (eds) Two-dimensional electrophoresis protocols. Humana Press, Totowa, pp 2–17
Speer R, Wulfkuhle JD, Liotta LA et al (2005) 3rd reverse-phase protein microarrays for tissue-based analysis. Curr Opin Mol Ther 7:240–245
Speers AE, Wu CC (2007) Proteomics of integral membrane proteins theory and application. Chem Rev 107(8):3687–3714
Tanaka K, Waki H, Ido Y et al (1988) Protein and polymer analyses up to m/z 100,000 by laser ionization time-of-flight mass spectrometry. Rapid Commun Mass Spectrom 2:151–151
Tseng TT, Tyler BM, Setubal JC (2009) Protein secretion systems in bacterial-host associations, and their description in the gene ontology. BMC Microbiol 9:S2
Unlu M, Morgan ME, Minden JS (1997) Difference gel electrophoresis: a single gel method for detecting changes in protein extracts. Electrophoresis 18:2071–2077
Vadivel AK (2015) Gel-based proteomics in plants: time to move on from the tradition. Front Plant Sci 6:369
Valcu CM, Junqueira M, Shevchenko A (2009) Comparative proteomic analysis of responses to pathogen infection and wounding in *Fagus sylvatica*. J Proteome Res 8:4077–4091
Veenstra TD, Smith RD (eds) (2003) Proteome characterization and proteomics. Academic, San Diego
Ventelon-Debout M, Delalande F, Brizard JP et al (2004) Proteome analysis of cultivar-specific deregulations of *Oryza sativa indica* and *O. sativa japonic*a cellular suspension undergoing rice yellow mottle virus infection. Proteomics 1:216–225
Wang L, Jiang W, Zhang Y et al (2013) Ax21-triggered immunity plays a significant role in rice defense against *Xanthomonas oryzae pv. oryzicola*. Phytopathology. https://doi.org/10.1094/PHYTO-12-12-0333-R
Washburn MP, Wolters D, Yates JR (2001) Large-scale analysis of the yeast proteome by multidimensional protein identification technology. Nat Biotechnol 19(3):242–247
Westermeier R (2006) Electrophoresis in practice. Wiley, Weinheim

White IR, Pickford R, Wood J et al (2004) Statistical comparison of silver and SYPRO ruby staining for proteomic analysis. Electrophoresis 17:3048–3054

Wilkins MR, Sanchez JC, Gooley AA et al (1996) Progress with proteome projects: why all proteins expressed by a genome should be identified and how to do it. Biotechnol Genet Eng Rev 13:19–50

Wittmann-Liebold H, Graack HR, Pohl T (2006) Two dimensional gel electrophoresis as tool for proteomics studies in combination with protein identification by mass spectrometry. Proteomics 17:4688–4703

Yajima W, Kav NN (2006) The proteome of the phytopathogenic fungus *Sclerotinia sclerotiorum*. Proteomics 6:5995–6007

Yang F, Melo-Braga MN, Larsen MR et al (2013) Battle through signaling between wheat and the fungal pathogen *Septoria tritici* revealed by proteomics and phosphoproteomics. Mol Cell Proteomics 12:2497–2508

Yang F, Li W, Derbyshire M et al (2015) Unraveling proteomics and Phosphoproteomics. Mol Cell Proteomics 12:2497–2508. incompatibility between wheat and the fungal pathogen *Zymoseptoria tritici* through apoplastic proteomics. BMC Genomics 16:362

Yates IJR, Gilchrist A, Howell KE et al (2005) Proteomics of organelles and large cellular structures. Nat Rev Mol Cell Biol 6(9):702–714

Zhou H, Ranish JA, Watts JD et al (2002) Quantitative proteome analysis by solid-phase isotope tagging and mass spectrometry. Nat Biotechnol 20(5):512–515

Zhou W, Eudes F, Laroche A (2006a) Identification of differentially regulated proteins in response to a compatible interaction between the pathogen *Fusarium graminearum* and its host, *Triticum aestivum*. Proteomics 6(16):4599–4609

Zhou W, Kolb FL, Riechers DE (2006b) Identification of proteins induced or upregulated by Fusarium head blight infection in the spikes of hexaploid wheat (*Triticum aestivum*). Genome 48(5):770–780

Zhu H, Bilgin M, Bangham R et al (2001) Global analysis of protein activities using proteome chips. Science 293:2101–2105

Zhu H, Bilgin M, Snyder M (2003) Proteomics. Annu Rev Biochem 1:783–812

Zhu W, Smith JW, Huang CM (2009) Mass spectrometry- based label-free quantitative proteomics. J Biomed Biotechnol 2010:Article ID: 840581

Zhu M, Simons B, Zhu N et al (2010) Analysis of abscisic acid responsive proteins in brassica Napus guard cells by multiplexed isobaric tagging. J Proteome 73(4):790–805

Zhu N, Zhu M, Dai S et al (2012) An improved isotope-coded affinity tag technology for thiol redox proteomics. J Integr OMICS 2(1):17–23

Zieske LR (2006) A perspective on the use of iTRAQ reagent technology for protein complex and profiling studies. J Exp Bot 57:1501–1508

Role of Pathogenesis-Related (PR) Proteins in Plant Defense Mechanism

12

Deepti Jain and Jitendra Paul Khurana

Abstract

Plant growth and development is often challenged by several abiotic and biotic stresses, such as drought, cold, salinity, wounding, heavy metals, and pathogen attacks, respectively. A plant responds to these threats by activating a cascade of genes, encoding different effectors, receptors, and signaling and protective molecules. Among all, the induction and accumulation of pathogenesis-related (PR) proteins in plants in response to these adverse conditions is very important as PR proteins are an indispensible component of innate immune responses in plants under biotic or abiotic stress conditions. The PR proteins protect the plants from further infection by not only accumulating locally in the infected and surrounding tissues but also in remote uninfected tissues. Induction of PRs has been reported from many plant species belonging to different families suggesting a general role for these proteins in adaptation to biotic or abiotic stress conditions. PR proteins are also involved in hypersensitive response (HR) or systemic acquired resistance (SAR) against infection. Thus, PR proteins have been defined as "proteins encoded by the host plant but induced only in pathological or related situations," the latter inferring situations of nonpathogenic origin. In this chapter, structure, biochemistry, source, regulation of gene expression, and role in defense mechanism of various pathogenesis-related proteins will be discussed.

Keywords

Abiotic stress · PR proteins · Stress response · Biotic stress · Pathogenesis

D. Jain (✉) · J. P. Khurana
Interdisciplinary Centre for Plant Genomics, Department of Plant Molecular Biology, Delhi University-South Campus, New Delhi 110021, Delhi, India
e-mail: deeptivjain@gmail.com

A. Singh, I. K. Singh (eds.), *Molecular Aspects of Plant-Pathogen Interaction*,
https://doi.org/10.1007/978-981-10-7371-7_12

12.1 Introduction

Plants, as sessile organisms, are often encountered by various abiotic and biotic stresses affecting their growth and agricultural yield. Plant stress tolerance and susceptibility are governed by a complex exchange of signals and responses collectively known by a general term, cellular stress response occurring under given environmental conditions in plants. So, the key difference between resistant and susceptible plants is the timely recognition of the invading pathogen or stress and the rapid and effective activation of host defense mechanisms.

Cellular stress response is a complex trait that happens due to the balanced coordination of physiological and biochemical alterations at the cellular and molecular level. These alterations could be the physical strengthening of the cell wall or through accumulation of various osmolytes, late embryogenesis abundant (LEA) and PR proteins. Physical strengthening of cell wall is often through lignification, suberization, and callose deposition. Cellular alterations mostly coupled with an efficient antioxidant system and prevent pathogen invasion by producing phytoalexins, phenolic compounds, and PR proteins. Various strategies acquired by plants during cellular stress responses serve the adaptive purpose of protecting a cell against unfavorable environmental conditions, both through short-term mechanisms that minimize acute damage to the overall cell's integrity and through long-term mechanisms which provide the cell a measure of pliancy against similar adverse conditions. For any organism sustainability, the cellular machinery must be activated in response to the various stresses, to ensure that the resources are used when required. Accordingly, plant cells have evolved to perceive different signals from their surroundings, to integrate them, and to respond by modulating the appropriate gene expression that may involve protein phosphorylation, ion fluxes, reactive oxygen species (ROS), and other singling events. A diverse array of plant protectants and defense genes get activated whose products include glutathione S-transferases (GST), peroxidases, proteinase inhibitors, cell wall proteins, hydrolytic enzymes (e.g., chitinases and β-1,3-glucanases), pathogenesis-related (PR) proteins, and phytoalexin biosynthetic enzymes, like phenylalanine ammonia lyase (PAL) and chalcone synthase (CHS) (Hammond-Kosack and Jones 1996). Among all, the synthesis and activation of pathogenesis-related (PR) proteins is very critical in response to any stress situation and/or invading pathogen.

During incompatible host-pathogen interactions, the plant's defensive responses restrict the damage caused by the pathogen. Subsequent infection by different types of pathogens is often limited by the defensive responses that is associated with a coordinated and integrated set of metabolic alterations. Further, various novel proteins are synthesized and induced which are collectively known as "pathogenesis-related proteins" (PRs). Pathogen-related proteins (PRs) have been defined as "proteins encoded by the host plant but induced only in pathological or related situations," the latter referring situations of nonpathogenic origin (Antoniw and Pierpoint 1978). PRs have not been identified because of their anti-pathogenic action, but solely because of their easily identification in infected plants. A number of PRs have been reportedly induced in many plant species belonging to various families suggesting a general protective role of PRs against biotic stress (Van Loon 1999). PRs not only accumulate locally in the infected tissue but are also induced systemically. Thus, PRs are associated with the development of systemic acquired resistance (SAR) or hypersensitive response (HR) against further infection by pathogenic fungi, bacteria, and viruses. HR is characterized by necrotic lesions resulting from localized host cell death at the site of infection (Goodman and Novacky 1994). Moreover, plants respond to pathogen infection by activating defense responses in uninfected parts of the plant. As a result, the whole plant develops resistance to subsequent infections. This systemic acquired resistance (SAR) is a generally occurring phenomenon and often confers broad-based resistance to a variety of different pathogens (Ryals et al. 1996; Delaney 1997) instigating the defensive capacity of plants in response to necrotizing infections. Over the decades, a number of reports depicted the role of different classes of PR proteins during abiotic and biotic stresses and their defense responses in plants; however, the mechanism of action is sparsely described. In this chapter, we will be briefly discussing the biochemistry, source, regulation of gene expression, and mode of action of PR proteins in defense mechanisms.

12.2 PR Proteins: An Overview

PR proteins comprise a huge family of proteins ubiquitous in the plant kingdom. Plant PR proteins were first identified and reported in tobacco plants infected by tobacco mosaic virus (Van Loon and Van Kammen 1970). Later, these proteins were reported in many different plant species. Most plant PR proteins share the common biochemical properties of being acid soluble, low molecular weight, and protease resistance (Leubner-Metzger and Meins 1999; Neuhaus 1999). PR proteins have similar functions, but depending on their isoelectric points, they may be acidic or basic proteins. Most acidic PR proteins are secreted in the extracellular spaces, whereas basic PR proteins are predominantly found in the vacuole (Legrand et al. 1987; Niki et al. 1998) through the signal peptide at C-terminus. However, such localization cannot be generalized for all PR proteins. Though PRs are most abundant in the leaves, they are detected in almost all plant organs including leaves, stems, roots, and flowers. Usually, acidic PRs are upregulated by various signaling molecules like salicylic acid (Yalpani et al. 1991; Sinha et al. 2014) and reactive oxygen species (Chamnongpol et al. 1998), while basic PRs are upregulated by gaseous phytohormone ethylene and methyl jasmonate (Xu et al. 1994) during pathogen attack. Apart from various environmental factors, there are certain internal developmental factors too that trigger the synthesis of these PR proteins.

Based on molecular mass, isoelectric point, localization, and biological activity, the PR proteins have been categorized into 17 families (Table 12.1), including β-1,3-glucanases, chitinases, thaumatin-like proteins, peroxidases, ribosome-inactivating proteins, thionins, nonspecific lipid transfer proteins, oxalate oxidase, and oxalate-oxidase-like proteins (Van Loon and Van Strien 1999) and numbered in the order in which they were discovered. When dealing with a stress-related sequence possibly related with PRs, it is necessary to gather information at both the nucleic acid and the protein level as homologies at the cDNA, or genomic level may be encountered without information on the expression of the encoded protein.

Table 12.1 Classification of pathogenesis-related proteins

Families	Properties	Example
PR-1	Antifungal	Tobacco PR-1a
PR-2	β-1,3-Glucanase	Tobacco PR-2
PR-3	Chitinase type I, II, IV, V, VI, VII	Tobacco P, Q
PR-4	Chitinase type I, II	Tobacco “R”
PR-5	Thaumatin-like	Tobacco S
PR-6	Proteinase inhibitor	Tomato inhibitor I
PR-7	Endoproteinase	Tomato P69
PR-8	Chitinase type III	Cucumber chitinase
PR-9	Peroxidase	Tobacco “lignin-forming peroxidase”
PR-10	Ribonuclease-like	Parsley “PR1”
PR-11	Chitinase, type I	Tobacco “class V” chitinase
PR-12	Defensin	Radish Rs-AFP3
PR-13	Thionin	Arabidopsis THI2.1
PR-14	Lipid transfer protein	Barley LTP4
PR-15	Oxalate oxidase	Barley OxOa (germin)
PR-16	Oxalate oxidase-like	Barley OxOLP
PR-17	Unknown	Tobacco PRp27

Though such sequences belong to the PR-type families, they are to be named PR-like proteins (PRLs) and cannot be considered to correspond to pathogen-induced PRs (Van Loon et al. 1994). Among these PRLs, chitinases and β-1,3-glucanases are two important hydrolytic enzymes that accumulate in many plant species after infection by different types of pathogens. These hydrolytic enzymes play the main role of defense reaction against fungal pathogen by degrading cell wall, because chitin and β-1,3-glucan are major structural components of the cell walls of many pathogenic fungi. Some of the tobacco PRs were characterized as chitinases and β-1,3-glucanases (Kauffmann et al. 1987) with potential antifungal activity suggested the group of PRs might be inhibiting pathogen growth and be responsible for the SAR. In spite of their common name, PR proteins show a great diversity in species specificity and in the mechanism of action and do not share any structural relationship among themselves.

PR proteins unveil diverse functions within the plant. Many PRs exhibit antifungal activity (Caruso et al. 1996) though few of the PR proteins also show antibacterial, insecticidal, nematicidal, and antiviral activity (Edreva 2005). PR proteins thus have a critical role in disease resistance, seed germination, and plant facilitation to adapt to the environmental stress.

12.2.1 Plant Chitinases

Chitinases (E.C. 3.2.1.14) are widely distributed across plant, animal, fungi, and bacteria kingdoms. These enzymes catalyze the cleavage of a bond between C1 and C4 of two consecutive N-acetyl-D-glucosamine monomers of chitin which is a common component of fungal cell walls and of the exoskeleton of arthropods (Bartnicki-Garcia 1968). Usually the plant chitinases are endo-chitinases capable of degrading chitin as well as inhibit fungal growth (Schlumbaum et al. 1986; Broekaert et al. 1988). Chitinases are either localized in the vacuole (Class I) or outside the cell (Class III) (Neuhaus et al. 1996). Many reports strongly indicated that chitinases, together with β-1,3 glucanases, play critical role in the plant defense response against fungal pathogens (Abeles et al. 1971).

Plant chitinases have been classified into seven classes, class I through VII, based on their primary structures. Certain isoforms of chitinases are induced by particular elicitors, and only few isoforms have antifungal activities, while some isoforms have shown another role like antifreeze activity (Sela-Buurlage et al. 1993; Yeh et al. 2000). Class I chitinases have a cysteine-rich N-terminal chitin-binding domain (CBD) that is homologous to havein, a chitin-binding lectin from the rubber tree (Suarez et al. 2001). Class II chitinases are similar to class I but they lack the N-terminal CBD. Class III chitinases are unique in structure and belong to the PR-8 family and family 18 of glycosyl-hydrolases. Class III chitinases are more closely related to the bacterial chitinases and generally have lysozyme activity. Class IV, V, VI, and VII chitinases belong to the PR-3 family of proteins (Meins et al. 1994).

Various biotic and abiotic factors have been known to induce chitinases in the plant. Various studies reported the inhibitory effect of plant chitinases on fungal growth by demonstrating on the growth of chitin-containing fungi (Mauch et al. 1988a, b). Various studies showed that chitinase expression is induced against phytopathogen systems, and resistant varieties have stronger upregulation than susceptible varieties in the sugar beet (Nielsen et al. 1992), wheat (Anguelova-Merhar et al. 2001), and tomato (Lawrence et al. 2000). Transformation of chitinase genes was performed in tobacco (Brogue et al. 1991), grapevine (Yamamoto et al. 2000), rice (Datta et al. 2001), and peanut (Rohini and Rao 2001), and enhanced disease resistance has been achieved. Overexpression studies have been made with chitinase genes, alone or together with β-1,3-glucanase genes in a number of plant species, and in most cases, the resulting transgenic plants exhibited enhanced levels of fungal disease resistance or delayed symptom development as compared to the control plants (Zhu et al. 1994; Jach et al. 1995; Jongedijk et al. 1995). However, several studies showed that plants transformed with either chitinase or β-1,3-glucanase gene alone did not exhibit resistance to certain pathogens or showed less resistance. Plant chitinases alone are unable to effectively degrade harder chitin structures of fungi as they usually affect only the hyphal tip, but when coexpressed with β-1,3-glucanase, these two enzymes act synergistically against fungal pathogen.

12.2.2 Plant β-1,3-Glucanases

Plant β-1,3-glucanases belong to the PR-2 family of pathogenesis-related proteins and reportedly play an important role in plant defense responses to pathogen infection. These enzymes have been identified across plants, yeasts, actinomycetes, bacteria, fungi, insects, and fish (Pan et al. 1989). These enzymes catalyze the cleavage of the β-1,3-glucosidic bonds in β-1,3-glucan (Simmons 1994) which is another major structural component of the cell walls of many pathogenic fungi (Adams 2004). β-1,3-Glucanases play an important role in plant defense and other physiological functions such as cell division and cell elongation (Fulcher et al. 1976), fruit ripening (Meins et al. 1992), pollen germination and tube growth (Meikle et al. 1991), fertilization (Ori et al. 1990), somatic embryogenesis (Helleboid et al. 2000), seed germination (Buchner et al. 2002), and flower formation (Akiyama et al. 2004). The role of plant β-1,3-glucanases as an important component of plant defense mechanisms against pathogens has been well documented (Legrand et al. 1987; Cordero et al. 1994). It has been postulated that β-1,3-glucanases hydrolyze fungal cell walls, which consequently causes the lysis of fungal cells when defending against fungi. On the pathogen encounter, β-1,3-glucanases also cause the formation of oligosaccharide elicitors, which elicit the production of other PR proteins or low molecular weight antifungal compounds, such as phytoalexins (Klarzynski et al. 2000).

β-1,3-Glucanase genes have been reported from a wide range of plant species, and many studies have shown that the synthesis of β-1,3-glucanases is stimulated by pathogen infections (Alonso et al. 1995; Roulin et al. 1997) and can change during plant development (Wyatt et al. 1991). Different plant species may have different β-1,3-glucanase genes, and a single plant species may have various copies of β-1,3-glucanase genes. Plant β-1,3-glucanases were classified into two major classes I and II and two minor classes based on their amino acid sequence, structural properties, and cellular localizations (Beerhues and Kombrink 1994). Generally, β-1,3-glucanases are stress regulated, but a few β-1,3-glucanases are exclusively developmentally regulated and do not show a stress-related regulation (Bucciaglia and Smith 1994; Sharma 2013). β-1,3-Glucanases are usually expressed at low concentration in plants, but when plants are challenged by fungal, bacterial, or viral pathogens, β-1,3-glucanases enzyme accumulate dramatically (Castresana et al. 1990; Lusso and Kuc 1995). Class I β-1,3-glucanases and class I chitinases showed synergistic effect in pathogen defense. Class I β-1,3-glucanase accumulated only at the site of tobacco mosaic virus (TMV) infection in tobacco plants, while class II and III β-1,3-glucanases accumulated both at the site of infection and systemically (Vögeli-Lange et al. 1994; Livne et al. 1997). Many reports showed that the transcript levels of glucanases accumulated after infected with pathogens, such as barley infected by powdery mildew (Ignatius et al. 1994), maize infected with *Aspergillus flavus* (Lozovaya et al. 1998), pepper infected with *Xanthomonas campestris* pv. *vesicatoria* and *Phytophthora capsici* (Jung and Hwang 2000), wheat infected with *Fusarium graminearum* (Li et al. 2001), chickpea infected with *Ascochyta rabiei* (Pass.) Labr. (Hanselle and Barz 2001), and peach infected with *Monilinia fructicola* (Zemanek et al. 2002). β-1,3-Glucanases and other PR protein induction in the plant can also occur due to some elicitors, including fungal β-glucan, chitin, chitosan, glycoproteins, and N-acetylchito oligosaccharides (Chang et al. 1992; Kaku et al. 1997), or by other factors, for example, salicylic acid-induced accumulation of mRNAs of class II and III β-1,3-glucanases in wild-type tobacco plants (Ward et al. 1991), abscisic acid (ABA) in tobacco (Rezzonico et al. 1998), and methyl jasmonate, ethylene, and gibberellin A3 in tomato seeds and leaves (Wu and Bradford 2003). Stress factors like wounding, drought, exposure to heavy metals, air pollutant ozone, and ultraviolet radiation can also upregulate β-1,3-glucanases in some plants (Thalmair et al. 1996; Fecht-Christoffers et al. 2003).

12.3 Role of PR Proteins in SAR and HR

PRs are most common in hypersensitive responses but appear to contribute to SAR also. An induced systemic resistance (ISR) can be induced by nonpathogenic rhizobacteria, considerably modified the relationships between necrotic lesion formation, PRs, and SAR. ISR induction with these rhizobacteria shows no symptoms in plants; however, this resistance is independent of the production of salicylic acid (SA) by the plant and is not associated with the accumulation of PRs (Pieterse et al. 1996; Van Loon et al. 1998). This indicates that plants can substantially enhance their defensive capacity against variety of pathogens in either SA-dependent or independent way (Pieterse and Van Loon 1999). SAR is mainly SA-dependent (Ryals et al. 1996) while ISR is SA-independent. Until now the mechanism involved in ISR has been unclarified. At least in *Arabidopsis*, similar to SAR, ISR depends on the functioning of the *npr1* gene, which in turn distinguishes ISR from the JA- and ethylene-dependent inducible defense response pathway effective against *Alternaria brassicicola*, which is independent of *npr1*. PRs are often associated with SAR, but not with ISR, which have led to hypothesize that PR accumulation is not a prerequisite for the induction of resistance, but they contribute to the protective state (Van Loon 1997). The JA- and ethylene-dependent pathway induced by, and effective against, *A. brassicicola* involves increase in SA, JA, and ethylene levels resulting in detection of PRs in the infected plants. The differential expression of various PRs determines the extent of the plant's response and its effectiveness to inhibit further infection. Recent report shows that SA-dependent expression of PR-1, PR-2, and PR-5 is required for increased protection against the biotrophic fungus *Peronospora parasitica* in *Arabidopsis*, whereas SA-independent but JA-dependent induction of PR-3 and PR-4 is associated with the induced resistance against the necrotrophic fungi *A. brassicicola* (Penninckx et al. 1996), *Botrytis cinerea* (Thomma et al. 1998), and *Fusarium oxysporum* f.sp. *matthiolae* (Bohlmann et al. 1998). These results suggest that PRs appear to contribute differentially to the induced resistance against different pathogens.

12.4 Signaling Involved in Pathogen-Induced Expression of PRs

The pathogen-activated PR gene expression plays a critical role in plant defense against pathogens. Regulation of PR gene expression has always been a highly active research area since the time PR proteins were discovered. However, the signaling behind the pathogen-induced PR gene expression is still poorly understood in plants. This is partly due to the complexity of environmental stimuli and stimulation by phytohormones that can induce the expression of various PR genes (Brederode et al. 1991).

12.5 Signals and Putative Receptors Involved in PR Gene Expression

During plant-pathogen interactions, a number of molecules derived from pathogens can serve as elicitors for PR gene induction such as chitin fragments and glucans from fungal cell wall, extracellular glycoproteins/peptides from few fungal species, oligosaccharides and harpins from bacteria, and Avr proteins derived from bacterial and fungal pathogens (Boller and Felix 1996). Though a large number of signals are known to induce PR gene expression, no receptors have been unambiguously established for these signal molecules. For example, β-glucan elicitor (GE) is released from *Phytophthora sojae* cell wall by β-1,3-glucanase from soybean and reportedly induces phytoalexin biosynthesis (Darvill and Albersheim 1984). Recently, a GE-binding protein (GEBP) has been purified from soybean whose antiserum partially inhibited the binding of GE to soybean membrane proteins and reduced the phytoalexin accumulation elicited by GE. Another set of elicitors is the polypeptide encoded by pathogen avirulence *(avr)* genes. Any pathogen containing a particular *avr* gene is recognized by the corresponding resistance *(R)* gene of the host plant and activates a variety of defense responses in the host, including increased PR gene expression. In the past decade, a number of *R* genes and *avr* genes have been isolated. For example, an elicitor encoded by the *avr9* gene from *Cladosporium fulvum*, a fungal pathogen of tomato, rapidly activated the transcription of glucanase and chitinase genes in plants carrying the cognate *R* gene *Cf9* (Ashfield et al. 1994; Wubben et al. 1996). Another fungal elicitor, NIP1 protein from the barley pathogen *Rhynchosporium secalis* is known to activate PR genes (Rohe et al. 1995). The only evidence that an Avr protein directly interacts with an R gene product comes from the study of bacterial speck disease in tomato. Tomato plants having *R* gene *Pto*, encoding cytoplasmic kinase, are resistant to the bacterial pathogen *Pseudomonas syringae* pv. *tomato* carrying the *avrPto* gene (Martin et al. 1993). The Pto and AvrPto proteins showed a highly specific association in yeast too (Scofield et al. 1996; Tang et al. 1996). The results confirmed Pto as the receptor for the AvrPto protein.

12.6 Different Pathways for PR Genes' Activation by Pathogens

Plants often exhibit increased production of reactive oxygen species (ROS), salicylic acid (SA), ethylene, and jasmonates upon infection by pathogens (Hammond-Kosack and Jones 1996; Yang et al. 1997). These molecules may serve as secondary signals to activate plant defense, and many of these reportedly work as inducers for PR gene expression. For example, SA induces acidic PR genes that are normally activated during SAR, whereas ethylene and jasmonates are known to induce proteinase inhibitors, defensin, thionin, and basic PR proteins (Brederode et al. 1991; Ward et al. 1991; Epple et al. 1995; Donnell, et al. 1996; Penninckx et al. 1996). However, the involvement of secondary messengers in the PR gene induction is uncertain in majority of studies. Cross talks are often common between signaling pathways mediated by these secondary messengers. The use of signaling pathway mutants would be supportive in clarifying the roles of these secondary messengers in plant defense responses against pathogen attacks.

12.7 Functions and Relevance of PR Expression in Disease Resistance

In the last decades, proteinase, peroxidase, ribonuclease, and lysozyme activities were assigned to PR-7, PR-9, PR-10, and PR-8, respectively. Also, membrane-permeabilizing functions are characteristic of defensins (PR-12), thiols (PR-13), lipid transfer proteins (LTPs, PR-14), and of osmotins and thaumatin-like proteins (PR-5). Multiple enzymatic, structural, and receptor functions are reported in germins (PR-15) and germin-like proteins (PR-16) (Van Loon and Van Strien 1999; Bernier and Berna 2001; Selitrennikoff 2001; Park et al. 2004a, b). Besides this, some PRs also exhibited antibacterial, insecticidal, nematicidal, and antiviral action, though an important common feature of most PRs is their antifungal effect. Their hydrolytic, proteinase inhibitory, and membrane-permeabilizing ability made them toxic to pathogens. Thus, hydrolytic enzymes (β-1,3-glucanases, chitinases, and proteinases) can effectively weaken and decompose fungal cell walls, containing glucans, chitin, and proteins, while PR-8 can damage gram-positive bacteria due to lysozyme activity (Van Loon and Van Strien 1999; Selitrennikoff 2001). The defensive functions of PRs against pathogens can be attributed to a number of their ingenious properties; their constitutive expression in seeds and plant organs, high fungitoxicity of seed osmotins and thaumatin-like proteins (Vigers et al. 1992; Abad et al. 1996), their accumulation in plant cell wall appositions formed against pathogen invasions (Jeun 2000; Jeun and Buchenauer 2001). Inspite of all studies and reports, the defensive mechanism of PR function against pathogen attack is still unclear. The protective role of PRs is supported by following evidences:

(a) *Transcript accumulation of PRs in pathogen-tolerant and susceptible plants.* Recently, the differential responses of resistant/susceptible plants were reported in tomato plants, inoculated with *Cladosporium fulvum* (Wubben et al. 1996), *Phytophthora infestans*-infected potato (Tonón et al. 2002), *Venturia inaequalis*-inoculated apple (Poupard et al. 2003), *Pseudomonas syringae*-infected grapevine (Robert et al. 2001), *Xanthomonas campestris* pv. *vesicatoria*, and TMV-Po-infected hot pepper (Park et al. 2004a, b).
(b) *The plants with high natural disease resistance constitutively express PRs.* This correlation has been proved in many pathosystems, such as apple-*Venturia*

inaequalis (Gau et al. 2004), tomato-*Alternaria solani* (Lawrence et al. 2000), and potato-*Phytophthora infestans* (Vleeshouwers et al. 2000).

(c) *Overexpressing PRs in transgenic plants results in increased resistance to pathogens.* Tobacco overexpressing *PR1a* gene showed increased tolerance to *Peronospora tabacina* and *Phytophthora parasitica* var. *nicotianae* (Alexander et al. 1993). Similarly, overexpression of thaumatin-like PR-5 in transgenic rice and orange plants showed increased tolerance to *Rhizoctonia solani* and *Phytophthora citrophthora*, respectively (Datta et al. 1999, Fagoaga et al. 2001); transgenic potato overexpressing PR-2 and PR-3 had improved resistance to *Phytophthora infestans* (Bachmann et al. 1998); transgenic carrot overexpressing PR-2 and PR-3 genes, coding for β-1,3-glucanase and chitinase, respectively, showed increased resistance to several fungal pathogens; and the transgenic tomato simultaneously expressing tobacco β-1,3-glucanase and chitinase genes had improved resistance to fungal pathogens (Melchers et al. 1998).

(d) *Accumulation of PRs in plants with locally or systemically induced resistance.* As discussed before, PRs are identified as markers of the systemic acquired resistance (SAR). SAR and the associated set of PRs are induced by different pathogens and various chemicals predominantly in a salicylic acid-dependent pathway. It is important to note that the direct role of PRs in disease resistance is being suggestive by their high expression in resistant or SAR-expressing plants, as well as transgenic resistant plants exhibiting high antimicrobial activity (Rauscher et al. 1999; Tonón et al. 2002; Anand et al. 2004).

12.8 Transcriptional Regulation of PR Gene Expression

The most active area in PR gene research is to study its transcriptional regulation. Several *cis*-regulatory elements mediating pathogen-induced PR gene expression have been identified and characterized through traditional promoter deletion analysis coupled with mutagenesis of putative regulatory elements, gain-of-function studies with synthetic promoters, and DNA-fingerprinting analysis (Yang et al. 1997; Rushton and Somssich 1998). Many of these elements are W-box (consensus TTGACC or TGAC-[N]x-GTCA), GCC-box (consensus AGCCGCC), MRE-like sequence (consensus A[A/C]C[A/T]A[A/C]C), G-box (consensus CACGTG), and SA-responsive element (SARE, with a consensus of TTCGACCTCC). Among all, the GCC-box and W-box have been extensively studied and have a wide role in PR gene regulation (Hart et al. 1993; Ohme-Takagi and Shinshi 1995). Defense responses mediated by ethylene are often associated with GCC-box and are known to confer ethylene-induced transcription of the tobacco *gln2* gene encoding a β-1,3-glucanase (Ohme-Takagi and Shinshi 1995). EREB clones (1–4) were identified in tobacco cDNA library with radiolabeled GCC-box as probe (Ohme-Takagi and Shinshi 1995). EREBP transcripts are induced by ethephon, a compound known to release ethylene upon its degradation that suggests that ethylene further induces the expression of EREBP genes. The EREBP-1 gene was reportedly induced by *Pseudomonas* bacteria and SA, suggesting a role of this gene in plant defense (Zhou et al. 1997; Horvath et al. 1998). The direct correlation of the EREBP proteins with a disease-resistance pathway was confirmed by the study of the signaling pathway mediated by the tomato gene *Pto* (Zhou et al. 1997). Phosphorylation also plays an important role in the activation of PR gene expression during pathogen attacks (Raz and Fluhr 1993). Another highly conserved *cis*-element, W-box, is present in parsley *PR1-1* and *PR1-2* (both encoding the PR1 protein), tobacco *CHN50* (encoding a class I basic chitinase), asparagus *AoPR1* (encoding the PR10 protein), potato *PR-10a* (encoding the PR10 protein), and maize *PRms* (encoding the PR1 protein). Besides PRs, W-box is also present in the promoter of other pathogen inducible genes such as the potato glutathione S-transferase gene *prp1* and the grape phytoalexin synthesis gene *Vst1*, suggesting a broader role for this element in pathogen-induced gene expression (Rushton and Somssich 1998). Three parsley cDNAs encoding W-box-binding proteins were identified by using a south-western screening (Rushton et al. 1996). These proteins are termed WRKY family proteins and contain the consensus sequence, WRKYGQK. Since the in vivo function of the cloned transcription factors is still to be worked upon, so rigorous tests on transgenic plants with altered expression of the transcription factor genes are required to establish their roles in PR gene expression and defense responses. In addition, it is necessary to answer few questions. How different signals affect PR gene expression? Do different signals converge on the same transcription factor? Do these transcription factors interact? Answering these questions help us in better understanding of cross talks between different signaling pathways.

12.9 Conclusions

PR proteins and their homologues are generally responsible for the defense against various stresses including pathogen attacks, wounding, use of chemicals, and pollutants. However, many PR proteins (members of PR 1, 2, 3, 4, 5, 8, 10, and 14 families) have demonstrated allergenicity, but the allergenicity is also guided by several environmental factors like the use of chemical inducers in agriculture and environmental pollutants. Recent reports have documented their critical importance as preservative agents in food industry and for producing disease-resistant plants by genetic engineering. Various studies have revealed that transgenic plants overexpressing PR genes mediate host plant resistance to phytopathogenic fungi. Such genetically modified (GM) plants with enhanced expression of PR proteins may be associated with increased allergenicity and toxicity, thus raising a serious question for their commercial acceptability, so different strategies are adopted to monitor the transformed crops for their allergenicity.

References

Abad LR et al (1996) Antifungal activity of tobacco osmotin has specificity and involves plasma membrane permeabilization. Plant Sci 118(1):11–23

Abeles FB et al (1971) Preparation and purification of glucanase and chitinase from bean leaves. Plant Physiol 47(1):129–134

Adams DJ (2004) Fungal cell wall chitinases and glucanases. Microbiology 150(7):2029–2035

Akiyama T et al (2004) Cloning, characterization and expression of OsGLN2, a rice endo-1, 3-β-glucanase gene regulated developmentally in flowers and hormonally in germinating seeds. Planta 220(1):129–139

Alexander D et al (1993) Increased tolerance to two oomycete pathogens in transgenic tobacco expressing pathogenesis-related protein 1a. Proc Natl Acad Sci 90(15):7327–7331

Alonso E et al (1995) Differential in vitro DNA binding activity to a promoter element of the gn1β-1, 3-glucanase gene in hypersensitively reacting tobacco plants. Plant J 7(2):309–320

Anand A et al (2004) Apoplastic extracts from a transgenic wheat line exhibiting lesion-mimic phenotype have multiple pathogenesis-related proteins that are antifungal. Mol Plant-Microbe Interact 17(12):1306–1317

Anguelova-Merhar VS et al (2001) β-1, 3-Glucanase and Chitinase activities and the resistance response of wheat to leaf rust. J Phytopathol 149(7–8):381–384

Antoniw JF, Pierpoint WS (1978) Purification of a tobacco leaf protein associated with resistance to virus infection [proceedings]. Biochem Soc Trans 6(1):248–250

Ashfield T et al (1994) Cf gene-dependent induction of a b-1, 3-glucanase promoter in tomato plants infected with Cladosporium fulvum. MPMI-Mol Plant Microbe Interact 7(5):645–656

Bachmann D et al (1998) Improvement of potato resistance to Phytophthora infestans by overexpressing antifungal hydrolases. 5th international workshop on pathogenesis-related proteins. Signaling pathways and biological activities. Aussois

Bartnicki-Garcia S (1968) Cell wall chemistry, morphogenesis, and taxonomy of fungi. Annu Rev Microbiol 22(1):87–108

Beerhues L, Kombrink E (1994) Primary structure and expression of mRNAs encoding basic chitinase and 1, 3-β-glucanase in potato. Plant Mol Biol 24(2):353–367

Bernier F, Berna A (2001) Germins and germin-like proteins: plant do-all proteins. But what do they do exactly? Plant Physiol Biochem 39(7):545–554

Bohlmann H et al (1998) Wounding and chemicals induce expression of the Arabidopsis thaliana gene Thi2. 1, encoding a fungal defense thionin, via the octadecanoid pathway. FEBS Lett 437(3):281–286

Boller T, Felix G (1996) Olfaction in plants: specific perception of common microbial molecules. In: Stacey G, Mullin B, Gresshoff PM (éds) Biology of plant-microbe interactions. International Society for Molecular Plant-Microbe Interactions, Knoxville, É.-U:1-8

Brederode FT et al (1991) Differential induction of acquired resistance and PR gene expression in tobacco by virus infection, ethephon treatment, UV light and wounding. Plant Mol Biol 17(6):1117–1125

Broekaert WF et al (1988) Comparison of some molecular, enzymatic and antifungal properties of chitinases from thorn-apple, tobacco and wheat. Physiol Mol Plant Pathol 33(3):319–331

Brogue K et al (1991) Transgenic plants with enhanced resistance to the fungal pathogen *Rhizoctonia solani*. Science 254(5035):1194–1197

Bucciaglia PA, Smith AG (1994) Cloning and characterization of tag 1, a tobacco anther β-1, 3-glucanase expressed during tetrad dissolution. Plant Mol Biol 24(6):903–914

Buchner P et al (2002) Characterization of a tissue-specific and developmentally regulated β-1, 3-glucanase gene in pea (*Pisum sativum*). Plant Mol Biol 49(2):171–186

Caruso C et al (1996) Structural and antifungal properties of a pathogenesis-related protein from wheat kernel. J Protein Chem 15(1):35–44

Castresana C et al (1990) Tissue-specific and pathogen-induced regulation of a *Nicotiana plumbaginifolia beta-1, 3-glucanase* gene. Plant Cell 2(12):1131–1143

Chamnongpol S et al (1998) Defense activation and enhanced pathogen tolerance induced by H2O2 in transgenic tobacco. Proc Natl Acad Sci 95(10):5818–5823
Chang M-M et al (1992) Molecular characterization of a pea *β-1, 3-glucanase* induced by *Fusarium solani* and chitosan challenge. Plant Mol Biol 20(4):609–618
Cordero MJ et al (1994) Differential expression and induction of chitinases and β1- 3-glucanases in response to fungal infection during germination of maize seeds. Mol Plant Microbe Interact 7:23–31
Darvill AG, Albersheim P (1984) Phytoalexins and their elicitors-a defense against microbial infection in plants. Annu Rev Plant Physiol 35(1):243–275
Datta K et al (1999) Over-expression of the cloned rice thaumatin-like protein (PR-5) gene in transgenic rice plants enhances environmental friendly resistance to *Rhizoctonia solani* causing sheath blight disease. Theor Appl Genet 98(6–7):1138–1145
Datta K et al (2001) Enhanced resistance to sheath blight by constitutive expression of infection-related rice chitinase in transgenic elite indica rice cultivars. Plant Sci 160(3):405–414
Delaney TP (1997) Genetic dissection of acquired resistance to disease. Plant Physiol 113(1):5
Donnell PJO et al (1996) Ethylene as a signal mediating the wound response of tomato plants. Science 274(5294):1914
Edreva A (2005) Pathogenesis-related proteins: research progress in the last 15 years. Gen Appl Plant Physiol 31(1–2):105–124
Epple P et al (1995) An *Arabidopsis thaliana thionin* gene is inducible via a signal transduction pathway different from that for pathogenesis-related proteins. Plant Physiol 109(3):813–820
Fagoaga C et al (2001) Increased tolerance to *Phytophthora citrophthora* in transgenic orange plants constitutively expressing a tomato pathogenesis related protein PR-5. Mol Breed 7(2):175–185
Fecht-Christoffers MM et al (2003) Effect of manganese toxicity on the proteome of the leaf apoplast in cowpea. Plant Physiol 133(4):1935–1946
Fulcher RG et al (1976) β-1, 3-glucans may be associated with cell plate formation during cytokinesis. Can J Bot 54(5–6):539–542
Gau AE et al (2004) Accumulation of pathogenesis-related proteins in the apoplast of a susceptible cultivar of apple *(Malus domestica cv. Elstar)* after infection by *Venturia inaequalis* and constitutive expression of PR genes in the resistant cultivar Remo. Eur J Plant Pathol 110(7):703–711
Goodman RN, Novacky AJ (1994) The hypersensitive reaction in plants to pathogens: a resistance phenomenon. Am Phytopathol Soc (APS)
Hammond-Kosack KE, Jones JD (1996) Resistance gene-dependent plant defense responses. Plant Cell 8(10):1773
Hanselle T, Barz W (2001) Purification and characterisation of the extracellular PR-2b β-1, 3-glucanase accumulating in different *Ascochyta rabiei*-infected chickpea (Cicer arietinum L.) cultivars. Plant Sci 161(4):773–781
Hart CM et al (1993) A 61 bp enhancer element of the tobacco *β-1, 3-glucanase B* gene interacts with one or more regulated nuclear proteins. Plant Mol Biol 21(1):121–131
Helleboid S et al (2000) Cloning of β-1, 3-glucanases expressed during *Cichorium* somatic embryogenesis. Plant Mol Biol 42(2):377–386
Horvath DM et al (1998) Four classes of salicylate-induced tobacco genes. Mol Plant-Microbe Interact 11(9):895–905
Ignatius SMJ et al (1994) Effects of fungal infection and wounding on the expression of chitinases and β-1, 3 glucanases in near-isogenic lines of barley. Physiol Plant 90(3):584–592
Jach G et al (1995) Enhanced quantitative resistance against fungal disease by combinatorial expression of different barley antifungal proteins in transgenic tobacco. Plant J 8(1):97–109
Jeun YC (2000) Immunolocalization of PR-protein P14 in leaves of tomato plants exhibiting systemic acquired resistance against *Phytophthora infestans* induced by pretreatment with 3-aminobutryic acid and preinoculation with tobacco necrosis virus. J Plant Dis Prot:352–367
Jeun YCH, Buchenauer H (2001) Infection structures and localization of the pathogenesis-related protein AP 24 in leaves of tomato plants exhibiting systemic acquired resistance against

Phytophthora infestans after pre-treatment with 3-aminobutyric acid or tobacco necrosis virus. J Phytopathol 149(3–4):141–154
Jongedijk E et al (1995) Synergistic activity of chitinases and β-1, 3-glucanases enhances fungal resistance in transgenic tomato plants. Euphytica 85(1–3):173–180
Jung HW, Hwang BK (2000) Pepper gene encoding a basic β-1, 3-glucanase is differentially expressed in pepper tissues upon pathogen infection and ethephon or methyl jasmonate treatment. Plant Sci 159(1):97–106
Kaku H et al (1997) N-acetylchitooligosaccharides elicit expression of a single (1→ 3)-β-glucanase gene in suspension-cultured cells from barley (Hordeum vulgare). Physiol Plant 100(1):111–118
Kauffmann S et al (1987) Biological function of pathogenesis-related' proteins: four PR proteins of tobacco have 1, 3-β-glucanase activity. EMBO J 6(11):3209
Klarzynski O et al (2000) Linear β-1, 3 glucans are elicitors of defense responses in tobacco. Plant Physiol 124(3):1027–1038
Lawrence CB et al (2000) Constitutive hydrolytic enzymes are associated with polygenic resistance of tomato to *Alternaria solani* and may function as an elicitor release mechanism. Physiol Mol Plant Pathol 57(5):211–220
Legrand M et al (1987) Biological function of pathogenesis-related proteins: four tobacco pathogenesis-related proteins are chitinases. Proc Natl Acad Sci U S A 84(19):6750–6754
Leubner-Metzger G, Meins Jr F (1999) 3 functions and regulation of plant β-(PR-2). Pathogenesis-related proteins in plants
Li WL et al (2001) Isolation and characterization of novel cDNA clones of acidic chitinases and β-1, 3-glucanases from wheat spikes infected by *Fusarium graminearum*. Theor Appl Genet 102(2–3):353–362
Livne B et al (1997) TMV-induced expression of tobacco β-glucanase promoter activity is mediated by a single, inverted, GCC motif. Plant Sci 130(2):159–169
Lozovaya VV et al (1998) β-l, 3-glucanase and resistance to *Aspergillus flavus* infection in maize. Crop Sci 38(5):1255–1260
Lusso M, Kuc J (1995) Evidence for transcriptional regulation of β-1, 3-glucanase as it relates to induced systemic resistance of tobacco to blue mold. Mol Plant-Microbe Interact 8(3):473–475
Martin GB et al (1993) Map-based cloning of a protein kinase gene conferring disease resistance in tomato. Sci-New York Wash 262:1432–1432
Mauch F et al (1988a) Antifungal hydrolases in pea tissue I. Purification and characterization of two chitinases and two β-1, 3-glucanases differentially regulated during development and in response to fungal infection. Plant Physiol 87(2):325–333
Mauch F et al (1988b) Antifungal hydrolases in pea tissue II. Inhibition of fungal growth by combinations of chitinase and β-1, 3-glucanase. Plant Physiol 88(3):936–942
Meikle PJ et al (1991) The location of (1→ 3)-β-glucans in the walls of pollen tubes of Nicotiana alata using a (1→ 3)-β-glucan-specific monoclonal antibody. Planta 185(1):1–8
Meins F Jr et al (1992) The primary structure of plant pathogenesis-related glucanohydrolases and their genes. Springer, Berlin
Meins F et al (1994) Plant chitinase genes. Plant Mol Biol Report 12(2):S22–S28
Melchers LS et al (1998) The utility of PR genes to develop disease resistance in transgenic crops. 5th international workshop on pathogenesis-related proteins. Signalling pathways and biological activities
Neuhaus JM (1999) Plant chitinases (pr-3, pr-4, pr-8, pr-11). Pathogenesis-related proteins in plants: 77–105
Neuhaus JM et al (1996) A revised nomenclature for chitinase genes. Plant Mol Biol Report 14(2):102–104
Nielsen KK et al (1992) An acidic class III chitinase in sugar beet: induction by *Cercospora beticola*, characterization, and expression in transgenic tobacco plants. Mol Plant-Microbe Interact: MPMI 6(4):495–506
Niki T et al (1998) Antagonistic effect of salicylic acid and jasmonic acid on the expression of pathogenesis-related (PR) protein genes in wounded mature tobacco leaves. Plant Cell Physiol 39(5):500–507

Ohme-Takagi M, Shinshi H (1995) Ethylene-inducible DNA binding proteins that interact with an ethylene-responsive element. Plant Cell 7(2):173–182
Ori N et al (1990) A major stylar matrix polypeptide (sp41) is a member of the pathogenesis-related proteins superclass. EMBO J 9(11):3429
Pan S-Q et al (1989) Direct detection of β-1, 3-glucanase isozymes on polyacrylamide electrophoresis and isoelectrofocusing gels. Anal Biochem 182(1):136–140
Park C-J et al (2004a) Molecular characterization of pepper germin-like protein as the novel PR-16 family of pathogenesis-related proteins isolated during the resistance response to viral and bacterial infection. Planta 219(5):797–806
Park CJ et al (2004b) Pathogenesis-related protein 10 isolated from hot pepper functions as a ribonuclease in an antiviral pathway. Plant J 37(2):186–198
Penninckx IA et al (1996) Pathogen-induced systemic activation of a plant defensin gene in *Arabidopsis* follows a salicylic acid-independent pathway. Plant Cell 8(12):2309–2323
Pieterse CMJ, van Loon LC (1999) Salicylic acid-independent plant defence pathways. Trends Plant Sci 4(2):52–58
Pieterse CM et al (1996) Systemic resistance in *Arabidopsis* induced by biocontrol bacteria is independent of salicylic acid accumulation and pathogenesis-related gene expression. Plant Cell 8(8):1225–1237
Poupard P et al (2003) A wound-and ethephon-inducible PR-10 gene subclass from apple is differentially expressed during infection with a compatible and an incompatible race of *Venturia inaequalis*. Physiol Mol Plant Pathol 62(1):3–12
Rauscher M et al (1999) PR-1 protein inhibits the differentiation of rust infection hyphae in leaves of acquired resistant broad bean. Plant J 19(6):625–633
Raz V, Fluhr R (1993) Ethylene signal is transduced via protein phosphorylation events in plants. Plant Cell 5(5):523–530
Rezzonico E et al (1998) Transcriptional down-regulation by abscisic acid of pathogenesis-related β-1, 3-glucanase genes in tobacco cell cultures. Plant Physiol 117(2):585–592
Robert N et al (2001) Molecular characterization of the incompatible interaction of *Vitis vinifera* leaves with *Pseudomonas syringae pv. pisi*: expression of genes coding for *stilbene synthase* and class 10 PR protein. Eur J Plant Pathol 107(2):249–261
Rohe M et al (1995) The race-specific elicitor, NIP1, from the barley pathogen, *Rhynchosporium secalis*, determines avirulence on host plants of the *Rrs1* resistance genotype. EMBO J 14(17):4168
Rohini VK, Rao KS (2001) Transformation of peanut *(Arachis hypogaea L.)* with tobacco chitinase gene: variable response of transformants to leaf spot disease. Plant Sci 160(5):889–898
Roulin S et al (1997) Expression of specific (1→ 3)-β-glucanase genes in leaves of near-isogenic resistant and susceptible barley lines infected with the leaf scald fungus *(Rhynchosporium secalis)*. Physiol Mol Plant Pathol 50(4):245–261
Rushton PJ, Somssich IE (1998) Transcriptional control of plant genes responsive to pathogens. Curr Opin Plant Biol 1(4):311–315
Rushton PJ et al (1996) Interaction of elicitor-induced DNA-binding proteins with elicitor response elements in the promoters of parsley *PR1* genes. EMBO J 15(20):5690
Ryals JA et al (1996) Systemic acquired resistance. Plant Cell 8(10):1809
Schlumbaum A et al (1986) Plant chitinases are potent inhibitors of fungal growth
Scofield SR et al (1996) Molecular basis of gene-for-gene specificity in bacterial speck disease of tomato. Science 274(5295):2063
Sela-Buurlage MB et al (1993) Only specific tobacco *(Nicotiana tabacum)* chitinases and β-1, 3-glucanases exhibit antifungal activity. Plant Physiol 101(3):857–863
Selitrennikoff CP (2001) Antifungal proteins. Appl Environ Microbiol 67(7):2883–2894
Sharma V (2013) Pathogenesis related defence functions of plant Chitinases and β-1, 3-Glucanases. Vegetos-An Int J Plant Res 26(2s):205–218
Simmons CR (1994) The physiology and molecular biology of plant 1, 3-β-D-glucanases and 1, 3; 1, 4-β-D-glucanases. Crit Rev Plant Sci 13(4):325–387

Sinha M et al (2014) Current overview of allergens of plant pathogenesis related protein families. Sci World J 2014:543195
Suarez V et al (2001) Substrate specificity and antifungal activity of recombinant tobacco class I chitinases. Plant Mol Biol 45(5):609–618
Tang X et al (1996) Initiation of plant disease resistance by physical interaction of AvrPto and Pto kinase. Science 274:5295–2060
Thalmair M et al (1996) Ozone and ultraviolet B effects on the defense-related proteins ß-1, 3-glucanase and chitinase in tobacco. J Plant Physiol 148(1):222–228
Thomma BPHJ et al (1998) Separate jasmonate-dependent and salicylate-dependent defense-response pathways in *Arabidopsis* are essential for resistance to distinct microbial pathogens. Proc Natl Acad Sci 95(25):15107–15111
Tonón C et al (2002) Isolation of a potato acidic 39 kDa β-1, 3-glucanase with antifungal activity against *Phytophthora infestans* and analysis of its expression in potato cultivars differing in their degrees of field resistance. J Phytopathol 150(4–5):189–195
Van Loon LC (1997) Induced resistance in plants and the role of pathogenesis-related proteins. Eur J Plant Pathol 103(9):753–765
Van Loon LC (1999) Occurrence and properties of plant pathogenesis-related proteins. In: Datta SK, Muthukrishnan S (eds) Pathogenesis-related proteins in plants, CRC press, Boca Raton, p 1–19
Van Loon LC, van Kammen A (1970) Polyacrylamide disc electrophoresis of the soluble leaf proteins from *Nicotiana tabacum var. "Samsun" and "Samsun NN"*. II. Changes in protein constitution after infection with tobacco mosaic virus. Virology 40(2):190–211
Van Loon LC, Van Strien EA (1999) The families of pathogenesis-related proteins, their activities, and comparative analysis of PR-1 type proteins. Physiol Mol Plant Pathol 55(2):85–97
Van Loon LC et al (1994) Recommendations for naming plant pathogenesis-related proteins. Plant Mol Biol Report 12(3):245–264
Van Loon LC et al (1998) Systemic resistance induced by rhizosphere bacteria. Annu Rev Phytopathol 36(1):453–483
Vigers AJ et al (1992) Thaumatin-like pathogenesis-related proteins are antifungal. Plant Sci 83(2):155–161
Vleeshouwers VGAA et al (2000) Does basal PR gene expression in *Solanum* species contribute to non-specific resistance to *Phytophthora infestans*? Physiol Mol Plant Pathol 57(1):35–42
Vögeli-Lange R et al (1994) Developmental, hormonal, and pathogenesis-related regulation of the tobacco class I β-1, 3-glucanase B promoter. Plant Mol Biol 25(2):299–311
Ward ER et al (1991) Coordinate gene activity in response to agents that induce systemic acquired resistance. Plant Cell 3(10):1085–1094
Wu C-T, Bradford KJ (2003) Class I chitinase and β-1, 3-glucanase are differentially regulated by wounding, methyl jasmonate, ethylene, and gibberellin in tomato seeds and leaves. Plant Physiol 133(1):263–273
Wubben JP et al (1996) Differential induction of chitinase and 1, 3-β-glucanase gene expression in tomato by *Cladosporium fulvum* and its race-specific elicitors. Physiol Mol Plant Pathol 48(2):105–116
Wyatt SE et al (1991) β-1, 3-glucanase, chitinase, and peroxidase activities in tobacco tissues resistant and susceptible to blue mould as related to flowering, age and sucker development. Physiol Mol Plant Pathol 39(6):433–440
Xu YI et al (1994) Plant defense genes are synergistically induced by ethylene and methyl jasmonate. Plant Cell 6(8):1077–1085
Yalpani N et al (1991) Salicylic acid is a systemic signal and an inducer of pathogenesis-related proteins in virus-infected tobacco. Plant Cell 3(8):809–818
Yamamoto T et al (2000) Transgenic grapevine plants expressing a rice chitinase with enhanced resistance to fungal pathogens. Plant Cell Rep 19(7):639–646
Yang Y et al (1997) Signal perception and transduction in plant defense responses. Genes Dev 11(13):1621–1639

Yeh S et al (2000) Chitinase genes responsive to cold encode antifreeze proteins in winter cereals. Plant Physiol 124(3):1251–1264

Zemanek AB et al (2002) Changes in β-1, 3-glucanase mRNA levels in peach in response to treatment with pathogen culture filtrates, wounding, and other elicitors. J Plant Physiol 159(8):877–889

Zhou J et al (1997) The Pto kinase conferring resistance to tomato bacterial speck disease interacts with proteins that bind a cis-element of pathogenesis-related genes. EMBO J 16(11):3207–3218

Zhu Q et al (1994) Enhanced protection against fungal attack by constitutive Co-expression of Chitinase and glucanase genes in transgenic tobacco. Bio/Technology 12:807–812

13 Antimicrobial Compounds and Their Role in Plant Defense

Anupama Razdan Tiku

Abstract

Plants are important nutrient source for several organisms like microbes, heterotrophic plants, insects as well as vertebrates. Even though they lack a proper defense mechanism like animals, still they have developed a mixture of chemicals which are mainly protein based and are used as a means of defense by detecting attacking organisms and preventing them from causing major damage. In order to protect themselves from these microbes like fungi, bacteria, etc., plant cells have developed the capability to identify attacking pathogens and use inducible defense mechanism by producing toxic chemicals or antimicrobial compounds in the form of pathogen-degrading enzymes and secondary metabolites involved with plant defense. Secondary metabolites generally are grouped into three major classes of chemicals, i.e. terpenoids, phenolic and alkaloids. Some of these antimicrobial compounds are constitutive in nature, i.e. they occur in biologically active forms in healthy plants, whereas other metabolites are inductive in nature. Glucosinolates and cyanogenic glycosides exist in inactive form and are activated as a response to attack by pathogen or tissue damage. These compounds are activated by release of plant enzymes at the time of breakdown of cells. Preformed antimicrobial compounds are termed as "phytoanticipin", while "phytoalexins" are those antimicrobial compounds which are synthesized (as a result of synthesis of enzymes) from precursors as a response to attack by pathogen. Preformed inhibitors are usually tissue specific and are mainly present in the outer layers of the cells of plant organs. These inhibitors are mostly successful against comprehensive range of probable pathogens, and specific virulent pathogens might circumvent the effect of these secondary metabolites by eluding them or by enduring or by detoxification. Most of these constitutive plant compounds show antifungal activity, e.g. phenols, phenolic

A. R. Tiku (✉)
Department of Botany, Ramjas College, University of Delhi, Delhi, India
e-mail: anupama.tiku@yahoo.com

A. Singh, I. K. Singh (eds.), *Molecular Aspects of Plant-Pathogen Interaction*,
https://doi.org/10.1007/978-981-10-7371-7_13

glycoside, unsaturated lactones, sulphur compounds, saponins, etc. "Phytoalexins" are the most considered antimicrobial plant defense compounds. These compounds are pathogen specific and therefore more effective in plant defense mechanism. Transcriptional and translational activities in a plant are prerequisite for the production of phytoalexins. Examples of these antimicrobial phytoalexins are scopoletin, camalexin, glucosinolates, etc. This chapter will mainly discuss the role of both phytoanticipins and phytoalexins as plant defense antimicrobial compounds and also their use as "antibiotic potentiators" and virulence attenuators along with their role in crop protection/phytoprotection.

Keywords
Defense mechanism · Secondary metabolites · Phytoalexins · Phenolics · Toxins and alkaloids

13.1 Introduction

All plants growing in natural habitats are surrounded by a large number of antagonists including variety of microbes, nematodes, insects and herbivores and various kinds of abiotic environmental stress which are major sources of hindrance for crop yield. Plants have evolved with a wide variety of inherent and stimulative defense mechanisms to shield themselves from their potential enemies (Cowan 1999). Inherent defense includes many preformed obstructions including cell wall, bark and cuticular wax to protect the plant from attack of pathogens. Plants use stimulative defense mechanism including production of toxic chemicals, enzymes which degrade pathogens and cell suicide to protect themselves from attacking pathogens. Research on the role of these chemicals acting as plant defense compounds took off in the mid of the nineteenth century, e.g. Muller and Borger introduced phytoalexins based on their observation of *Phytophthora infestan* infection in potato tubers. Plant cells showed inhibition towards the pathogen by producing a chemical which reacted hypersensitively and was named as phytoalexin (Jeandet et al. 2014a, b; Jeandet 2015).

Plant compounds are identified as primary metabolites and secondary metabolites.

Primary metabolites (sugars, amino acids, nucleic acids) are involved in growth, development or reproduction, whereas secondary metabolites serve as toxic chemicals or defense-related proteins (Freeman and Beattie 2008). Plants produce diverse varieties of secondary metabolites using resultants of primary metabolisms including enzymes involved in various metabolic pathways and major biomolecules of the cell (Table 13.1) (Fig. 13.1). These are used to protect plants against invading microbial pathogens on account of their toxic nature. Some of them are as well involved in protection of plants from abiotic stress (such as UV-B exposure) and also used for communication between plants and other organisms (including attracting useful organisms like pollinators) or hostile interactions (such as restraint against pathogens and herbivores) (Schafer and Wink 2009).

Table 13.1 Plants containing antimicrobial compounds

Common name	Scientific name	Compound	Class	Activity
Allspice	*Pimenta dioica*	Eugenol	Essential oil	General
Aloe	*Aloe barbadensis*, *Aloe vera*	Latex	Complex mixture	*Corynebacterium*, *Salmonella*, *Streptococcus*, *S. aureus*
Apple	*Malus sylvestris*	Phloretin	Flavonoid derivative	General
Ashwagandha	*Withania somnifera*	Withaferin A	Lactone	Bacteria, fungi
Bael tree	*Aegle marmelos*	Essential oil	Terpenoid	Fungi
Barberry	*Berberis vulgaris*	Berberine	Alkaloid	Bacteria, protozoa
Basil	*Ocimum basilicum*	Essential oils	Terpenoids	*Salmonella*, bacteria
Bay	*Laurus nobilis*	Essential oils	Terpenoids	Bacteria, fungi
Betel pepper	*Piper betle*	Catechols, eugenol	Essential oils	General
Black pepper	*Piper nigrum*	Piperine	Alkaloid	Fungi, *Lactobacillus*, *Micrococcus*, *E. coli*, *E. faecalis*
Blueberry	*Vaccinium* spp.	Fructose	Monosaccharide	*E. coli*
Brazilian pepper tree	*Schinus terebinthifolius*	Terebinthine	Terpenoids	General
Buchu	*Barosma betulina*	Essential oil	Terpenoid	General
Burdock	*Arctium lappa*		Polyacetylene, tannins, terpenoids	Bacteria, fungi, viruses
Buttercup	*Ranunculus bulbosus*	Protoanemonin	Lactone	General
Caraway	*Carum carvi*		Coumarins	Bacteria, fungi, viruses
Cascara sagrada	*Rhamnus purshiana*	Tannins	Polyphenols	Viruses, bacteria, fungi
			Anthraquinone	
Cashew	*Anacardium*	Salicylic acids	Polyphenols	*P. acnes* bacteria, fungi
	Pulsatilla			
Ceylon cinnamon	*Cinnamomum verum*	Essential oils, others	Terpenoids, tannins	General
Chamomile	*Matricaria chamomilla*	Anthemic acid	Phenolic acid	*M. tuberculosis*, *S. typhimurium*, *S. aureus*, helminths

(continued)

Table 13.1 (continued)

Common name	Scientific name	Compound	Class	Activity
Chaparral	*Larrea tridentata*	Nordihydroguaiaretic acid	Lignan	Skin bacteria
Chilli peppers, paprika	*Capsicum annuum*	Capsaicin	Terpenoid	Bacteria
Clove	*Syzygium aromaticum*	Eugenol	Terpenoid	General
Coca	*Erythroxylum coca*	Cocaine	Alkaloid	Gramnegative and grampositive cocci
Cranberry	*Vaccinium* spp.	Fructose	Monosaccharide	Bacteria
		Other		
Dill	*Anethum graveolens*	Essential oil	Terpenoid	Bacteria
Eucalyptus	*Eucalyptus globulus*	Tannin	Polyphenol	Bacteria, viruses
Fava bean	*Vicia faba*	Fabatin	Thionin	Bacteria
Gamboge	*Garcinia hanburyi*		Resin	General
Garlic	*Allium sativum*	Allicin, ajoene	Sulfoxide	General
			Sulfated terpenoids	
Ginseng	*Panax notoginseng*		Saponins	*E. coli*, *Sporothrix schenckii*, *Staphylococcus*, *Trichophyton*
Glory lily	*Gloriosa superba*	Colchicine	Alkaloid	General
Goldenseal	*Hydrastis canadensis*	Berberine, hydrastine	Alkaloids	Bacteria, *Giardia duodenalis*, trypanosomes
				Plasmodia
Gotu kola	*Centella asiatica*	Asiaticoside	Terpenoid	*M. leprae*
Grapefruit peel	*Citrus paradisi*		Terpenoid	Fungi
Green tea	*Camellia sinensis*	Catechin	Flavonoid	General
				Shigella
				Vibrio
				S. mutans
				Viruses
Hemp	*Cannabis sativa*	βResercyclic acid	Organic acid	Bacteria and viruses
Henna	*Lawsonia inermis*	Gallic acid	Phenolic	*S. aureus*

Hops	*Humulus lupulus*	Lupulone, humulone	Phenolic acids	General
Horseradish	*Armoracia rusticana*		Terpenoids	General
(Japanese) herb	*Rabdosia trichocarpa*	Trichorabdal A	Terpene	*Helicobacter pylori*
Legume (West Africa)	*Millettia thonningii*	Alpinumisoflavone	Flavone	*Schistosoma*
Lemon balm	*Melissa officinalis*	Tannins	Polyphenols	Viruses
Lemon verbena	*Aloysia triphylla*	Essential oil	Terpenoid	Ascaris
Licorice	*Glycyrrhiza glabra*	Glabrol	Phenolic alcohol	*S. aureus*, *M. tuberculosis*
Mountain tobacco	*Arnica montana*	Helenalins	Lactones	General
Oak	*Quercus rubra*	Tannins	Polyphenols	General
		Quercetin (available commercially)	Flavonoid	
Olive oil	*Olea europaea*	Hexanal	Aldehyde	General
Onion	*Allium cepa*	Allicin	Sulfoxide	Bacteria, *Candida*
Oregon grape	*Mahonia aquifolium*	Berberine	Alkaloid	*Plasmodium*
				Trypanosomes, general
Pau d'arco	*Tabebuia*	Sesquiterpenes	Terpenoids	Fungi
Papaya	*Carica papaya*	Latex	Mix of terpenoids, organic acids, alkaloids	General
Pasqueflower	*Anemone pulsatilla*	Anemonins	Lactone	Bacteria
Peppermint	*Mentha piperita*	Menthol	Terpenoid	General
Periwinkle	*Vinca minor*	Reserpine	Alkaloid	General
Peyote	*Lophophora williamsii*	Mescaline	Alkaloid	General
Poppy	*Papaver somniferum*	Opium	Alkaloids and others	General
Purple prairie clover	*Petalostemum*	Petalostemumol	Flavonol	Bacteria, fungi
Quinine	*Cinchona* sp.	Quinine	Alkaloid	*Plasmodium* spp.

(continued)

Table 13.1 (continued)

Common name	Scientific name	Compound	Class	Activity
Rauvolfia, chandra	*Rauvolfia serpentina*	Reserpine	Alkaloid	General
Rosemary	*Rosmarinus officinalis*	Essential oil	Terpenoid	General
Sainfoin	*Onobrychis viciifolia*	Tannins	Polyphenols	Ruminal bacteria
Savory	Satureja montana	Carvacrol	Terpenoid	General
Senna	*Cassia angustifolia*	Rhein	Anthraquinone	*S. aureus*
Cabbage	*Brassica oleracea*	Quercetin	Flavonoids	General
Rape seed	*B. rapa*	Cyanidin	Hydroxycinnamic acid	
Nabicol	*B. napus*	p-coumaric acid		
		Ferulic acid		
Broad bean	*Vicia faba*	Quinolizidine	Alkaloids	Bacteria
		Piperidine	Amines	Fungi
		Pyridine		
		Indolizidine		
Wild barley	*Hordeum chilense*	Hydroxamic acid	Aglucones	Pests
	H. brevisubulatum			Pathogens
	H. bulbosum			
Grapes	*Vitis vinifera*	Gallic acid	Polyphenols	Generals
		Catechin		
Makoi	*Solanum nigrum*	Luteolin	Flavonoids	Bacteria
		Oleuropein glycoside	Alkaloids	Fungi
		Veremivirine		
		Myristic acid		

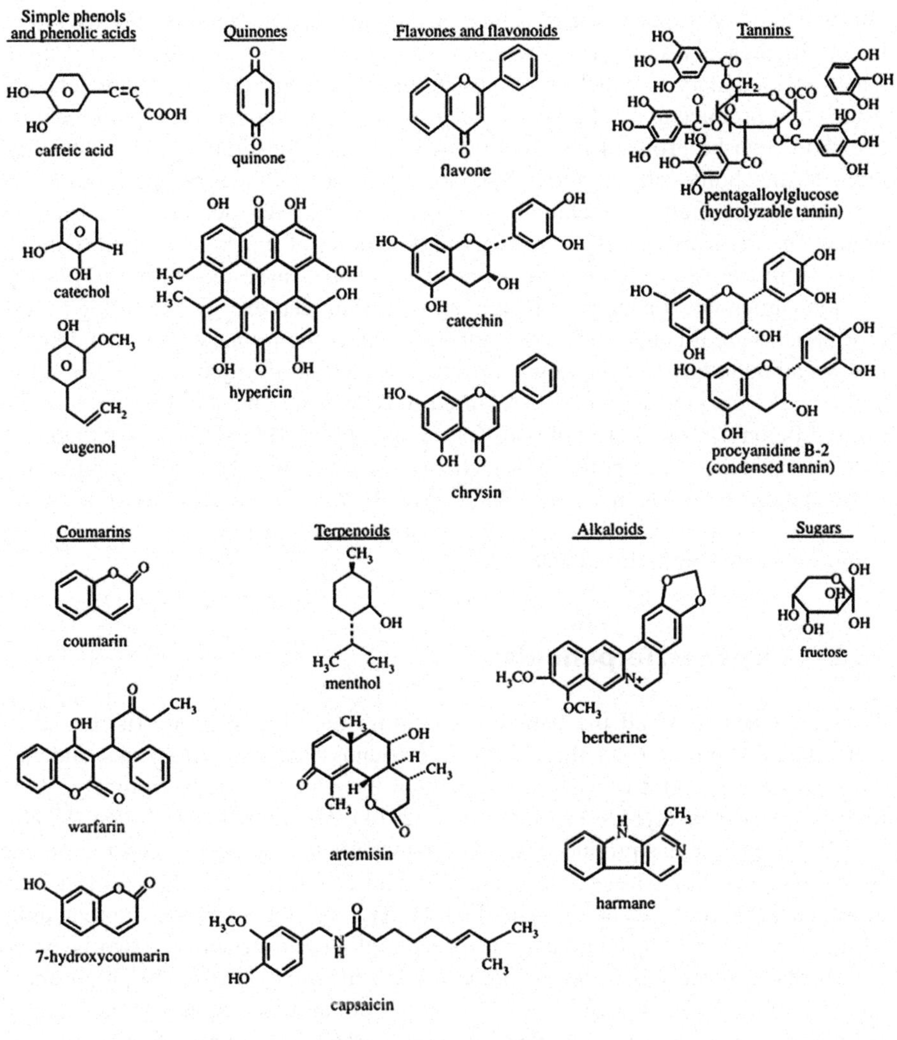

Fig. 13.1 Structures of common antimicrobial plant chemicals

The three major types of secondary metabolites are terpenes/terpenoids, phenolics and nitrogen- and sulphur-containing compounds produced mostly from common amino acids. Terpenes are formed from *5-C isopentanoid* units, which are toxic in nature and hence discourage many herbivores from eating the plants. Phenolics produced mainly from *shikimic* acid pathway are vital to protect the plants from invading microbial pathogens and fungi. Their defensive roles have been confirmed by in vitro experiments conducted by changing expression of secondary metabolites of plants using modern molecular methods (Mes et al. 2000; Van Etten et al. 2000). Studies have shown that around 200,000 secondary metabolites produced by plants are part of biochemical protective systems of plants, evolved over millions of years during which the respective plants and their foes have existed together (Wink 1999). Due to high energy and nutrient requirements required for production and maintenance of secondary metabolites, plants normally wait till the time pathogens are detected to produce toxic chemicals/defense-related proteins. Hence, plant defense chemicals can be grouped into two types which are constitutive metabolites and induced metabolites. Constitutive metabolites are also known as "prohibitins" or phytoanticipins. Induced metabolites, also known as phytoalexins, are formed to protect the plant from infection involving synthesis of enzymes (complex molecules) from simple molecules (Van Etten et al. 1994; Grayer and Harborne 1994).

Phytoanticipins consume large amount of carbon and energy exhibiting fitness cost under natural conditions. They are accepted as the primary chemical defense that a pathogen has to overcome, whereas production of phytoalexin takes up to 2–3 days, as enzyme synthesis is required for converting their precursors into the desired defense protein (Hammond-Kosack and Jones 1996). Moreover the same type of secondary metabolites/toxic chemicals exists in plants belonging to the same species or taxonomically related species. We will discuss possible types/groups of secondary metabolites acting as antimicrobial compounds used in plant defense mechanism in this chapter.

13.2 Terpenes/Terpenoids

They exist in almost all the plants representing the biggest group of secondary metabolites. There are more than 22,000 compounds described as terpenes, and they are grouped together based on their common biosynthetic origin from glycolytic intermediates or acetyl-CoA (Mazid et al 2011). Hydrocarbon isoprene (C5H8) which is a volatile gas produced during photosynthesis can protect cell membrane from damage due to extreme temperature or light conditions. It is the simplest form of terpenoid. Terpenoids are classified on the basis of number of the isoprene units which are used in their construction, for example, monoterpenoids (two isoprene units), sesquiterpenoids (three isoprene units), diterpenoids (four isoprene units) and triterpenoids (six isoprene units). Terpenes are assumed to be intricate part of plant defense mechanism in the form of toxins and feeding restraint to many insects and herbivores (Gershenzon and Croteau 1991). They are also active against bacteria, fungi and protozoa, for example, pyrethrins, menthol, camphor, farnesol and artemisinins. Capsaicin present in chilli peppers has bactericidal properties. Below several examples will be discussed from five major subclasses of terpenes.

13.2.1 Monoterpenes (C_{10})

Monoterpenes along with its derived compounds are vital representatives of insect toxicity, for example, pyrethroids/pyrethrins (monoterpene esters) exist in the leaves and flowers of chrysanthemum species presenting robust insecticidal reactions (neurotoxin) towards insects like bees, beetles, wasps, moths, etc. They are key elements of commercial insecticides. Monoterpenoids as well as sesquiterpenoids are the key elements of highly volatile compounds and essential oils, which are the reason for the fragrance of plants producing them. They are insect toxins and prevent bacterial and fungal infections in the plant. For example, mint plants (*Mentha* spp.) produce huge quantity of monoterpenes (menthone and menthol), formed and stored in glandular trichrome present in epidermal cells. In gymnosperms (conifers) like fir, lime and pine, monoterpenes collect in resin ducts of needles, twigs and trunk primarily in the form of alpha-pinene, beta-pinene, limonene and myrcene which are potent insect repellents effecting bark beetles and other serious pests of conifer species (Turlings et al. 1995).

13.2.2 Sesquiterpenes (C_{15})

Many sesquiterpenes have been reported till now, and the part they play in plant defense is mainly of anti-herbivore representatives belonging to Asteraceae family classified by a lactone ring with five members (a cyclic ester) having tough feed repelling properties affecting several insects and herbivores.

ABA is a well-known example of sesquiterpene playing important role as a plant hormone. In cotton four different sesquiterpenoid phytoalexins, desoxyhemigossypol, hemigossypol, desoxy-6-methoxygossypol and 6-methoxygossypol (Garas and Waiss 1986), help the plant to fight against fungus *Verticillium dahliae*. Resistant varieties of tobacco with *Phytophthora* had induction of sesquiterpenoid phytoalexins like capsidiol, rishitin, etc.

13.2.3 Diterpenes (C_{20})

Abietic acid, a significant diterpene, mostly exists in resin canals of tree trunks of legumes and pine trees. Toxic resin serves as chemical deterrent against the feeding insects which pierce these canals. Phorbol (an ester) is another example of diterpene which is present in members of Euphorbiaceae and proves to be a skin irritant as well as toxic for feeding herbivores. Gossypol (Fig. 13.2) which has tough antifungal and antibacterial characteristics is produced by cotton (*Gossypium hirsutum*) (Bennett and Wallsgrove 1994).

13.2.4 Triterpenes (C_{30})

Triterpene structures are quite similar to plant and animal steroidal hormones and sterols. The milkweeds generate numerous bitter-tasting glycosides, i.e. sterols which protect the plants against insects and herbivores. Limonoid is a bitter substance in citrus fruits and serves as anti-herbivore compounds present in family Rutaceae. Azadirachtin (Fig. 13.3), a complex limonoid, is present in *Azadirachta indica*, which restrains some insects from feeding on the plant and also protects the

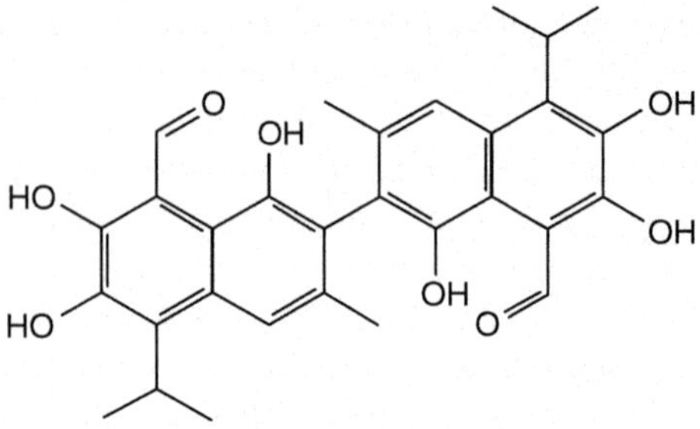

Fig. 13.2 Structure of diterpene gossypol

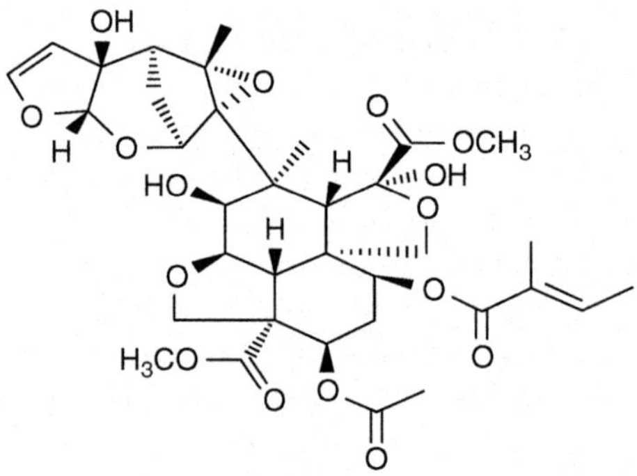

Fig. 13.3 Structure of triterpene azadirachtin

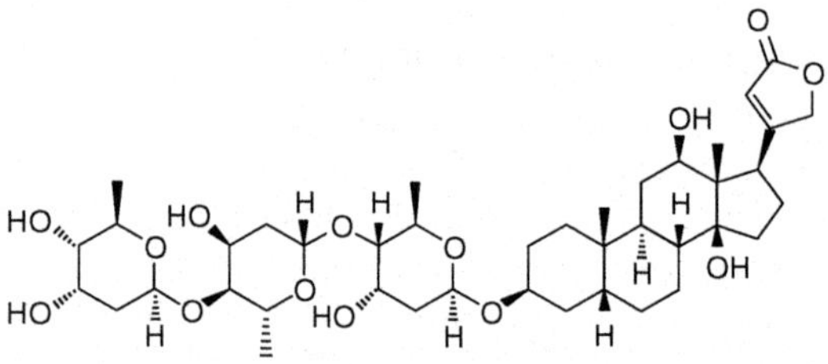

Fig. 13.4 Structure of triterpene digoxin

plant from microbial infections (Mordue and Blackwell 1993). Foxglove (*Digitalis purpurea*) also produces glycosides known as digitoxin and digoxin (Fig. 13.4).

13.2.5 Polyterpenes $(C_5)_n$

These are high molecular weight terpenes, for example, rubber has many polyterpenes showing antimicrobial properties. Rubber is found in long vessels known as laticifers made up of 1500–15,000 isopentenyl units having nearly all C-C double bonds with cis (Z) configuration. Another example is gutta rubber which has its double bonds in trans (E) configuration (Eisner and Meinwald 1995). Few polyterpenes (tetraterpenes) effect growth, e.g. hormone gibberellic acid, and also contribute to red, yellow and orange pigments (carotenoids).

13.3 Saponins

Saponins exist in large concentration in healthy plants exhibiting strong antifungal activity. These molecules have been found to be determinants of plant resistance to fungal attack (Osbourn 1996). These compounds also exhibit other properties like piscicidal, insecticidal and molluscicidal. Saponins are triterpenoids with attached sugar group, i.e. glycosylated triterpenoids which exist in membranes of cell of several species of plants. Having detergent properties they disrupt the membranes of cells of attacking fungal pathogens. Saponins are classified into three major groups based on the types of aglycone which are triterpenoid, a steroid or a steroidal glycoalkaloid. Saponins exist in the form of *triterpenoid saponins* mostly in dicotyledonous plants and also in few monocots, while *steroid saponins* exist mostly in monocots, for example, plants belonging to family Agavaceae, Liliaceae, Dioscoreaceae, etc. Saponins in the form of *saponin digitoxin* mostly exist in dicots, for example, foxglove. *Avena* has both the types mentioned above (Price et al. 1987). Saponins in the form of steroidal glycoalkaloids exist mostly in Solanaceae family (potato, tomato, etc.) along with Liliaceae family. Saponins found in oats and tomato and their function in defense of plants against phytopathogenic fungi have been studied in detail (Osbourn 1996).

Some vital saponins involved in antimicrobial activity are avenacins and avenacosides (mainly present in oats and related species like *Arrhenatherum elatius*). Avenacins exist only in roots, while avenacosides exist in roots as well as shoots. Avenacosides which are inactive biologically are transformed into antifungal monodesmosidic saponin 26-desglucoavenacosides. Oats are able to resist root-infecting fungus because of the presence of triterpenoid avenacin saponins. *Gaeumannomyces graminis* var. *tritici* is not able to infect oat plant even though it wreaks havoc to wheat and barley plant because of the presence of avenacins (Fig. 13.5a).

One of the major saponin, α-tomatine (Fig. 13.5b), a monodesmosidic steroidal glycoalkaloid, exists in its biologically active form in a healthy plant. It safeguards the tomato plants from both *Verticillium albo-atrum* (Pegg and Woodward 1986) and vascular wilt fungi *Fusarium oxysporum* f.sp. *lycopersici* (Smith and Machardy 1982). Saponins prove to be toxic for fungi as a result of the ability of saponins to associate with sterols present in membrane leading to formation of pores (Price et al. 1987; Fenwick et al. 1992). Actions of saponins like α-tomatine are dependent upon pH, but some fungi render the saponins ineffective by altering the pH at the infection site. Plants have developed a mechanism to safeguard themselves from their own saponins by placing them in the vacuole or in other organelles, whose membranes can avoid lysis because of low or altered sterol composition.

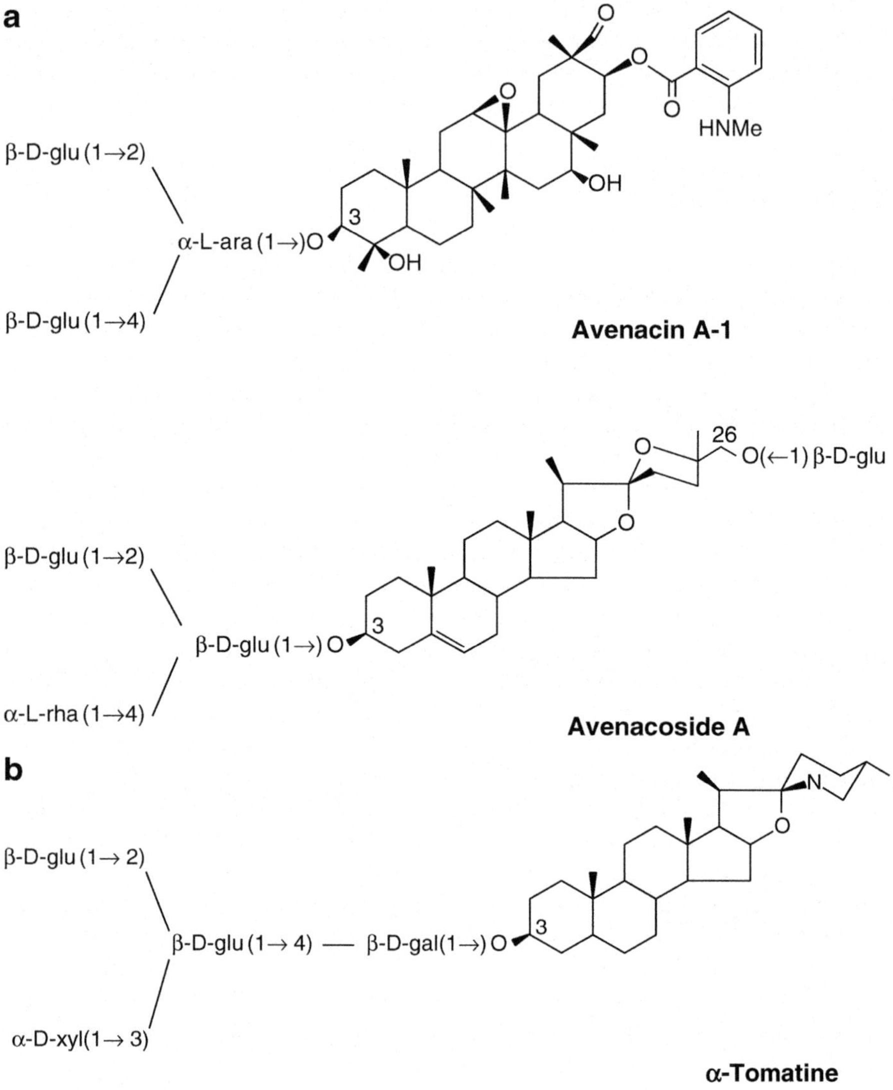

Fig. 13.5 (**a**) Structure of avenacin saponins present in oat (**b**) Structure of saponin present in tomato

13.4 Phenolics

Phenolics are a big class of secondary metabolites formed by plants to safeguard themselves against pathogens. They are created mostly via shikimic acid and malonic acid pathways in plants, comprising extensive variety of protective metabolites such as furanocoumarins, lignin, anthocyanins, flavonoids, tannins and

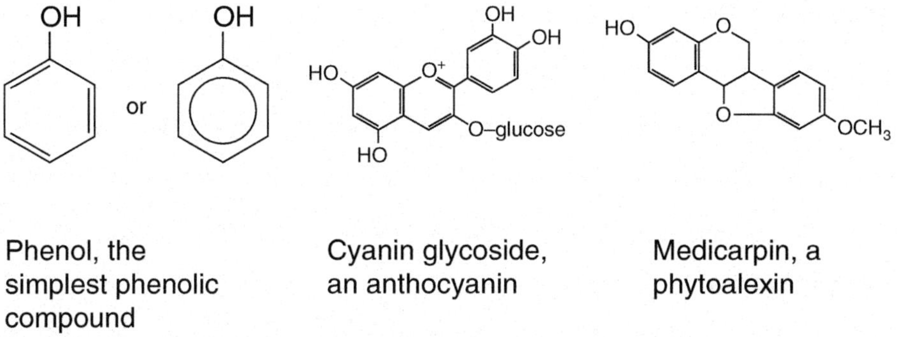

Fig. 13.6 Structure of important phenolics

phytoalexins. These secondary products play a vital part in plant defense mechanism against threats like fungi, bacteria and nematodes as well as restraint herbivores from feeding the plant (Mazid et al. 2011). They contain phenol which is a hydroxyl functional group on an aromatic ring named phenol (Fig. 13.6).

13.4.1 Coumarin

Simple phenolic compounds made up of fused benzene and α-pyrone rings occur extensively in vascular plants. They protect various plants in many ways against threats from insects and fungi. They are mostly derived through shikimic acid pathway and are involved actively against wide range of microbes (Brooker et al. 2008). These cyclic compounds behave as natural pesticides for plants and represent a point from where the exploration of new derivatives which possess a range of enhanced antifungal activity can start. Derivatives of halogenated coumarin operate efficiently against fungal growth, for example, 7-hydroxylated simple coumarin works against *Orobanche cernua* preventing its penetration and germination in host vascular system. Some coumarins also show high defense activity against soilborne plant pathogenic fungi (Brooker et al. 2008). Hydroxycinnamic acid, related to coumarins, shows inhibition towards gram-positive bacteria. Phytoalexins which are hydroxylated derivatives of coumarins formed in carrots show antifungal activity.

13.4.2 Furanocoumarin

A wide variety of plants produce furanocoumarins which are phenolic compounds. Plants belonging to family of Apiaceae usually produce these compounds. These phenolic compounds become toxic only when activated by light (UV-A), and they get integrated into the pyrimidine bases of DNA of vertebrate and invertebrate herbivore causing rapid cell death due to blocked transcription and repair mechanism. Psoralen is a basic furanocoumarin used against fungi (Fig. 13.7).

Fig. 13.7 Structure of furanocoumarin psoralen

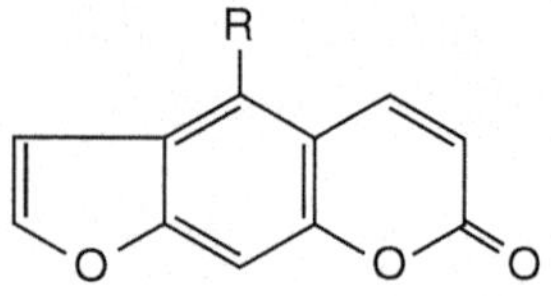

13.4.3 Lignin

Branched polymer of phenylpropanoid groups exists primarily in the secondary cell walls of plants. Three different types of alcohols which are coniferyl, coumaryl and sinapyl oxidize to free radicals (ROS) by a ubiquitous plant enzyme – peroxidase – which reacts randomly as well as simultaneously to produce lignin. The reactive proportions of these three monomeric units in lignin fluctuate among species, plant organs and even different layers of single cell wall (Lewis and Yamamoto 1990). These phenolic monomers are present hundred or thousand times in the structure of lignin and constitute the main component of wood. It is insoluble, rigid and virtually indigestible therefore serving as exceptional physical wall against attack by pathogens. The physical toughness and chemical durability of lignin causes it to be impossible for herbivores and insect pathogens to digest it. Lignification prevents the development of pathogens and microbes and is formed frequently during infection or wounding.

13.4.4 Flavonoids

Flavonoids, which are one of the biggest classes of plant phenolics, perform diverse roles from pigmentation to defense mechanism of plants. Flavones and flavonols are two main groups of flavonoids present in flowers. They protect the cells from harmful UV-B radiation by accumulating in epidermal layers of stems and plants (Lake et al. 2009). Anthocyanins as well as colourful water-soluble flavonoid pigments are formed by plants to protect their foliage from UV damage. They are hydroxylated phenolic substances existing as C_6_C_3 unit connected to an aromatic ring, produced by plants in response to microbial infection. Because of their capability to complex with extracellular and soluble proteins and bacterial cell wall proteins, they can act against extensive range of microorganisms. Lipophilic flavonoids, often pathogen specific in their toxicity, disrupt cellular structure, microbial membranes and pathogen metabolism. An example is catechin, a flavonoid occurring in leaves of oblong green tea. Tea leaves showing antimicrobial activity have mixture of catechin compounds working against various bacteria and microorganisms, for example, *Vibrio cholerae*, *Streptococcus mutans*, *Shigella*, etc. Phloretin existing in some varieties of apples displays action against many varieties of microorganisms. Galangin which is produced from *Helichrysum aureonitens* (a perennial herb) is effective against extensive array of gram-positive bacteria as well as microbes and fungi.

13.4.5 Isoflavonoids

Isoflavonoids play a vital role in development of plant as well as defense mechanism. Their precursor is naringenin which is a flavanone intermediate, present in plants. They play a vital part in the formation of nitrogen-fixing nodules by symbiotic rhizobia and are secreted by legumes (Sreevidya et al. 2006). Phytoalexins are isoflavonoids, produced in response to pathogen attack having antibiotic and antifungal characteristics. These pathogen-specific, toxic compounds unsettle cellular structure and metabolism of pathogen, for example, medicarpin which is formed by alfalfa (*Medicago sativa*), rishitin also a toxin formed in tomatoes as well as potatoes (family Solanaceae) and camalexin formed by *Arabidopsis thaliana.*

13.4.6 Tannins

Tannins are group of polymeric phenolic substances which have the property of astringency which gives the ability to tan leather or precipitate gelatin from solution. They exist mostly in various plant parts like roots, bark, wood, fruits and leaves. Tannins are grouped into two types, hydrolyzable and condensed tannins. Hydrolyzable tannins are based on gallic acid (multiple aster of D-glucose), while flavonoid monomers are precursors of condensed one which are also known by the alias "proanthocyanidins". Condensed flavan derivatives in plant woody tissues and polymerization of quinone units are also known to form tannins. These phenolics are water soluble and mainly stored in vacuoles of cell. Tannins combine with salivary proteins and digestive enzymes of insects and herbivorous animals making the insect protein inactive. A herbivore or an insect having large intake of tannins does not gain weight and eventually dies. Tannins are capable of binding proteins, for example, protocatechuic and chlorogenic acids playing a vital role in disease resistance in certain plants. In onions smudge disease caused by *Colletotrichum circinans* (fungus) and growth of other fungi is restricted due to formation of these two tannins. Some tannin like chlorogenic acid is oxidized into effective quinones showing antifungal properties in case of some disease-resistant plants (Cowan 1999).

13.5 Nitrogen-Containing Secondary Metabolites

Nitrogen-containing secondary metabolites including alkaloids, cyanogenic glycosides and nonprotein amino acids are synthesized from common amino acids and are of significant interest due to their anti-herbivore defense role in plants.

13.5.1 Alkaloids

Alkaloids are the largest family of bitter-tasting nitrogenous compounds (Fig. 13.8) present in many vascular plants (20%) (Hegnauer 1988) mostly in herbaceous dicots and some gymnosperms and monocots. Pyrrolizidine alkaloids (Pas) are quite toxic and help in defense against infection caused by microbes and attack by herbivores. They are usually produced from few common amino acids like lysine, tyrosine, aspartic acid and tryptophan (Pearce et al. 1991). Cocoa, coffee and tea contain caffeine which is an alkaloid. These are toxic to both insects and fungi. Nicotine, an alkaloid, is formed by tobacco plant roots which is further transported to leaves of tobacco plant and is stored in vacuoles. Atropine, another neurotoxin and cardiac stimulant alkaloid, is formed by lethal nightshade plant (*Atropa belladonna*) and is highly toxic in large quantities. Capsaicin formed by members of genus *Capsicum* also has antimicrobial properties and works actively in plant defense mechanism. Their action involve effect on nervous system especially the chemical transmitters, membrane transport system, protein synthesis and miscellaneous enzyme activities (Creelman and Mullet 1997). Shikimic acid is a precursor for indole and its derivatives, amino acid tryptophan and its derivatives (psychedelic compounds, dimethyltryptamine), many alkaloids and other aromatic metabolites (Figs.13.9 and 13.10) which play an important part in resistance against microbes and fungi and nematodes.

13.5.2 Cyanogenic Glycosides

Hydrogen cyanide (HCN), a lethal chemical that halts cellular respiration in aerobic organisms, is produced by breaking down of cyanogenic glycosides (very deadly

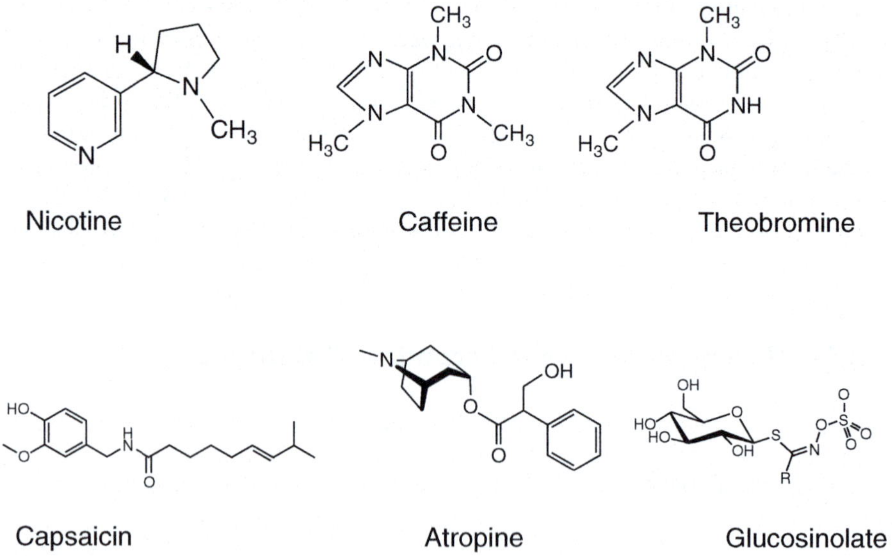

Fig. 13.8 Structure of common alkaloids

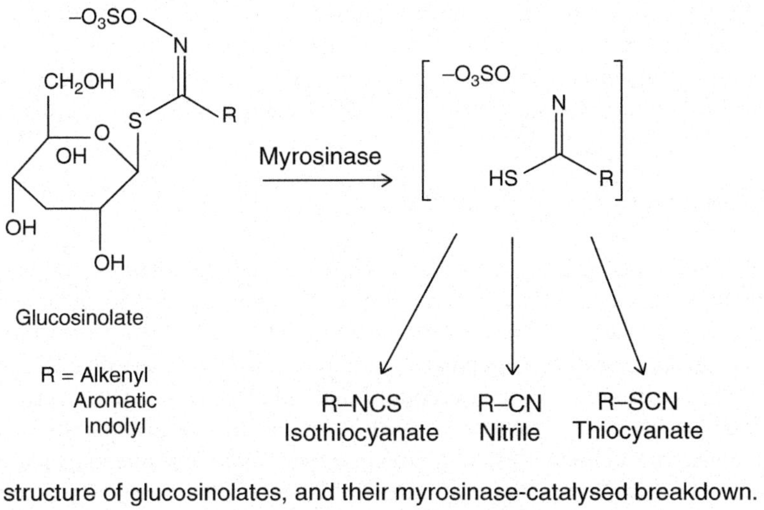

Fig. 13.9 Structure of glucosinolates and their myrosinase - catalysed breakdown

Fig. 13.10 Structure of phytoalexin capsidiol

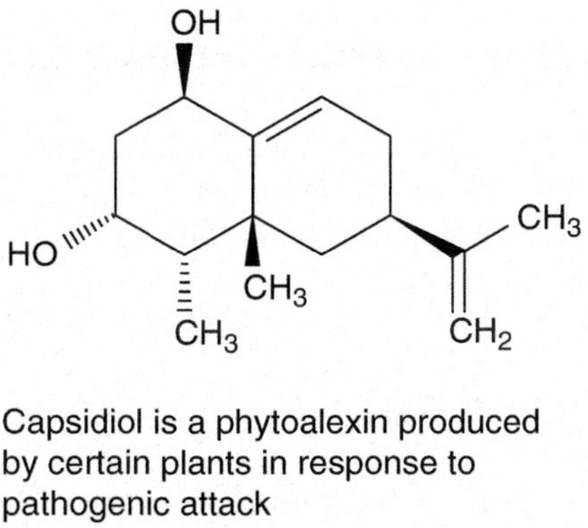

nitrogenous compounds). When the herbivores feed on the plants that produce cyanogenic glycosides, the plant tissue gets damaged which leads to mixing of enzyme (including glycosidases and hydroxynitrile lyases) and substrate which are stored in separate sections of the healthy plant, producing HCN. HCN binds itself to the Fe-containing heme group of cytochrome oxidase as well as other respiratory enzymes causing cellular respiration poisoning. Cyanogenic glycosides are mainly present in families Gramineae, Rosaceae and Leguminosae (Seigler 1991). Seeds of almond, apricot, cherries and peaches contain a common glycoside called amygdalin, whereas *Sorghum bicolor* contains dhurrin (Poulton 1990). Cassava has a long shelf life because of the presence of cyanogenic glycosides. Lima bean (*Phaseolus lunatus* L.) is a classic example of a plant which may be researched to understand defense mechanism in plants showing inducible indirect anti-herbivore properties achieved by producing VOC, i.e. volatile organic compounds (Ballhorn et al. 2009).

Among members of mustard family (Brassicaceae), mustard oil glycosides are present which when broken down by thioglucosidase enzymes produce cyanide gas. Cyanogenic glycosides are the best example where defenses are produced only in reaction to damage to plant tissue or attack by pathogen like "phytoalexins".

13.5.3 Nonprotein Amino Acids

Several plant species have rare amino acids known as nonprotein amino acids that do integrate into proteins but are also present as free forms and act as shielding defensive substance (Johnson et al. 1989). For example, azetidine-2-carboxylic acid and canavanine block the synthesis or uptake of protein amino acids. They are close analogs of arginine and proline. After consumption, canavanine (NPA) is detected by herbivore enzyme that sticks arginine to arginine tRNA molecule and gets assimilated into protein in place of arginine. Resulting protein is a nonutility one due to its faulty tertiary structure or disrupted catalytic site (Rosenthal 1991). Plants which produce nonprotein amino acids do not get harmed themselves due to its toxicity, but they gain defense against a wide array of pathogenic microbes, insects and herbivorous animals.

13.6 Sulphur Containing Secondary Metabolites

These metabolites directly or indirectly defend plants against pathogen microbes. Defensins, thionins, GSH, phytoalexins and GSL are the examples of sulphur containing secondary metabolites (Saito 2004; Grubb and Abel 2006; Halkier and Gershenzon 2006).

13.6.1 GSH

GSH is the main form of organic S which is present as soluble fraction in plants. GSH gets amassed in the plant after it is attacked by fungi. Main cellular sections of the plant can carry between 3 and 10 mM of GSH. GSH displays properties of an antioxidant as well as reducing agent in an effort to diminish the oxidative stress (Kang and Kim 2007).

GSH targets and detoxifies xenobiotic and cytotoxins in vacuoles. For synthesis of GSH and other phytochelatins involved in heavy metal detoxification, enzymes are produced inside specialized epidermal cells like trichomes.

13.6.2 GSL (Glucosinolates)

GSL increases the resistance of higher plants to parasites, predators as well as competitors. They are a group of N and S with low molecular mass and holding plant glycosides. The breakdown products which are released in the form of protective volatile substances exhibit lethal or repelling effects (DeVos and Jander 2009), for example, allyl sulfoxides in *Allium* (Leustek and Saito 1999) and mustard oil glycosides in Cruciferae. Myrosinase catalyses the volatiles coming from GSL (Fig. 13.10) cleaving glucose from its bond with S atom. Resulting product (aglycone) reorganizes the loss of sulphate giving pungent as well as chemically reactive products, which includes isothiocyanates (R-N=C=S), thiocyanates and nitriles. These toxic and feed repelling products protect the plants from herbivores (Geu-Flores et al. 2009). It prevents various fungal diseases including light leaf spot (*Pyrenopeziza brassicae*), sclerotina stem rot (*Sclerotinia sclerotiorum*) and alternaria (*Alternaria brassicae*) from affecting the plants. The GSL become effective only when it comes in contact with myrosinase (a plant enzyme) which happens only if plant tissue gets damaged. (In healthy plants both enzymes and glucosinolate substrates are separated from each other due to compartmentalization.) The activity of enzymes involved in antioxidant defense mechanism may get affected by isothiocyanates, and defense of cell from DNA damage and GST activity may get affected because of detoxification from xenobiotic (Lipka et al. 2010). These three hydrolysis products are toxic for a wide array of fungi, and the breakdown products' nature depends upon the glucosinolate structure, myrosinase enzyme existing, the species of plant and variety of factors such as protein cofactors, temperature, metal ion concentrations and pH. Resistance of cabbage to *P. parasitica* and resistance of Indian mustard and oilseed rape to *L. maculans* have been associated with high glucosinolate levels. Breakdown products of glucosinolate also have effect on several nonpathogens of *Brassica*. These compounds are used as naturally available fungicides to regulate several pest harvest pathogens and cereal diseases related to fruits and vegetables (Mari et al. 1993).

13.6.3 Phytoalexin

Plants synthesize antimicrobial and often antioxidative elements de novo. These then accumulate quickly at infected areas affected by pathogen (bacteria or fungi) in order to limit the spread of invading microbes (Fig. 13.10). Organic phytoalexins produced by most of the plant families have a diverse chemistry, and mostly they are grouped under the classes including alkaloids, glycosteroids and terpenoids. The examples include isoflavonoids of Leguminosae and sesquiterpenoids of Solanaceae. S metabolites are only produced in Cruciferae plant family (Harborne 1999; Gross et al. 1994; Monde et al. 2000). The phytoalexins which are produced in the plants have a toxic effect on the attacking pathogens because they are responsible for rupturing the walls of the cell, delaying maturation, disrupting metabolism and preventing pathogen reproduction. The plant mutant which is incapable of producing phytoalexin experiences widespread presence of pathogens compared to wild plants which produce phytoalexin. Production of phytoalexins comes under systemic acquired resistance (SAR) or long-term resistance, which involves communication between the damaged tissue and the rest of the plant as well as usage and synthesis of plant hormones including salicylic acid, abscisic acid, ethylene and jasmonic acid. The entire mechanism involves genes that transcribe enzyme involved in synthesis of phytoalexins. Natural phenols such as isoflavonoids, polyphenols as well as associated substances have very important part to play in defense of plant, for example, in *Vitis vinifera* (grape) trans-resveratrol (a phytoalexin) plays a role in plant defense against fungal pathogens including *Botrytis cinerea*, and another phytoalexin *delta*-viniferin is present in grapevine formed during fungal infection by *Plasmopara viticola*. Sakuranetin is a flavonoid type of phytoalexin found in *Polymnia fruticosa* and rice against the pathogen *Pyricularia oryzae*. 6-Methoxymellein (a dihydroisocoumarin) is synthesized in carrot slices by UV-c gene against the pathogen *Botrytis cinerea* (Kurosaki and Nishi 1983). Phytoalexin danielone exists in papaya fruit displaying extraordinary antifungal activity against *Colletotrichum gloeosporioides* (Echeverri et al. 1997). Stilbenes are a phytoalexin produced in *Eucalyptus sideroxylon* during pathogen attacks.

Allicin with the structure 3-hydroxy-5-methoxy-6-methyl-2-pentyl-4H-pyran-4-one is the first phytoalexin isolated from garlic (Illic et al. 2011). Allicin is another phytoalexin obtained from garlic. It is an organosulfur compound formed when alliinase enzyme converts alliin into allicin which also gives aroma to the fresh garlic. Allicin being an unstable compound quickly changes into a series of other sulphur-containing compounds such as diallyl disulphide having antibacterial, antifungal, antiviral or antiprotozoal activity. Other examples of associations of phytoalexin induction with the resistance in plants are displayed in Table 13.2. Production and function of phytoalexins for disease resistance can be restored or enhanced through genetic engineering techniques by introducing simple genetic constructs carrying phytoalexin synthesizing genes, for example, in grapevine, phytoalexin "resveratrol" is synthesized by stilbene synthase (STS) gene. STS genes (Vst1 and Vst2) from grapevine were transferred to tobacco through genetic transformation; and when the assays were performed in transformed plants, they showed higher resistance to *B. cinerea* (Hain et al. 1993). STS gene responsible for phytoalexin production was later transformed from different sources into many plant systems like *Arabidopsis*, papaya, tomato, wheat, barley, rice and alfalfa conferring resistance to various pathogens (Jeandet et al. 2013). In alfalfa, overexpression of enzyme isoflavone 7-O-methyltransferase played a critical part in biosynthesis of phytoalexin "maiackiain" which was associated with resistance of the plant to *Phoma medicaginis He and Dixon (2000).* Transformation of soya bean hairy root with both AhRS3 (peanut resveratrol synthase 3) and ROMT (resveratrol-0-methyltransferase) gene resulted in the resistance of transformed plant against *Rhizoctonia salami* (Zernova et al. 2014). Cytokinin overexpression in tobacco plant resulted in surge of resistance of plant against the pathogen *Phytophthora syringae* due to up-regulated synthesis of capsidiol and scopoletin, the two phytoalexins which we involved in the process (Grosskinsky et al. 2011). Similarly in the case of *Arabidopsis*, production of camalexin as well as resistance to *B. cinerea* was severally affected because of the mutation occurring in two MAP kinases (MPK3 and MPK6) (Ren et al. 2008). Some genes which function as phytoalexin biosynthesis regulators have also been identified, for example, overexpression of Rac protein in rice leads to accumulation of phytoalexin momilactone (Ono et al. 2001) and resistance of plant against bacterial blight. Overexpression of non-expressor of pathogenesis-related gene 1 induced the biosynthesis of phytoalexin gossypol (Parkhi et al. 2010).

Table 13.2 Phytoalexin induction and resistance to plant pathogens

Plant	Pathogen	Phytoalexin
Grapevine (*Vitis* spp.)	*Plasmopara viticola*	Viniferins
Grapevine (*Vitis* spp.)	*Botrytis cinerea*	Viniferins
Carnation (*Dianthus* spp.)	*Fusarium oxysporum*	Dianthalexin
		Methoxydianthramide S
Pea (*Pisum sativum*)	*Pseudomonas syringae*	Pisatin
Chickpea (*Cicer arietinum*)	*Ascochyta rabiei*	Medicarpin
		Maakiain
Citrus spp.	*Phytophthora citrophthora*	Scoparone
Oat (*Avena sativa*)	*Puccinia coronata*	Avenalumins

13.7 Proteins and Enzymes

Several plants and their seeds have protein that impedes pathogen and best enzymes after developing complexes that would block active sites or modify enzyme conformations. These proteins have small structure and mostly consist of amino acid cysteine. They include proteinase inhibitors, lectins, anhydrase inhibitors and defensins. Unlike simple chemicals like phenolics, alkaloids, terpenoids, etc. they need large amount of energy and resources of plant for their production. They are usually produced in significant concentration only after the attack by pest or pathogen. On activation these defensive protein and enzymes would effectively impede the growth of pathogens including fungi, bacteria as well as insect herbivores and nematodes.

13.7.1 Defensins

Small cysteine-rich protein displaying comprehensive antimicrobial activity was first isolated from endosperm of barley (*Hordeum vulgare*) and wheat (*Triticum aestivum*). They are best characterized in seeds but are also present in other plant tissues like leaves, pods, tubers, fruits, barks, etc. They display extensive array of biological activities and impede the spread of wide-ranging fungi and bacteria (Thomma et al. 2002). They also impede digestive protein in herbivores. The exact mechanism of their action is still under consideration; it looks like that they attack molecular targets in the pathogens' plasma membrane. They disrupt cellular ion balance by forming new membrane pores as well as impede pre-existing ion channels.

13.7.2 Digestive Enzyme Inhibitors

Digestive enzyme inhibitors inhibit the digestion and disrupt the nutrient absorption by herbivores. Alpha-amylase inhibitor proteins bind to amylase enzyme impeding digestion of starch in legumes. Glycoproteins and non-enzymatic proteins like lectins bind to carbohydrates displaying various properties, for example, in invertebrates they cause lumping of blood cells and in insects they disrupt digestion (Peumans and Van-Damme 1995). Ricin combines lectin molecule with an N-glycoside hydrolase to produce a very powerful toxin in *Ricinus*. It impedes protein synthesis in animal cells.

13.7.3 Protease Inhibitors

Protease inhibitors impede digestive enzymes like chymotrypsin and trypsin. They are produced by plants as a defense mechanism to protect them from attack by herbivores. They have wide distribution but mainly seen in legumes, members of Solanaceae and grasses.

13.7.4 Hydrolytic Enzymes

When pathogens attack plants, some of the plants in order to defend themselves produce hydrolytic enzymes. These enzymes act upon pathogenic fungi by degrading their cell walls in the extracellular spaces where the enzymes get accumulated. For example, the degradation of chitin in cell wall of fungi is catalysed by chitinases. The degradation of glycosidic linkages in glucans existing in several oomycete (water moulds) cell walls is catalysed by glucanases. Another example of a hydrolytic enzyme with the ability to degrade cell walls of bacteria is lysozymes.

In this review we have discussed the utility of secondary metabolites occurring in plant in their defense system. We have tried our best to portray the role of different chemical compounds produced by plants itself against various plant pathogens. Several factors are involved in the interaction of a species of plants with its respective pest and non-pest and hence quite complex. Synthesis and accumulation of these compounds occur in tissues which are young and developing like leaves as well as in tissues of reproduction like seeds and flowers, thereby protecting young plant tissues. Some of the antimicrobial secondary metabolites are preformed, while others are induced by infection itself (phytoalexins, cyanogenic glycosides, glucosinolates, etc.). Organic farming can be made sustainable by identifying and properly using the natural chemical compounds to combat the pathogens, thereby eliminating or at least minimizing the use of fungicides in agriculture. Further research is required for development of natural pesticides. The genes required for producing these valuable defensive compounds can be isolated and then synthesized in bulk for the crop plants to be reengineered metabolically. This will make the plants more resistant towards threats posed by microbial pathogens, various herbivores and several environmental stresses.

References

Ballhorn DJ, Kautz S, Heil M, Hegeman AD (2009) Cyanogenesis of wild lima bean (Phaseolus lunatus L.) is an efficient direct defence in nature. Plant Signal Behav 4(8):735–745

Bennett RN, Wallsgrove RM (1994) Secondary metabolites in plant defence mechanisms. New Phytol 127:612–633

Brooker N, Windorski J, Blumi E (2008) Halogenated coumarins derivatives as novel seed protectants. Commun Agric Appl Biol Sci 73(2):81–89

Cowan MM (1999) Plant products as antimicrobial agents. Clin Microbiol Rev 12(4):564

Creelman RA, Mullet JE (1997) Biosynthesis and action of jasmonates in plants. Ann Rev Plant Physiol Plant Mol Biol 48:355–381

DeVos M, Jander G (2009) Myzus persicae (Green peach aphid) salivary components induce defence responses in Arabidopsis thaliana. Plant Cell Environ 32(11):1548–1560

Echeverri F, Torres F, Quinones W, Cardona G, Archbold R, Roldan J, Brito I, Luis JG, LahlouU EH (1997) Danielone a phytoalexin from papaya fruit. Phytochemistry 44(2):255256. INIST: 2558881 (http://cat.inist.fr/?aModele=afficheN&cpsidt=2558881)

Eisner T, Meinwald J (1995) Chemical ecology: the chemistry of biotic interaction. In: (eds). National Academy Press, Washington, DC

Fenwick GR, Price KR, Tsukamota C, Okubo K (1992) Saponins. In: DMello JP, Duffus CM, Duffus JH (eds) Toxic substances in crop plants. Royal Society of Chemistry, Cambridge, pp 285–327

Freeman BC, Beattie GA (2008) An overview of plant defences against pathogens and herbivores. Plant Health Instr 0226-01.

Garas NA, Waiss AC (1986) Differential accumulation of antifungal sesquiterpenoids in cotton stems inoculated with Verticillium dahliae. Phytopathology 76:1011–1017

Gershenzon J, Croteau R (1991) Terpenoids. In: Rosenthal GA, Berenbaum MR (eds) Herbivores their interaction with secondary plant metabolites, Vol I: the chemical participants, 2nd edn. Academic, San Diego, pp 165–219

Geu-Flores F, Olsen CE, Halkier BA (2009) Towards engineering glucosinolates into non-cruciferous plants. Planta 229(2):261–270

Grayer RJ, Harborne JB (1994) A survey of antifungal compounds from higher plants 1982–1993. Phytochemistry 37:19–42

Gross D, Porzel A, Schmidt J (1994) Phytoalexine mit Indolstruktur aus Kohlrabi [Phytoalexins with indole structure from Kohlrabi]. Z Naturforsch 49(5–6):281–285

Grosskinsky DK, Naseem M, Abdelmoshem UA, Plickert N, Engelke T, Griebel T, Zeier J, Novak O, Strand M, Pfeifhofer H et al (2011) Cytokinins mediate resistance against Pseudomonas syringae in tobacco through increased antimicrobial phytoalexin synthesis independent of salicylic acid signaling. Plant Physiol 157:815–830

Grubb C, Abel S (2006) Glucosinolate metabolism and its control. Trends Plant Sci 11:89–100

Hain R, Reif HJ, Krause E, Langebartels R, Kindl H, Vornam B, Wiese W, Schmelzer E, Schreier P, Stöcker R et al (1993) Disease resistance results from foreign phytoalexin expression in a novel plant. Nature 361:153–156

Halkier BA, Gershenzon J (2006) Biology and biochemistry of glucosinolates. Ann Rev Plant Biol 57:303–333

Hammond-Kosack KE, Jones JDG (1996) Resistance gene-dependent plant defence responses. Plant Cell 8:1773–1791

Harborne JB (1999) The comparative biochemistry of phytoalexin induction in plants. Biochem Syst Ecol 27:335–367

He XZ, Dixon RA (2000) Genetic manipulation of isoflavone 7-O-methyltransferase enhances biosynthesis of 4′-O-methylated isoflavonoid phytoalexins and disease resistance in alfalfa. Plant Cell 12(9):1689–1702

Hegnauer R (1988) Biochemistry, distribution and taxonomic relevance of higher plant alkaloids. Phytochemistry 27:2423–2427

Ilic Dusica P, Nikolic Ljubisa B, Stanojevic Ljiljana P, Nikolic Vesna D, Stankovic Mihajlo Z, Milorad C (2011) Allicin and related compounds: biosynthesis, synthesis and pharmacological activity(PDF). Facta Universitatis 9(1):9–20. https://doi.org/10.2298/FUPCT1101009I
Jeandet P (2015) Phytoalexins: current progress and future prospects. Molecules 20:2770–2774
Jeandet P, Clément C, Courot E, Cordelier S (2013) Modulation of phytoalexin biosynthesis in engineered plants for disease resistance. Int J Mol Sci 14:14136–14170
Jeandet P, Clément C, Courot E (2014a) Resveratrol production at large scale using plant cell suspensions. Eng Life Sci. https://doi.org/10.1002/elsc.201400022
Jeandet P, Hébrard C, Deville MA, Cordelier S, Dorey S, Aziz A, Crouzet J (2014b) Deciphering the role of phytoalexins in plant-microorganism interactions and human health. Molecules 19:18033–18056
Johnson R, Narvaez J, An G, Ryan C (1989) Expression of proteinase inhibitors I and II in transgenic tobacco plants: effects on natural defence against Manduca sexta larvae. Proc Natl Acad Sci U S A 86:9871–9875
Kang SY, Kim YC (2007) Decursinol and decursin protect primary cultured rat cortical cells from glutamate-induced neurotoxicity. J Pharm Pharmacol 59(6):863–870
Kurosaki F, Nishi A (1983) Isolation and antimicrobial activity of the phytoalexin 6methoxymellein from cultured carrot cells. Phytochemistry 22(3):669. https://doi.org/10.1016/S00319422(00)869599
Lake JA, Field KJ, Davey MP, Beerling DJ, Lomax BH (2009) Metabolomic and physiological responses reveal multi-phasic acclimation of Arabidopsis thaliana to chronic UV radiation. Plant Cell Environ 32(10):1377–1389
Leustek T, Saito K (1999) Sulfate transport and assimilation in plants. Plant Physiol 120:637–643
Lewis NG, Yamamoto E (1990) Lignin: occurrence, biogenesis and biodegradation. Ann Rev Plant Physiol Plant Mol Biol 41:455–496
Lipka U, Fuchs R, Kuhns C, Petutschnig E, Lipka V (2010) Live and let die-Arabidopsis non-host resistance to powdery mildews. Eur J Cell Biol 89(2):194–199
Mari M, Lori R, Leoni O, Marchi A (1993) In vim activity of glucosinolate-derived isothiocyanates against postharvest fruit pathogens. Ann Appl Biol 123:155–164
Mazid M, Khan TA, Mohammad F (2011) Role of secondary metabolites in defence mechanisms of plants. Biol Med 3(2):232–249
Mes JJ, Van Doorn AA, Wijbrandi J, Simons G, Cornelissen BJC, Haring MA (2000) Expression of the Fusarium resistance gene I-2 colocalizes with the site of fungal containment. Plant J 23:183–193
Monde K, Osawa SM, Harada N, Takasugi M, Suchy M, Kutschy P, Dzurilla M, Balentova E (2000) Synthesis and absolute stereochemistry of a cruciferous phytoalexin, (−)-Spirobrassinin. Chem Lett 598:886–887
Mordue AJ, Blackwell A (1993) Azadirachtin: an update. J Insect Physiol 39:903–924
Ono E, Wong HL, Kawasaki T, Hasegawa M, Kodama O, Shimamoto K (2001) Essential role of the small GTPase Rac in disease resistance of rice. Proc Natl Acad Sci U S A 98:759–764
Osbourn AE (1996) Preformed antimicrobial compounds and plant defences against fungal attack. Plant Cell 8:1821–1831
Parkhi V, Kumar V, Campbell LM, Bell AA, Shah J, Rathore KS (2010) Resistance against various fungal pathogens and reniform nematode in transgenic cotton plants expressing Arabidopsis NRP1. Transgenic Res 19:959–975
Pearce G, Strydom D, Johnson S, Ryan CA (1991) A polypeptide from tomato leaves induces wound inducible proteinase inhibitor proteins. Science 253:895–898
Pegg GF, Woodward S (1986) Synthesis and metabolism of a-tomatine in tomato isolines in relation to resistance to Verticillium albo-atrum. Physiol MOI Plant Pathol 28:187–201
Peumans WJ, Van Damme EJM (1995) Lectins as plant defence proteins. Plant Physiol 109:347–342
Poulton JE (1990) Cyanogenesis in plants. Plant Physiol 94:401–405
Price KR, Johnson IT, Fenwick GR (1987) The chemistry and biological significance of saponins in food and feeding stuffs. Crit Rev Food Sci Nutr 26:27–133

Ren D, Liu Y, Yang KY, Han L, Mao G, Glazebrook J, Zhang S (2008) A fungal-responsive MAPK cascade regulates phytoalexin biosynthesis in Arabidopsis. Proc Natl Acad Sci U S A 105:5638–5643

Rosenthal GA (1991) The biochemical basis for the deleterious effects of L-canavanine. Phytochemistry 30:1055–1058

Saito K (2004) Sulfur assimilatory metabolism. The long and smelling road. Plant Physiol 136:2443–2450

Schafer H, Wink M (2009) Medicinally important secondary metabolites in recombinant microorganisms or plants: progress in alkaloid biosynthesis. Biotechnol J 4(12):1684–1703

Seigler DS (1991) Secondary metabolites and plant systematic. In: Conn EE (ed) The biochemistry of plants, Secondary plant products, vol 7. Plenum, New York/London, pp 139–176

Smith CA, MacHardy WE (1982) The significance of tomatine in the host response of susceptible and resistant tomato isolines infected with two races of Fusarium oxysporum f. sp. lycopersici. Phytopathology 72:415–419

Sreevidya VS, Srinivasa RC, Rao C, Sullia SB, Ladha JK, Reddy PM (2006) Metabolic engineering of rice with soyabean isoflavone synthase for promoting nodulation gene expression in rhizobia. J Exp Bot 57(9):1957–1969

Thomma BPHJ, Cammue BPA, Thevissen K (2002) Plant defenses. Planta 216(2):193–202

Turlings TCJ, Loughrin JH, Mccall PJ, Roese USR, Lewis WJ, Tumlinson JH (1995) How caterpillar-damaged plants protect themselves by attracting parasitic wasps. Proc Natl Acad Sci U S A 92:4169–4174

Van Etten HD, Mansfield JW, Bailey JA, Farmer EE (1994) Two classes of plant antibiotics: phytoalexins versus "phytoanticipins". Plant Cell 6:1191–1192

Wink M (1999) Functions of plant secondary metabolites and their exploitation in biotechnology, Annual plant reviews, vol 3. CRC Press, Boca Raton

Zernova OV, Lygin AV, Pawlowski ML, Hill CB, Hartman GL, Widholm JM, Lozovaya VV (2014) Regulation of plant immunity through modulation of phytoalexin synthesis. Molecules 19:7480–7496

Explorations of Plant's Chemodiversity: Role of Nitrogen-Containing Secondary Metabolites in Plant Defense

14

Sanjay Kumar Singh

Abstract

In nature, plants are surrounded by a number of biotic and abiotic environmental stresses. Biotic ecosystems contain a wide variety of bacteria, viruses, fungi, nematodes, mites, insects, mammals, and other herbivorous animals, greatly responsible for heavy reduction in crop productivity. Henceforth, to cope up from these biotic stresses, the plant defense mechanism increasingly requires the availability of large numbers of phytochemicals. Chemodiversity in plants offers a valuable source; for example, nitrogen-containing secondary metabolites, previously regarded as waste products, are now recognized for their resistant activity against herbivores, pests, pathogens, and diseases. In this chapter, I have described the increasing role of nitrogen-containing secondary metabolites during plant defense. These metabolites impose their effects by acting as deterrence/antifeedant, toxicity, or precursors to physical defense systems. Many specialized herbivores and pathogens do not merely circumvent the deterrent or toxic effects of secondary metabolites but actually utilize these compounds as host recognition signals and/or nutrients. This is true for both cyanogenic glucosides and glucosinolates which are discussed in detail. Their biochemical and molecular mechanism of action is compared and contrasted.

Keywords

Secondary metabolites · Plant defense · Pathogen · Herbivores

S. K. Singh (✉)
Kentucky Tobacco Research and Development Center, University of Kentucky, Lexington, KY, USA
e-mail: sanjaysingh@uky.edu

A. Singh, I. K. Singh (eds.), *Molecular Aspects of Plant-Pathogen Interaction*,
https://doi.org/10.1007/978-981-10-7371-7_14

14.1 Introduction

The term chemodiversity, generally, leaves aside larger molecules, which involve in vital primary metabolic functions and form the majority of the organic body mass of living beings. Thus, small molecules that often have a defensive or offensive signaling function mainly contribute to the chemodiversity. Since the beginning, humans have utilized the plants, one of the most prolific sources of biochemical diversity, for its own benefits. Since ancient times, plants have provided mankind with cures for health problems and continue to be the most capable pool of bioactive chemicals for the development of modern drugs (Dias et al. 2012; Cragg and Newman 2013; Harvey et al. 2015). More than 20,000 natural molecules have been studied so far, and numerous have been used as novel anticancer, antibiotic, anti-inflammatory or anti-pain agents, etc. In the previous few decades, plants have turned into a critical source for the discovery of novel and unique pharmaceutical compounds (Cordell 2000; Farnsworth 1988; Newman et al. 2000). Plants are reported to have high chemodiversity including more than 21,000 alkaloids, 700 nonprotein amino acids (NPAAs), 200 cyanogenic glycosides (CGs) and glucosinolates, >20,000 terpenoids, >10,000 polyphenols, >1500 polyacetylenes and fatty acids, 750 polyketides, and 200 carbohydrates (Wink 2008, 2013; Theis and Lerdau 2003).

Approximately 450 million (M) years ago, plants began to inhabit the terrestrial earth during the mid-Ordovician period and over the subsequent 40 M years spread across the earth surface. The evolution of species-specific metabolic systems from core metabolic pathways of aquatic ancestors was one of the reasons behind the success of early land plants, as they were able to synthesize the structurally and functionally diverse chemicals to cope with frequent biotic and abiotic ecological pressures (Weng et al. 2012). Several of these chemicals, such as cuticular components and phenolic compounds, are universal in all land plants and, therefore, provide indispensable physical and chemical protection against desiccation and UV radiation (Fig. 14.1). Other classes of specialized metabolites, including those that contribute to plant-specific flavors, colors, and scents, frequently occur in a lineage-specific manner and play specialized roles for the host species in their natural habitat (Weng et al. 2012). Present knowledge of secondary metabolism and its evolution in the plant has been primarily driven by studying of angiosperms or flowering plants, ranging from well-studied model species, such as rice and *Arabidopsis* (Romeo 2004; D'Auria and Gershenzon 2005), to the reference species including medicinal plants with remarkable pharmaceutical properties, e.g., *Vinca minor*, *Catharanthus roseus*, and *Rauvolfia serpentina* (Facchini and De Luca 2008; De Luca et al. 2012; Patra et al. 2013). These studies revealed massive chemical diversity in flowering plants and provide deep insight on their widespread speciation and global domination over the last 170 M years following the Permian-Triassic extinction event (Wikström et al. 2001). The vast expansion of plant chemodiversity associated with secondary metabolites reflects the tremendous adaptability of land-dwelling plants. For example, plant hormones regulate various aspects of plant growth and development in response to environmental cues, whereas phenolic and waxy cuticles act as UV protectant and prevent excessive water loss. Plant polymers

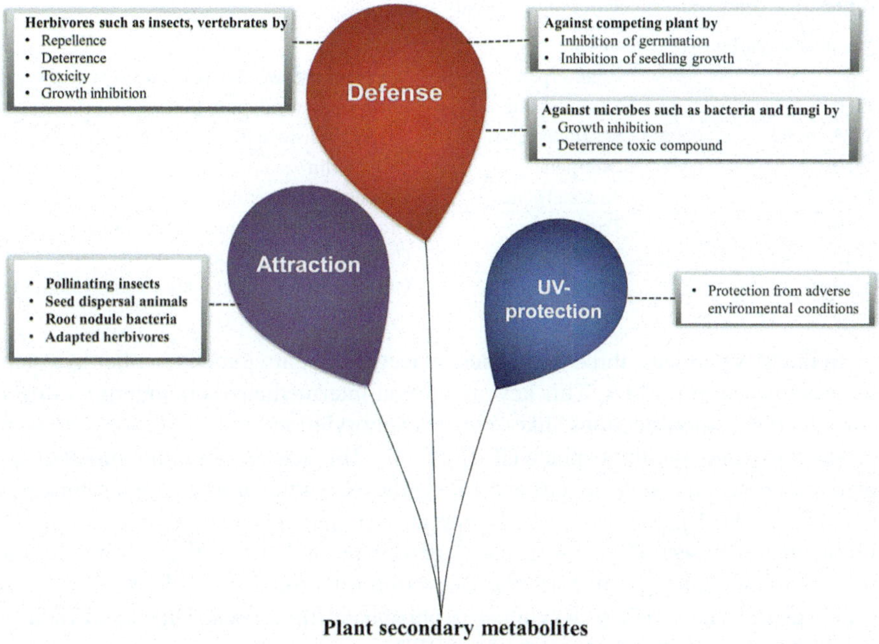

Fig. 14.1 Functional diversity of plant secondary metabolites

including lignin and sporopollenin provide mechanical support, gamete protection, and wound healing. New metabolic pathways continuously arose throughout terrestrial plant evolution, resulting in a contemporary collection of secondary metabolites. Therefore, some of these specialized metabolites are common across various taxonomic groups, while others were found in some limited species.

14.2 Secondary Metabolites Are Divided into Three Major Groups

On the basis of their chemical nature, plant secondary metabolites can be divided into three chemically distinct groups: terpenes, phenolics, and nitrogen-containing compounds.

14.2.1 Terpenes

Terpenes (also known as terpenoids) constitute the largest class of secondary metabolites. Plants and other natural sources are reported to produce more than 30,000 terpenoids (Bohlmann et al. 1998).

In plants, terpenes are biosynthesized in at least two different pathways. The main and well-studied biosynthetic route is known as the mevalonic acid (MA) pathway.

Table 14.1 Important molecules of terpenoids

Number of carbon	Name	Example
C5	Hemiterpene	Isoprene, prenol, isovaleric acid
C10	Monoterpene	Limonene, eucalyptol, pinene
C15	Sesquiterpene	ABA (abscisic acid)
C20	Diterpene	Gibberellin
C25	Sesterterpenes	Ophiobolin A, ceroplastol
C30	Triterpene	Brassinosteroids, squalene, lanosterol
C40	Tetraterpene	Carotenoids, lycopene
C>40	Polyterpenes	Ubiquinones, rubber, cytokonines, vitamin E

In the MA pathway, three molecules of acetyl-CoA are joined together in a stepwise manner to form MA. This key six-carbon intermediate then undergoes different chemical modifications like pyrophosphorylation and decarboxylation to produce isopentenyl diphosphate (IPP). Finally, IPP acts as a building block of terpenes. The second route of terpene biosynthesis is known as methylerythritol-4-phosphate (MEP) pathway, which operates in plastids (Tholl and Lee 2011; Lichtenthaler 1999). Glyceraldehyde-3-phosphate and two carbon atoms derived from pyruvate condense to form the five-carbon intermediate, 1-deoxy-d-xylulose 5-phosphate. The 1-deoxy-d-xylulose 5-phosphate further rearranged and reduced to MEP, which eventually converted into IPP.

Terpenes are the structurally diverse class of secondary metabolites from hemi- to polyterpenes (Table 14.1). All terpenes are originated from the union of five-carbon elements (also referred to as C5 units) that have the branched carbon skeleton of isopentane. The basic structural elements of terpenes are also known as isoprene, and, thus, terpenes are sometimes also called as isoprenoids. The terpenes can be classified in different groups on the basis of a number of C5 units they comprised of (Table 14.1). For instance, 10-carbon terpenes, which contain two C5 units, are called monoterpenes, while 15-carbon terpenes (three C5 units) are sesquiterpenes. In spite of structural similarities, terpenes can be synthesized in different compartments in the cell. For instance, nowadays it is believed that sesquiterpenes and triterpenes are synthesized through the cytosolic MA pathway, whereas mono-, di-, and tetraterpenes are derived from the chloroplastic MEP pathway (Thimmappa et al. 2014).

Terpenes have roles in both primary and secondary metabolism. Certain terpenes have been well studied for their functions in plant growth or development and therefore can be considered as primary rather than secondary metabolites. For instance, the gibberellins, an important group of phytohormones which are essential for numerous growth and developmental processes in plants including seed germination, leaf expansion, stem elongation, pollen maturation, trichome development, and the induction of flowering (Achard and Genschik 2009), are diterpenes. Brassinosteroids, also a class of plant hormones with growth-regulating functions such as activation of the cell cycle during seed germination (Zadvornova et al. 2005), control of cell cycle progression (González-García et al. 2011), and induction of exaggerated growth of hydroponically grown plants (Arteca and Arteca

2001), are derived from triterpenes. Terpenes are toxins and also act as a feeding deterrent to many herbivorous insects and mammals (Gershenzon and Croteau 1992). For instance, pyrethroids, a monoterpene ester reported from *Chrysanthemum* species, show remarkable insecticidal activity (Mori 2012). Monoterpenes accumulate in resin ducts found in the needles, twigs, and trunk of conifers, such as Douglas-fir, lodgepole pine, *Pinus contorta*, *Picea engelmannii*×*glauca*, and *Abies lasiocarpa*×*bifolia*, and are toxic to numerous insects, including bark beetles, a serious pest of conifer species throughout the planet (Trapp and Croteau 2001).

Essential oils, which lend a characteristic odor to their foliage, are mixtures of volatile monoterpenes and sesquiterpenes. Essential oils have been broadly used for bactericidal, virucidal, fungicidal, insecticidal, medicinal, and cosmetic applications (Isman 2000). Recently they are also used in pharmaceutical, sanitary, cosmetic, agricultural, and food industries (Holley and Patel 2005). *Mentha piperita*, *Citrus limon*, *Ocimum basilicum*, and *Salvia officinalis* are some well-known plants that contain essential oils. Essential oils are frequently found in glandular hairs and serve to repel the potential herbivores even before they take a trial bite. Caryophyllene, a sesquiterpene, is a common constituent of the essential oil of numerous plants including *Piper nigrum* and *Syzygium aromaticum*. Caryophyllene is known to possess anti-inflammatory, antimicrobial, anticarcinogenic, antibiotic, antioxidant, and local anesthetic properties (Legault et al. 2013; Kuwahata et al. 2012; Lee et al. 2005).

14.2.2 Phenolic Compounds

Plants produce a large variety of secondary metabolites that contain a phenol group: one or more hydroxyl functional groups on benzene rings (Randhir et al. 2004). These substances are classified as phenolic compounds or phenolics. The structures of these phenolics may range from simple phenolic molecule to complex high-molecular-weight polymer (Velderrain-Rodriguez et al. 2014). Phenolic compounds are found in nearly all the plant kingdom and located in nearly all plant parts. Main classes of phenolic compounds reported in higher plants are given in Table 14.2.

Shikimic acid and malonic acid are two basic pathways involve in the biosynthesis of phenolic compounds in plants. The shikimic acid pathway is involved in biosynthesis of most plant phenolics. Shikimic acid pathway converts simple carbohydrate precursors derived from glycolysis and the pentose phosphate pathway (PPP) into the three aromatic amino acids: phenylalanine, tyrosine, and tryptophan. Phenylalanine acts as a precursor of biosynthesis of most abundant classes of secondary phenolic compound in the plant.

Phenolic compounds play a vital role in growth and reproduction of plants, providing protection against pathogens and herbivores (Bravo 1998). Phenolic compounds are also involve in providing the color and sensory characteristics of fruits and vegetables (Alasalvar et al. 2001), in absorbing harmful ultraviolet (UV) radiation, and in reducing the growth of nearby competing plants. Phenolic compounds also have a wide range of physiological properties, such as antiallergenic, antiatherogenic, anti-inflammatory, antimicrobial, cardioprotective, and vasodilatory

Table 14.2 Main classes of phenolic compounds in higher plants

Classes and subclasses	Examples of specific compounds	Natural sources
Non-flavonoid compounds		
Phenolic acids	Hydroxybenzoic acids; hydroxycinnamic acids	*Macrotyloma uniflorum*
Benzoic acids	Gallic acid; protocatechuic acid	Quercus infectoria, *Hibiscus sabdariffa, Vitex agnus-castus*
	4-hydroxybenzoic acid	
Hydroxycinnamic acid	Coumaric acid; caffeic acid; ferulic acid; sinapic acid	*Arachis hypogaea, Eucalyptus globulus, Citrus limon*
Hydrolyzable tannins	Pentagalloylglucose	*Rhus chinensis*
Stilbenes	Resveratrol	*Fallopia japonica*
Lignans	Secoisolariciresinol; matairesinol; lariciresinol; pinoresinol	*Linum usitatissimum, Sesamum indicum*
Flavonoid compounds		
Condensed tannins or proanthocyanidins	Procyanidin, prodelphinidins	*Vitis vinifera*
Anthocyanidins	Pelargonidin; cyanidin; malvidin	*Geranium dissectum, Philodendron bipinnatifidum*
Flavanols	Catechins; gallocatechins	*Uncaria rhynchophylla, Camellia sinensis*
Flavanones	Naringenin; hesperetin	Citrus × paradisi, *Mentha aquatica*
Flavones	Apigenin; luteolin	*Petroselinum crispum, Apium graveolens, Ambrosia psilostachya*
Flavonols	Kaempferol; quercetin; myricetin	*Aloe vera, Coccinia grandis*
Isoflavones	Daidzein; genistein; glycitein	*Pueraria mirifica*

effects (Benavente-Garcia et al. 2000; Manach et al. 2005; Middleton et al. 2000; Puupponen-Pimiä et al. 2001; Samman et al. 2001).

Lignin is formed from three different phenylpropanoid alcohols, namely, coniferyl, coumaryl, and sinapyl. The physical toughness of lignin acts as a herbivore deterrent, while its chemical durability makes it relatively indigestible to herbivore and insect pathogens (Lattanzio et al. 2006; Rosenthal and Berenbaum 2012). The flavonoids, one of the largest classes of plant phenolics, are involved in pigmentation and defense (Treutter 2005). Tannins, a mainly constituent of woody plants, are general toxins that significantly reduce the growth and survivorship of many herbivores and also act as feeding repellents (Barbehenn and Peter Constabel 2011). Protocatechuic acid prevents smudge in onions, a disease caused by the fungus *Colletotrichum circinans*, and prevents spore germination and growth of other fungi as well (Kakkar and Bais 2014).

14.2.3 Nitrogen-Containing Compounds

A large number of plant secondary metabolites have nitrogen as part of their structure. They are synthesized from common amino acids. Nitrogen-containing secondary metabolites can be categorized into four categories: alkaloids, cyanogenic glycosides, glucosinolates, and nonprotein amino acids.

14.2.3.1 Alkaloids

Alkaloids are typically defined as plant-derived pharmacologically active basic compounds, which synthesized from amino acids and may contain one or more heterocyclic nitrogen atoms. The alkaloids are an extremely heterogeneous group of more than 15,000 nitrogen-containing secondary metabolites. The alkaloids include more than 150 families and found in around 20% of the vascular plant species. Alkaloids in plants are common in families of seed-bearing vascular plants or angiosperms, e.g., Magnoliaceae, Solanaceae, Papaveraceae, Leguminosae, Ranunculaceae, Rubiaceae, and Apocynaceae. The alkaloidal plant species may contain single or multiple alkaloids. For example, *Catharanthus roseus* contains 130 terpenoid indole alkaloids, including anticancerous vinblastine, and their synthesis can be regulated by multiple pathways (van Der Heijden et al. 2004; Patra et al. 2013). The alkaloids can accumulate in a different part of the plants including leaf, epidermal and hypodermal cells, bundle sheaths, and latex vessels. Alkaloids are usually synthesized from one of a few common amino acids, such as lysine, tyrosine, or tryptophan. However, the basic carbon skeleton of some alkaloids may contain a component derived from the terpene pathway also. Table 14.3 lists the major alkaloid types, their amino acid precursors, and natural plant sources. Alkaloids usually occur as salts of organic acids, such as acetic, malic, lactic, citric, and oxalic, in plants, while some basic alkaloids, like nicotine, also occur freely in nature (Ramawat et al. 2009). Very often, the alkaloids are biosynthesized in a particular plant organ but accumulate in another. For example, in tobacco, nicotine is synthesized in roots but is translocated to and stored in leaves (Shoji et al. 2000; Yazaki 2005; Morita et al. 2009). The alkaloids may be divided into three subclasses: proto-alkaloids, true alkaloids, and atypical alkaloids. Proto-alkaloids and true alkaloids are directly derived from amino acids, while atypical alkaloids are derived from sources other than amino acids, e.g., terpenoid-containing alkaloids.

14.2.3.1.1 Proto-alkaloids

These are nitrogen-containing alkaloids which originated from amino acids. Proto-alkaloids include mescaline, adrenaline, and ephedrine.

14.2.3.1.2 True Alkaloids

These alkaloids, generally, contain a heterocyclic ring with nitrogen, derived from amino acids and always basic in nature. These alkaloids are toxic and normally present in plants as salts of organic acids, e.g., nicotine, morphine, and codeine.

Table 14.3 Example of some true alkaloids and their natural sources

Alkaloid class	Example	Natural occurrence	Biosynthetic precursor
Pyrrolidine	Stachydrine, hygrine	*Erythroxylum coca*,	Aspartate
		Leonurus japonicus	
Piperidine	Coniine, piperine, solenopsin	*Piper nigrum*	Lysine
		Psilocaulon absimile	
		Petrosimonia monandra	
		Conium maculatum	
Tropane	Atropine, racemic, hyoscyamine	*Atropa belladonna*	Aspartate
		Hyoscyamus niger	
		Mandragora officinarum	
Isoquinoline	Papaverine, narcotine, berberine	*Papaver somniferum*	Tyrosine
		Argemone mexicana	
Quinolizidine	Lupinine	*Lupinus albus*	Lysine
Indole	Reserpine, ergatomine	*Ipomoea violacea*	Tryptophan
		Turbina corymbosa	
Pyrrolizidine	Heliotridine	*Adenostyles alliariae*	Aspartate
		Cordia myxa	

14.2.3.1.3 Atypical Alkaloids

These are alkaloid-like compounds that do not derive from amino acids. The atypical alkaloids include terpene-like alkaloids, steroid-like alkaloids, and purine-like alkaloids such as caffeine, theobromine, ephedrine, colchicine, erythromycin, and taxol. These are less commonly found in nature.

14.2.3.2 Cyanogenic Glycosides (CGs)

CGs are a group of nitrile-containing plant secondary metabolites that produce cyanide following their enzymatic breakdown. There are approximately 25 known CGs which occur in at least 2600 plant species, such as members of Fabaceae, Rosaceae, Leguminosae, Linaceae, and Compositae family, of which a number of species are used as food including apples, apricots, cherries, peaches, plums, quinces, cassava, peas, beans, barley, and sorghum (Eisler 1991; Haque and Bradbury 2002; Ganjewala et al. 2010; Vetter 2000). Chemically, CGs are glycosides of α-hydroxynitriles which are stored in cell vacuoles (Vetter 2000; Fleming 1999). The CG content in plant discourages feeding by insects and other herbivores. Most of the CGs are believed to be derived from L-valine, L-isoleucine, L-leucine, L-phenylalanine, L-tyrosine, and cyclopentenyl-glycine, a nonprotein amino acid. In plants, CG biosynthesis occurs in three steps (Vetter 2000). In the first step, two successive N-hydroxylations of amino group of parent amino acid are catalyzed by an enzyme of cytochrome P450 family which, finally, converted into aldoxime. The second step includes conversion of aldoxime into cyanohydrin by another cytochrome P450 enzyme. In the final step, cyanohydrins get glycosylated by a soluble enzyme

UDP-glucosyltransferase. CGs play pivotal roles in organization of chemical defense system in plants and in plant-insect interactions (Zagrobelny et al. 2004).

14.2.3.3 Glucosinolates

Glucosinolates (also known as mustard oil glycosides) are the second class of glycoside after CGs. Glucosinolates are sulfur- and nitrogen-containing plant secondary metabolites common in the agriculturally important Brassicaceae family. Glucosinolates degrade to produce the compounds responsible for the smell and taste of vegetables such as cabbage, broccoli, and radishes, which act as toxin and herbivore repellents. More than 130 glucosinolates have been identified in plants (Radojčić Redovniković et al. 2008). The glucosinolate biosynthesis comprises three steps: amino acid chain elongation, conversion of the amino acid moiety to the glucosinolate core structure, and subsequent side chain modifications. The structural diversity of glucosinolates arises from side chain elongation of the amino acid precursors and from various secondary modifications including oxidation, desaturation, hydroxylation, methoxylation, sulfation, and glucosylation. Most glucosinolates in the member of the Brassicaceae are synthesized from methionine that is modified by the sequential addition of one to nine additional methylene groups to its side chain (Graser et al. 2000). Glucosinolates are stored in the intact plant discretely from the enzymes (myrosinase) that hydrolyze them, and they are brought into contact with the hydrolyzing enzymes only when the plant is crushed because of wounding and insect or pathogen attack. Loss of cellular integrity triggers the binary glucosinolate-myrosinase system and causes the generation of thioglucose, sulfate, and an unstable intermediate which spontaneously rearranges into several degradation products which can include nitriles, epithionitriles, isothiocyanates, oxazolidine-2-thiones, and thiocyanates (Radojčić Redovniković et al. 2008).

14.2.3.4 Nonprotein Amino Acids (NPAAs)

There are common 20 amino acids, also referred to as protein amino acids, which are incorporated into proteins by plants and animals. Nonetheless, several plants also contain unusual amino acids, called NPAA, that are not incorporated into proteins. Instead, these NPAAs are present in the free form and act as defensive molecules. Many NPAAs are very similar in structure to protein amino acids and, therefore, have similar properties. NPAAs can mimic the behavior of standard amino acids and, thus, can act as metabolic antagonists or inhibitors. For instance, canavanine and azetidine-2-carboxylic acid have structure much like that of arginine and proline, respectively. About 900 NPAAs have been isolated from plants. Of these, some 250 are found, particularly, within a small subset of plant families including the Hippocastanaceae, Leguminosae, Sapindaceae, Aceraceae, and Cucurbitaceae (Wink 2011).

14.2.4 Role of Nitrogen-Containing Secondary Metabolites in Plant Defense

Plants have a range of defense mechanisms, which occur soon after the pathogen attack that leads to the formation of a wide range of phytochemicals and by-products including nitrogen-containing secondary metabolites. These chemicals help the plant to respond to the incompatible interaction and finally help them to cope up with adverse conditions (Dixon 2001).

14.2.4.1 Alkaloids

Alkaloids are a diverse group of secondary metabolites with a variety of targets and biological activities including interference with neurotransmitters, disruption of DNA replication, and inhibition of protein synthesis (Mithöfer and Boland 2012). Alkaloids are produced by a large number of higher plant species and mostly involved in defense-related functions such as inhibition of competitors and herbivore deterrents (Roberts 2013). The inhibitory effects of alkaloids on glycosidase and trehalose metabolism deter herbivores, and the capability to quench singlet reactive oxygen confers protection against this toxic photosynthetic by-product (Mithöfer and Boland 2012; González-Lamothe et al. 2009). Alkaloids also act as phytoanticipins and phytoalexins and, naturally protect the plants from disease (González-Lamothe et al. 2009). The α-tomatine, for example, is a spirosolane-type alkaloid that occurs in tomato plants and possesses antimicrobial, antifungal, and anti-inflammatory activities (Friedman 2002; Chiu and Lin 2008; Ito et al. 2007; Morrow et al. 2004; Simons et al. 2006; Thorne et al. 1985). Several potently antibacterial alkaloids have been identified in the different classes of alkaloid including indole, indolizidine, isoquinoline, aaptamine, piperazine, quinoline, quinolone, aaptamine-indole, bisindole, and indole-quinoline in plants like *Zanthoxylum tetraspermum*, *Prosopis glandulosa*, *Clausena heptaphylla*, and *Teclea afzelii* (Maneerat et al. 2012; Chakraborty et al. 1995a, b; Samoylenko et al. 2009; Nissanka et al. 2001; Iwasa et al. 2001; Kuete et al. 2008; Wang et al. 2013).

Alkaloids have toxic and repellent effects on a wide range of generalist herbivores in order to reduce or prevent damage to plants (van Dam et al. 1995; Hartmann 1999; Hartmann and Ober 2000; Ober 2003). Sugar-mimic alkaloids act as inhibitors of several sugars and glycosidase-metabolizing enzymes leading to toxic effects on the insect. Morus species are a good example of plants that contain sugar-mimic alkaloids. Leaves exude of *Morus* species rich in sugar-mimic alkaloids, 1,4-dideoxy-1,4-imino-d-arabinitol and 1-deoxynojirimycin, which are toxic to the *Samia ricini* (also known as eri silkworm), a generalist herbivore, but not to the domesticated silkworm, *Bombyx mori*, a mulberry specialist (Hirayama et al. 2007). Yasuda et al. (2002) reported 13 sugar-mimic alkaloids from the pods of *Angylocalyx pynaertii*, a member of Leguminosae (Yasuda et al. 2002). The nature of toxicity and target of plant alkaloid can be diversified but frequently involves in cell signaling disruption (Mithöfer and Boland 2012). Sanguinarine ((13-methyl[1,3]benzodioxolo[5,6-c]-1,3-dioxolo[4,5]phenanthridinium), a benzophenanthridine alkaloid, mainly found in the Papaveraceae family, which includes *Sanguinaria canadensis*, *Argemone mexicana*, and *Chelidonium majus*, is shown to have antioxidant, antitumor, antibacterial, and anti-inflammatory properties (Chaturvedi et al. 1997). Sanguinarine is also reported to suppress cyclooxygenase, lipoxygenase, cholinesterase, Na^{+}/K^{+}-ATPase, cAMP- and Ca^{2+}-dependent protein kinase, NF-κB activation, nitric oxide synthase, and mitogen-activated protein kinase phosphatase-1 activities (Jeng et al. 2007; Vavrečková et al. 1996; Ulrichová et al. 1983; Seifen et al. 1979; Wang et al. 1997; Chaturvedi et al. 1997; Huh et al. 2006; Vogt et al. 2005). Sanguinarine inhibits choline acetyltransferase, an enzyme that catalyzes the biosynthesis of the neurotransmitter acetylcholine, and, finally affect neurotransmission. Nicotine, mostly found in leaves of *Nicotiana* species, binds to nicotinic acetylcholine receptors and blocks or displaces the endogenous neurotransmitters. Nicotine acts as either an agonist or antagonist targeting nicotinic acetylcholine receptors in insects, causing continual stimulation of the parasympathetic nervous system which finally leads to paralysis and death of insect (Dewey and Xie 2013).

Toxic effects of plant alkaloids on bacterial and fungal activities have been shown in a number of studies. Quinolizidine alkaloids (QAs) which frequently occur in members of Fabaceae family, like *Lupinus*, *Baptisia*, *Thermopsis*, *Genista*, *Cytisus*, *Echinosophora*, and *Sophora*, are involved in plant protection against insect pests (Philippi et al. 2015; Wang et al. 2000; Zhao et al. 1998). QAs extracted from *Lupinus angustifolius* and *Genista vuralii* have shown to have antibacterial properties (Erdemoglu et al. 2007, 2009). The antifungal properties of alkaloids also have been proved for several plant-associated fungi by bioassay experiments (Wippich and Wink 1985; Ma et al. 1999; Zhao et al. 1998; Zhou et al. 2003). The antifungal alkaloids are reported from different plants, such as *Corydalis incisa*, *Corydalis ambigua*, *Dictamnus dasycarpus*, and *Veratrum taliense*, which are reported to be effective against a wide range of phytopathogenic fungi including *Cladosporium cucumerinum*, *Erysiphe graminis*, *Cladosporium herbarum*, *Phytophthora capsici*, and *Rhizoctonia cerealis*.

14.2.4.1.1 Cyanogenic Glycosides

CGs can act as a defense molecule both against herbivory and phytopathogens. In general, an inverse correlation is frequently reported between the degree of herbivore pressure and the CG content in plant (Schappert and Shore 1999; Gleadow and Woodrow 2000; Ballhorn 2011). Dhurrin (4-hydroxymandelonitrile-β-d-glucoside) is a well-studied CG, reported to be present in several plant species including *Sorghum bicolor.* Dhurrin acts as an oviposition activator for the pests such as *Atherigona soccata* and *Chilo partellus* (Alborn et al. 1992). Efficient hydrolysis of dhurrin and, subsequent, release of cyanide are essential to deter insect herbivory in *Sorghum bicolor* (Krothapalli et al. 2013). Larvae of *Phyllotreta nemorum* eat 80% less tissue of the dhurrin-overproducing transgenic *Arabidopsis* plant compared to wild-type (Tattersall et al. 2001). The CG content, the rate of HCN release, and the susceptibility of the attacker to HCN are three main factors which determine the effectiveness of CGs against attackers (Ballhorn et al. 2005; Kadow et al. 2012). Many organisms, including humans, have mechanisms to detoxify and excrete HCN; therefore, HCN poisoning occurs only when the rate of detoxification is lesser than the rate of intake. Depending on the insect species, CGs can act both as feeding deterrents or phagostimulants. For instance, CG acts as a feeding stimulant for *Spodoptera eridania* larvae as it prefers to graze on CG-containing plants, such as *Phaseolus lunatus*, and grows better when cyanide is present in their diet (Brattsten et al. 1983). In contrast, *Prunus dulcis* plants with a high concentration of CGs are resistant to larvae of *Capnodis tenebrionis* (Malagon and Garrido 1990). Ellsbury et al. (1992) studied the variation in feeding damage to *Trifolium repens* (white clover) by larvae of *Hypera postica* (alfalfa weevils) (Ellsbury et al. 1992). They found that larvae of *Hypera postica* preferred leaflets of *Trifolium repens* with less or no CG content. Although all CGs have a potential danger through the production of HCN, there are differences in the sensitivity of different animal species. CG content of *Prunus padus*, also known as bird cherry, triggers the anorexia, weakness, depression, stupor, circling, bruxism, excessive salivation, and tenesmus in herbivores which, finally, leads to death (Sargison et al. 1996). CGs are also reported to have the antifungal properties. For instance, CGs can inhibit the growth of some fungi, such as *Magnaporthe oryzae* (also known as blast fungus), in dose-dependent manner (Seo et al. 2011).

CGs can be harmful to human also. Different types of CGs may be found in various cyanogenic food plants, for example, taxiphyllin in bamboo shoots and linamarin and lotaustralin in cassava (Organization 2013). The tubers of cassava which is used as staple food in many tropical countries, such as the Pacific Island countries, Latin America, Africa, and regions of Asia, contain high levels of CGs. Although traditional tuber processing methods, such as grating, grinding, soaking, and drying, caused the removal or degradation of a major fraction of the CGs present in cassava tubers. However, partial paralysis of the limbs caused by chronic cyanide poisoning is still widespread in cassava-eating regions. Tropical ataxic neuropathy and konzo are some health-related issues that can be caused by continuous dietary exposure to CGs (Tylleskär et al. 1992; Ernesto et al. 2002; Oluwole et al. 2000).

14.2.4.1.2 Glucosinolates

Most of the glucosinolates in plants are involved in responses to external or environmental stimuli. Glucosinolates are also involved in communicating and activating a variety of information relating to plant defense against insects, bacteria, and fungi. Depending on developmental stage and environmental condition, glucosinolate pattern varies between species and ecotypes as well as between and within individual plants. Environmental conditions such as temperature and light (Hasegawa et al. 2000; Engelen-Eigles et al. 2006), changes in nutritional status (Kaur et al. 1990; Underhill et al. 1980), biotic (e.g., fungal infection and insect damage), and abiotic (e.g., wounding) (Halkier and Gershenzon 2006; del Carmen et al. 2013) stress can alter the glucosinolate profile significantly. A change of the glucosinolate profile by several environmental factors has supported the idea regarding possible roles of glucosinolates in the plant defense against insects, herbivores, and microbial pathogens.

Glucosinolates and their hydrolysis products evidently act as mediators in plant-insect interactions. Glucosinolates can function as general poison and deterrent for generalist insects. Glucosinolates in *Brassica* show growth inhibition or feeding deterrence to a wide range of general herbivores such as birds, land slugs, and generalist insects (Giamoustaris and Mithen 1995, 1996). Martin and Müller (2007) found that *Sinapis alba* (white mustard) respond to *Athalia rosae* (turnip sawfly) damage by systematically accumulating higher levels of glucosinolates and, thus, apparently increasing their resistance (Martin and Müller 2007). An increase in short-chain aliphatic methylsulfinyl glucosinolates in *Arabidopsis thaliana* in response to both specialist and generalist phloem-feeding aphids is also known (Mewis et al. 2005). *Brassica napus* lines with higher glucosinolate content are also reported to have less damage in response to generalists such as pigeons and slugs (Giamoustaris and Mithen 1995). *Brassica juncea* with high glucosinolate concentrations is less prone to damage caused by both crucifer specialist, *Plutella xylostella*, and the generalist, *Spodoptera eridania* (Li et al. 2000). Moreover, insect herbivore feeding may substantially increase the levels of glucosinolates in plants. In *Arabidopsis*, comparison of glucosinolate accumulation and expression of glucosinolate biosynthetic genes in wild-type and mutant lines affected in defense signaling indicated that feeding of the aphid generalist *Myzus persicae (Sulzer)*, the aphid specialist *Brevicoryne brassicae* (L.), and the *Spodoptera exigua Hübner*, a lepidopteran generalist, can increase the accumulation of aliphatic glucosinolate content (Mewis et al. 2006). The plant also alters the nature of glucosinolates in affected area to deter the herbivores. For instance, *Myzus persicae* feeds on *Arabidopsis* and causes an overall decrease in glucosinolate content, but the production of 4-methox yindol-3-ylmethylglucosinolate is induced. This altered composition of glucosinolate, finally, acts as a deterrent for herbivores (Kim and Jander 2007).

The role of glucosinolates in defense against pathogens is not well studied like for herbivores. However, there are several reports indicating glucosinolate and its hydrolysis products can be toxic to bacteria and fungi (Smolinska et al. 2003; Mari et al. 2002; Li et al. 1999). *Brassica* crops are used as a break crop. The glucosinolates and their hydrolysis products secreted from *Brassica* canola and Indian mustard show inhibitory effects on soilborne fungal pathogen, *Gaeumannomyces graminis* var. *tritici*, which causes take-all of wheat (Angus et al. 1994). The 4-methylsulphinylbutyl isothiocyanate, a glucosinolate-derived isothiocyanates, is reported to have broad spectrum of antimicrobial activity. Growth of wide range of the fungi, such as *Alternaria brassicicola*, *Plectosphaerella cucumerina*, *Botrytis cinerea*, *Fusarium oxysporum*, and *Peronospora parasitica*, and bacteria, like *Erwinia carotovora* and *Pseudomonas syringae*, is inhibited by the presence of 4-methylsulphinylbutyl isothiocyanate (Tierens et al. 2001). Also, tryptophan-derived indole glucosinolates are reported to enhance the resistance of *Arabidopsis thaliana* against fungi like *Plectosphaerella cucumerina* and *Phytophthora brassicae* (Sanchez-Vallet et al. 2010; Schlaeppi et al. 2010).

Additionally, exogenous treatment of phytohormones like jasmonic acid (JA) and salicylic acid (SA), key signal regulators of plant defenses, to the plant also alters the glucosinolate profile which, again, proves the role of glucosinolates in plant defense. Previous studies showed that exogenous JA application can induce the accumulation of indole glucosinolate content in white mustard and oilseed rape (Bodnaryk 1994; Doughty et al. 1995). In addition, SA application is also reported to alter glucosinolate accumulation in oilseed rape (Kiddle et al. 1994). The hydrolysis products of glucosinolate have negative effects on vertebrates too. A diet highly rich in glucosinolates can cause the growth depression, poor palatability, decreased food efficiency, hypertrophy and hyperplasia of the thyroid, and liver lesions and necrosis in vertebrates (Anilakumar et al. 2006).

14.2.4.1.3 Nonprotein Amino Acids

NPAAs are commonly found in plants. NPAAs are present in widely consumed animal foods also. For instance, *Medicago sativa* is rich in canavanine, while *Lens culinaris*, a widely used edible pulse, contains homoarginine. In plants, NPAAs possess different roles including antiherbivory, antimicrobial, and allelochemical activity. The NPAA can protect the producer plants against stress, microorganisms, plants, insects, or higher animals including human (Bell 2003; McSweeney et al. 2008). NPAAs exert their toxicity in several ways. Some block the synthesis or uptake of protein amino acids, while others can be misincorporated into proteins and, finally, lead to production of nonfunctional proteins.

The protein-synthesizing machinery of plants that produce NPAAs can discriminate between protein and NPAAs, and, therefore, they are not susceptible to the toxicity of NPAAs. For instance, *Convallaria majalis* produces an analog of the protein amino acid L-proline known as L-azetidine-2-carboxylic acid. Although *Convallaria majalis* can differentiate the L-proline and L-azetidine-2-carboxylic acid, it can be easily misincorporated in proteins of *Vigna aureus*, which does not synthesize azetidine-2-carboxylic acid, and strongly inhibit the growth of germinating seedlings (Fowden 1963).

14.2.4.1.4 Aliphatic NPAAs

β-methylamino-L-alanine (BMAA) is a derivative of the alanine with a methylamino group on the side chain. BMAA is produced by the cyanobacteria in root nodules of cycads and has potent neurotoxic properties. BMAA is also accumulated in the seeds of cycads and causes amyotrophic lateral sclerosis/parkinsonism-dementia (ALS/P-D) (Steele and Guzman 1987; Ince and Codd 2005). ALS is a rare group of progressive neurological disorders that mainly involve the neurons responsible for controlling voluntary muscle movements such as chewing, walking, and breathing. Dencichine (β-N-oxalyl-l-α,β-diaminopropionic acid) is a hemostatic agent present in widely used traditional Chinese medicinal herbs, such as Panax species and *Lathyrus sativus*. Dencichine is a neuro-excitatory NPAA which causes the motor neuron disease, neurolathyrism, a condition with acute neurotoxic symptoms such as the inability to stand, neck stiffening, and head retraction (Campbell et al. 1993). Canavanine, an arginine analog, is synthesized in some leguminous plants (Bell et al. 1978) and plays a pivotal role in plant chemical defense against insects (Rosenthal 2001). Canavanine functions as an allelopathic chemical and inhibits plant growth (Nakajima et al. 2001). Incorporation of canavanine in place of arginine produces structurally aberrant proteins which exhibit altered protein conformation and impaired function in insects, such as *Manduca sexta* and *Heliothis virescens* (Rosenthal and Dahlman 1986; Berge et al. 1986). Animals fed on seeds of canavanine-containing plants developed hematological and serological abnormalities and induce antibody-mediated autoimmune phenomena (Bell 2003). Indospicine is a hepatotoxic NPAA found in *Indigofera* plant species. It accumulates as the free amino acid in the tissues (like muscle) of grazing animals including the horse and acts as a competitive inhibitor of arginase and causes reproductive losses and severe to mild liver disease (Fletcher et al. 2015). Djenkolic acid commonly found in *Archidendron pauciflorum* causes djenkolism, an acute kidney malfunction (Bunawan et al. 2014; Bell 2003). L-methionine sulfoximine, seleno-cystathionine, selenomethionine, and dl-phosphinothricin are examples of other NPAAs of plant origin that are involved in plant defense (Bell 2003; Shaw et al. 1999; Schrauzer 2000; Kitajima and Chiba 2013; Tardito et al. 2012).

14.2.4.1.5 NPAAs with Aromatic Skeletons

Plants produce several NPAAs with aromatic skeletons, such as L-3,4-dihydroxyphenylalanine (L-DOPA) and m-tyrosine, that are involved in plant defense. L-DOPA is a compound with strong allelopathic activity. It is found in leaves and seeds of *Mucuna pruriens* (velvet bean) that has a nutritional quality similar to the soybean (Nishihara et al. 2005). L-DOPA acts as a precursor of many alkaloids, such as catecholamines and melanin, which are released into soils and inhibit the growth of nearby plants. L-DOPA is an important secondary metabolite for chemical defense against herbivores in plants (Huang et al. 2011; Van Alstyne et al. 2006). Plants with high L-DOPA content are less prone to attack of small mammals or insects (Rehr et al. 1973). It is also a key chemical involving in sclerotization and melanization of insects which finally affects the development and immunity of insects (Gallot et al. 2010; Andersen 2010). The L-DOPA acts as a herbicide and suppresses the growth of several weed species such as *Sinapis arvensis*, *Cirsium arvense*, *Papaver rhoeas*, and *Lamium amplexicaule* (Topal and Kocaçalişkan 2006). *m*-Tyrosine is an example of another NPAA with aromatic skeletons with phytotoxic properties. It is exuded from the roots of fine fescue grasses and inhibits the growth of a wide range of neighboring plant and, therefore, grants a competitive advantage to fescue grasses (Bertin et al. 2007; Huang et al. 2012). The toxicity of *m*-tyrosine is due to its misincorporation into cellular protein in place of protein amino acid phenylalanine (Gurer-Orhan et al. 2006; Klipcan et al. 2009). The *m*-tyrosine can also prevent the growth of bacteria including *Escherichia coli* and *Bacillus* species (Smith et al. 1964; Aronson and Wermus 1965).

14.2.4.1.6 NPAAs with Cyclic and Heterocyclic Skeletons

The 5-hydroxytryptophan (5-HTP) is found in the seeds of *Griffonia simplicifolia* and has been associated with the insecticidal properties (Janzen et al. 1977). Homoproline, a lysine-derived NPAA, is a critical regulator of systemic acquired resistance (SAR) and basal immunity to bacterial infection in plants including *Arabidopsis thaliana* and *Nicotiana tabacum* (Navarova et al. 2012; Vogel-Adghough et al. 2013). Homoproline signals the plants for effective biosynthesis of defense signal SA, accumulation of the phytoalexin camalexin, and expression of defense-related genes. Mimosine and its derivatives (α-amino-β-(3-hydroxy-4-oxo-1,4-dihydropyridin-1-yl)-propanoic acid), found in a leguminous *Leucaena leucocephala* (Xuan et al. 2006), have a strong herbicidal impact on several plants namely *Brassica rapa* and *Phaseolus vulgaris* (Xuan et al. 2006, 2016). Mimosine has insecticidal (Ishaaya et al. 1991) properties also and can inhibit the growth of first-instar larvae of *Tribolium castaneum.* β-(Isoxazolin-5-on-2-yl)-alanine (BIA), found in *Pisum*, *Lens*, *Lathyrus*, and *Vicia* plant species (Lambein et al. 1990), is a potent growth inhibitor of several eukaryotic organisms, such as yeasts; unicellular green algae; phytopathogenic fungi, such as *Botrytis cinerea*, *Pythium ultimum*, and *Rhizoctonia solani*; and higher plants (Schenk et al. 1991).

References

Achard P, Genschik P (2009) Releasing the brakes of plant growth: how GAs shutdown DELLA proteins. J Exp Bot 60(4):1085–1092

Alasalvar C, Grigor JM, Zhang D, Quantick PC, Shahidi F (2001) Comparison of volatiles, phenolics, sugars, antioxidant vitamins, and sensory quality of different colored carrot varieties. J Agric Food Chem 49(3):1410–1416

Alborn H, Stenhagen G, Leuschner K (1992) Biochemical selection of sorghum crop varieties resistant to sorghum shoot fly (*Atherigona soccata*) and stem borer (*Chilo partellus*): role of allelochemicals. In: Rizvi SJH, Rizvi V (eds) Allelopathy: basic and applied aspects. Springer, Dordrecht, pp 101–117

Andersen SO (2010) Insect cuticular sclerotization: a review. Insect Biochem Mol Biol 40(3):166–178

Angus JF, Gardner PA, Kirkegaard JA, Desmarchelier JM (1994) Biofumigation: isothiocyanates released frombrassica roots inhibit growth of the take-all fungus. Plant Soil 162(1):107–112

Anilakumar K, Khanum F, Bawa A (2006) Dietary role of glucosinolate derivatives: a review. J Food Sci Technol-Mysore 43(1):8–17

Aronson JN, Wermus GR (1965) Effects of m-tyrosine on growth and sporulation of Bacillus species. J Bacteriol 90(1):38–46

Arteca JM, Arteca RN (2001) Brassinosteroid-induced exaggerated growth in hydroponically grown Arabidopsis plants. Physiol Plant 112(1):104–112

Ballhorn DJ (2011) Constraints of simultaneous resistance to a fungal pathogen and an insect herbivore in lima bean (*Phaseolus lunatus* L.) J Chem Ecol 37(2):141–144

Ballhorn DJ, Lieberei R, Ganzhorn JU (2005) Plant cyanogenesis of *Phaseolus lunatus* and its relevance for herbivore–plant interaction: the importance of quantitative data. J Chem Ecol 31(7):1445–1473

Barbehenn RV, Peter Constabel C (2011) Tannins in plant-herbivore interactions. Phytochemistry 72(13):1551–1565

Bell EA (2003) Nonprotein amino acids of plants: significance in medicine, nutrition, and agriculture. J Agric Food Chem 51(10):2854–2865

Bell EA, Lackey JA, Polhill RM (1978) Systematic significance of canavanine in the Papilionoideae (faboideae). Biochem Syst Ecol 6(3):201–212

Benavente-Garcia O, Castillo J, Lorente J, Ortuño A, Del Rio JA (2000) Antioxidant activity of phenolics extracted from *Olea europaea* L. leaves. Food Chem 68(4):457–462

Berge MA, Rosenthal GA, Dahlman DL (1986) Tobacco budworm, *Heliothis virescens* [Noctuidae] resistance to l-canavanine, a protective allelochemical. Pestic Biochem Physiol 25(3):319–326

Bertin C, Weston LA, Huang T, Jander G, Owens T, Meinwald J, Schroeder FC (2007) Grass roots chemistry: meta-tyrosine, an herbicidal non-protein amino acid. Proc Natl Acad Sci 104(43):16964–16969

Bodnaryk RP (1994) Potent effect of jasmonates on indole glucosinolates in oilseed rape and mustard. Phytochemistry 35(2):301–305

Bohlmann J, Meyer-Gauen G, Croteau R (1998) Plant terpenoid synthases: molecular biology and phylogenetic analysis. Proc Natl Acad Sci U S A 95(8):4126–4133

Brattsten LB, Samuelian JH, Long KY, Kincaid SA, Evans CK (1983) Cyanide as a feeding stimulant for the southern army worm, *Spodoptera eridania*. Ecol Entomol 8(2):125–132

Bravo L (1998) Polyphenols: chemistry, dietary sources, metabolism, and nutritional significance. Nutr Rev 56(11):317–333

Bunawan NC, Rastegar A, White KP, Wang NE (2014) Djenkolism: case report and literature review. Int Med Case Rep J 7:79–84

Campbell CG, Mehra RB, Agrawal SK, Chen YZ, Abd El Moneim AM, Khawaja HIT, Yadov CR, Tay JU, Araya WA (1993) Current status and future strategy in breeding grasspea (*Lathyrus sativus*). Euphytica 73(1):167–175

Chakraborty A, Chowdhury BK, Bhattacharyya P (1995a) Clausenol and clausenine—two carbazole alkaloids from *Clausena anisata*. Phytochemistry 40(1):295–298

Chakraborty A, Saha C, Podder G, Chowdhury BK, Bhattacharyya P (1995b) Carbazole alkaloid with antimicrobial activity from *Clausena heptaphylla*. Phytochemistry 38(3):787–789

Chaturvedi MM, Kumar A, Darnay BG, Chainy GB, Agarwal S, Aggarwal BB (1997) Sanguinarine (pseudochelerythrine) is a potent inhibitor of NF-κB activation, IκBα phosphorylation, and degradation. J Biol Chem 272(48):30129–30134

Chiu FL, Lin JK (2008) Tomatidine inhibits iNOS and COX-2 through suppression of NF-kappaB and JNK pathways in LPS-stimulated mouse macrophages. FEBS Lett 582(16):2407–2412

Cordell GA (2000) Biodiversity and drug discovery—a symbiotic relationship. Phytochemistry 55(6):463–480

Cragg GM, Newman DJ (2013) Natural products: a continuing source of novel drug leads. Biochim Biophys Acta 1830(6):3670–3695

D'Auria JC, Gershenzon J (2005) The secondary metabolism of Arabidopsis thaliana: growing like a weed. Curr Opin Plant Biol 8(3):308–316

De Luca V, Salim V, Atsumi SM, Yu F (2012) Mining the biodiversity of plants: a revolution in the making. Science 336(6089):1658–1661

del Carmen M-BM, Moreno DA, Carvajal M (2013) The physiological importance of glucosinolates on plant response to abiotic stress in Brassica. Int J Mol Sci 14(6):11607–11625

Dewey RE, Xie J (2013) Molecular genetics of alkaloid biosynthesis in *Nicotiana tabacum*. Phytochemistry 94:10–27

Dias DA, Urban S, Roessner U (2012) A historical overview of natural products in drug discovery. Metabolites 2(2):303–336

Dixon RA (2001) Natural products and plant disease resistance. Nature 411(6839):843–847

Doughty KJ, Kiddle GA, Pye BJ, Wallsgrove RM, Pickett JA (1995) Selective induction of glucosinolates in oilseed rape leaves by methyl jasmonate. Phytochemistry 38(2):347–350

Eisler R (1991) Cyanide hazards to fish, wildlife, and invertebrates: a synoptic review. U.S. Department of the Interior, Fish and Wildlife Service, Laurel

Ellsbury M, Pederson G, Fairbrother T (1992) Resistance to foliar-feeding *Hyperine weevils* (coleoptera: Curculionidae) in cyanogenic white clover. J Econ Entomol 85(6):2467–2472

Engelen-Eigles G, Holden G, Cohen JD, Gardner G (2006) The effect of temperature, photoperiod, and light quality on gluconasturtiin concentration in watercress (*Nasturtium officinale* R. Br.) J Agric Food Chem 54(2):328–334

Erdemoglu N, Ozkan S, Tosun F (2007) Alkaloid profile and antimicrobial activity of *Lupinus angustifolius* L. alkaloid extract. Phytochem Rev 6(1):197–201

Erdemoglu N, Ozkan S, Duran A, Tosun F (2009) GC-MS analysis and antimicrobial activity of alkaloid extract from *Genista vuralii*. Pharm Biol 47(1):81–85

Ernesto M, Cardoso AP, Nicala D, Mirione E, Massaza F, Cliff J, Haque MR, Bradbury JH (2002) Persistent konzo and cyanogen toxicity from cassava in northern Mozambique. Acta Trop 82(3):357–362

Facchini PJ, De Luca V (2008) Opium poppy and Madagascar periwinkle: model non-model systems to investigate alkaloid biosynthesis in plants. Plant J 54(4):763–784
Farnsworth NR (1988) Screening plants for new medicines. Biodiversity 1:83–97
Fleming FF (1999) Nitrile-containing natural products. Nat Prod Rep 16(5):597–606
Fletcher M, Al Jassim R, Cawdell-Smith A (2015) The occurrence and toxicity of Indospicine to grazing animals. Agriculture 5(3):427
Fowden L (1963) Amino-acid analogues and the growth of seedlings. J Exp Bot 14(3):387–398
Friedman M (2002) Tomato glycoalkaloids: role in the plant and in the diet. J Agric Food Chem 50(21):5751–5780
Gallot A, Rispe C, Leterme N, Gauthier J-P, Jaubert-Possamai S, Tagu D (2010) Cuticular proteins and seasonal photoperiodism in aphids. Insect Biochem Mol Biol 40(3):235–240
Ganjewala D, Kumar S, Devi S, Ambika K (2010) Advances in cyanogenic glycosides biosynthesis and analyses in plants: a review. Acta Biologica Szegediensis 54(1):1–14
Gershenzon J, Croteau R (1992) Terpenoids. In: Herbivores: their interactions with secondary plant metabolites, The chemical participants, vol 1, 2nd edn. Academic, San Diego, pp 165–219
Giamoustaris A, Mithen R (1995) The effect of modifying the glucosinolate content of leaves of oilseed rape (*Brassica napus ssp. oleifera*) on its interaction with specialist and generalist pests. Ann Appl Biol 126(2):347–363
Giamoustaris A, Mithen R (1996) Genetics of aliphatic glucosinolates. IV. Side-chain modification in Brassica oleracea. Theor Appl Genet 93(5–6):1006–1010
Gleadow RM, Woodrow IE (2000) Polymorphism in cyanogenic glycoside content and cyanogenic β-glucosidase activity in natural populations of Eucalyptus cladocalyx. Funct Plant Biol 27(7):693–699
González-García M-P, Vilarrasa-Blasi J, Zhiponova M, Divol F, Mora-García S, Russinova E, Caño-Delgado AI (2011) Brassinosteroids control meristem size by promoting cell cycle progression in *Arabidopsis* roots. Development 138(5):849–859
González-Lamothe R, Mitchell G, Gattuso M, Diarra M, Malouin F, Bouarab K (2009) Plant antimicrobial agents and their effects on plant and human pathogens. Int J Mol Sci 10(8):3400
Graser G, Schneider B, Oldham NJ, Gershenzon J (2000) The methionine chain elongation pathway in the biosynthesis of glucosinolates in *Eruca sativa* (Brassicaceae). Arch Biochem Biophys 378(2):411–419
Gurer-Orhan H, Ercal N, Mare S, Pennathur S, Orhan H, Heinecke JW (2006) Misincorporation of free m-tyrosine into cellular proteins: a potential cytotoxic mechanism for oxidized amino acids. Biochem J 395(2):277–284
Halkier BA, Gershenzon J (2006) Biology and biochemistry of glucosinolates. Annu Rev Plant Biol 57:303–333
Haque MR, Bradbury JH (2002) Total cyanide determination of plants and foods using the picrate and acid hydrolysis methods. Food Chem 77(1):107–114
Hartmann T (1999) Chemical ecology of pyrrolizidine alkaloids. Planta 207(4):483–495
Hartmann T, Ober D (2000) Biosynthesis and metabolism of pyrrolizidine alkaloids in plants and specialized insect herbivores. In: Leeper FJ, Vederas JC (eds) Biosynthesis: aromatic polyketides, isoprenoids, alkaloids. Springer, Berlin/Heidelberg, pp 207–243
Harvey AL, Edrada-Ebel R, Quinn RJ (2015) The re-emergence of natural products for drug discovery in the genomics era. Nat Rev Drug Discov 14(2):111–129
Hasegawa T, Yamada K, Kosemura S, Yamamura S, Hasegawa K (2000) Phototropic stimulation induces the conversion of glucosinolate to phototropism-regulating substances of radish hypocotyls. Phytochemistry 54(3):275–279
Hirayama C, Konno K, Wasano N, Nakamura M (2007) Differential effects of sugar-mimic alkaloids in mulberry latex on sugar metabolism and disaccharidases of Eri and domesticated silkworms: enzymatic adaptation of *Bombyx mori* to mulberry defense. Insect Biochem Mol Biol 37(12):1348–1358
Holley RA, Patel D (2005) Improvement in shelf-life and safety of perishable foods by plant essential oils and smoke antimicrobials. Food Microbiol 22(4):273–292

Huang T, Jander G, de Vos M (2011) Non-protein amino acids in plant defense against insect herbivores: representative cases and opportunities for further functional analysis. Phytochemistry 72(13):1531–1537

Huang T, Rehak L, Jander G (2012) Meta-tyrosine in *Festuca rubra ssp. commutata* (Chewings fescue) is synthesized by hydroxylation of phenylalanine. Phytochemistry 75:60–66

Huh J, Liepins A, Zielonka J, Andrekopoulos C, Kalyanaraman B, Sorokin A (2006) Cyclooxygenase 2 rescues LNCaP prostate cancer cells from sanguinarine-induced apoptosis by a mechanism involving inhibition of nitric oxide synthase activity. Cancer Res 66(7):3726–3736

Ince PG, Codd GA (2005) Return of the cycad hypothesis – does the amyotrophic lateral sclerosis/parkinsonism dementia complex (ALS/PDC) of Guam have new implications for global health? Neuropathol Appl Neurobiol 31(4):345–353

Ishaaya I, Hirashima A, Yablonski S, Tawata S, Eto M (1991) Mimosine, a nonprotein amino acid, inhibits growth and enzyme systems in *Tribolium castaneum*. Pestic Biochem Physiol 39(1):35–42

Isman MB (2000) Plant essential oils for pest and disease management. Crop Prot 19(8):603–608

Ito S, Ihara T, Tamura H, Tanaka S, Ikeda T, Kajihara H, Dissanayake C, Abdel-Motaal FF, El-Sayed MA (2007) Alpha-Tomatine, the major saponin in tomato, induces programmed cell death mediated by reactive oxygen species in the fungal pathogen *Fusarium oxysporum*. FEBS Lett 581(17):3217–3222

Iwasa K, Moriyasu M, Tachibana Y, Kim H-S, Wataya Y, Wiegrebe W, Bastow KF, Cosentino LM, Kozuka M, Lee K-H (2001) Simple isoquinoline and benzylisoquinoline alkaloids as potential antimicrobial, antimalarial, cytotoxic, and anti-HIV agents. Bioorg Med Chem 9(11):2871–2884

Janzen DH, Juster HB, Arthur Bell E (1977) Toxicity of secondary compounds to the seed-eating larvae of the bruchid beetle *Callosobruchus maculatus*. Phytochemistry 16(O):223–227

Jeng J-H, Wu H-L, Lin B-R, Lan W-H, Chang H-H, Ho Y-S, Lee P-H, Wang Y-J, Wang J-S, Chen Y-J (2007) Antiplatelet effect of sanguinarine is correlated to calcium mobilization, thromboxane and cAMP production. Atherosclerosis 191(2):250–258

Kadow D, Voß K, Selmar D, Lieberei R (2012) The cyanogenic syndrome in rubber tree Hevea brasiliensis: tissue-damage-dependent activation of linamarase and hydroxynitrile lyase accelerates hydrogen cyanide release. Ann Bot 109:1253. mcs057

Kakkar S, Bais S (2014) A review on protocatechuic acid and its pharmacological potential. ISRN Pharmacol 2014:9

Kaur S, Gupta SK, Sukhija PS, Munshi SK (1990) Accumulation of glucosinolates in developing mustard (*Brassica juncea* L.) seeds in response to sulphur application. Plant Sci 66(2):181–184

Kiddle GA, Doughty KJ, Wallsgrove RM (1994) Salicylic acid-induced accumulation of glucosinolates in oilseed rape (*Brassica napus* L.) leaves. J Exp Bot 45(9):1343–1346

Kim JH, Jander G (2007) *Myzus persicae* (green peach aphid) feeding on Arabidopsis induces the formation of a deterrent indole glucosinolate. Plant J 49(6):1008–1019

Kitajima T, Chiba Y (2013) Selenomethionine metabolism and its toxicity in yeast. Biomol Concepts 4(6):611–616

Klipcan L, Moor N, Kessler N, Safro MG (2009) Eukaryotic cytosolic and mitochondrial phenylalanyl-tRNA synthetases catalyze the charging of tRNA with the meta-tyrosine. Proc Natl Acad Sci U S A 106(27):11045–11048

Krothapalli K, Buescher EM, Li X, Brown E, Chapple C, Dilkes BP, Tuinstra MR (2013) Forward genetics by genome sequencing reveals that rapid cyanide release deters insect herbivory of *Sorghum bicolor*. Genetics 195(2):309–318

Kuete V, Wansi JD, Mbaveng AT, Kana Sop MM, Tadjong AT, Beng VP, Etoa FX, Wandji J, Meyer JJM, Lall N (2008) Antimicrobial activity of the methanolic extract and compounds from *Teclea afzelii* (Rutaceae). S Afr J Bot 74(4):572–576

Kuwahata H, Katsuyama S, Komatsu T, Nakamura H, Corasaniti MT, Bagetta G, Sakurada S, Sakurada T, Takahama K (2012) Local peripheral effects of β-Caryophyllene through CB 2 receptors in neuropathic pain in mice. Pharmacol Pharm 3(04):397

Lambein F, Kuo Y-H, Ikegami F, Murakoshi I (1990) Toxic and non-toxic nonprotein amino acids in the Vicieae. In: Amino acids. Springer, pp 21–28

Lattanzio V, Lattanzio VM, Cardinali A (2006) Role of phenolics in the resistance mechanisms of plants against fungal pathogens and insects. Phytochemistry Adv Res 661:23–67

Lee S-J, Umano K, Shibamoto T, Lee K-G (2005) Identification of volatile components in basil (*Ocimum basilicum* L.) and thyme leaves (*Thymus vulgaris* L.) and their antioxidant properties. Food Chem 91(1):131–137

Legault J, Côté P-A, Ouellet S, Simard S, Pichette A (2013) Iso-caryophyllene cytotoxicity induced by lipid peroxidation and membrane permeabilization in L-929 cells. J Appl Pharma Sci 3:25

Li Y, Kiddle G, Bennett RN, Wallsgrove RM (1999) Local and systemic changes in glucosinolates in Chinese and European cultivars of oilseed rape (Brassica nap us L.) after inoculation with Sclerotinia sclerotiorum (stem rot). Ann Appl Biol 134(1):45–58

Li Q, Eigenbrode SD, Stringam GR, Thiagarajah MR (2000) Feeding and growth of *Plutella xylostella* and *Spodoptera eridania* on *Brassica juncea* with varying glucosinolate concentrations and myrosinase activities. J Chem Ecol 26(10):2401–2419

Lichtenthaler HK (1999) The 1-deoxy-d-xylulose-5-phosphate pathway of isoprenoid biosynthesis in plants. Annu Rev Plant Physiol Plant Mol Biol 50:47–65

Ma W, Fukushi Y, Tahara S (1999) Fungitoxic alkaloids from *Hokkaido Corydalis* species. Fitoterapia 70(3):258–265

Malagon J, Garrido A (1990) Relation between cyanogenic glycosides content and the resistance to *Capnodis tenebrionis* (L.) in stone fruits. Boletín de Sanidad Vegetal, Plagas 16(2):499–503

Manach C, Mazur A, Scalbert A (2005) Polyphenols and prevention of cardiovascular diseases. Curr Opin Lipidol 16(1):77–84

Maneerat W, Phakhodee W, Ritthiwigrom T, Cheenpracha S, Promgool T, Yossathera K, Deachathai S, Laphookhieo S (2012) Antibacterial carbazole alkaloids from *Clausena harmandiana* twigs. Fitoterapia 83(6):1110–1114

Mari M, Leoni O, Iori R, Cembali T (2002) Antifungal vapour-phase activity of allyl-isothiocyanate against *Penicillium expansum* on pears. Plant Pathol 51(2):231–236

Martin N, Müller C (2007) Induction of plant responses by a sequestering insect: relationship of glucosinolate concentration and myrosinase activity. Basic Appl Ecol 8(1):13–25

McSweeney C, Collins E, Blackall L, Seawright A (2008) A review of anti-nutritive factors limiting potential use of *Acacia angustissima* as a ruminant feed. Anim Feed Sci Technol 147(1):158–171

Mewis I, Appel HM, Hom A, Raina R, Schultz JC (2005) Major signaling pathways modulate Arabidopsis glucosinolate accumulation and response to both phloem-feeding and chewing insects. Plant Physiol 138(2):1149–1162

Mewis I, Tokuhisa JG, Schultz JC, Appel HM, Ulrichs C, Gershenzon J (2006) Gene expression and glucosinolate accumulation in *Arabidopsis thaliana* in response to generalist and specialist herbivores of different feeding guilds and the role of defense signaling pathways. Phytochemistry 67(22):2450–2462

Middleton E, Kandaswami C, Theoharides TC (2000) The effects of plant flavonoids on mammalian cells: implications for inflammation, heart disease, and cancer. Pharmacol Rev 52(4):673–751

Mithöfer A, Boland W (2012) Plant defense against herbivores: chemical aspects. Annu Rev Plant Biol 63:431–450

Mori T (2012) Pyrethroids: from Chrysanthemum to modern industrial insecticide, vol 314. Springer, Berlin

Morita M, Shitan N, Sawada K, Van Montagu MCE, Inzé D, Rischer H, Goossens A, Oksman-Caldentey K-M, Moriyama Y, Yazaki K (2009) Vacuolar transport of nicotine is mediated by a multidrug and toxic compound extrusion (MATE) transporter in *Nicotiana tabacum*. Proc Natl Acad Sci 106(7):2447–2452

Morrow WJ, Yang YW, Sheikh NA (2004) Immunobiology of the Tomatine adjuvant. Vaccine 22(19):2380–2384

Nakajima N, Hiradate S, Fujii Y (2001) Plant growth inhibitory activity of L-Canavanine and its mode of action. J Chem Ecol 27(1):19–31

Navarova H, Bernsdorff F, Doring AC, Zeier J (2012) Pipecolic acid, an endogenous mediator of defense amplification and priming, is a critical regulator of inducible plant immunity. Plant Cell 24(12):5123–5141

Newman DJ, Cragg GM, Snader KM (2000) The influence of natural products upon drug discovery. Nat Prod Rep 17(3):215–234

Nishihara E, Parvez MM, Araya H, Kawashima S, Fujii Y (2005) L-3-(3,4-Dihydroxyphenyl)alanine (L-DOPA), an allelochemical exuded from velvetbean (*Mucuna pruriens*) roots. Plant Growth Regul 45(2):113–120

Nissanka APK, Karunaratne V, Bandara BMR, Kumar V, Nakanishi T, Nishi M, Inada A, Tillekeratne LMV, Wijesundara DSA, Gunatilaka AAL (2001) Antimicrobial alkaloids from Zanthoxylum tetraspermum and caudatum. Phytochemistry 56(8):857–861

Ober D (2003) Chapter nine chemical ecology of alkaloids exemplified with the pyrrolizidines. Recent Adv Phytochem 37:203–230

Oluwole OS, Onabolu AO, Link H, Rosling H (2000) Persistence of tropical ataxic neuropathy in a Nigerian community. J Neurol Neurosurg Psychiatry 69(1):96–101

Patra B, Schluttenhofer C, Wu Y, Pattanaik S, Yuan L (2013) Transcriptional regulation of secondary metabolite biosynthesis in plants. Biochim Biophys Acta (BBA) – Gene Regul Mech 1829(11):1236–1247

Philippi J, Schliephake E, Jürgens H-U, Jansen G, Ordon F (2015) Feeding behavior of aphids on narrow-leafed lupin (*Lupinus angustifolius*) genotypes varying in the content of quinolizidine alkaloids. Entomol Exp Appl 156(1):37–51

Puupponen-Pimiä R, Nohynek L, Meier C, Kähkönen M, Heinonen M, Hopia A, Oksman-Caldentey KM (2001) Antimicrobial properties of phenolic compounds from berries. J Appl Microbiol 90(4):494–507

Radojčić Redovniković I, Glivetić T, Delonga K, Vorkapić-Furač J (2008) Glucosinolates and their potential role in plant. Period Biol 110(4):297–309

Ramawat KG, Dass S, Mathur M (2009) The chemical diversity of bioactive molecules and therapeutic potential of medicinal plants. In: Ramawat KG (ed) Herbal drugs: ethnomedicine to modern medicine. Springer, Berlin/Heidelberg, pp 7–32

Randhir R, Lin Y-T, Shetty K (2004) Stimulation of phenolics, antioxidant and antimicrobial activities in dark germinated mung bean sprouts in response to peptide and phytochemical elicitors. Process Biochem 39(5):637–646

Rehr SS, Janzen DH, Feeny PP (1973) L-dopa in legume seeds: a chemical barrier to insect attack. Science 181(4094):81–82

Roberts MF (2013) Alkaloids: biochemistry, ecology, and medicinal applications. Springer, New York

Romeo J (2004) Secondary metabolism in model systems: recent advances in phytochemistry, vol 38. Elsevier, San Diego

Rosenthal GA (2001) L-Canavanine: a higher plant insecticidal allelochemical. Amino Acids 21(3):319–330

Rosenthal GA, Berenbaum MR (2012) Herbivores: their interactions with secondary plant metabolites, Ecological and evolutionary processes, vol 2. Academic, San Diego

Rosenthal GA, Dahlman DL (1986) L-Canavanine and protein synthesis in the tobacco hornworm *Manduca sexta*. Proc Natl Acad Sci U S A 83(1):14–18

Samman S, Sandström B, Toft MB, Bukhave K, Jensen M, Sørensen SS, Hansen M (2001) Green tea or rosemary extract added to foods reduces nonheme-iron absorption. Am J Clin Nutr 73(3):607–612

Samoylenko V, Ashfaq MK, Jacob MR, Tekwani BL, Khan SI, Manly SP, Joshi VC, Walker LA, Muhammad I (2009) Indolizidine, antiinfective and antiparasitic compounds from Prosopis glandulosa Torr. Var. glandulosa. Planta Med 75(04):48

Sanchez-Vallet A, Ramos B, Bednarek P, López G, Piślewska-Bednarek M, Schulze-Lefert P, Molina A (2010) Tryptophan-derived secondary metabolites in Arabidopsis thaliana confer non-host resistance to necrotrophic *Plectosphaerella cucumerina* fungi. Plant J 63(1):115–127

Sargison ND, Williamson DS, Duncan JR, McCance RW (1996) Prunus padus (bird cherry) poisoning in cattle. Vet Rec 138(8):188
Schappert PJ, Shore JS (1999) Cyanogenesis, herbivory and plant defense in *Turnera ulmifolia* on Jamaica. Ecoscience 6(4):511–520
Schenk SU, Lambein F, Werner D (1991) Broad antifungal activity of β-isoxazolinonyl-alanine, a non-protein amino acid from roots of pea (*Pisum sativum* L.) seedlings. Biol Fertil Soils 11(3):203–209
Schlaeppi K, Abou-Mansour E, Buchala A, Mauch F (2010) Disease resistance of Arabidopsis to *Phytophthora brassicae* is established by the sequential action of indole glucosinolates and camalexin. Plant J 62(5):840–851
Schrauzer GN (2000) Selenomethionine: a review of its nutritional significance, metabolism and toxicity. J Nutr 130(7):1653–1656
Seifen E, Adams RJ, Riemer RK (1979) Sanguinarine: a positive inotropic alkaloid which inhibits cardiac Na+, K+−ATPase. Eur J Pharmacol 60(4):373–377
Seo S, Mitsuhara I, Feng J, Iwai T, Hasegawa M, Ohashi Y (2011) Cyanide, a coproduct of plant hormone ethylene biosynthesis, contributes to the resistance of rice to blast fungus. Plant Physiol 155(1):502–514
Shaw CA, Bains JS, Pasqualotto BA, Curry K (1999) Methionine sulfoximine shows excitotoxic actions in rat cortical slices. Can J Physiol Pharmacol 77(11):871–877
Shoji T, Yamada Y, Hashimoto T (2000) Jasmonate induction of putrescine N-methyltransferase genes in the root of *Nicotiana sylvestris*. Plant Cell Physiol 41(7):831–839
Simons V, Morrissey JP, Latijnhouwers M, Csukai M, Cleaver A, Yarrow C, Osbourn A (2006) Dual effects of plant steroidal alkaloids on *Saccharomyces cerevisiae*. Antimicrob Agents Chemother 50(8):2732–2740
Smith LC, Ravel JM, Lax SR, Shive W (1964) The effects of phenylalanine and tyrosine analogs on the synthesis and activity of 3-deoxy-d-arabino-heptulosonic acid 7-phosphate synthetases. Arch Biochem Biophys 105:424–430
Smolinska U, Morra MJ, Knudsen GR, James RL (2003) Isothiocyanates produced by Brassicaceae species as inhibitors of *Fusarium oxysporum*. Plant Dis 87(4):407–412
Steele JC, Guzman T (1987) Observations about amyotrophic lateral sclerosis and the parkinsonism-dementia complex of Guam with regard to epidemiology and etiology. Can J Neurol Sci J Can Sci Neurol 14(3 Suppl):358–362
Tardito S, Chiu M, Franchi-Gazzola R, Dall'Asta V, Comi P, Bussolati O (2012) The non-proteinogenic amino acids l-methionine sulfoximine and dl-phosphinothricin activate mTOR. Amino Acids 42(6):2507–2512
Tattersall DB, Bak S, Jones PR, Olsen CE, Nielsen JK, Hansen ML, Høj PB, Møller BL (2001) Resistance to an herbivore through engineered cyanogenic glucoside synthesis. Science 293(5536):1826–1828
Theis N, Lerdau M (2003) The evolution of function in plant secondary metabolites. Int J Plant Sci 164(S3):S93–S102
Thimmappa R, Geisler K, Louveau T, O'Maille P, Osbourn A (2014) Triterpene biosynthesis in plants. Annu Rev Plant Biol 65(1):225–257
Tholl D, Lee S (2011) Terpene specialized metabolism in *Arabidopsis thaliana*. Arabidopsis Book Am Soc Plant Biol 9:e0143
Thorne HV, Clarke GF, Skuce R (1985) The inactivation of herpes simplex virus by some Solanaceae glycoalkaloids. Antivir Res 5(6):335–343
Tierens KFM-J, Thomma BPHJ, Brouwer M, Schmidt J, Kistner K, Porzel A, Mauch-Mani B, Cammue BPA, Broekaert WF (2001) Study of the role of antimicrobial glucosinolate-derived isothiocyanates in resistance of Arabidopsis to microbial pathogens. Plant Physiol 125(4):1688–1699
Topal S, Kocaçalişkan I (2006) Allelopathic effects of DOPA against four weed species. DPU Fen Bil Enst Dergisi 11:27–32
Trapp SC, Croteau RB (2001) Genomic organization of plant terpene synthases and molecular evolutionary implications. Genetics 158(2):811–832

Treutter D (2005) Significance of flavonoids in plant resistance and enhancement of their biosynthesis. Plant Biol (Stuttg) 7(6):581–591

Tylleskär T, Rosling H, Banea M, Bikangi N, Cooke R, Poulter N (1992) Cassava cyanogens and konzo, an upper motoneuron disease found in Africa. Lancet 339(8787):208–211

Ulrichová J, Walterová D, Preininger V, Slavik J, Lenfeld J, Cushman M, Šimánek V (1983) Inhibition of acetylcholinesterase activity by some isoquinoline alkaloids. Planta Med 48(06):111–115

Underhill E, Chisholm M, Steck W (1980) (E)-5,(Z)-7-Dodecadienal, a sex pheromone component of the western tent caterpillar, *Malacosoma californicum* (Lepidoptera: Lasiocampidae). The Canadian Entomologist 112(06):629–631

Van Alstyne KL, Nelson AV, Vyvyan JR, Cancilla DA (2006) Dopamine functions as an antiherbivore defense in the temperate green alga *Ulvaria obscura*. Oecologia 148(2):304–311

van Dam NM, Vuister LWM, Bergshoeff C, de Vos H, van Der Meijden E (1995) The "raison D'être" of pyrrolizidine alkaloids in *Cynoglossum officinale*: deterrent effects against generalist herbivores. J Chem Ecol 21(5):507–523

van Der Heijden R, Jacobs DI, Snoeijer W, Hallard D, Verpoorte R (2004) The Catharanthus alkaloids: pharmacognosy and biotechnology. Curr Med Chem 11(5):607–628

Vavrečková C, Gawlik I, Müller K (1996) Benzophenanthridine alkaloids of Chelidonium majus; I. Inhibition of 5-and 12-lipoxygenase by a non-redox mechanism. Planta Med 62(05):397–401

Velderrain-Rodriguez GR, Palafox-Carlos H, Wall-Medrano A, Ayala-Zavala JF, Chen CYO, Robles-Sanchez M, Astiazaran-Garcia H, Alvarez-Parrilla E, Gonzalez-Aguilar GA (2014) Phenolic compounds: their journey after intake. Food Funct 5(2):189–197

Vetter J (2000) Plant cyanogenic glycosides. Toxicon 38(1):11–36

Vogel-Adghough D, Stahl E, Návarová H, Zeier J (2013) Pipecolic acid enhances resistance to bacterial infection and primes salicylic acid and nicotine accumulation in tobacco. Plant Signal Behav 8(11):e26366

Vogt A, Tamewitz A, Skoko J, Sikorski RP, Giuliano KA, Lazo JS (2005) The benzo [c] phenanthridine alkaloid, sanguinarine, is a selective, cell-active inhibitor of mitogen-activated protein kinase phosphatase-1. J Biol Chem 280(19):19078–19086

Wang BH, Lu ZX, Polya GM (1997) Inhibition of eukaryote protein kinases by isoquinoline and oxazine alkaloids. Planta Med 63(06):494–498

Wang SF, Liu AY, Ridsdill-Smith TJ, Ghisalberti EL (2000) Role of alkaloids in resistance of yellow lupin to red-legged earth mite Halotydeus destructor. J Chem Ecol 26(2):429–441

Wang X, Wu X, Zhang Y, Tian YA (2013) Distributed spatial data integrated management prototype for geological applications. In: 2013 21st international conference on geoinformatics, 20–22 June 2013, pp 1–7

Weng J-K, Philippe RN, Noel JP (2012) The rise of chemodiversity in plants. Science 336(6089):1667–1670

Wikström N, Savolainen V, Chase MW (2001) Evolution of the angiosperms: calibrating the family tree. Proc R Soc Lond Ser B Biol Sci 268(1482):2211–2220

Wink M (2008) Plant secondary metabolism: diversity, function and its evolution. Nat Prod Commun 3(8):1205–1216

Wink M (2011) Annual plant reviews, biochemistry of plant secondary metabolism. Wiley, Oxford

Wink M (2013) Evolution of secondary metabolites in legumes (Fabaceae). S Afr J Bot 89:164–175

Wippich C, Wink M (1985) Biological properties of alkaloids. Influence of quinolizidine alkaloids and gramine on the germination and development of powdery mildew, *Erysiphe graminis f. sp. hordei*. Experientia 41(11):1477–1479

World Health Organization (2013) Cyanogenic glycosides. Toxicological evaluation of certain food additives and naturally occurring toxicants, WHO Food Additive Series 30. WHO, Geneva

Xuan T, Elzaawely A, Deba F, Fukuta M, Tawata S (2006) Mimosine in Leucaena as a potent bioherbicide. Agron Sustain Dev 26(2):89

Xuan TD, Minh TN, Khanh TD (2016) Isolation and biological activities of 3-hydroxy-4 (1H)-pyridone. J Plant Interact 11(1):94–100

Yasuda K, Kizu H, Yamashita T, Kameda Y, Kato A, Nash RJ, Fleet GW, Molyneux RJ, Asano N (2002) New sugar-mimic alkaloids from the pods of Angylocalyx pynaertii. J Nat Prod 65(2):198–202

Yazaki K (2005) Transporters of secondary metabolites. Curr Opin Plant Biol 8(3):301–307

Zadvornova Y, Alekseichuk G, Laman N, Khripach V, Grut S (2005) Effect of brassinosteroids on activation of the cell cycle during germination of *Brassica oleracea* L. seeds. Doklady Natsional'noi Akademii Nauk Belarusi 49:70–73

Zagrobelny M, Bak S, Rasmussen AV, Jørgensen B, Naumann CM, Møller BL (2004) Cyanogenic glucosides and plant–insect interactions. Phytochemistry 65(3):293–306

Zhao B, Grant G, Langevin D, MacDonald L (1998) Deterring and inhibiting effects of quinolizidine alkaloids on spruce budworm (Lepidoptera: Tortricidae) oviposition. Environ Entomol 27(4):984–992

Zhou C-X, Liu J-Y, Ye W-C, Liu C-H, Tan R-X (2003) Neoveratalıne A and B, two antifungal alkaloids with a novel carbon skeleton from *Veratrum taliense*. Tetrahedron 59(30):5743–5747

Check for updates

Plant Cell Wall: A Simple Physical Barrier or a Complex Defense Modulator – Exploring Its Dynamic Role at Plant-Fungus Interface

15

Sumanti Gupta and Amit Roy

Abstract

Plants are continuously threatened by many pathogens, among which fungal pathogen accounts for a measureable large quantity. Understanding of plant-fungal interaction is constantly coevolving along with the evolution of both the interacting partners. According to the previous scientific literature, many fungi are associated with a single host. Present chapter is focused at elaborating the role of plant cell wall in participating in the interaction. Host cell wall is the outermost barrier which any pathogen has to breach for successful invasion and establishment. Now the question is what is the role of host cell wall in regulating the pathogen's infiltration or restriction? Present study explains the structural dynamism of cell wall which is believed to have functional relevance and is dependent on the behavior of the infecting fungi. Moreover, cell wall is also known to elicit immune signals that brings about transcriptional reprogramming and helps in mounting defense against attacking fungi. Additionally, cell wall-mediated responses trigger expressions of many antimicrobials which are regulated by hormone signaling pathways. This study also sheds light on the impact of plant-fungal association on tritrophic interactions with other beneficial and pathogenic biotic components. But the role of host cell wall while dealing with multiple interacting partners is still elusive. Thus, the knowledge of cell wall glycobiology is expected to proceed further based on researches conducted at natural microenvironments.

S. Gupta (✉)
Department of Botany, Rabindra Mahavidyalaya, Tarakeswar, Hooghly, India
e-mail: sumantigupta@gmail.com

A. Roy
Department of Entomology, College of Agriculture, University of Kentucky, Lexington, KY, USA

Present Address: Faculty of Forestry and Wood Sciences, EXTEMIT-K, Czech University of Life Sciences, Kamýcká 1176, Prague 6, Suchdol 165 21, Czech Republic

A. Singh, I. K. Singh (eds.), *Molecular Aspects of Plant-Pathogen Interaction*,
https://doi.org/10.1007/978-981-10-7371-7_15

Keywords
Plant-fungal interaction · Host cell wall · Elicitation · Defense signaling · Tritrophic interactions

15.1 Introduction: Cell Wall Versus Immunity

The tug-of-war between plants and pathogens is endless. In order to defend oneself, both the interacting partners constantly opt for appropriate counter-reactive measures that are temporally and spatially regulated (Jones and Dangl 2006). But is this dynamic interplay between plants and its invading pathogens completely understood? This is undoubtedly a million dollar question with a number of answers that are also coevolving with the up gradation of understanding of the plant-pathogen interaction chemistry.

However, there are few conceptions that have received unanimous scientific endorsements. The pathogens, being aggressive in nature, are believed to make the initial attempt of overpowering their host. The pathogen's selection of host may or may not be specific. But, pathogens usually possess some conserved pathogen-associated molecular patterns or microbe-associated molecular patterns (PAMPs/MAMPs) which are readily recognized by the pattern recognition receptors (PRRs) of the hosts. The PAMPs/MAMPs and PRR recognition event leads to activation of downstream defense reactions termed as pattern-triggered immunity (PTI) (Tsuda and Katagiri 2010). Sudden imbalance of ionic concentrations, increase in calcium ionic influxes, activation of MAP (mitogen-associated protein) kinases, phosphorylation of some targeted proteins, and secretion of antimicrobial compounds all leading to cell wall reinforcement of host at the site of attempted penetration are said to be the features of PTI (Buchanan et al. 2015). Additionally, pathogen invasion induces the generation of some host components referred to as danger-associated molecular patterns (DAMPs) such as callose, glucans, fructans, etc. which are recognized as nonself and trigger similar set of downstream signals known as danger-triggered immunity (DTI) (Boller and Felix 2009). Both PTI and DTI provide basal resistance that non-specifically restrict a large number of pathogens at the site of penetration. But, a few diplomatic invaders evade the restrictions imposed by PTI/DTI and secrete specific effector proteins which lead to effector-triggered susceptibility (ETS). As counter-defensive measure, hosts secrete effector-specific R proteins (resistance proteins) that directly or in combination with decoy/guardees interact with pathogen effectors and activate effector-triggered immunity (ETI). The features of ETI not only overlap with those of PTI but, being of much greater amplitude, compensate for the shortfalls of the effect of PTI (Thomma et al. 2011). However, the pathogen has to infringe through the initial physical barrier of host cell wall in order to successfully establish itself within the host (Malinovsky et al. 2014). Thus it is logical to comment that cell wall has roles in allowing or restricting the invader. Present study shall explore the role of plant cell wall at plant-fungus interface.

15.2 Structural Complexity Versus Functional Relevance

Cell wall is the initial physical obstacle that a pathogen has to break for successful establishment within the host. Thus, the structural complexity undoubtedly regulates the function of the cell wall against phytopathogenic fungi. Occasionally the cell wall is covered with a protective layer called cuticle that the fungi breaks through by secreting cell wall-degrading enzymes which also act as important virulence factors (Nuhse 2012). The load-bearing cellulose microfibrils are known to provide the structural integrity of the primary cell wall. They are synthesized by multimeric complexes made up of cellulose synthase catalytic subunits (CESAs) that are well characterized in *Arabidopsis thaliana* (Endler and Persson 2011). CESAs comprise of two subfamilies that synthesize cellulose in both primary (CESA 1, 3, 6) and secondary (CESA 4, 7, 8) cell walls with some crossovers that has still not been characterized (Endler and Persson 2011). However, perturbation of cell wall integrity due to fungal invasion which is diplomatic in case of biotrophic attack and destructive in case of necrotrophic attack, is often compensated by increased lignifications and downstream resistance responses offering the fact that structural cohesiveness of the outermost barrier is always under the alert surveillance of plant defense armory (Caño-Delgado et al. 2003; Hernández-Blanco et al. 2007). The structural integrity is also maintained by the hemicelluloses that are composed of beta 1–4-linked backbone of mannose, glucose, or xylose. They help in reinforcing the interactions with both cellulose and lignin (Endler and Persson 2011). Among several hemicelluloses, xylans are found to be predominant in secondary cell walls where the beta 1–4 xylose residues are often substituted by arabinosyl residues that are esterified with ferulic acid groups. These ferulate esters aid the cross-linking of xylans with lignins ultimately imparting enhanced structural rigidity (Harris and Stone 2008). Xylans are degraded by fungal xylanases and recognized as potential PAMPS by host PRRs in *Lycopersicum esculentum* (Ron and Avni 2004). In addition to feruloylation, xylans can also undergo acetylation and methylation that are specifically targeted by some fungal xylan degrading enzymes. Degraded xylan molecules perceived as DAMPS are known to trigger downstream defense signals (Pogorelko et al. 2013).

Pectin, a hetero complex polysaccharide, serves as the matrix of primary cell wall. Pectic backbones are comprised of either homogalacturonan (HGA) or rhamnogalacturonan. Among these, HGA appears to be more relevant in defense signaling since they undergo specific time-dependent chemical substitutions like methyl esterification and de-esterification at C6 position and acetylation at C2–C3 positions. Such chemical modifications provide a fine tune balance to the host plant according to its prevailing local requirements by forming calcium-mediated cross-linked gels and helping in cell adhesion (Cabrera et al. 2008). HGA is cleaved by fungal polygalacturonases (PG) into oligogalacturonide (OG) fragments that either aid pathogenic entry or act as nonself DAMPs and impart resistance during compatible and incompatible interaction, respectively (Galletti et al. 2009). Interestingly, studies on endogenous pectin methyl esterases (PMEs) showed that they hold dual roles in degrading host HGA into OGs during fungal penetration and also trigger host resistance response by aiding the formation of gel structures to bolster damaged cell walls (Bethke et al. 2014). Such counterintuitive reports support the fact that apart from giving a structural integrity, the role of pectin in defense is quite dubious and still demands extensive case-specific experimentation to come to generalized conceptualizations. Lignin is the prime phenolic polymer of secondary cell walls that increases mechanical strength and water impermeability (Albersheim et al. 2010). Lignin not only provides structural firmness but also actively regulates defense responses by regulating phenylpropanoid pathway and the production of antimicrobial phytoalexins, stilbenes, coumarins, and flavonoids. Besides the key stress hormone that modulates defense, salicylic acid (SA) is also produced from the same biosynthetic pathway that produces lignin polymers (Lozovaya et al. 2007). But, the random incorporation of lignin monolignols to form complete lignin polymers, whether has any ecological implication to meet the protective demands of the host against attacking fungi, is still not clearly understood.

15.3 Plant-Fungus Interface

Immediately after contact the interaction between plant and fungus starts that largely depends on the topology, chemical composition, and molecular features of the contact areas of both the associated partners.

15.3.1 Fungal Penetration Strategy

15.3.1.1 Spore Properties

Spore texture, composition, and molecular properties influence the attachment, adhesion, and penetration of the pathogenic fungus within the plant. On the other side, most of the plants are coated with protective waxes that affect fungal attachment, adhesion, and invasion. Biotrophic fungi of barley *Blumeria graminis* perceive signals from the host surface and direct the growth of germ tube. Germination takes place only after the spore becomes proximate with the host surface stratum (Nielsen et al. 2000). Spore attachment of necrotrophic pathogen of wheat *Stagonospora nodorum* takes place immediately after 30 s and depends on conidial and host surface-released glycoproteins (Newey et al. 2007). In case of hemibiotrophic pathogen of bean *Colletotrichum lindemuthianum*, the spore outer surface consists of fibrillar porous layer abundant in carbohydrates that are required for attachment to hydrophobic layers and sense hard textured surface needed for appressoria formation (Rawlings et al. 2007). Besides in blast fungi *Magnaporthe grisea*, the fungal-released proteins of extracellular matrix (ECM) play a crucial factor in spore attachment, germ tube, and appressoria formation (Inoue et al. 2007). Additionally, the conidial cell wall composition specially beta 1–3 glucan content is also important for pathogenicity as reported in *Alternaria brassicicola* (Joubert et al. 2011) suggesting that cell wall integrity is central to protection of fungal pathogens against host antimicrobial compounds.

15.3.1.2 Mode of Penetration

Following attachment, the germination of the fungal spore and penetration within the host interior appear to be a hallmark achievement for the successful fungus which occurs in different ways. Few fungi develop specialized structures that are capable of piercing the plant cuticular layer and cell wall of epidermal cells, while most of the others rely on their secreted cell wall-degrading enzymes (CWDE) that digest the host cell wall and facilitate their invasion (Łaźniewska et al. 2012). In a handful of cases, involvement of both specialized structures and CWDE occurs. In case of *Magnaporthe grisea*, it develops highly melanized dome-shaped appressoria that exerts tremendous physical force due to high turgor pressure and forces into the host epidermal cells (Choi et al. 2011). *Rhizotonia solani* forms complex appressorial structures known as infection cushions (Pannecoucque and Höfte 2009). CWDE of fungi comprises of xylanases, exogalacturonases, pectin methylesterases, endoglucanases, and polysaccharide deacetylases (Carapito et al. 2008). However the biotrophic fungi and necrotrophic fungi differ strategically in using their CWDEs. The biotrophs employ CWDE for loosening the cell wall of the host and allowing the fungi stealthily, while necrotrophs apply brute force and engage their CWDEs in totally digesting the cell wall of the host (Spanu and Kämper 2010). The function of CWDEs is largely dependent on the pH of the host cell sap (Niture and Pant 2007). Again some fungi like *Fusarium oxysporum* f.sp. *ciceri* and *Puccinia striiformis* f.sp. *tritici* rely on the natural openings such as breaches within root hairs, stomata, etc. of the host for entry (Gupta et al. 2009; Moldenhauer et al. 2006).

15.4 Role of Cell Wall in Elicitation

Previous researchers have well documented the role of cell wall in elicitation during fungal infection (Malinovsky et al. 2014). The following section shall brief how the elicitation mediated by cell wall changes trigger the defense signaling in the infected host plant.

15.4.1 Role of Cell Surface Chemistry and Topology

Hydrophobicity of host cell surface is undoubtedly perceived by most pathogenic fungi for successful attachment which is followed by germination of infecting structures and establishment of the pathogen within the host. Now the cell wall's hydrophobic nature is largely imparted by the presence of the protective layer cuticle. Cuticle is composed of polyester cutin or cutan and epi and intracuticular waxes. Besides triterpenoids and phenyl propanoids are also present (Nawrath 2006). The structure and chemical composition of cuticle differ according to the microenvironment and the nature of plant-fungal interaction as studied in *Arabidopsis thaliana*-fungi interaction (Kurdyukov et al. 2006). On the contrary, different fungal pathogen has developed different breaching mechanism which is capable of using the phylloplane cuticle topology and chemistry for self-sustenance and pathogenicity. Even then, cuticle is known to hold dual roles where in one hand they serve as protective layer and prevent fungal invasion, and on the other hand, their altered structure helps in release of some antimicrobials that lead to resistance against the attacking fungi (Mang et al. 2009).

Cell wall-associated trichomes also have prominent roles in plant-pathogen recognition. Similar to cuticle they also have twofold function but of disparate nature. They are known to trap fungal spores and secrete specialized exudates rich in secondary metabolites having antifungal activity (Nonomura et al. 2009). Besides, studies on *Nicotiana tabacum* have revealed the expression of several pathogenesis-related proteins (PR5 and PR14) and lipid transfer protein (LTP) in leaf trichomes during *Peronospora tabacina* attack (Harada et al. 2010). Conversely, the spatial organization and topology of trichomes are sensed by some group of fungi like *Fusarium graminearum* and *Colletotrichum acutatum* infecting strawberry and *Arabidopsis thaliana*, respectively (Salazar et al. 2007; Skadsen and Hohn 2004). Moreover, the high density of trichomes increase the humidity of the host surface resulting in colonization of many pathogenic fungi that mark specific weak areas in and around the trichomes as probable entry gates (Calo et al. 2006).

15.4.2 Role of Wall-Degrading Enzyme Inhibitors

Fungal CWDEs are important molecules which they categorically use to promote their growth and establishment *in planta* (Łaźniewska et al. 2012). In opposition, host plants also activate immune responses that restrict the devastations caused by fungal CWDEs. Degradation of the primary component of cell wall pectin, the homogalacturonan by polygalacturonases (PGs), results in release of fragments of oligogalacturonides (OGs) that are important DAMP molecules. Studies on *Arabidopsis thaliana* during *Botrytis cinerea* attack revealed enhanced expression of defense proteins like PAD3 (phytoalexin deficient 3) and PGIP (polygalacturonase-inhibiting protein) due to release of OGs from infected cell wall of host (Galletti et al. 2009). Additionally there exist many other CWDE inhibitors of infected hosts that have been reviewed by Lagaert et al. (2009). Pepper pectin methylesterase inhibitor protein (CaPME11) exhibits antifungal activity against *Fusarium oxysporum* f.sp. *matthiole, Alternaria brassicicola, and Botrytis cinerea* (An et al. 2008). Ongoing researches on fungal enzyme verses host inhibitor projects a constant evolution of both the interacting cognate proteins that are subjected to dynamic ecological selection pressure (Beliën et al. 2007).

15.4.3 Host Cell Wall Reinforcement Strategies

The most common counter response of the infected host against attempted penetration of the fungal pathogen is to reinforce its cell wall and prevent further fungal ingress. Reinforcement is achieved by deposition of a number of compounds such as callose, phenolics, lignin, cellulose, pectin, suberin lipids, hydroxyproline-rich glycoproteins (HRGPs), and peroxidases (Schmelzer 2002). The nature and chemical constituent of the deposition is dependent on a particular pathosystem and is temporally regulated. For example, *Septoria tritici* infecting wheat induces callose deposition, whereas accumulation of syringly rich lignins is found in wheat during *Puccinia graminis* f.sp. *tritici* attack (Shetty et al. 2009; Menden et al. 2007). Callose (beta 1–3 glucan) deposition is one of the most important reinforcement molecule deployed by the host. It leads to papillae formation at the site of attempted penetration and restricts a wide range of fungi. The key enzyme of callose synthesis is callose synthase PMR4. PMR4 is known to regulate salicylic acid and PR1 protein expression responses via NPR1 expression in *Arabidopsis thaliana* (Dong et al. 2008). Interestingly, in *Cicer-Fusarium oxysporum* f.sp. *ciceri* interaction, the degradation of callose was found to be linked to enhanced susceptibility indicating that callose deposition and/or degradation is often governed by the invading fungus also that diplomatically reprograms the host machinery for its self-sustenance (Gupta et al. 2010).

Lignins are also known to form cell wall apposition and provide penetration resistance in wheat during powdery mildew attack (Bhuiyan et al. 2009). Lignin content is also regulated by phenyl propanoid pathway and jasmonic acid (Taheri and Tarighi 2010). HRGPs are also important cell wall strengthening compounds that are released in the apoplast in the form of monomers that undergo cross-linking with the help of H_2O_2 and class III peroxidases, thus forming thick-walled network of extensin. These extensins not only serve as anchor for host cell wall lignification but also cause agglutination of the infecting fungi (Almagro et al. 2009) (Fig. 15.1).

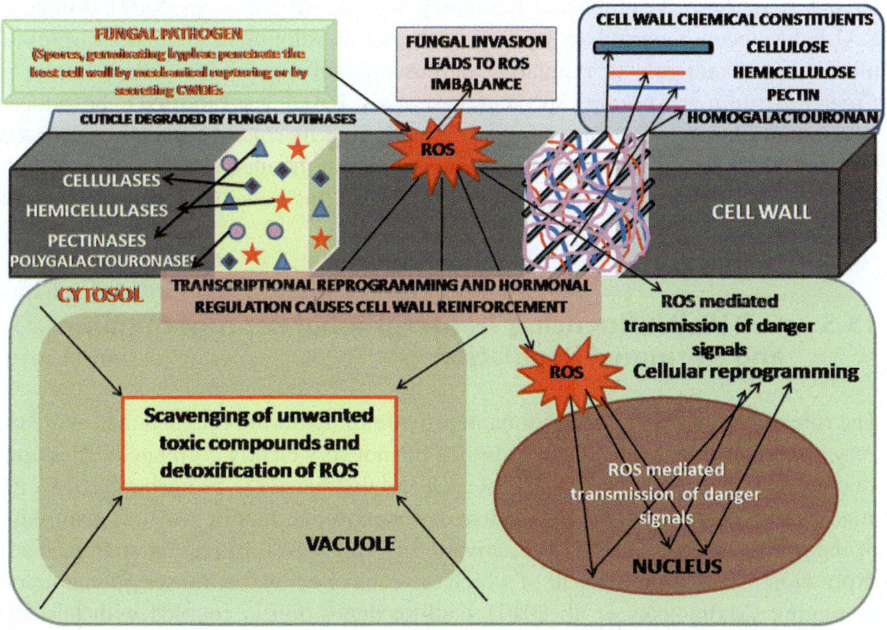

Fig. 15.1 Schematic diagram representing the role of cell wall in regulating plant defense against pathogenic fungi

15.4.4 Role of Peptide Hormones

Peptide hormones are a class of new entrants in the list of plant hormones. Till date few peptide hormones have been identified that are known to regulate diverse roles starting from controlling cell differentiation and growth to inducing defense signaling (Ryan et al. 2002). In the present section, light will be shed on the role of systemin and rapid alkalinization factor (RALF) in controlling defense in particular. Systemin was initially identified in tomato leaves as a result of systemic wound response during *Manduca sexta* pest infestation (Pearce et al. 1991). Follow-up researches have identified many systemins from other solanaceous as well as non-solanaceous member (Ryan et al. 2002). But, their structure and sequence homology greatly differ across different plant species suggesting a need to look through as, why nature has retained the functional conservation of systemin, in spite of its structural dissimilarity across diverse species. Wounding is a response that is deeply associated with the architectural change in the cell wall. Fungal penetration definitely induces wounding, although there has been no report of plant systemins associated with fungal penetration till date, which demands extensive research ahead on this field.

RALFs were reported from tobacco known to cause rapid alkalinization of the medium (Pearce et al. 2001). This medium alkalinization is reported to induce MAP kinases in vitro in tobacco and tomato cell cultures (Pearce et al. 2001). Systemins also induce medium alkalinity, but RALFs are identified to be larger components having signaling role other that regulating defense (Ryan et al. 2002). Although RALF has been reported to have growth and development regulatory roles in tobacco, its exact role in regulating defense is still not proven. Interestingly, in *Cicer-Fusarium oxysporum* f.sp. *ciceri* case study, RALF was reported to be induced in resistant plants during fungal invasion predicting its role in somehow adversely affecting the fungal spread and establishment within the incompatible plant (Gupta et al. 2009). However, the exact mode of action of this novel molecule needs to be elaborated.

15.5 Cell Wall Influencing Transcriptional Reprogramming and Hormonal Regulation

The role of cell wall in transcriptional reprogramming during fungal attack is indeed very important. Although it is well implied but not very clearly correlated in previous studies, that the cell wall being the external and first barrier to be encountered by the attacking fungi surely transmits a sense of danger to its interior parts. Transmission of the danger alarms leads to the activation of cell wall fortification strategies and expression of antimicrobials all of which are controlled by the transcriptional reprogramming (Malinovsky et al. 2014). Callose deposition is coupled with late PTI responses. Besides, attempted penetration of fungal pathogen causes an obvious imbalance in cellular ionic concentrations thus culminating in generation and accumulation of reactive oxygen species (ROS) at the site of invasion (Kobayashi et al.

2007). This ROS accumulation causes downstream transcriptional upregulation leading to calcium spiking, MAP kinase, and calcium-dependent protein kinase expression (CDPK), which are influenced by the changes in cell wall architecture and integrity (Boudko 2012; Boudsocq et al. 2010). Among the genes related with cell wall defense, penetration genes (*PEN*) are most widely characterized. *PEN1* contains a SNARE (soluble N-ethylmaleimide-sensitive factor attachment protein receptor) domain and encodes membrane-associated syntaxin SYP121/PEN1. It is essential for papillae formation and helps in delivering the cell wall reinforcement material to the appropriate site of attack (Assaad et al. 2004). *PEN2* gene encodes glycoside hydrolase that participates in peroxidase-mediated antifungal production, while *PEN3* gene-encoding ABC (ATP-binding cassette) transporters deliver the compound to the necessary site of attempted penetration (Stein et al. 2006).

Although indirect, the role of cell wall in modulating hormonal expression during fungal attack is also noteworthy. Many fungi like *Fusarium* sp. rely on natural opening like stomata or breaches adjoining root hairs for entering the host. Thus, in order to prevent fungal entry, the host guard cell wall modifies accordingly, decreases the osmotic potential, and initiates abscisic acid (ABA)-mediated stomatal closure (Ton et al. 2009). Besides, callose synthase that affects the deposition of callose is known to trigger NPR1-mediated SA response in *Arabidopsis thaliana* (Dong et al. 2008). Similarly lignins that are products of phenyl propanoid pathway are regulated by jasmonic acid pathway (Taheri and Tarighi 2010).

15.6 Plant-Fungus Interaction and Its Impact on Other Biotic Interactions at Different Trophic Levels: Is the Role of Host Cell Wall Defined During Multiple Interactions?

Fungus is one of the most commonly occurring plant-associated organisms with a considerable diversity comprising 120,000 described species. According to Hawksworth's (1991) extrapolation, there are 1.5 million fungal species on earth, most of which consume plant matter. The ratio of fungi to plant species is 5:1 considering approximately 300,000 plant species on earth. Thus, it is likely that a plant is associated with more than five fungi (Arnold et al. 2007; Jumpponen and Jones 2010). With such expansive species diversity, it is understandable that plant-fungal interactions are dynamic and ecologically complex (Southworth 2012).

It is presumed that all plants in the ecosystem are symbiotic with fungal endophytes that reside in plant tissues (Petrini 1996). These symbiotic association aids the host plant in gaining fitness benefits such as conferring biotic and abiotic stress tolerance and inducing growth and development. For example, class 2 fungal endophytes play a crucial role in plant adaptation to various abiotic stresses such as drought, salinity, and temperature (Redman et al. 2001; Márquez et al. 2007; Rodriguez et al. 2009, 2008). However, in the present section, we confine ourselves to fungi-mediated biotic interactions only. From a plant's perspective, fungi can mediate both beneficial and harmful interactions. In the following section, we shall discuss few of such naturally occurring tritrophic interactions mediated by fungi (Fig. 15.2).

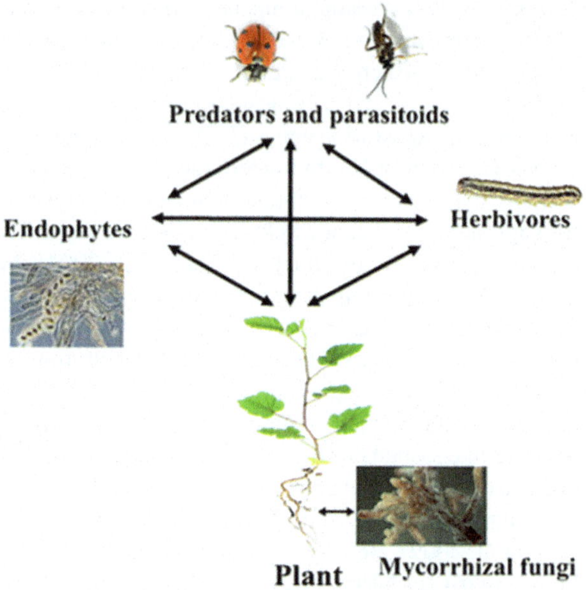

Fig. 15.2 Simplified pictorial representation of interaction dynamics of plant and its associated organisms including insects, fungus (harmful + beneficial), and parasite/predators (Model adapted from Lamit and Gehring 2012)

15.6.1 Fungi as Mediators of Symbiotic Trophic Associations

Plants endorse different mutualistic associations where they provide nutritional or structural resources to their partners and obtain essential components related to their growth. Otherwise, there is a tremendous competition of utilizable carbon sources in the soil among soil organisms. Plants often produce excess photosynthate in presence of adequate nitrogen and release some part of it into the rhizosphere (Hikosaka 2005). These nutritional conditions provide opportunity for symbiosis between plants and fungi; plants require nitrogen, and fungi need accessible carbon, which they are unable to scavenge efficiently from soil (Smith et al. 2009). There are several examples of plant-fungus symbiotic interactions. Many of these fungal symbiotic partners, including mycorrhizal fungi, are obligate in nature and are therefore unable to survive without a plant host. In this symbiotic association, fungus receives carbon, in form of plant-derived sugar, and in return, they provide their host with the limiting soil nutrients such as nitrogen and/or phosphate. Several gene expression and transporter-staining experiments provide evidence for such mutual interactions (Fellbaum et al. 2012; Jones et al. 2009; Balestrini and Lanfranco 2006; Karandashov and Bucher 2005; Guether et al. 2009).

Metarhizium is an insect pathogenic, root colonizing fungus. They are not obligate biotrophs like mycorrhizal fungi. Their survival also depends on the reciprocal nutrient exchange with a host plant, whereby the fungus receives carbon in exchange for the insect-derived nitrogen (Hajek and St. Leger 1994; Behie et al. 2017; Behie and Bidochka 2014). They infect soilborne insects and transfer the insect-derived nitrogen to the host plants via fungal hyphae for the plant-derived photosynthate (Behie et al. 2017). This serves as a classic example of a secondary interaction between plant and insect mediated by fungus where insects become the prey and plants are provided with insect-derived nitrogen by the entomopathogenic fungus.

Occasionally, fungus benefits the host plant through cross-compartment signaling in both directions such as shoot to root and vice versa. For instance, aboveground inoculation with an incompatible race of the hemibiotrophic fungus *Fusarium oxysporum* (Foc race 1) on Cavendish banana enhances resistance of the host plant to a compatible strain of fungus (Foc race 4) when the strain is subsequently inoculated to the roots (Wu et al. 2013). The enhanced resistance, shoot to root SAR (systemic acquired resistance), is associated with the higher levels of salicylic acid (SA) and other defense-related genes including PR-1 (pathogenesis-related protein). Conversely, the hemibiotrophic fungus, *Colletotricum graminicola*, strongly suppresses the same pathogen growth when it is inoculated in the leaves after 6 days of preinoculation of roots in maize plants. Transcriptional studies confirmed the significant increase in the expression of ABA (abscisic acid), SA, and other genes associated with the biosynthesis of benzoxazinoids (DIMBOA) and PR proteins until 4 dpi (days post inoculation) (Balmer et al. 2013) indicating an active root to shoot SAR. Moreover, Leath et al. (Leath and Byers 1977) reported reduced pea aphid densities on the forage legumes infected by *Fusarium* in roots.

Interestingly, plant-fungal interaction can also modulate herbivore population by attracting the natural enemies. Plant volatile compounds facilitate the host searching by the natural enemies (Turlings et al. 1991a, b; Kessler and Baldwin 2001). The plant-emitted chemical signals may originate from various sources like the host plant, herbivore-damaged plant, fungus-infected plants, etc.. For example, infection by the fungus *Alternaria brassicae* on *Brassica rapa* (mustard) seedlings induces release of volatile cues from glucosinolate degradation (Doughty et al. 1996). The white mold fungus, *Sclerotium rolfsii*, infected *Arachis hypogaea* (groundnut) plant releases E-4, 8-dimethyl-1, 3, 7-nonatrine, methyl salicylate, etc. (Cardoza et al. 2002). Interestingly, some herbivores find ways to escape natural enemies by evolving mechanisms to reduce volatile cues from the damaged plants. Hence, in addition to the host plants, the parasitoids feeding upon herbivores need to depend on other sources for volatile cues to locate their prey from faraway. Fungus-infected plants not only provide valuable cucs to herbivores for host finding and oviposition, but they also emit volatiles that are highly attractive to parasitoids. Studies even revealed that there is a difference between the volatile emission profiles from the host plant after compatible or incompatible interactions with the fungal pathogens (Huang et al. 2003). Moreover, pathogen in combination with herbivory shows a different emission profile from the host (Cardoza and Tumlinson 2006), which may serve as a valuable cue for the natural enemies such as parasitoid wasps like *Cotesia marginiventris*. For instance, the volatile blend produced by beet armyworm (BAW) and white mold-infected groundnut plant contains methyl salicylate and fungus-produced 3-octanone, in addition to all the volatiles produced from the healthy groundnut plant alone (Cardoza et al. 2002). This infected groundnut plant exposed to BAW plants is more attractive to *C. marginiventris* (Cardoza et al. 2003) compare to healthy plants.

15.6.2 Fungi as Inducers of Pathogenic Trophic Associations

There are a number of pathogenic fungi that induce or facilitate other harmful interactions with plants (i.e., promote egg laying of herbivores) in addition to the damage they cause. In other words, when a pathogen attacks, the plant's response to those attacks affects the oviposition preference of different herbivores. There are many reports of higher oviposition of herbivores on pathogen-infected host plants in different pathosystems. *Helicoverpa armigera* (Hübner) adult females were reported to deposit more eggs on tomato leaves inoculated with root fungal endophyte, *Acremonium strictum* (Jallow et al. 2008). Similarly, stem-boring weevil, *Apiononopordi* (Coleoptera: Apionidae), prefers *Puccinia punctiformis* (rust fungus, Uredinales)-infected shoots of its host gray-green perennial herb, *Cirsium arvense* (Asteraceae) (Friedli and Bacher 2001). Abreha et al. (2015) showed that susceptible potato cultivar after fungus infection is preferred by the generalist moth, *Spodoptera littoralis* (Boisduval) (Lepidoptera: Noctuidae). Females were found to lay more eggs on the *Phytophthora infestans* inoculated potato plant over uninoculated ones indicating a preference of *Spodoptera* females for diseased plants over healthy plants. There was no significant difference on the larval performance after feeding on the infested leaves. However, white mold fungus, *Sclerotium rolfsii*-infected peanut plants are more preferred by BAW, *Spodoptera exigua*. BAW grows better on the mold-infected plants due to higher level of soluble sugars and lower levels of soluble phenolics (Cardoza et al. 2003). *Phytophthora plurivora* (root pathogen) enhances the performance of gypsy moth (*Lymantria dispar*). However, herbivores do not always prefer to lay eggs on leaves infected with pathogens. For example, BAW avoided oviposition on powdery mildewed (*Podosphaera pannosa*) leaves of *Rosa chinensis* (Yang et al. 2013), leaf beetle *Phaedon cochleariae* does not prefer to oviposit on *Alternaria brassicae*-infested chinese cabbage (Rostas and Hilker 2002), and *Cassida rubiginosa*, leaf-feeding beetle, prefers laying eggs on healthy plants over *Phoma destructiva*-inoculated *Cirsium arvense* (Kruess 2002).

There is another interesting dimension in the plant-fungus-insect interaction where insect-fungus mutualistic relationships help them in successful colonization in host plant, for example, bark beetles (Coleoptera: Curculionidae, Scolytinae) ectosymbioses with fungi (Harrington 2005; Six 2012) for successful colonization on the host plant (Fig. 15.3), to be more specific, the western pine beetle (*Dendroctonus brevicomis LeConte*) symbiosis with two symbiotic fungal partners, *Entomocorticium* sp. (Basidiomycota) and *Cerato cystiopsis brevicomi* (Ascomycota) (Paine and Birch 1983; Hsiau and Harrington 1997). The two fungi carried in a prothoracic mycangium were found only in female beetles. During tree colonization, females inoculate the tree with their symbiotic fungi and oviposit in the tree's phloem layer. Larvae feed and grow initially on a combination of fungi and phloem. However, in the second instar, larvae start feeding on the nutrient-poor bark (Miller and Keen 1960), which is hypothesized to be mediated by the symbiotic fungi. This transition is crucial for larval survival, as they require both bark and phloem in the diet for development (Valiev et al. 2009). This is an example of

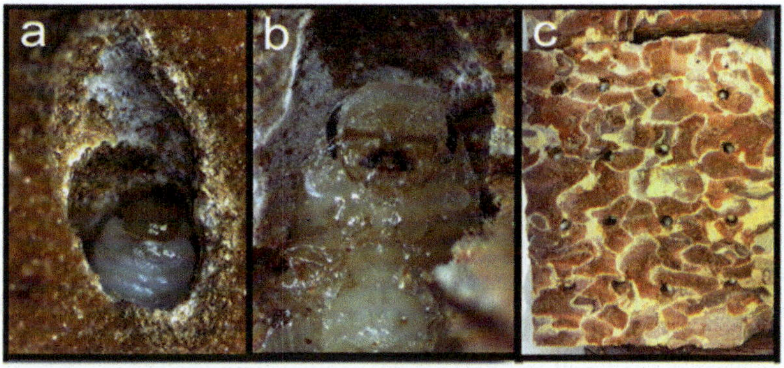

Fig. 15.3 Plant-fungus-insect tritrophic interaction. (**a**) Western pine beetle feed on symbiotic fungi during early phase of larval development on ponderosa pine bark. Fungi are visible as white mass in larval tunnel. (**b**) Fungal spores line the pupal chamber for incorporation into the beetle mycangia after metamorphosis. (**c**) Pseudo-chambers created by Bracewell et al. showed one pupa in each chamber (Adapted with permission from Bracewell and Six 2015)

symbiosis between insect and fungus since the fungi gain transportation and entry to the host trees by insects while, in return, the fungi provide nutritional benefit to the growing larvae.

Thus, plant-fungal-herbivore interactions are highly variable with no broad generalizations at present. However, these ecologically significant findings demonstrate the impact of phytopathogen (i.e., fungi)-induced alteration in plant chemistry (including volatile emission profile) influencing the plant-insect interactions, not only at the second trophic level but also the third level.

But, in order to mediate such tritrophic interactions, the plant cell wall definitely has a specific role in allowing different combinations of these multiple agents (symbiotic/pathogenic fungi along with beneficial insect/harmful herbivore) within itself. How the host cell wall perceives and channelizes the entire signaling that regulates the entry or exit of multiple organisms simultaneously or in sequence within the resident host is still a gross mystery. But, since penetration and invasion are a critical event for establishment of a symbiont or pathogen, the role of host cell wall is fundamental, which needs to be extensively studied that too under natural ecological microenvironment.

15.7 Challenges of Cell Wall Biology Ahead

According to the thumb rule of nature, cell wall is a universal barrier, which is sufficient to restrict a wide range of fungal pathogens. But what still remains unanswered are:

1. How does the cell wall of a particular host restrain so many wide range of fungal pathogens that constantly try to threaten it?

2. The PAMP/DAMP versus PRR recognition is primarily cell membrane centric and under transcriptional regulation. How does cell wall specifically function during PTI/ETI signaling?
3. It is believed that the initial fungal penetration is manifested within host interior by changes in ROS level. Is the same signal transmitted systemically to distant parts? What roles do the distant cell walls have? Do they also have any memory that helps in priming?
4. How is the chemical glycobiology of cell wall regulated according to functional demands more importantly during gradual invasion and establishment of the fungal pathogen? Does the role of host cell wall become redundant after pathogen establishes itself within the host?
5. How does the cell wall of a host cater to multiple invaders?

Apart from the above, there probably lie many more questions in scientific minds that are likely to receive satisfactory answers as the study of cell wall biology proceeds. On the contrary, many more queries will also come up in the near future that shall pave the path of new and novel researches ahead.

Acknowledgment Authors are grateful to Dr. Ryan Bracewell, University of Montana, USA, for providing the beetle pictures.

References

Abreha KB, Alexandersson E, Vossen JH, Anderson P, Andreasson E (2015) Inoculation of transgenic resistant potato by Phytophthora infestans affects host plant choice of a generalist moth. Plos One 10(6):e0129815. https://doi.org/10.1371/ journal.pone.0129815.

Albersheim P, Darvill A, Roberts K, Sederoff R, Staehelin A (2010) Plant cell walls. Garland Sci 154:483

Almagro L, Gómez Ros L, Belchi-Navarro S, Bru R, Ros Barceló A, Pedreno M (2009) Class III peroxidases in plant defence reactions. J Exp Bot 60:377–390

An SH, Sohn KH, Choi HW, Hwang IS, Lee SC, Hwang BK (2008) Pepper pectin methylesterase inhibitor protein CaPMEI1 is required for antifungal activity, basal disease resistance and abiotic stress tolerance. Planta 228:61–78

Arnold AE, Henk DA, Eells RL, Lutzoni F, Vilgalys R (2007) Diversity and phylogenetic affinities of foliar fungal endophytes in loblolly pine inferred by culturing and environmental PCR. Mycologia 99:185–206

Assaad FF, Qiu J-L, Youngs H, Ehrhardt D, Zimmerli L, Kalde M, Wanner G, Peck SC, Edwards H, Ramonell K (2004) The PEN1 syntaxin defines a novel cellular compartment upon fungal attack and is required for the timely assembly of papillae. Mol Biol Cell 15:5118–5129

Balestrini R, Lanfranco L (2006) Fungal and plant gene expression in arbuscular mycorrhizal symbiosis. Mycorrhiza 16:509–524

Balmer D, Papajewski DV, Planchamp C, Glauser G, Mauch-Mani B (2013) Induced resistance in maize is based on organ-specific defence responses. Plant J 74:213–225

Behie SW, Bidochka MJ (2014) Ubiquity of insect-derived nitrogen transfer to plants by endophytic insect-pathogenic fungi: an additional branch of the soil nitrogen cycle. Appl Environ Microbiol 80:1553–1560

Behie SW, Moreira CC, Sementchoukova I, Barelli L, Zelisko PM, Bidochka MJ (2017) Carbon translocation from a plant to an insect-pathogenic endophytic fungus. Nat Commun 8:14245

Beliën T, van Campenhout S, van Acker M, Robben j, Courtin c M, Delcour j A, Volckaert G (2007) Mutational analysis of endoxylanases XylA and XylB from the phytopathogen Fusarium graminearum reveals comprehensive insights into their inhibitor insensitivity. Appl Environ Microbiol 73:4602–4608

Bethke G, Grundman RE, Sreekanta S, Truman W, Katagiri F, Glazebrook J (2014) Arabidopsis pectin methylesterases contribute to immunity against pseudomonas syringae. Plant Physiol 164:1093–1107

Bhuiyan NH, Selvaraj G, Wei Y, King J (2009) Gene expression profiling and silencing reveal that monolignol biosynthesis plays a critical role in penetration defence in wheat against powdery mildew invasion. J Exp Bot 60:509–521

Boller T, Felix G (2009) A renaissance of elicitors: perception of microbe-associated molecular patterns and danger signals by pattern-recognition receptors. Annu Rev Plant Biol 60:379–406

Boudko DY (2012) Molecular basis of essential amino acid transport from studies of insect nutrient amino acid transporters of the SLC6 family (NAT-SLC6). J Insect Physiol 58:433–449

Boudsocq M, Willmann MR, Mccormack M, Lee H, Shan L, He P, Bush J, Cheng S-H, Sheen J (2010) Differential innate immune signalling via Ca2+ sensor protein kinases. Nature 464:418–422

Bracewell RR, Six DL (2015) Experimental evidence of bark beetle adaptation to a fungal symbiont. Ecol Evol 5:5109–5119

Buchanan BB, Gruissem W, Jones RL (2015) Biochemistry and molecular biology of plants. Wiley, Somerset

Cabrera JC, Boland A, Messiaen J, Cambier P, van Cutsem P (2008) Egg box conformation of oligogalacturonides: the time-dependent stabilization of the elicitor-active conformation increases its biological activity. Glycobiology 18:473–482

Calo L, García I, Gotor C, Romero LC (2006) Leaf hairs influence phytopathogenic fungus infection and confer an increased resistance when expressing a Trichoderma α-1, 3-glucanase. J Exp Bot 57:3911–3920

Caño-Delgado A, Penfield S, Smith C, Catley M, Bevan M (2003) Reduced cellulose synthesis invokes lignification and defense responses in Arabidopsis thaliana. Plant J 34:351–362

Carapito R, Hatsch D, Vorwerk S, Petkovski E, Jeltsch J-M, Phalip V (2008) Gene expression in fusarium graminearum grown on plant cell wall. Fungal Genet Biol 45:738–748

Cardoza YJ, Tumlinson JH (2006) Compatible and incompatible Xanthomonas infections differentially affect herbivore-induced volatile emission by pepper plants. J Chem Ecol 32:1755

Cardoza YJ, Alborn HT, Tumlinson JH (2002) In vivo volatile emissions from peanut plants induced by simultaneous fungal infection and insect damage. J Chem Ecol 28:161–174

Cardoza YJ, Teal PE, Tumlinson JH (2003) Effect of peanut plant fungal infection on oviposition preference by Spodoptera exigua and on host-searching behavior by Cotesia marginiventris. Environ Entomol 32:970–976

Choi J, Kim KS, Rho H-S, Lee Y-H (2011) Differential roles of the phospholipase C genes in fungal development and pathogenicity of Magnaporthe oryzae. Fungal Genet Biol 48:445–455

Dong X, Hong Z, Chatterjee J, Kim S, Verma DPS (2008) Expression of callose synthase genes and its connection with Npr1 signaling pathway during pathogen infection. Planta 229:87–98

Doughty KJ, Blight MM, Bock CH, Fieldsend JK, Pickett JA (1996) Release of alkenyl isothiocyanates and other volatiles from Brassica rapa seedlings during infection by Alternaria brassicae. Phytochemistry 43:371–374

Endler A, Persson S (2011) Cellulose synthases and synthesis in Arabidopsis. Mol Plant 4:199–211

Fellbaum CR, Gachomo EW, Beesetty Y, Choudhari S, Strahan GD, Pfeffer PE, Kiers ET, Bücking H (2012) Carbon availability triggers fungal nitrogen uptake and transport in arbuscular mycorrhizal symbiosis. Proc Natl Acad Sci 109:2666–2671

Friedli J, Bacher S (2001) Mutualistic interaction between a weevil and a rust fungus, two parasites of the weed Cirsium arvense. Oecologia 129:571–576

Galletti R, de Lorenzo G, Ferrari S (2009) Host-derived signals activate plant innate immunity. Plant Signal Behav 4:33–34

Guether M, Neuhäuser B, Balestrini R, Dynowski M, Ludewig U, Bonfante P (2009) A mycorrhizal-specific ammonium transporter from Lotus japonicus acquires nitrogen released by arbuscular mycorrhizal fungi. Plant Physiol 150:73–83

Gupta S, Chakraborti D, Rangi RK, Basu D, Das S (2009) A molecular insight into the early events of Chickpea (Cicer arietinum) and Fusarium oxysporum f. sp. ciceri (Race 1) interaction through cDNA-AFLP analysis. Phytopathology 99:1245–1257

Gupta S, Chakraborti D, Sengupta A, Basu D, Das S (2010) Primary metabolism of chickpea is the initial target of wound inducing early sensed Fusarium oxysporum f. sp. ciceri race I. PLoS One 5:e9030

Hajek A, St. Leger R (1994) Interactions between fungal pathogens and insect hosts. Annu Rev Entomol 39:293–322

Harada E, Kim J-A, Meyer AJ, Hell R, Clemens S, Choi Y-E (2010) Expression profiling of tobacco leaf trichomes identifies genes for biotic and abiotic stresses. Plant Cell Physiol 51:1627–1637

Harrington TC (2005) Ecology and evolution of mycophagous bark beetles and their fungal partners. Insect-Fungal Assoc 1:22

Harris PJ, Stone BA (2008) Chemistry and molecular organization of plant cell walls, Biomass Recalcitrance. Blackwell, Oxford, pp 61–93

Hawksworth DL (1991) The fungal dimension of biodiversity: magnitude, significance, and conservation. Mycol Res 95:641–655

Hernández-Blanco C, Feng DX, Hu J, Sánchez-Vallet A, Deslandes L, Llorente F, Berrocal-Lobo M, Keller H, Barlet X, Sánchez-Rodríguez C (2007) Impairment of cellulose synthases required for Arabidopsis secondary cell wall formation enhances disease resistance. Plant Cell 19:890–903

Hikosaka K (2005) Leaf canopy as a dynamic system: ecophysiology and optimality in leaf turnover. Ann Bot 95:521–533

Hsiau PTW, Harrington TC (1997) Ceratocystiopsis brevicomi sp. nov., a mycangial fungus from Dendroctonus brevicomis (Coleoptera: Scolytidae). Mycologia 89:661–669

Huang J, Cardoza YJ, Schmelz EA, Raina R, Engelberth J, Tumlinson JH (2003) Differential volatile emissions and salicylic acid levels from tobacco plants in response to different strains of Pseudomonas syringae. Planta 217:767–775

Inoue K, Suzuki T, Ikeda K, Jiang S, Hosogi N, Hyong G-S, Hida S, Yamada T, Park P (2007) Extracellular matrix of Magnaporthe oryzae may have a role in host adhesion during fungal penetration and is digested by matrix metalloproteinases. J Gen Plant Pathol 73:388–398

Jallow MF, Dugassa-Gobena D, Vidal S (2008) Influence of an endophytic fungus on host plant selection by a polyphagous moth via volatile spectrum changes. Arthropod Plant Interact 2:53–62

Jones JD, Dangl JL (2006) The plant immune system. Nature 444:323–329

Jones DL, Nguyen C, Finlay RD (2009) Carbon flow in the rhizosphere: carbon trading at the soil–root interface. Plant Soil 321:5–33

Joubert A, Bataille-Simoneau N, Campion C, Guillemette T, Hudhomme P, Iacomi-Vasilescu B, Leroy T, Pochon S, Poupard P, Simoneau P (2011) Cell wall integrity and high osmolarity glycerol pathways are required for adaptation of Alternaria brassicicola to cell wall stress caused by brassicaceous indolic phytoalexins. Cell Microbiol 13:62–80

Jumpponen A, Jones K (2010) Seasonally dynamic fungal communities in the Quercus macrocarpa phyllosphere differ between urban and nonurban environments. New Phytol 186:496–513

Karandashov V, Bucher M (2005) Symbiotic phosphate transport in arbuscular mycorrhizas. Trends Plant Sci 10:22–29

Kessler A, Baldwin IT (2001) Defensive function of herbivore-induced plant volatile emissions in nature. Science 291:2141–2144

Kobayashi M, Ohura I, Kawakita K, Yokota N, Fujiwara M, Shimamoto K, Doke N, Yoshioka H (2007) Calcium-dependent protein kinases regulate the production of reactive oxygen species by potato NADPH oxidase. Plant Cell 19:1065–1080

Kruess A (2002) Indirect interaction between a fungal plant pathogen and a herbivorous beetle of the weed Cirsium arvense. Oecologia 130:563–569

Kurdyukov S, Faust A, Nawrath C, Bär S, Voisin D, Efremova N, Franke R, Schreiber L, Saedler H, Métraux J-P (2006) The epidermis-specific extracellular BODYGUARD controls cuticle development and morphogenesis in Arabidopsis. Plant Cell 18:321–339

Lagaert S, Beliën T, Volckaert G (2009) Plant cell walls: protecting the barrier from degradation by microbial enzymes. Semin Cell Dev Biol 20:1064–1073. Elsevier

Lamit LJ, Gehring CA (2012) Dynamic interplay in a multivariate world: case studies in mycorrhizal and endophytic fungal interactions with herbivores, Biocomplexity Plant-Fungal Interact. Wiley-Blackwell, Oxford, pp 185–204

Łaźniewska J, Macioszek VK, Kononowicz AK (2012) Plant-fungus interface: the role of surface structures in plant resistance and susceptibility to pathogenic fungi. Physiol Mol Plant Pathol 78:24–30

Leath K, Byers R (1977) Interaction of Fusarium root rot with pea aphid and potato leafhopper feeding on forage legumes. Phytopathology 67:226–229

Lozovaya VV, Lygin AV, Zernova OV, Ulanov AV, Li S, Hartman GL, Widholm JM (2007) Modification of phenolic metabolism in soybean hairy roots through down regulation of chalcone synthase or isoflavone synthase. Planta 225:665–679

Malinovsky FG, Fangel JU, Willats WG (2014) The role of the cell wall in plant immunity. Front Plant Sci 5:178

Mang HG, Laluk KA, Parsons EP, Kosma DK, Cooper BR, Park HC, Abuqamar S, Boccongelli C, Miyazaki S, Consiglio F (2009) The Arabidopsis RESURRECTION1 gene regulates a novel antagonistic interaction in plant defense to biotrophs and necrotrophs. Plant Physiol 151:290–305

Márquez LM, Redman RS, Rodriguez RJ, Roossinck MJ (2007) A virus in a fungus in a plant: three-way symbiosis required for thermal tolerance. Science 315:513–515

Menden B, Kohlhoff M, Moerschbacher BM (2007) Wheat cells accumulate a syringyl-rich lignin during the hypersensitive resistance response. Phytochemistry 68:513–520

Miller JM, Keen FP (1960) Biology and control of the western pine beetle: a summary of the first fifty years of research. US Department of Agriculture, Washington, DC

Moldenhauer J, Moerschbacher B, van der Westhuizen A (2006) Histological investigation of stripe rust (Puccinia striiformis f. sp. tritici) development in resistant and susceptible wheat cultivars. Plant Pathol 55:469–474

Nawrath C (2006) Unraveling the complex network of cuticular structure and function. Curr Opin Plant Biol 9:281–287

Newey LJ, Caten CE, Green JR (2007) Rapid adhesion of Stagonospora nodorum spores to a hydrophobic surface requires pre-formed cell surface glycoproteins. Mycol Res 111:1255–1267

Nielsen KA, Nicholson RL, Carver TL, Kunoh H, Oliver RP (2000) First touch: an immediate response to surface recognition in conidia of Blumeria graminis. Physiol Mol Plant Pathol 56:63–70

Niture SK, Pant A (2007) Production of cell wall-degrading enzymes by a pH tolerant estuarine fungal isolate Fusarium moniliforme NCIM1276 in different culture conditions. World J Microbiol Biotechnol 23:1169–1177

Nonomura T, Xu L, Wada M, Kawamura S, Miyajima T, Nishitomi A, Kakutani K, Takikawa Y, Matsuda Y, Toyoda H (2009) Trichome exudates of Lycopersicon pennellii form a chemical barrier to suppress leaf-surface germination of Oidium neolycopersici conidia. Plant Sci 176:31–37

Nuhse TS (2012) Cell wall integrity signaling and innate immunity in plants. Front Plant Sci 3:280

Paine T, Birch M (1983) Acquisition and maintenance of mycangial fungi by Dendroctonus brevicomis LeConte (Coleoptera: Scolytidae). Environ Entomol 12:1384–1386

Pannecoucque J, Höfte M (2009) Interactions between cauliflower and Rhizoctonia anastomosis groups with different levels of aggressiveness. BMC Plant Biol 9:95

Pearce G, Strydom D, Johnson S, Ryan CA (1991) A polypeptide from tomato leaves induces wound-inducible proteinase inhibitor proteins. Science 253:895–899
Pearce G, Moura DS, Stratmann J, Ryan CA (2001) RALF, a 5-kDa ubiquitous polypeptide in plants, arrests root growth and development. Proc Natl Acad Sci 98:12843–12847
Petrini O (1996) Ecological and physiological aspects of host-specificity in endophytic fungi. In: Endophytic fungi in grasses and woody plants. APS, St Paul, pp 87–100
Pogorelko G, Lionetti V, Fursova O, Sundaram RM, Qi M, Whitham SA, Bogdanove AJ, Bellincampi D, Zabotina OA (2013) Arabidopsis and Brachypodium distachyon transgenic plants expressing Aspergillus nidulans acetylesterases have decreased degree of polysaccharide acetylation and increased resistance to pathogens. Plant Physiol 162:9–23
Rawlings SL, O'connell RJ, Green JR (2007) The spore coat of the bean anthracnose fungus Colletotrichum lindemuthianum is required for adhesion, appressorium development and pathogenicity. Physiol Mol Plant Pathol 70:110–119
Redman RS, Dunigan DD, Rodriguez RJ (2001) Fungal symbiosis from mutualism to parasitism: who controls the outcome, host or invader? New Phytol 151:705–716
Rodriguez RJ, Henson J, VAN Volkenburgh E, Hoy M, Wright L, Beckwith F, Kim Y-O, Redman RS (2008) Stress tolerance in plants via habitat-adapted symbiosis. ISME J 2:404–416
Rodriguez R, White J Jr, Arnold AE, Redman R (2009) Fungal endophytes: diversity and functional roles. New Phytol 182:314–330
Ron M, Avni A (2004) The receptor for the fungal elicitor ethylene-inducing xylanase is a member of a resistance-like gene family in tomato. Plant Cell 16:1604–1615
Rostas M, Hilker M (2002) Asymmetric plant-mediated cross-effects between a herbivorous insect and a phytopathogenic fungus. Agric For Entomol 4:223–231
Ryan CA, Pearce G, Scheer J, Moura DS (2002) Polypeptide hormones. Plant Cell 14:S251–S264
Salazar SM, Castagnaro AP, Arias M, Chalfoun N, Tonello U, Díaz Ricci JC (2007) Induction of a defense response in strawberry mediated by an avirulent strain of Colletotrichum. Eur J Plant Pathol 117:109–122
Schmelzer E (2002) Cell polarization, a crucial process in fungal defence. Trends Plant Sci 7:411–415
Shetty NP, Jensen JD, Knudsen A, Finnie C, Geshi N, Blennow A, Collinge DB, Jørgensen HJL (2009) Effects of β-1, 3-glucan from Septoria tritici on structural defence responses in wheat. J Exp Bot 60:4287–4300
SIX DL (2012) Ecological and evolutionary determinants of bark beetle – fungus symbioses. Insects 3:339–366
Skadsen RW, Hohn TM (2004) Use of fusarium graminearum transformed with gfp to follow infection patterns in barley and Arabidopsis. Physiol Mol Plant Pathol 64:45–53
Smith FA, Grace EJ, Smith SE (2009) More than a carbon economy: nutrient trade and ecological sustainability in facultative arbuscular mycorrhizal symbioses. New Phytol 182:347–358
Southworth D (2012) Defining complex interactions between plants and fungi. In: Biocomplexity plant–fungal interact. Wiley-Blackwell, Ames, pp 205–213
Spanu P, Kämper J (2010) Genomics of biotrophy in fungi and oomycetes – emerging patterns. Curr Opin Plant Biol 13:409–414
Stein M, Dittgen J, Sánchez-Rodríguez C, Hou B-H, Molina A, Schulze-Lefert P, Lipka V, Somerville S (2006) Arabidopsis PEN3/PDR8, an ATP binding cassette transporter, contributes to nonhost resistance to inappropriate pathogens that enter by direct penetration. Plant Cell 18:731–746
Taheri P, Tarighi S (2010) Riboflavin induces resistance in rice against Rhizoctonia solani via jasmonate-mediated priming of phenylpropanoid pathway. J Plant Physiol 167:201–208
Thomma BP, Nürnberger T, Joosten MH (2011) Of PAMPs and effectors: the blurred PTI-ETI dichotomy. Plant Cell 23:4–15
Ton J, Flors V, Mauch-Mani B (2009) The multifaceted role of ABA in disease resistance. Trends Plant Sci 14:310–317
Tsuda K, Katagiri F (2010) Comparing signaling mechanisms engaged in pattern-triggered and effector-triggered immunity. Curr Opin Plant Biol 13:459–465

Turlings T, Tumlinson J, Eller F, Lewis W (1991a) Larval-damaged plants: source of volatile synomones that guide the parasitoid Cotesia marginiventris to the micro-habitat of its hosts. Entomol Exp Appl 58:75–82

Turlings TC, Tumlinson JH, Heath RR, Proveaux AT, Doolittle RE (1991b) Isolation and identification of allelochemicals that attract the larval parasitoid, Cotesia marginiventris (Cresson), to the microhabitat of one of its hosts. J Chem Ecol 17:2235–2251

Valiev A, Ogel ZB, Klepzig KD (2009) Analysis of cellulase and polyphenol oxidase production by southern pine beetle associated fungi. Symbiosis 49:37–42

Wu Y, Yi G, Peng X, Huang B, Liu E, Zhang J (2013) Systemic acquired resistance in Cavendish banana induced by infection with an incompatible strain of Fusarium oxysporum f. sp. cubense. J Plant Physiol 170:1039–1046

Yang FZ, Li Y, Yang B (2013) The inhibitory effects of rose powdery mildew infection on the oviposition behaviour and performance of beet armyworms. Entomol Exp Appl 148:39–47

The manufacturer's authorised representative in the EU is Springer Nature Customer Service Centre GmbH, Europaplatz 3, 69115 Heidelberg, Germany. If you have any concerns regarding our products, please contact ProductSafety@springernature.com

Printed and bound by CPI Group (UK) Ltd, Croydon, CR0 4YY
01/10/2026
02225871-0008